THE DISABILITY BIOETHICS READER

The Disability Bioethics Reader is the first introduction to the field of bioethics presented through the lens of critical disability studies and the philosophy of disability.

Introductory and advanced textbooks in bioethics focus almost entirely on issues that disproportionately affect disabled people and that centrally deal with becoming or being disabled. However, such textbooks typically omit critical philosophical reflection on disability. Directly addressing this omission, this volume includes 36 chapters, most appearing here for the first time, that cover key areas pertaining to disability bioethics, such as:

- state-of-the-field analyses of modern medicine, bioethics, and disability theory
- health, disease, and the philosophy of medicine
- issues at the edge- and end-of-life, including physician-aid-in-dying, brain death, and minimally conscious states
- enhancement and biomedical technology
- invisible disabilities, chronic pain, and chronic illness
- implicit bias and epistemic injustice in health care
- disability, quality of life, and well-being
- race, disability, and healthcare justice
- connections between disability theory and aging, trans, and fat studies
- prenatal testing, abortion, and reproductive justice.

The Disability Bioethics Reader, unlike traditional bioethics textbooks, also engages with decades of empirical and theoretical scholarship in disability studies—scholarship that spans the social sciences and humanities—and gives serious consideration to the history of disability activism.

Joel Michael Reynolds is Assistant Professor of Philosophy and Disability Studies at Georgetown University, Senior Research Scholar in the Kennedy Institute of Ethics, Senior Advisor to The Hastings Center, and core faculty in Georgetown's Disability Studies Program. Reynolds is author of *The Life Worth Living: Disability, Pain, and Morality* (University of Minnesota Press), the founder of *The Journal of Philosophy of Disability*, and co-founder of the *Oxford Studies in Disability, Ethics, and Society* book series from Oxford University Press.

Christine Wieseler is Assistant Professor of Philosophy at California State Polytechnic University, Pomona. Wieseler is author of articles published in *Hypatia, IJFAB: International Journal of Feminist Approaches to Bioethics*, and *Social Philosophy Today* as well as chapters in two edited book collections.

THE DISABILITY BIOETHICS READER

Edited by Joel Michael Reynolds and Christine Wieseler

Routledge
Taylor & Francis Group

NEW YORK AND LONDON

Cover image: Rhonda K. Rayman

First published 2022
by Routledge
605 Third Avenue, New York, NY 10158

and by Routledge
4 Park Square, Milton Park, Abingdon, Oxon, OX14 4RN

Routledge is an imprint of the Taylor & Francis Group, an informa business

© 2022 Taylor & Francis

The right of Joel Michael Reynolds and Christine Wieseler to be identified as the authors of the editorial material, and of the authors for their individual chapters, has been asserted in accordance with sections 77 and 78 of the Copyright, Designs and Patents Act 1988.

Library of Congress Cataloging-in-Publication Data
A catalog record for this title has been requested

ISBN: 978-0-367-22002-0 (hbk)
ISBN: 978-0-367-22003-7 (pbk)
ISBN: 978-1-003-28948-7 (ebk)

DOI: 10.4324/9781003289487

Typeset in Bembo
by codeMantra

CONTENTS

PART VI
Issues at the Edge and End of Life

PART VII
Disability, Difference, and Health Care

PART VIII
Intellectual and Mental Disabilities

TABLES

FIGURES

NOTES ON CONTRIBUTORS

Sean Aas is Assistant Professor of Philosophy at Georgetown University, a senior research scholar at the Kennedy Institute of ethics, and a Greenwall Foundation faculty scholar. His primary areas of research are bioethics and social and political philosophy. Dr. Aas is especially interested in questions about boundaries and the basis of our rights in bodies, especially the questions around rights in prostheses, transplants, and other atypical parts. Much of his work focuses on disability, especially disability as social construct, disability and political egalitarianism, and disability and health.

Nicole D. Agaronnik is a medical student at Harvard Medical School in Boston, Massachusetts.

Ron Amundson is Professor emeritus at University of Hawaii Hilo and an elected Fellow of the American Association for the Advancement of Science. His interests include evolutionary biology and philosophy of disability, and he is best known for his 2005 book *The Changing Role of the Embryo in Evolutionary Thought* (Cambridge University Press).

Alexandre Baril is Associate Professor in the School of Social Work at the University of Ottawa. His work, carried out from an intersectional perspective, is at the crossroads of gender, queer, trans, disability/crip/mad studies, critical gerontology, and critical suicidology. His passion and commitment to equity and diversity issues have earned him several awards, including the Canadian Disability Studies Association Tanis Doe Francophone Award in 2020 for his research, teaching, and activism on disability. A prolific author, he has published in journals such as *Hypatia*, *Feminist Review*, *TSQ: Transgender Studies Quarterly*, *Sexualities*, *DSQ: Disability Studies Quarterly*, *Journal of Literary & Cultural Disability Studies*, *Disability & Society*, and *Somatechnics*.

Harold Braswell is Associate Professor of healthcare ethics and a Bicentennial Fellow at Saint Louis University. His research attempts to develop a disability studies approach on bioethical topics at the end of life. He is the author of *The Crisis of US Hospice Care* (Johns Hopkins University Press, 2019) and is currently at work on his second book, *Inhospitable: How Housing Discrimination Shapes the Way We Die*.

Teresa Blankmeyer Burke is Professor of Philosophy at Gallaudet University. Her research interests include bioethics, philosophy of disability, and Deaf philosophy. She currently serves as faculty administrator for faculty development in the office of the provost at Gallaudet University, as co-editor of the *Journal of Philosophy of Disability*, and as chair of the American Philosophical Association's (APA) Task Force on Diversity and Inclusion. She also served as Disability Representative on the APA Committee on Inclusiveness in the Profession from 2008 through 2017, and was acting chair of that committee and member of the APA board of officers in 2015–2016. She is a member of the APA Committee on the Status of Disabled People in the Profession, which was established in 2020.

Havi Carel is Professor of Philosophy at the University of Bristol. She has recently completed a Wellcome Trust Senior Investigator Award funded project, the Life of Breath. She received the Health Humanities' Inspiration Award 2018 for this work. Her third monograph, *Phenomenology of Illness*, was published by Oxford University Press in 2016. She was selected as a "Best of Bristol" lecturer in 2016. Havi is the author of *Illness* (2008, 2013, 2018), shortlisted for the Wellcome Trust Book Prize, and of *Life and Death in Freud and Heidegger* (2006). She is the co-editor of *Health, Illness and Disease* (2012), *New Takes in Film-Philosophy* (2010), and *What Philosophy Is* (2004).

Eli Clare is an American writer, activist, educator, and speaker. His work focuses on queer, transgender, and disability issues. Clare was one of the first scholars to popularize the bodymind concept. He has written two books on creative non-fiction, *Brilliant Imperfection: Grappling with Cure* (2017) and *Exile and Pride: Disability, Queerness, and Liberation* (1999, 2009, 2015); and has written a collection of poetry, *The Marrow's Telling: Words in Motion* (2007). His work has been published in many periodicals and anthologies.

Elizabeth Dietz is a PhD candidate in the Biology and Society program at Arizona State University, where their research and teaching take up questions about reproductive justice, the governance of queer and trans medicine, and informed consent in biomedicine.

John H. Evans is the Tata Chancellor's Chair in social sciences, professor of sociology, associate dean of the Social Sciences, and co-director of the Institute for Practical Ethics at the University of California, San Diego. He specializes in examining debates that involve religion and science in the public sphere, as well as using social science to contribute to humanistic and ethical debates. His most recent books are *Morals Not Knowledge: Recasting the Contemporary U.S. Conflict between Religion and Science* (University of California Press, 2018) and *The Human Gene Editing Debate* (2020, Oxford University Press).

Joseph J. Fins is the E. William Davis, Jr., MD Professor of Medical Ethics, Professor of Medicine and Chief of the Division of Medical Ethics at Weill Cornell Medical College, and Visiting Professor of Law at Yale Law School. He is the author of *Rights Come to Mind: Brain Injury, Ethics, and the Struggle for Consciousness* (Cambridge University Press, 2015).

Erica Hua Fletcher is a Postdoctoral Research Fellow in Veteran Resilience and Recovery at UCLA's Semel Institute, in affiliation with the VA of Greater Los Angeles. She also serves as co-chair of the Society for Medical Anthropology's Anthropology and Mental Health Interest Group. Her interdisciplinary research in the health humanities examines the contributions of contemporary mental health patient rights groups in their efforts to democratize, politicize, and ultimately shape community mental health care in the USA. Her work has been published in *Culture, Medicine and Psychiatry*, *Journal of Medical Humanities*, and *Emotion, Space & Society*.

Rosemarie Garland-Thomson is a bioethicist, author, educator, humanities scholar, and thought leader in disability justice and culture. She is professor emerita of English and bio-ethics at Emory University, a founding co-director of Emory's Disability Studies Initiative, and a fellow of and senior advisor to the Hastings Center. Named by Utne Reader as one of "50 Visionaries Who Are Changing Your World," she is most recently co-editor of *About Us: Essays from the Disability Series of the New York Times* and is the author of *Staring: How We Look* and several other books.

Laura Guidry-Grimes is Assistant Professor in the Department of Medical Humanities and Bioethics in the University of Arkansas for Medical Sciences (UAMS) with a secondary ap-pointment in Psychiatry. She serves as a clinical ethicist at UAMS Medical Center and Arkan-sas Children's Hospital, which involves consulting on complex cases, rounding with different services, providing ethics education, and working on policies and ethical climate. Her research focuses on psychiatric ethics, disability bioethics, vulnerability in clinical settings, and the in-teraction between institutional structures and agency. She co-authored *Basics of Bioethics*, *Fourth Edition* (Routledge, 2020) with Robert M. Veatch.

Lauren Guilmette is Assistant Professor of Philosophy at Elon University. Her areas of research specialization include affect theory, twentieth-century continental philosophy, feminist philos-ophy, queer theory, and applied ethics geared to service-learning and community engagement. Her areas of competency and interest include philosophy of disability and critical disability studies, critical philosophy of race, social and political philosophy, and the history of philosophy.

Anita Ho is Associate Professor in bioethics at both the University of British Columbia and the University of California, San Francisco, as well as the regional director of ethics (North-ern California) for Providence St. Joseph Health. An international scholar with a unique combined academic training and experience in philosophy, clinical/organizational ethics, public health, and business, Ho is the author of more than 70 publications. Her current re-search focuses on ethical dimensions of utilizing innovative and artificial intelligence tech-nologies in health care, disability and equity in health care, supportive decision making, and end-of-life care decisions.

Lisa I. Iezzoni is Professor of Medicine at Harvard Medical School. Dr. Iezzoni has spent more than three decades conducting health services research focusing on two primary areas: risk ad-justment methods for predicting cost and clinical outcomes of care, and healthcare experiences and outcomes of persons with disabilities. Her most recent book is *Making Their Days Happen: Personal Assistance Services Supporting People with Disability Living in Their Homes and Communities* (Temple University Press, 2022).

Eva Feder Kittay is Distinguished Professor of Philosophy emerita at Stony Brook University. Her primary interests include feminist philosophy, ethics, social and political theory, metaphor, and the application of these disciplines to disability studies. She has received numerous awards, including a Guggenheim Fellowship and NEH Fellowship, and the Lebowitz Prize for philo-sophical achievement and contribution from the American Philosophical Association and Phi Beta Kappa.

Erin Gentry Lamb is Carl F. Asseff, MD, MBA, JD, Designated Professor in Medical Humanities, Associate Professor of Bioethics and faculty lead of the humanities pathway at the Case Western Reserve University School of Medicine. While trained in the field of literature, her research interests include aging and ageism, death and dying, disability, bioethics, and healthcare and social justice. She co-edited *Research Methods in the Health Humanities* (Oxford, 2019) and her work appears in forums such as *The Journal of Medical Humanities*, *The Health and Humanities Reader*, the *Encyclopaedia of Health Humanities*, and *Keywords in Health Humanities*.

Lydia Nunez Landry is an independent scholar and disability justice activist. She is co-founder of Gulf Coast ADAPT and a board member of Not Dead Yet, both grassroots disability advocacy organizations. Her work focuses on ableism in mainstream feminism and the systemic abuse and neglect of disabled and older people forced into institutions and the abolition of these practices. She holds a degree in social work from the University of Houston-Clear Lake.

Hil Malatino is Assistant Professor in the Department of Women's, Gender, and Sexuality Studies and a research associate in the Rock Ethics Institute of Penn State University. Malatino's research and teaching draws upon trans and intersex studies, critical sexuality studies, transnational feminisms, disability studies, and medical ethics to theorize how experiences of violence, trauma, and resilience play out in intersex, trans, and gender non-conforming lives.

Kevin Mintz, PhD, is a T32 Postdoctoral Fellow in the Ethical, Legal, and Social Implications of Genetics and Genomics. He received his Ph.D. from the Department of Political Science at Stanford University in 2019. He also holds an AB in Government from Harvard College, and an MSc in Political Theory from the London School of Economics and Political Science. Prior to returning to Stanford, Kevin was a Postdoctoral Fellow in The Department of Bioethics at the National Institutes of Health. His research focuses on disability bioethics, research ethics, business ethics, and the degree to which genetics should be used to construct social or political identities. His work has appeared in a variety of academic journals and newspapers, including *Pediatrics, The Hastings Center Report,* and the *Los Angeles Times.*

Anna Mollow is an independent scholar whose work focuses on disability studies, fat studies, critical race studies, feminism, and queer studies. She is the co-editor, with Robert McRuer, of *Sex and Disability* (Duke UP, 2012) and the co-editor, with Merri Lisa Johnson, of *DSM-CRIP* (Social Text Online, 2013). Her essays have appeared in *African American Review, Hypatia: Journal of Feminist Philosophy, The Journal of Literary and Cultural Disability Studies, WSQ: Women's Studies Quarterly, MELUS: Multi-Ethnic Literature of the United States, Bitch magazine, Autostraddle, Everyday Feminism,* and *Huffington Post.* She is currently completing a book manuscript titled *The Disability Drive.*

David M. Peña-Guzmán is Associate Professor at the School of Humanities and Liberal Studies in San Francisco State University. He received his PhD in Philosophy from Emory University in 2015 and was a postdoctoral fellow at the Berman Institute of Bioethics in Johns Hopkins University from 2016 to 2017. He is the author of *When Animals Dream: The Hidden World of Animal Consciousness* (Princeton University Press, forthcoming), and a co-author of *Chimpanzee Rights: The Philosophers' Brief* (Routledge, 2018). His research has also appeared in *Animal Sentience, Animal Law, Hypatia, Foucault Studies,* and *International Journal of Feminist Approaches to Bioethics.*

Andrea J. Pitts is Associate Professor of Philosophy at the University of North Carolina, Charlotte. They are the author of *Nos/Otras: Gloria E. Anzaldúa, Multiplicitous Agency, and Resistance*

(SUNY Press 2021), and co-editor with Mariana Ortega and José Medina of *Theories of the Flesh: Latinx and Latin American Feminisms, Transformation, and Resistance* (Oxford University Press 2020) and *Beyond Bergson: Examining Race and Colonialism through the Writings of Henri Bergson* with Mark Westmoreland (SUNY Press 2019).

Tia Powell holds the Trachtenberg chair in Bioethics at Albert Einstein College of Medicine, where she is professor of epidemiology and psychiatry and director of the Montefiore Einstein Center for Bioethics. She currently chairs a committee for the National Academy of Medicine to recommend the next decade of social science research to support those with dementia and their caregivers. She graduated magna cum laude from Harvard-Radcliffe College, and AOA from Yale Medical School. Her book, *Dementia Reimagined: Building a Life of Joy and Dignity from Beginning to End*, was published by Penguin Random House in 2019.

Alison Reiheld is Associate Professor of Philosophy at Southern Illinois University—Edwardsville. Her research focuses on the way that clinical structures and habits of practice render patients vulnerable, some groups more than others. These include members of racialized groups subject to stigma, disabled persons, fat persons, members of groups defined by gender and sexuality, and those with multiple group membership. Professor Reiheld was previously Director of Women's Studies at SIUE (2015–2019) and principal researcher on the Trans Health Ethics Project at SIUE.

Michael Rembis is Director of the Center for Disability Studies and Associate Professor in the Department of History at the University of Buffalo (SUNY). Rembis has authored and edited many books, articles, and book chapters, including *Defining Deviance: Sex, Science, and Delinquent Girls, 1890–1960*; *Disability Histories,* co-edited with Susan Burch; *The Oxford Handbook of Disability History,* co-edited with Catherine Kudlick and Kim Nielsen; and *Disabling Domesticity*. In 2012, Rembis and co-editor Kim Nielsen launched the Disability Histories book series with University of Illinois Press. His research interests include the history of institutionalization, mad people's history, the history of psychiatry, and the history of eugenics.

Joel Michael Reynolds is Assistant Professor of Philosophy and disability studies at Georgetown University, senior research scholar in The Kennedy Institute of Ethics, senior advisor to The Hastings Center, and core faculty in Georgetown's Disability Studies Program. He is the founder of *The Journal of Philosophy of Disability* and the co-founder of the *Oxford Studies in Disability, Ethics, and Society* book series from Oxford University Press. Reynolds is the author or co-author of over three dozen journal articles, book chapters, and encyclopedia entries spanning the fields of biomedical ethics, philosophy of disability, social epistemology, and twentieth-century European philosophy as well as the author of *The Life Worth Living: Disability, Pain, and Morality* (The University of Minnesota Press, 2022).

Jackie Leach Scully is Professor of bioethics and director of the Disability Innovation Institute at the University of New South Wales, Australia. Her work examines the ethical implications of a range of biomedical and life science innovations for disabled people's lives, and explores their moral responses. Author of the pioneering *Disability Bioethics: Moral Bodies, Moral Difference* (Rowman & Littlefield 2008), she is a Fellow of the Academy of Social Sciences, the Royal Academy of Arts, the Royal Society of New South Wales, and the Hastings Center, and is currently Editor of the *International Journal of Feminist Approaches to Bioethics*.

Emma Sheppard is lecturer in sociology at Coventry University in the UK. Her research is on embodied experiences of chronic pain, fatigue, and crip time.

Anita Silvers (1940–2019) was Professor and former Chair of the Department of Philosophy at San Francisco State University. She contracted polio as a child, leaving her with partial quadriplegia. She became a highly regarded advocate for disability rights as well as "an institution in professional philosophy." Silvers received the inaugural California Faculty Association Human Rights Award (1989), the Quinn Prize for service to Philosophy (2009, from the APA), the Lebowitz Prize for Philosophical Achievement and Contribution (2013, from Phi Beta Kappa), and the Wang Family Excellence Award for contributions to the California State University system (2017).

Joseph A. Stramondo is Associate Professor of Philosophy, and director of the Institute for Ethics and Public Affairs at San Diego State University, co-president of the Society for Disability Studies, and sits on the board of the Society for Philosophy and Disability. Stramondo specializes in philosophy of disability and bioethics.

Desiree Valentine is Assistant Professor of Philosophy at Marquette University and a 2022-23 Laurance S. Rockefeller Fellow of the Center for Human Values at Princeton University. Valentine's work lies at the intersections of critical philosophy of race, critical disability theory, feminist philosophy, and bioethics. She is presently working on a project exploring the intersections of racism and ableism, and her work has appeared in *Critical Philosophy of Race*, *Journal of Speculative Philosophy*, *Bioethics*, *Journal of Philosophy of Disability*, and *Puncta: Journal of Critical Phenomenology*.

David Wasserman has been on the Department of Bioethics Faculty since January 2013. Previously, he was Director of Research at the Center for Ethics, Yeshiva University. He is a fellow of The Hastings Center and has written extensively on ethical issues in biotechnology, neuroscience, disability, reproduction, genetics, and health care. He has co-authored two books: *Disability, Difference, Discrimination: Perspectives on Justice in Bioethics and Public Policy*, with Anita Silvers and Mary Mahowald (1998), and *Debating Procreation*, with David Benatar (2015).

Christine Wieseler is Assistant Professor of Philosophy at California State Polytechnic University, Pomona. Wieseler is the author of a number of journal articles and book chapters, including "The Desexualization of Disabled People as Existential Harm and the Importance of Ambiguity" in *Normality, Abnormality, and Pathology in Merleau-Ponty* edited by Susan Bredlau and Talia Welsh; "Epistemic Oppression and Ableism in Bioethics" in *Hypatia*; and "Missing Phenomenological Accounts: Disability Theory, Body Integrity Identity Disorder; and Being an Amputee" in *IJFAB: International Journal of Feminist Approaches to Bioethics*.

Robert A. Wilson is Professor of Philosophy at the University of Western Australia, having previously taught at Queen's University (1992–1996), the University of Illinois, Urbana-Champaign (1996–2001), the University of Alberta (2000–2017), and La Trobe University (2017–2019). He was born in Broken Hill, New South Wales and grew up there and in Perth, Western Australia. He has authored or edited seven books, most recently The Eugenic Mind Project (MIT Press, 2018), led the Canadian-based project working with eugenics survivors who produced the website EugenicsArchive.ca, and co-directed and produced the documentary film Surviving Eugenics (2015) available at that site.

Yolonda Wilson is Associate Professor of Healthcare Ethics, Philosophy, and African-American studies at St. Louis University. Her research interests include bioethics, social and political philosophy, race theory, and feminist philosophy. Wilson is currently developing an account of justice that gives specific requirements for racial justice in health care at the end of life.

DISABILITY BIOETHICS

Introduction to *The Disability Bioethics Reader*

Joel Michael Reynolds and Christine Wieseler

The field of bioethics emerged against a tumultuous backdrop. The not-so-distant history of state-sponsored eugenics in Britain, the USA, and Germany loomed large as highly publicized biomedical events in the middle of the 20th century sparked novel public awareness of the relationship between medicine and ethics (or the lack thereof) across the globe. These included the Tuskegee Study and Guatemala Experiments, Jim Crow medical care, and the development of unprecedented life-sustaining technologies ranging from ventilators to artificial heart valves. Yet, even in its infancy, bioethics was not one field, but many. That is still true today and, in many ways, even more so. "Bioethics" includes academics, medical practitioners of every stripe, policy and public health experts, and, increasingly, scientists whose research ranges across the life sciences. If one considers the impact of modern biomedicine on contemporary life, this should not be surprising. Practices of health care are at once scientific—finding empirical answers, saving lives—and political—concerned with constraints like resource distribution and inflected by a host of legal, social, and political considerations.

In light of this, the questions bioethicists ask are expectedly broad and touch nearly every aspect of life: what purpose does the practice of medicine serve? What does it mean to care for another person? Or for groups of people, non-human animals, land, and the earth itself? Is life sacred? How does one define need, harm, risk, and benefit? What counts as life? Who owns life? Who decides where life begins and ends? Who decides whose care can be withheld or limited and whose cannot? What medical research should be conducted and how, when, where, with whom, and by whom? By exploring questions such as these, *The Disability Bioethics Reader* introduces you to the field of bioethics. Unlike other bioethics readers, however, you will learn about core issues in the field through the lens of, in the light of, research in philosophy of disability and critical disability studies (Wieseler 2015; Davis 2016; Hall 2019; Cureton and Wasserman 2020; Reynolds and Burke 2021).

To explain why we take this approach and to give context to its historical import, let us return to the claim that bioethics is not one field. "Bioethics" is an umbrella term spanning an interdisciplinary, intradisciplinary, and transdisciplinary cluster of inquiries; it is loosely tied together by moral questions that arise in the study of organic life and the many fields, domains, and industries that investigate, engage with, and seek to act upon life's processes. This cluster

DOI: 10.4324/9781003289487-1

is massive, for it includes any and all investigations of ethical, social, and political issues that arise from the ever-growing intersection of biomedicine, health-related industries, and contemporary life. The range of topics bioethicists study is thus staggering—from the use of human embryonic materials in basic research to the moral status and treatment of chimpanzees or other non-human animals to the proper procedures for vaccine distribution at local, national, and international levels; from the ethical issues facing surgeons, general practitioners, or NICU nurses to the biosecurity risks posed by changes to multilateral geopolitical instruments like NATO to the legal implications of copyright law for therapeutics developed via the contemporary tools of genome editing such as CRISPR-Cas9. The list goes on and on. And it's not just the field of bioethics that is diverse—it is also the professions in which it is practiced or applied and which it supports. There are academic bioethicists, typically people with PhDs whose job entails some combination of teaching, research, and service at a college or university. There are also clinical bioethicists who work in hospitals, of which there are two primary types: people with PhDs in fields such as philosophy or sociology who work alongside doctors and nurses, and people with core training in medicine who have gained expertise in bioethics through programs, seminars, certification, or other sorts of additional educational accreditation and training. Given this breadth in the field of bioethics or, rather, breadth in the many fields and practices that make up what falls under the umbrella term "bioethics," we will begin by defining what we mean by "disability bioethics."

I Disability Bioethics

In her ground-breaking 2008 work, *Disability Bioethics: Moral Bodies, Moral Difference*, Jackie Leach Scully writes,

> Disability ethics, like feminist ethics, is a form of ethical analysis consciously and conscientiously attentive to the experience of being/having a "different" embodiment. Where feminist ethics' concern is with the non-normativity introduced by gendered bodies, however, disability ethics looks at the embodied effects of impairment.
>
> *(11)*

Scully makes clear in that book that disability bioethics involves more than merely paying attention to certain pockets of empirical research. It must also involve active uptake and integration of the analysis, understanding, and interpretation of the lived experiences of disabled people. Furthermore, and as we argue in more detail below, it must also involve an ethical commitment to *centering*—not just noting—those experiences. Tellingly, in her 2013 book, *Bioethics and Disability: Toward a Disability-Conscious Bioethics*, Alicia Ouellette makes a related argument, tying it more directly to longstanding concerns over bias against disabled people.

> Bioethicists who dismiss the disability perspective are making a mistake…If the evidence [demonstrates] that the story of disability need not be the story of tragedy and that biased and disproven assumptions about life with disability are at play in medical decision-making, then it is the business of bioethics to work with disability experts to figure that out and to work to eliminate that bias.
>
> *(2011, 69)*

Combining Scully and Ouellette's insights, one arrives at the idea that disability bioethics involves a combination of empirical and non-empirical commitments.[1]

In this spirit, *disability bioethics* for us refers to bioethical inquiry that involves, at minimum, the following three aspects:

1   A critical relationship to common narratives and "common sense" claims concerning disability.
2   Theory and practice rooted in *critical* disability scholarship, with an emphasis placed on testimony by and work from disabled people as well as an emphasis placed on participatory models of research and practice.[2]
3   Inquiry committed to increasing justice and equity for people with disabilities.

Put differently, the last qualification means: disability bioethics is grounded in *disability justice* (Brown 2011, Piepzna-Samarasinha 2018; Lewis 2021). This is contentious, for it raises the question of how one distinguishes between the academic and the activist. Without getting into the weeds of that longstanding debate (see Stramondo forthcoming), if one engages in bioethical inquiry *without* the aim of increasing justice and equity for people with disabilities (not to mention other oppressed groups), we think one is engaged in a very different sort of project than that of this volume and the sort of work we hope becomes mainstream in "mainstream bioethics." That claim might lead one to wonder: why do we hold onto the term "disability bioethics" and not instead push for a change to what counts as "bioethics" itself sans qualification? It is our hope that eventually it will be a given that "mainstream" bioethics takes the approach we are here advocating, but in the meantime, we find it helpful for pedagogical, research, and other purposes to use a distinct term that picks out, among other things, the three aspects detailed above.

II Appreciating the Need for Disability Bioethics

Bioethicists who don't "do" disability bioethics often share four types of unexamined assumptions about disability. First, their positions tend to be in line with the medical model of disability, which narrowly and solely conceptualizes disability at the level of an individual's body (Reynolds 2022). Second, bioethicists too often endorse, implicitly or explicitly, what Elizabeth Barnes terms *bad-difference* views of impairment, which hold that impairment inherently and inevitably reduces quality of life (QOL) and would do so even apart from the removal of social factors such as stigmatization, induced poverty, inaccessibility, and, in a word, inequity (Goering 2008; Barnes 2016; Campbell and Stramondo 2017; Scuro 2018; Amundson 2022). Third, bioethicists' arguments too often involve biological determinism, biological reductionism, and/or defenses of strong objectivity (Wieseler 2016; Amundson 2022). When bioethicists rely on such suspect accounts, it leads to fundamentally inaccurate conceptions of human bodies and lives. Fourth, bioethicists too often assume that disabled experiences can be analogized from able-bodied experiences and thus do not regularly draw upon evidence grounded in the lived experiences of people with disabilities (Miserando 2003; Landry 2022).

This leads to bioethical discussions about disability that (i) in fact conflict with the lived experiences, perspectives, and interests of disabled people, (ii) fail to take into account the troubled historical relationship between medical practice and people with disabilities, and (iii) ignore the larger social, cultural, and political forces that oppress people with disabilities and thereby ignore a core component of social determinates of health for this large and varied population. Insofar as the perspectives of disability rights advocates and disabled people remain marginal within everything from introductory textbooks to scholarly monographs in bioethics, the state of scholarship across the field further compounds this issue. This creates serious problems, for it can lead students, medical professionals, educators, and scholars to think that disability, and the relationship of disability studies to bioethics more specifically, is an apolitical issue and that research concerning disability in general and the lived experiences of disability in particular are not essential to bioethical inquiry.

III The Stakes of Disability Bioethics

The difference between doing disability bioethics and doing non-disability bioethics (or "regular" bioethics) has high stakes. If one is a clinical bioethicist, it will impact what, how, and whether one will communicate with patients and clinicians about complex ethical issues. It will also impact how and whether one treats patients in a number of respects, including vis-à-vis certain diagnoses, prognoses, treatment options and plans, and referrals. If one is an academic bioethicist, it will impact what, how, and whether one argues in scholarly journals about issues that often have significant practical, real-world consequences. Unfortunately, though, insights, methodologies, and research from disability activism and scholarship have too often been simply ignored, not taken seriously, or misrepresented across bioethics' 50-plus-year history.[3] This is especially frustrating given the fact that central debates in bioethics, as well as in public health, focus upon issues that disproportionately affect disabled people. Topics including euthanasia, physician aid-in-dying, pre-implantation genetic diagnosis (PGD), prenatal testing, selective abortion, enhancement, resource allocation, and emergency rationing—among many others—are *all* premised on shared and implicit assumptions regarding disability, especially in relationship to QOL.

In short, doing disability bioethics *well* involves careful consideration of where one turns for data, how one interprets and otherwise reflectively analyzes that data, who is at the table—and brought to it—with respect to decision-making at basic, clinical, and translational levels, and the values upon which one's research more generally is anchored. Biomedical research, treatment, and engagement with Autistic people[4] is a powerful example here—some people, and far too many medical professionals, think that by listening to the parents of Autistic people, they are thereby plugged into the disability community. That is false and can be damagingly so. Some of those who think this way aren't even aware of the Autistic Self Advocacy Network (ASAN) or the very serious concerns regarding organizations like Autism Speaks. An essential aspect of disability bioethics involves research. If you are a clinician or clinical bioethicist who wants or needs, given the demands of one's job, to learn about X (understanding this as a random variable indicating a particular impairment), where do you go for information? If your first answer is "a genetics textbook," this should cause pause. Such a resource might be useful, but it should not be definitive and, we think, it should certainly not be the *only* source to consult. One's patient living with X or who identifies as X is, before all else, a *person*, and research that focuses upon and takes up their life as they in fact experience it is essential to any number of clinical considerations relating to their care and health outcomes.

The majority of bioethicists and medical professionals today are able-bodied people, and a number of studies have suggested that nondisabled people are likely to estimate the QOL of disabled people to be much lower, on average, than disabled people themselves report. Healthcare professionals are not immune from this phenomenon. In a recent survey of 714 practicing physicians in the USA published in *Health Affairs* by Iezzoni et al., 82.4% report that people with significant disability have worse quality of life than nondisabled people (2021). This judgment directly conflicts with decades of social scientific research suggesting that people with significant disability, just as with non-significant disability, experience similar, not lower levels of QOL as nondisabled people. That study also reports that only 40.7% of the physicians surveyed expressed confidence in their ability to provide the same quality of care to disabled patients as they do to nondisabled patients. These findings are worrisome on a number of fronts, but especially with respect to the following three implications: (1) there is a substantial discrepancy between how physicians conceive of the relationship between QOL and significant disability and how disabled people in fact experience it, (2) there may be a link between this discrepancy and the quality as well as the equity of care of disabled patients, and (3) this issue on the whole has not improved in a statistically significant way for decades.

Just consider how this issue played out during the COVID-19 crisis. In a December 2020 National Public Radio (NPR) piece, investigative reporter Joseph Shapiro detailed the story of a woman with intellectual disabilities who sought medical care at the start of the COVID-19 pandemic in the small, rural town of Pendleton in Oregon, USA. She needed a ventilator, but her physician denied it, citing her "low quality of life." He asked her to sign a form that would allow the hospital to deny her further care. After threats of lawsuits, this woman was transferred to another hospital where proper care was offered, and she recovered. Oregon Senator Sara Gelser told NPR: "Nothing happened to that hospital. Nothing happened to that physician...the health authority confirmed that, in fact, that was a coerced do-not-intubate order, they confirmed it happened... but there was no sanction." Shapiro further reports, "the state records that NPR obtained show other people with disabilities were denied coronavirus tests or treatment when they showed up at hospitals with symptoms." Such blatant cases of discrimination on the basis of disability have been so widespread during the COVID-19 pandemic across the globe that the United Nations put out guidelines to try and mitigate the problem (2020). The Iezzoni et al. study adds fuel to the larger body of evidence that people with disabilities do in fact receive worse—prejudicial, or otherwise inequitable—forms of health care compared to than their nondisabled counterparts (Reynolds & Peña-Guzmán 2019). This is yet another piece of evidence for the necessity of disability bioethics as an approach to—we would hope *the primary* approach to—doing and learning about bioethics.

Furthermore, we highlight this study to note that engagement with disability bioethics is not just a demand relative to education and the academy. As an applied field tied to one *of the larger economic sectors of most industrialized nations*, bioethics impacts healthcare education, practice, and policy in countless ways. It is in light of such pressing issues about the state of bioethics education and scholarship and their real-world impact that we have developed *The Disability Bioethics Reader*.

IV Language and Content

A note on language: we encouraged contributing authors to use their preferred nomenclature concerning disability instead of attempting to make it homogenous across the volume. There is no consensus (globally, nationally, and even at more narrow levels like "across disability studies" or "across disability activism") concerning the terms "disabled people" vs. "people with disabilities," etc. There is generally consensus, however, that euphemisms like "uniquely abled" or "differently abled" are misguided. In short, people use different terms for different reasons, reasons that are often context dependent. We ourselves purposely switch between "disabled people" and "people with disabilities" to signal this plurality of views.

We agree with Eli Clare's statement in *Brilliant Imperfection: Grappling with Cure* that "trigger warnings are in essence tools for self-care and collective care" (2017, xx). In terms of content warnings, we recognize that some of the topics into which this reader delves are likely to be difficult. Although we cannot anticipate how individual readers will respond, we expect that the following topics may be triggering: discussion of eugenics, psychiatric hospitalization, medicalized abuse, epistemic and other forms of oppression, and suicide, among others. Chapter titles and abstracts generally provide information that can assist you in determining when you may need self-care and/or collective care.

V The Design of *The Disability Bioethics Reader*

This reader is designed for use in undergraduate and graduate courses in colleges and universities, medical school education, continuing medical education courses, and other continuing medical education credit opportunities. Though we intend it to be of interest in any and all bioethics-related courses, it will be of special interest for teachers approaching bioethics from various critical perspectives, including feminist philosophy, critical philosophy of race, gender and sexuality studies, feminist science studies, and science and technology studies. We have been blessed by the fact that this project has been collaborative from the beginning. We solicited, received, and incorporated constructive feedback from a wide range of disability activists, disability studies scholars, disability-conscious healthcare professionals, and others all the way from the project's inception to its final production and concerning everything from its framework to specific topics to authors to include. We placed priority on selecting authors with relevant lived experience where possible and on the inclusion of multiple chapters exploring tensions and opportunities for future engagement between disability bioethics and other fields of study (e.g., critical race theory, feminist bioethics, fat studies, trans studies, aging studies, and animal studies). We have had the great joy of working with and learning from our exceptional authors, together shaping chapters, topics, and arguments through lively conversation, deliberation, and constructive disagreement.

All of this being said, a reader such as this should be seen, we hope, as a snapshot of a wide range of living, ongoing, and complex research projects. This volume does not represent all that is going on in disability bioethics. *Far from it!* As anyone who has ever edited a volume of this size knows, the original plan and shape has transformed over the last many years as we have gotten and responded to feedback, as authors (and thereby topics) unfortunately had to drop out or happily dropped in, and as constraints of the real world played out as they always do.

We welcome and look forward to feedback from teachers, students, practitioners, and others about your experience of using this reader. Just as with disability justice, disability bioethics is an evolving enterprise. We cannot say with confidence where disability bioethics is headed, but we are thrilled for you to be part of the journey and are so excited for whatever lies ahead. The idea for this volume started thanks to an experience we had over and over again: colleagues from around the globe asking us what readings to include on disability in their bioethics classes. So, to all of those friends and colleagues, and to their students, here is just a taste of what's out there.[5]

Notes

1 In addition to Scully and Oullette, for further research and arguments concerning how to think about "disability bioethics" and also what that term should mean, both definitionally and programmatically, see Shakespeare (2014); Stramondo (2016); Parens (2017); Garland-Thomson and Iezzoni (2021); Garland-Thomson (2022). As the very existence of and diverse content in this volume makes abundantly clear, that list of references just touches the surface of the many ways in which "disability bioethics" has been building (even if unevenly and contestedly) since at least the 1990s.
2 When we say "critical" disability scholarship, we mean research that does not simply report statistics or present humanistic or social scientific analyses *of* disabled people, but research that instead seriously engages testimony, activism, and scholarship by disabled people and, further, that treats as its interpretive north star how people with disabilities in fact experience and understand their lives.

3 There is some anachronism at play in this claim since the field of disability studies didn't really materialize until the late 1970s/early 1980s (expectedly, the rise of modern disability *activism* predates the academic field; in the USA and UK, it is typically dated to the mid-late 1960s/early 1970s). The field of bioethics is usually said to originate in the late 60s; see Evans, "A Critical History of Bioethics," this volume. Still, by the heyday of bioethics—keeping in mind that it did not fully "take off" in certain respects until the mid-1980s or so—the field of disability studies and the presence of disability activists on the national stage were established. For example, the national *Society for Disability Studies* (renamed to that in 1986) was established in 1982 and the international journal *Disability & Society* (renamed to that in 1994) was established in 1986.

4 We use the language of "Autistic people" and "Autistic person" following Lydia X. Z. Brown. See "Identity-First Language," Autistic Self-Advocacy Network. https://autisticadvocacy.org/about-asan/identity-first-language/. Accessed September 1, 2020. This post clarifies: "ASAN intern Lydia Brown originally published this article August 4, 2011 on their blog *Autistic Hoya* under the title 'The Significance of Semantics: Person-First Language: Why It Matters.'"

5 Acknowledgments: The editors would first like to thank all our fantastic contributors as well as Andy Beck at Routledge for his constant support for this project. We individually thank Nancy Berlinger, Tom Cole, Elizabeth Dietz, Kristie Dotson, Nick Evans, Rosemarie Garland-Thomson, Lauren Guilmette, Josephine Johnston, Gregory E. Kaebnick, Eva Feder Kittay, Alex Levine, Becca Longtin, Erik Parens, David Peña-Guzmán, Diane Price-Herndl, Gaile Pohlhaus, Rhonda Rayman, Jennifer Scuro, Millie Solomon, Gail Weiss, Gregor Wolbring, and Rachel Zacharias. Joel would like to thank his students at the University of Massachusetts Lowell and Georgetown University, who provided feedback on some of the chapters that ended up in this volume. He is also grateful for everyone at The Hastings Center for their wisdom and support during the germination of this project. We want to give a huge thanks to Ari Watson, who tirelessly helped get the project over the finish line. Thank you, Ari! Special thanks to Laura Guidry-Grimes for truly pivotal feedback on and suggestions for this introduction, its framing, and how we describe and think about "disability bioethics" more generally.

References

Amundson, Ron. 2022 [2005]. "Disability, Ideology, and Quality of Life: A Bias in Biomedical Ethics." This volume.

Barnes, Elizabeth. 2016. *The Minority Body*. New York: Oxford University Press.

Brown, Lydia X. Z. 2011. See "Identity-First Language," *Autistic Self-Advocacy Network*. https://autisticadvocacy.org/about-asan/identity-first-language/. Accessed Sep. 1, 2020.

Campbell, Stephen M., and Joseph A. Stramondo. 2017. "The Complicated Relationship of Disability and Well-Being." *Kennedy Institute of Ethics Journal* 27(2): 151–184. https://doi.org/10/gf96rz.

Clare, Eli. 2017. *Brilliant Imperfection: Grappling with Cure*. Durham, NC: Duke University Press.

Cureton, Adam, and David Wasserman. 2020. *The Oxford Handbook of Philosophy and Disability*. New York: Oxford University Press.

Davis, Lennard J. 2016. *The Disability Studies Reader*. 5th ed. New York: Routledge.

Garland-Thomson, Rosemarie. 2022 [2017]. "Disability Bioethics: From Theory to Practice." In *Disability Bioethics Reader*, edited by Joel Michael Reynolds and Christine Wieseler, pp. XX. New York; London: Routledge.

Garland-Thomson, Rosemarie and Lisa I. Iezzoni. 2021. "Disability Cultural Competence for All as a Model." *The American Journal of Bioethics* 21(9): 26–28, https://doi.org/10.1080/15265161.2021.1958652

Goering, Sara. 2008. "'You Say You're Happy, But…': Contested Quality of Life Judgments in Bioethics and Disability Studies." *Bioethical Inquiry* 5: 125–135. https://doi.org/10.1007/s11673-007-9076-z

Hall, Melinda C. 2019. "Critical Disability Theory", *The Stanford Encyclopedia of Philosophy*. Edward N. Zalta, ed., https://plato.stanford.edu/archives/win2019/entries/disability-critical/.

Iezzoni, L. I., S. R. Rao, J. Ressalam, D. Bolcic-Jankovic, N. D. Agaronnik, K. Donelan, T. Lagu, and E. G. Campbell (2021). "Physicians' Perceptions of People with Disability and Their Health Care." *Health Affairs* 40(2): 297–306. https://doi.org/10.1377/hlthaff.2020.01452.

Landry, Lydia Nunez. 2022. "Chronic Illness, Well-being, and Social Values." In *Disability Bioethics Reader*, edited by Joel Michael Reynolds and Christine Wieseler, pp. 156–169. New York; London: Routledge.

Lewis, Talila "TL". "January 2021 Working Definition of Ableism (Developed in Community with Disabled Black and Other Negatively Racialized People, Especially Dustin Gibson)". https://www.talilalewis.com/blog/january-2021-working-definition-of-ableism.

Miserando, Christine. 2003. "The Spoon Theory." Butyoudontlooksick.com. https://butyoudont-looksick.com/articles/written-by-christine/the-spoon-theory/

Ouellette, Alicia. 2011. *Bioethics and Disability: Toward A Disability-Conscious Bioethics*. New York: Cambridge University Press.

Parens, Erik. 2017. "Choosing Flourishing: Toward a More "Binocular" Way of Thinking about Disability." *Kennedy Institute of Ethics Journal* 27(2): 135–150. https://doi.org/10.1353/ken.2017.0013.

Piepzna-Samarasinha, Leah Lakshmi. 2018. *Care Work: Dreaming Disability Justice*. Vancouver: Arsenal Pulp Press.

Reynolds, Joel Michael. 2022. "Theories of Disability." In *Disability Bioethics Reader*, edited by Joel Michael Reynolds and Christine Wieseler, pp. XX. Routledge.

Reynolds, Joel Michael and David Peña-Guzmán. 2019. "The Harm of Ableism: Medical Error and Epistemic Injustice." *Kennedy Institute of Ethics Journal* 29(3): 205–242. https://doi.org/10.1353/ken.2019.0023.

Reynolds, Joel Michael and Teresa Blankmeyer Burke. 2021. "Introducing the Journal of the Philosophy of Disability." *The Journal of Philosophy of Disability* 1: 1–10.

Scully, Jackie Leach. 2008. *Disability Bioethics: Moral Bodies, Moral Difference, Feminist Constructions*. Lanham, MD: Rowman and Littlefield.

Scuro, Jennifer. 2018. *Addressing Ableism: Philosophical Questions via Disability Studies*. Lanham, MD: Lexington Books.

Shakespeare, Tom. 2014. *Disability Rights and Wrongs Revisited*. 2nd ed. London: Routledge.

Shapiro, Joseph. 2020. "Oregon Hospitals Didn't Have Shortages. So Why Were Disabled People Denied Care?" NPR. https://www.npr.org/2020/12/21/946292119/oregon-hospitals-didnt-have-shortages-so-why-were-disabled-people-denied-care. Accessed 9.5.21.

Stramondo, Joseph A. Forthcoming. "How Disability Activism Advances Disability Bioethics." *Ethical Theory and Moral Practice*.

Stramondo, Joseph A. 2016. "Why Bioethics Needs a Disability Moral Psychology." Hastings Center Report 46: 22–30. https://doi.org/10.1002/hast.585.

United Nations. April 29, 2020. UN Human Rights Office of the High Commissioner. "Covid-19 and The Rights of Persons with Disabilities: Guidance." https://www.ohchr.org/Documents/Issues/Disability/COVID-19_and_The_Rights_of_Persons_with_Disabilities.pdf.

Wieseler, Christine. 2015. "Thinking Critically about Disability in Biomedical Ethics Courses." *American Association of Philosophy Teachers Studies in Pedagogy* 1: 82–97. https://doi.org/10.5840/aaptstudies20159236

Wieseler, Christine. 2016. "Objectivity as Neutrality, Nondisabled Ignorance, and Strong Objectivity in Biomedical Ethics." *Social Philosophy Today* 32: 85–106. https://doi.org/10.5840/socphiltoday201692933

Suggested Further Reading

Albrecht, Gary L. 2006. *Encyclopedia of Disability*. 5 vols. Thousand Oaks, CA: Sage.

Bauman, H. Dirksen L. 2014. *Deaf Gain Raising the Stakes for Human Diversity*. Minneapolis: University of Minnesota Press.

Clare, Eli. 2015. *Exile and Pride: Disability, Queerness, and Liberation*. 2nd ed. Durham, NC: Duke University Press.

Kaposy, Chris. 2018. *Choosing Down Syndrome*. Cambridge: MIT Press.

Wendell, Susan. 1996. *The Rejected Body: Feminist Philosophical Reflections on Disability*. New York: Routledge.

PART I

History, Medicine, and Disability

1

A SHORT HISTORY OF MODERN MEDICINE AND DISABILITY

Michael Rembis

People living in ancient Greece and Rome, medieval China, and the Western Hemisphere before the arrival of Columbus all developed ways of knowing and treating the things that ail most humans. In this chapter, I will consider the history of medicine, disability, and madness from the eighteenth century through the twenty-first century in the USA. During these years, medical knowledge and practice moved away from tradition and custom and people's homes and storefronts to hospitals, clinics, laboratories, and universities. "Regular" or allopathic physicians in the, most of whom were white men until the mid-twentieth century, wrestled the authority to know and to treat away from homeopaths, midwives, and the local barber/surgeon and bonesetter. Home remedies and patent medicines, as well as more "heroic" treatments such as purging and bloodletting, gave way to the formalized, institutionalized, and increasingly inaccessible medical interventions familiar to most twenty-first-century observers. This transformation – the rise of modern medicine – took place slowly, over decades, and involved social and cultural battles and changing economic structures as much as it did scientific or technological breakthroughs or advancements in medical knowledge.

As the nineteenth century turned into the twentieth century, the regular physicians, or "regulars," as they became known, worked to expand their purview and their power and control over what constituted medical knowledge and treatment. They formalized and institutionalized diagnostic labels and disability categories, and medicalized and pathologized what had been considered relatively rare, but not entirely uncommon misfortunes or human differences, such as physical ailments, epilepsy, blindness, deafness, and madness. New ways of knowing led to new ways of treating humans whom physicians increasingly classified as defective, deficient, deformed, insane, or disabled. Although many physicians resisted early twentieth-century eugenics, an equal, if not greater, number of them actively sought to eliminate "defective" humans through various medical procedures, including sterilization, institutionalization, and euthanasia. At the very least, physician-eugenicists sought to educate people to make "wise" reproductive choices. Although eugenics would persist, in altered forms, into the twenty-first century, after World War I, physicians increasingly sought to rehabilitate disabled Americans, to make them "normal" and able to "reenter" "normal" society. This combination of eugenics and the medical/rehabilitation model came to dominate thinking on disability and madness by the

DOI: 10.4324/9781003289487-3

mid-twentieth century, and although it has faced opposition from the disability rights movement and the mad people's movement, it remains a powerful force in the twenty-first century.

Modern medicine and disability share a troubled past. On the one hand, advancements in modern medicine such as vaccines, prosthetics, surgery, pharmaceuticals, and other technologies have helped disabled people in places like the USA live longer, healthier lives than at any other time in human history. On the other hand, disabled people have had to suffer the consequences of living medicalized, pathologized lives shaped by stigma, objectification, devaluation, and, in most cases, figurative or literal removal from society. Disabled and mad people have been forced to endure objectifying, permanently damaging, and in some cases lethal, medical experiments and treatments. Although a lot has changed in the wake of the social movements of the 1960s and 1970s, and the passage of the Americans with Disabilities Act (ADA) in 1990, disabled and mad people continue to be forced into second-class citizenship, characterized by exceedingly high rates of un- or underemployment, poor educational outcomes, and severe and persistent poverty, which is rooted at least in part in the idea that they are medical and biological misfits and liabilities. This chapter will trace the troubled history of modern medicine and disability in four parts, from the colonial period through the early twenty-first century.

I Early Healthcare, Colonial–1860

The roots of modern medicine did not take hold in the USA until the mid-nineteenth century. Before that time, colonial settlers, African slaves, and Native peoples relied on their own, often widely divergent, ways of tending to the sick, the injured, and the distracted or distressed. Among colonial settlers, a small group of well-educated white male physicians attempted to assert their professional authority in the late-colonial period, but they met resistance from their fellow colonists who increasingly looked upon any attempts to assert social authority or to obscure access to knowledge and information with suspicion. As the sociologist Paul Starr has noted, medicine remained "domestic" for almost 100 years between 1760 and 1860 (Starr 2017).

Throughout the colonial period and well into the nineteenth century, medicine was practiced in the home or on the plantation, usually by women, who were considered the primary caregivers within households. Some women, such as midwives, who possessed special knowledge and skills, extended their services beyond their own homes and families to their neighbors and communities. Women practitioners lived and worked among a broad range of male healers, surgeons, and druggists, who relied on everything from diet, exercise, water, and medicinal plants to poisonous minerals, purging, bloodletting, blistering, and prolonged restraint, to intervene in the lives of sick, disabled, and mad people.

Historical documents related to medical practice and disability or madness for the colonial and early national period are scant, but they suggest that people who survived smallpox and other diseases and became disabled, people who experienced disabling injuries, people who had sensory impairments, and people who were distracted or distressed lived within their communities as long as they were able to work or contribute in informal ways to the local economy, and were not considered violent. In many instances, class, race, gender, and social context were equally if not more important than embodiment in determining the lived experiences of disabled and mad people (Boster 2013; Groce 1985; Nielsen 2012). It appears that some disabled people, like the deaf community living on Martha's Vineyard, thrived (Groce 1985). There is also evidence, however, that some disabled and mad people were shut away in cellars, confined to a single room, chained to walls and floors, denied adequate food, clothing, and shelter, and cast out of their communities.

Hospitals and insane asylums, as they were called, were rare during the colonial and early national period. Both hospitals and asylums, with the exception of a few private asylums, held poor or "indigent" people who did not have family, friends, or people in the community to care for them. One of the nation's oldest hospitals, Bellevue Hospital in New York City, began in 1736 as an almshouse. Bellevue moved to its current location on Manhattan's lower east side in 1795 (Oshinsky 2017). From the beginning, almshouses, hospitals, and asylums held disabled and mad people. The Pennsylvania Hospital, which was founded in 1752, was intended for the "reception and relief" of "Lunaticks" and other "distemper'd and sick Poor" residents of the province. The first asylum exclusively dedicated to holding "ideots, lunatics, and other persons of unsound minds" opened in Virginia in 1773 (Nielsen 2012). By the early 1840s, there were 21 public and private asylums scattered throughout 26 states. Although the numbers remained widely disputed, especially with respect to race, the 1840 census, the first to enumerate insanity, as well as other disabilities, counted between 8,272 and 17,181 "Insane and Idiotic" people among a total population of approximately 17 million – for a ratio of either one in 2,062 or one in every 900 Americans (Rembis forthcoming). It is safe to assume, based on anecdotal and other evidence, that many more Americans lived with physical and sensory disabilities and madness within their communities.

As the 1840 census indicates, the relationship between disability, madness, and medicine remained ill-defined during the decades between the founding of the nation and the outbreak of the Civil War. Ideas about disability and madness remained both complex and dynamic during the decades between the founding of the nation and the outbreak of the Civil War. Defenders of slavery, for example, created categories of madness which they assigned exclusively to enslaved people. Proponents of the slave system also argued that Black people living in the north suffered disproportionately from "insanity" and other ailments. Most people considered insanity a "brain disease." Yet its causes remained a fluid mix of social and biological factors that always had their roots in dominant understandings of gender, race, and class. What twenty-first-century observers would call developmental or intellectual disabilities were similarly ill-defined during the early nineteenth century. Literacy varied widely by region and often depended upon one's gender, race, and class. Compulsory education did not become widespread until the turn of the twentieth century and the classrooms that existed in the early nineteenth century were ungraded. What was legally referred to as mental "competency" was almost always defined by an individual's ability to fulfill their social roles or work for wages. Though conditions such as epilepsy had been known for centuries, they were also not well understood. People living with what some twenty-first-century observers might call "seizure disorders" often found themselves incarcerated in asylums for the "insane." Originally thought to be "dumb," deaf and blind people became the focus of some educational reformers during the early nineteenth century. Yet it would take more than a century for them to achieve full citizenship, and during that time their experiences as disabled people would become increasingly medicalized. Finally, in an age before antibiotics and effective surgical techniques, many Americans lived with all sorts of undifferentiated physical disabilities that incapacitated them to varying degrees at numerous points throughout their lives.

More than anything, early nineteenth-century Americans seemed concerned with managing the consequences that disability and madness posed for families, communities, and the state. Disabled Revolutionary War veterans who applied for government pensions, for example, had their disabilities measured in terms of their capacity to conduct manual labor. Many disabled Revolutionary War veterans who received a pension continued to fulfill their role as husband, father, and head of household (Blackie 2014). After the Revolution, sensory impairments and "idiocy" increasingly became a matter of concern among educators, not medical practitioners.

Blind, deaf, deaf/blind, and "idiots" who could afford it attended newly created residential schools and institutes, where they worked with teachers to build the knowledge and skills thought to be necessary to become productive and responsible citizens of the new nation. Some students fared well in these settings. They formed bonds, built careers, and created American Sign Language (Edwards 2012). Others returned home or were left to drift through society (Richards 2014). Some students remained residents at the schools for most if not all of their lives. In addition to those with intellectual or sensory impairments, people with physical ailments managed as well as they could on their own, or with the help of family and friends, to work and to fulfill other important familial and social roles.

Although their authority remained challenged and their ability to treat largely ineffective, regular physicians continued to try to consolidate their authority throughout the decades before the Civil War. In October 1844, physician-administrators of asylums formed the first professional medical organization in the country. Thirteen prominent asylum superintendents came together in Philadelphia and formed the Association of Medical Superintendents of American Institutions for the Insane (AMSAII). Almost three years later, in May 1847, physicians formed the American Medical Association (AMA). Both organizations worked within a wide and diverse medical marketplace to establish standards in the credentialing of physicians and in the treatment of patients and asylum inmates. These regular physicians continued to face resistance, however, from an American public reluctant to accept their exclusive claim to medical knowledge and professional expertise. Women physicians, Thomsonians (followers of Samuel Thompson), homeopaths, osteopaths, water healers, housewives, and enslaved healers, among other practitioners, remained part of the American medical landscape, even as the number of hospitals and asylums, and the ranks of regular physicians, continued to grow.

The asylum system continued to expand, and communities continued to build hospitals and almshouses, but they were not looked upon by most Americans as places of refuge or healing. They were avoided, and, in most instances, used only as a last resort. The carnage of the Civil War combined with changing demographics and a growing industrial economy would significantly alter the relationship between medicine and disability in America.

II The Rise of the Regulars, 1860–1950

The most destructive military conflict in American history, the Civil War, increased the visibility of certain types of disability and hastened changes that were beginning to take place within medical practice during the 1840s and 1850s. The Union alone constructed 200 hospitals and treated more than one million patients by the war's end in 1865 (Rosenberg 1995). Physicians performed 60,000 surgeries during the war, three-quarters of which were amputations. Nearly 10% of the war's 476,000 wounded soldiers endured an amputation. Between 1861 and 1873, the federal government issued 150 patents for artificial limbs. It issued the first patent for a wheelchair in 1869. In addition to physical disabilities, the war produced mental distress in both combatants and noncombatants. Although all madness cannot be attributed to the war, the federal census showed a 100% increase in insanity from 1850 to 1870. By 1880, there were about 91,000 insane people in a population of 50 million, or one in every 554 Americans. The asylum system, like the hospital system, expanded significantly during and after the war. In 1880, there were about 41,000 people incarcerated in 75 public and 40 private asylums scattered across 38 states (Rembis forthcoming).

While the Civil War produced intense and relatively rapid changes in the history of disability, madness, and modern medicine, large demographic and economic shifts played an equally important role in producing change over time. Among the most important of these shifts was the emancipation of about 4 million enslaved people. At the same time, the American population increased from 17 million in 1840 to 50 million in 1880. Most of these new arrivals came from Ireland and Germany. Forced by famine, poverty, religious violence, and economic dislocation to leave their homelands, they came to the USA to build a new life. Between 1880 and 1920, the nation's population doubled, from 50 million to 100 million. Unlike earlier arrivals, immigrants entering the USA around the turn of the twentieth century came from southern and eastern Europe. Although many of these "new" immigrants lived in rural areas, many more lived in towns and cities and worked in America's growing industries. By 1920, for the first time, more Americans (51%) lived in cities than on farms. When they became sick or injured, the growing American population increasingly turned to "regular" physicians and hospitals for treatment and relief.

Although a diverse range of healers continued to practice medicine in the decades following the Civil War, regular physicians used organizations such as the AMSAII and the AMA, as well as an expanding network of hospitals, asylums, medical schools, clinics, and institutes, to enhance their power and authority in American society. The development of new technologies such as anesthesia (1846), antiseptic (1880) and shortly thereafter aseptic procedures, "germ theory" (after 1880), the x-ray (1895) and lab-based or scientific medicine made it more difficult for people to practice medicine in the home. While patent medicines and home remedies persisted well into the twentieth century, medical knowledge and training became more exclusive, and medical practice moved outside the home in the decades after the Civil War.

By the second decade of the twentieth century, regular physicians had solidified their position of authority within America's increasingly racialized and segregated healthcare landscape. The number of hospitals in the USA increased from 178 in 1872 to more than 4,000 in 1910, and the percentage of physicians affiliated with hospitals increased from about 14% to nearly 50% during the same time (Starr 2017). The number of public and private asylums increased from 115 in 1880 to 521 in 1920. The AMSAII, which became the American Medico-Psychological Association in 1893, renamed itself the American Psychiatric Association (APA) in 1921. The number of regular physicians and hospitals would continue to expand over the course of the twentieth century. By the early twenty-first century, the USA had approximately 880,000 board-certified medical doctors and 6,146 hospitals.

The advent of scientific medicine in the late nineteenth and early twentieth centuries brought with it new ways of thinking about and treating disability and madness. Ideas of illness and recovery, and normal and abnormal functioning and behavior, took hold in the decades surrounding the turn of the twentieth century. Regular physicians increasingly intervened in the lives of disabled and mad people with the intent of "curing" them or returning them to "normal." Some of the interventions employed by physicians were considered drastic, even by contemporary physicians. They included surgical interventions into the bodies of mad people, such as the removal of teeth and large sections of the colon, or the mutilation of their genitals, as well as metrazol, insulin, and electric shock, and lobotomy (Braslow 1997; Scull 2005). Those disabled or mad people who could not be "fixed," especially those who were poor or people of color, became part of America's growing institutionalized population, which included people with epilepsy, people considered feebleminded or mentally defective, and mad and physically disabled people.

Although not all physicians believed in it, the new science of eugenics, which became popular during the early twentieth century, provided both ideological and scientific justifications for the removal of disabled and mad people from society, for medical interventions such as sterilization, and, in rare instances, for "euthanasia." The idea that disabled and mad people were both abnormal and unfit and had either to be cured or removed from society and prevented from reproducing became the dominant ideology during the first half of the twentieth century. While most eugenicists argued that "unfit" individuals could be found in any group of humans, the policies and practices they helped to institutionalize unevenly affected women, racial/ethnic minorities, and poor people. Sterilization, for example, persisted well into the twentieth century and after the Supreme Court decision in *Buck v. Bell* in 1927 was increasingly used on poor white women and women of color. By the early 1970s, approximately 67,000 Americans had been sterilized. By the mid-1950s, the USA had about 600 public and private institutions, which held approximately 560,000 disabled and mad citizens and cost taxpayers hundreds of millions of dollars per year. These systems of intervention and removal were undergirded by what disability rights activists would call a "medical model of madness and disability".

Equally important in the rise of modern medicine and the "medical model" was the concept of rehabilitation. Initially applied to veterans of World War I through the Soldiers Rehabilitation Act of 1918, Congress expanded rehabilitation to include injured (mostly white male) workers with the passage of the Smith-Fess or Civilian Rehabilitation Act of 1920. Historically, disabled veterans were entitled to federal pensions for the duration of their lives, and when they died, surviving spouses could claim their benefits. With the passage of legislation in the wake of World War I, veterans would no longer receive pensions. And both veterans and civilians would receive vocational training and assistance, with the goal of returning them to the labor market (Linker 2011). From the outset, federal and state rehabilitation programs were exclusionary. They did not include or instead offered inferior services to Black veterans (Lawrie 2014). Because few women served in the military and because the types of labor most women performed (i.e., reproductive and domestic labor) were not covered under the Civilian Rehabilitation Act, they were also excluded from most rehabilitation programs. Federal and state rehabilitation programs expanded after World War II, but they did not become entirely inclusive (Puaca 2017). As a result, divisions emerged between veterans and mostly white, male "employable" disabled people, and those disabled and mad people marked for institutionalization and removal from society.

III Deinstitutionalization and the Social Model, 1950–2000

By the mid-1950s, the medical industry oversaw a vast network of institutions that spanned the country. State-funded institutions held approximately 560,000 people with intellectual and developmental disabilities, epilepsy, mad people, and people with sensory and physical disabilities. Medical practitioners at the time advised families living with disabilities and madness to institutionalize their disabled or mad relatives and to try to forget them. Common medical wisdom held that if a disabled or mad person could not be cured or rehabilitated and if medical professionals considered them incapable of experiencing the benefits of education and independent labor or fulfilling the gendered role of wife and mother, they would be an emotional and financial drain on their family and a source of public shame.

Medical authority was not absolute, however. Beginning as far back as the early nineteenth century, mad people and their allies spoke out in an effort to reform the nation's growing asylum system (Rembis forthcoming). By the end of the century, blind, deaf, and deaf/blind people were forming their own organizations to advocate for their rights. During the 1930s, people with physical and sensory disabilities came together to pressure the Roosevelt administration to include disabled people in New Deal work programs (Longmore and Goldberger 2000). In the decades after World War II, parents began speaking out in behalf of their intellectually and developmentally disabled children. By the 1960s and early 1970s, survivors of polio and people with other types of disabilities, as well as mad people, began forging their own movements for civil rights, independent living, and the deinstitutionalization of disabled and mad people.

Although their beliefs did not always align with one another, from the mid-nineteenth century on, disabled and mad people found allies among journalists, lawyers, family members, and physicians and other healthcare workers, with whom they organized to reform, and later close, many state institutions. The first American asylum exposés appeared in the 1830s. They were so common by the 1870s that one reporter called them, "a somewhat hackneyed theme" (Rembis forthcoming). An ex-inmate of an asylum, Clifford Beers, and his physician and reformer colleagues led the early twentieth-century mental hygiene movement, which had as one of its goals reforming the nation's asylums. With the help of reporters, conscientious objectors to World War II exposed life inside America's institutions to the world in the late 1940s (Taylor 2009). By the time the investigative reporter, Geraldo Rivera, entered the Willowbrook State School on Staten Island in New York City in 1972, disabled and mad people and their allies had been speaking out about the conditions within institutions for more than a century (Rembis 2019).

After World War II, disabled and mad people and their allies wanted to do more than expose the conditions within institutions and work for reform. They wanted to move disabled people out of institutions into the community. A 1950 study, counted 88 local parent groups – most of which had been formed between 1946 and 1950 – with 19,300 members in 19 states who were advocating for their disabled children to live in the community. These local groups came together and formed the National Association for Retarded Children (NARC). By 1960, the NARC had 681 local affiliates and a membership of 62,000 people, most of whom were parents, dedicated to finding care and education for their disabled children outside of large state institutions (Castles 2004).

Beginning in the 1960s, young disabled people, primarily in places like Berkeley, California, New York City, and the University of Illinois, began speaking out for their right to a mainstream education, including post-secondary education, and their right to live and work independently in the community (Patterson 2012). About the same time, mad people who considered themselves ex-patients or ex-inmates of asylums and survivors of psychiatric diagnosis and treatment began the anti-psychiatry and mad people's liberation movements. People with intellectual and developmental disabilities, and their allies, as well as disabled and mad people became part of a broad disability rights movement that by the end of the 1960s was gaining momentum and increasing recognition from the federal government.

In their efforts to secure their rights, activists developed the social model of disability, which was a direct response to decades of institutionalization, neglect, and abuse, and the long history of the medicalization and pathologization of disability and madness. Put simply, the social model of disability made a critical distinction between impairment (body) and disability (society). It rooted disabled people's devalued and marginalized status in societal barriers that disabled them, not in any individual deficit. Disability studies scholars would later refer to this system of exclusion as "ableism." They argued that ableism and ableist attitudes were present in all societies that were built by and for nondisabled people (Goodley 2011, 2014). Beginning with the Section 504 sit-ins in April 1977, and continuing through the passage of the ADA in 1990, disability rights activists used the social model of disability to counter the older and more deeply entrenched medical model of disability (Heumann 2020). They argued that structural barriers within the built environment and long-standing discriminatory practices needed to be eliminated from society to ensure the equal rights of all people with disabilities.

While many mad people allied themselves with the disability rights movement, they maintained a more critical stance toward modern medicine, and specifically psychiatry, which led some mad activists and Mad Studies scholars to critique the social model of disability. Mad people argued that the social model, which separated impairment and disability, and left impairment largely untheorized, perpetuated medicalized ways of knowing disability. Under the social model of disability, the idea that something was an impairment was not challenged by activists or theorists. It was simply set aside, and framed as difference, while they focused on eliminating external barriers to the integration of disabled people into society. Mad activists who challenged psychiatric diagnosis – what they called psychiatrization – could not, and did not, leave impairment untouched. Instead, they, along with feminists and queer and critical race theorists, challenged disability rights activists and Disability Studies scholars to think in more nuanced ways about impairment, leading to more robust critiques of the medical model of disability (LeFrançois et al. 2013; Russo et al. 2016; Spandler et al. 2015).

IV Post-2000

With the passage of dozens of federal laws, including the Americans with Disabilities Act (1990), and court decisions such as the Willowbrook Consent Decree (1975) and the Supreme Court ruling in Olmstead v. L. C. (1999), most disabled Americans were able to live in the community and attend the same schools as their nondisabled peers by the early twenty-first century (Rembis 2019). Yet disabled people remained much less likely than nondisabled to complete their K-12 education, earn a post-secondary degree, and get a job. Unemployment rates remained high for disabled people, especially for those people living with intellectual and developmental disabilities and madness or distress. Homelessness and incarceration in jails and prisons became a serious concern for people living with mental health problems, especially people of color. Jails in places like New York City, Chicago, and Los Angeles, and the police, became part of the medicalized, and often punitive and carceral, response to madness during the early twenty-first century (Rembis 2014). Disabled and mad activists with their allies have done a lot to challenge the authority of the medical model and to make the world more accessible, but a lot of work remains for them to achieve full citizenship.

Despite the rising influence of the social model of disability, medicalized ways of knowing and treating disability and madness have persisted into the twenty-first century. Public funding for community-based services and supports, and programs such as vocational rehabilitation, remain medicalized. The biomedical model of madness has become increasingly dominant since the third revision of psychiatry's Diagnostic and Statistical Manual of Mental Disorders (DSM) in 1980. Despite mounting opposition, which includes high-ranking physicians who have spoken out against the fifth revision of the DSM, the biomedical model shows no signs of abating. It seems that with the mapping of the human genome at the end of the twentieth century, physicians and government officials have become more concerned with discovering the genetic or molecular causes of various human conditions, and with creating even more individualized interventions, than they are with attending to the important role that social and economic inequities play in determining health disparities. The advent of genetically tailored, individualized medicine, which involves the search for "biomarkers" of madness, as well as the development of new reproductive technologies, including in vitro fertilization, prenatal diagnosis, and selective abortion, have raised new questions and new concerns for disabled and mad activists, and scholars working in Disability Studies and Mad Studies. It seems that the often contentious, but in some ways necessary, relationship between disability, madness, and modern medicine will persist well into the future.

References

Blackie, Daniel. 2014. "Disability, Dependency, and Family in the Early United States." In *Disability Histories*, edited by Susan Burch and Michael Rembis, 17–34. Urbana: University of Illinois Press.

Boster, Dea H. 2013. *African American Slavery and Disability: Bodies, Property, and Power in the Antebellum South, 1800–1860*. New York: Routledge.

Braslow, Joel T. 1997. *Mental Ills and Bodily Cures: Psychiatric Treatment in the First Half of the Twentieth Century*. Berkeley: University of California Press.

Castles, Katherine. 2004. "'Nice, Average Americans': Postwar Parents' Groups and the Defense of the Normal Family." In *Mental Retardation in America: A Historical Reader*, edited by Steven Noll and James W. Trent, pp. 351–370. New York: New York University Press.

Edwards, R. A. R. 2012. *Words Made Flesh: Nineteenth-Century Deaf Education and the Growth of Deaf Culture*. New York: New York University Press.

Goodley, Dan. 2011. *Disability Studies: An Interdisciplinary Introduction*. Los Angeles, CA: SAGE.

———. 2014. *Dis/Ability Studies: Theorising Disablism and Ableism*. Abingdon, Oxon; New York: Routledge.

Groce, Nora Ellen. 1985. *Everyone Here Spoke Sign Language: Hereditary Deafness on Martha's Vineyard*. Cambridge, MA: Harvard University Press.

Heumann, Judith E. 2020. *Being Heumann: An Unrepentant Memoir of a Disability Rights Activist*. Boston, MA: Beacon Press.

Lawrie, Paul R. D. 2014. "'Salvaging the Negro': Race, Rehabilitation, and the Body Politic in World War I, America 1917–1924." In *Disability Histories*, edited by Susan Burch and Michael Rembis, pp. 321–344. Urbana: University of Illinois Press.

LeFrançois, Brenda A., Robert Menzies, and Geoffrey Reaume. 2013. *Mad Matters: A Critical Reader in Canadian Mad Studies*. Toronto: Canadian Scholars' Press.

Linker, Beth. 2011. *War's Waste: Rehabilitation in World War I America*. Chicago: University of Chicago Press.

Longmore, Paul K., and David Goldberger. 2000. "The League of the Physically Handicapped and the Great Depression: A Case Study in the New Disability History." *The Journal of American History* 87, no. 3 (December 1): 888–922. doi:10.2307/2675276.

Nielsen, Kim E. 2012. *A Disability History of the United States*. Boston, MA: Beacon Press.

Oshinsky, David. 2017. *Bellevue: Three Centuries of Medicine and Mayhem at America's Most Storied Hospital*. New York: Anchor Books.

Patterson, Lindsey. 2012. "Points of Access: Rehabilitation Centers, Summer Camps, and Student Life in the Making of Disability Activism, 1960–1973." *Journal of Social History* 46, no. 2 (December 1): 473–499. doi:10.1093/jsh/shs099.

Puaca, Laura Micheletti. 2017. "The Largest Occupational Group of all the Disabled: Homemakers with Disabilities and Vocational Rehabilitation in Postwar America." In *Disabling Domesticity*, edited by Michael Rembis, pp. 73–102. New York: Palgrave Macmillan.

Rembis, Michael. Forthcoming. *Writing Mad Lives in the Age of the Asylum*. New York: Oxford University Press.

———. 2014. "The New Asylums: Madness and Mass Incarceration in the Neoliberal Era." In *Disability Incarcerated*, edited by Liat Ben-Moshe, Chris Chapman, and Allison Carey, pp. 139–159. New York: Palgrave.

———. 2019. *Disability: A Reference Handbook*. Contemporary World Issues Series Santa Barbara, CA: ABC-CLIO.

Richards, Penny. 2014. "Thomas Cameron's 'Pure and Guileless Life,' 1806–1870: Affection and Developmental Disability in a North American Family." In *Disability Histories*, edited by Michael Rembis and Susan Burch, pp. 35–57. Urbana: University of Illinois Press.

Rosenberg, Charles E. 1995. *The Care of Strangers: The Rise of America's Hospital System*. Baltimore, MD: Johns Hopkins University Press.

Russo, Jasna, and Angela Sweeney. 2016. *Searching for a Rose Garden: Challenging Psychiatry, Fostering Mad Studies*. Monmouth: PCCS Books.

Scull, Andrew T. 2005. *Madhouse: A Tragic Tale of Megalomania and Modern Medicine*. New Haven, CT: Yale University Press.

Spandler, Helen, Jill Anderson, and Bob Sapey. 2015. *Madness, Distress and the Politics of Disablement*. Bristol: Policy Press.

Starr, Paul. 2017. *The Social Transformation of American Medicine: The Rise of a Sovereign Profession and the Making of a Vast Industry*. New York: Basic Books.

Taylor, Steven J. 2009. *Acts of Conscience: World War II, Mental Institutions, and Religious Objectors*. Syracuse, NY: Syracuse University Press.

Further Reading

Burch, Susan, and Michael Rembis, eds. 2014. *Disability Histories.* Urbana: University of Illinois Press.
Longmore, Paul K., and Catherine J. Kudlick. 2016. *Telethons: Spectacle, Disability, and the Business of Charity.* New York: Oxford University Press.
Rembis, Michael. Forthcoming. *Writing Mad Lives in the Age of the Asylum.* New York: Oxford University Press.
Stern, Alexandra. *Telling Genes: The Story of Genetic Counseling in America.* 2012. Baltimore, MD: Johns Hopkins University Press.
Taylor, Steven J. 2009. *Acts of Conscience: World War II, Mental Institutions, and Religious Objectors.* Syracuse, NY: Syracuse University Press.

2

EUGENICS, DISABILITY, AND BIOETHICS

Robert A. Wilson

I The Eugenic Touchstone

Despite having peaked in worldwide influence approximately 100 years ago and having few unbridled advocates since 1945, eugenics continues to serve as a touchstone for a number of core debates and issues within disability bioethics today. Eugenics is both a set of ideas and a historical social movement anchored in the thought that we should use scientific knowledge and technology to selectively direct the traits of future generations of people. In short, the overarching aim of eugenics is to bring about *intergenerational* human improvement. As a project that takes itself to be ameliorative, to make things better, eugenics rests on a distinction between some notion of better and worse traits and, seemingly, better and worse sorts of people.

I begin by saying enough about eugenics to explain why disability is central to eugenics (Section II). I then elaborate on why cognitive disability has played and continues to play a special role in eugenics and in thinking about moral status (Section III) before identifying three reasons why eugenics remains a live issue in contemporary bioethics (Section IV). After a reminder of the connections between Nazi eugenics, medicine, and bioethics (Section V), I return to take up two more specific clusters of issues at the intersection of eugenics, disability, and bioethics. These concern questions of life, death, and reproductive value (Section VI) and the value of standpoint theory and epistemology for understanding some of the tensions between bioethics and disability in light of a shared eugenic past (Section VII).

II Eugenic Traits and Disability

Eugenics is typically thought of as a set of ideas that began in the writings of Francis Galton in the nineteenth century, developed as a social movement in the first half of the twentieth century, and ended with the atrocities of the Nazi regime, including the murder of millions of people deemed to be of inferior stock. There are limitations to this received view of the history of eugenics, including neglecting the perspectives of survivors and creating the false impression that eugenics is only of historical interest. The recent sterilization of Latina and African-American women in the California prison system and of girls and women with intellectual disabilities in Australia underscores the limitations of this "eugenics past" view (Wilson

DOI: 10.4324/9781003289487-4

2018, ch.1). Recent defences of eugenics by philosophers and bioethicists reinforce the naiveté of such a view (Wilson 2019; see also Section IV).

Galton's eugenics arose within a broader context in which evolutionary thinking had been adapted to social transformation and change, with forms of artificial selection occupying centre stage in the opening chapters of Charles Darwin's *On the Origin of Species by Means of Natural Selection* in 1859. Darwin's classic "one long argument" for natural selection begins with an extended analogy between the power of artificial selection, directed by human agency and applied to farming animal stocks and plant species, and the idea of selection without such direction: natural selection.

This analogy and focus on human improvement can create the impression that early eugenic thought was chiefly directed at what later would be called *positive* eugenics—the selection of desirable traits to be passed down to future generations. Yet the development of eugenics in North America around the *eugenic family studies* in the 1870s with their focus on so-called degenerate families should remind us that negative eugenics was an integral part of eugenic thinking from that outset. So while the distinction between negative and positive eugenics is useful for some purposes, the two are closely entwined historically as well (I would suggest) conceptually.

Galton brought two general ideas forcefully together that go far beyond the bare-bones idea that eugenics is a project of human improvement. These are the idea that *human reproductive value is unevenly distributed* both within and across human populations and the idea that *we can direct the constitution of human populations over generational time* by harnessing the insights of science and technology. Galton defines the term "eugenics" as "[t]he science of improving stock, not only by judicious mating, but whatever tends to give the more suitable races or strains of blood a better chance of prevailing over the less suitable than they otherwise would have had" (Galton 1883, 25n).

The so-called science of eugenics sought to provide two ways of achieving this goal of improving stock. First, eugenics provided the means to distinguish those of higher quality reproductive value from those of lower quality reproductive value; second, it provided the means to guide, constrain, and even shape human populations to promote higher quality people in future generations. Intelligence testing was co-opted to play this first role early on in the twentieth century, following the advance it made through the introduction of what later became a standardized test by Binet. Policies of eugenic sexual sterilization and immigration restriction laws were determinate forms of eugenic policies that guided, constrained, and shaped future populations that many states and nations adopted in that century's first half (Wilson 2018, ch.2).

To say that human reproductive potential is unevenly distributed within and across human populations is a euphemistic way of expressing the idea that some people have traits that make them more valuable as hereditary contributors to future generations, whereas other people have traits that make them less valuable in this respect. That is because the traits themselves have differential value to human society and, further, are assumed to be heritable.

Such traits are what have been called *eugenic traits*: traits that are the basis for policies, laws, and practices that differentially treat people thought to have them. Eugenic traits include both valued traits, such as the artistic and scientific talents that Galton began with, and devalued traits, such as low intelligence, criminality, and sexual promiscuity. Historically, the most common eugenic traits that served as the basis for the practice of eugenic sterilization in at least North America were feeblemindedness, mental deficiency, epilepsy, and relatively indeterminate forms of mental illness, such as insanity (Wilson 2018, ch.3). Here many eugenic traits that are devalued are some form or another of what we categorize as *disabilities*, particularly cognitive and psychiatric disabilities.

This is true not only of the eugenic traits drawn on in practices such as sterilization but of those articulated in the policies and laws governing immigration restriction in the late nineteenth and early twentieth centuries. Immigration has historically been strongly racialized, treating peoples with specific national, linguistic, or cultural shared backgrounds as races. Although the restriction of immigration based on country or region of origin of the immigrant was racialized in countries such as the USA, Canada, and Australia, that racialization itself involved disabling characterizations. Umbrella terms, such as "the unfit", "the "dependent, defective, and delinquent", and "the feebleminded", all indexed disability and were used liberally in describing potential immigrants as racialized others (Dolmage 2011).

This indexing of disability via an encompassing expression is perhaps clearest in Harry Laughlin's attempt to establish the term "socially inadequate person" as the central organizing concept in his model eugenic sterilization law. Such a person "fails chronically in comparison with normal persons, to maintain himself or herself as a useful member of the organized social life of the state" (Laughlin 1922, 446). Most of the ten socially inadequate classes of people that Laughlin specifies designate putative sorts of people with disabilities: the feeble-minded, insane, epileptic, blind, deaf, and deformed. Laughlin then simply extends this conceptualization of state-level sterilization policy to federal-level immigration policy by advocating for the sterilization of all immigrants who are "potential parents of socially inadequate offspring" (451).

III Personhood, Cognitive Disability, and Moral Status

Cognitive or intellectual disability looms large in the history of eugenics, chiefly because the most prominent cluster of eugenic traits—feeble-mindedness or mental deficiency and their determinate forms—either were explicitly named in eugenic laws or policies, or were assumed to underlie or accompany other explicitly identified eugenic traits, such as epilepsy or criminality. The same is true of mental illness and its varieties, such as insanity or psychosis.

Within bioethics, especially that dimension to it informed most directly by moral philosophy, cognitive disability and mental illness have also been of particular importance among conditions viewed as diminishing health and well-being. This is in part because of their relationship to the concept of *personhood* that has featured in discussions of moral status and personal identity and their interaction with questions about life, death, and quality of life. Conceptualized as an inherited state or condition acquired early in life, cognitive disability in particular has been taken to compromise one's moral status as a person.

Consider John Locke's claims in *An Essay Concerning Human Understanding*. What makes for the same person over time is not simply identity in substance or continuous life, but some kind of *psychological continuity*. This is what makes persons morally distinctive for Locke, such that they can be the subjects of moral praise and blame. Locke's views continue a long philosophical tradition of defining persons in terms of their cognitive capacities and holding that such capacities differentiate humans from non-human animals and plants. Locke and the neo-Lockeans who occupy a central place in contemporary discussions of moral status and personal identity emphasize in particular the significance of the ability to remember, plan, or, more generally, bear psychological connections both to the past and the future.

If these are the basis for rights-imbuing moral status, then individuals lacking those capacities—either because they have never acquired them or because they have lost them—are seen to have a reduced moral status. One way in which the claim that individuals lacking certain capacities have a reduced moral status has featured in discussion of the relationship between cognitive disability and bioethics and moral philosophy is through what Eva Kittay calls "leveling by intrinsic properties" (Kittay 2017), where those lacking the intrinsic cognitive capacities that mark personhood have the reduced moral status of non-human animals with comparable intrinsic properties. The centrality of a broad range of cognitive abilities to views of what make human life valuable is also reflected in bioethically informed decisions about the prolongation of life through medical intervention and the corresponding withdrawal of such life-sustaining treatments and voluntary active and physician-assisted suicide.

IV Eugenics Is Alive in Contemporary Bioethics

As a field associated with medicine and health care, bioethics has developed only in the last 50 years, well after the demise of eugenics as a widespread social movement at the end of the Second World War. It is common to view bioethics, at least in the USA, as arising in reaction to knowledge of the atrocities of Nazi eugenics, particularly to how "medicine went mad" under that reign (Caplan 1992). Nonetheless, there are at least three further reasons why bioethics has intervened in the link between eugenics and disability and why eugenics seeps into contemporary disability bioethics.

The first is that those working on the history of eugenics, on the sociology of medical technologies, and, more explicitly, on disability have warned of the possibility of a new eugenics. Here so-called *newgenic* practices and policies primarily operate via individual choice over the use of reproductive technologies, rather than compulsory state mandate. Yet, critics argue, they ultimately achieve outcomes comparable to those of the eugenic past. Such technologies, including in vitro fertilization and prenatal screening with selective termination, provide one way to influence the traits that individuals in the next generation have and so mediate the selective, intergenerational control at the heart of eugenics. Which traits should we select for or select against, when technologies create choices for us here? The deployment of such reproductive technologies also raises questions about the nature of parenting, reproductive autonomy, and what constitutes human well-being and quality of life. These are all questions that bioethicists have grappled with and that are topics of ongoing discussion in disability bioethics.

The second is that some prominent contemporary philosophers and bioethicists have sought to explore forms of eugenics in a more favourable light. Here they respond with seeming bravado to the challenge that the label "eugenics" itself signals a no-go zone, displaying at times what might be thought of as a kind of philosophical shock-jockery. For example, the principle of procreative beneficence, which holds (roughly) that parents have an obligation to select the best possible child they can, has been defended within bioethics as part of such a rejuvenated form of eugenics (Savulescu 2001), suitably distanced from unacceptable forms that eugenics might take; see Stramondo (2018) and Barker and Wilson (2019) for critical discussion with respect to disability. These philosophical explorations might be seen more charitably as aiming to sift the worthy wheat at the core of eugenics from the dehumanizing chaff that is mixed together with it as a result of a contingent and regretful association of eugenics with its Nazification. As Michael Selgelid (a former, long-standing editor of the journal *Monash Bioethics Review*) says circumspectly, "The fact that the previous practice of eugenics was bad does not imply that eugenics, per se, is necessarily an altogether bad thing or that a better future eugenics would not be possible" (Selgelid 2014, 6).

Third and finally, from the standpoint of many people with disabilities, eugenics does not feel that distant from their lived experience (Garland-Thomson 2012). Whether or not people like Selgelid profess an enthusiastic view of a possible eugenic future, such discussion is itself a red flag from the perspective of those with disabilities, especially for those who were the explicit focus of past eugenic policies, practices, and laws. Since eugenics seems to them in all its variations a project aimed at eliminating people *like them*, identifying a possible "better future eugenics" misses the political forest for the theoretical trees (Garland-Thomson 2012).

V Nazi Eugenics, Medical Ethics, and the Birth of Bioethics

In Germany, eugenics or "racial hygiene" antedates the rise of the Nazis in the 1930s, though the Nazis did come to tailor their specific eugenic laws in light of developments in the USA, including Laughlin's model sterilization law. As in North America, physicians came to play important roles in Nazi eugenics, including in the passage and implementation of the original Nazi sterilization law. This law came into effect in January 1934 and led to the sterilization of over 400,000, chiefly those deemed to have one or more of the disabilities specified in the law. Doctors also actively directed a programme of supposed mercy killings focused initially on psychiatrically diagnosed patients. This programme, known as the T4 programme, began in 1939 in German hospitals and was extended by law in 1941 into concentration camps, killing over 200,000 people on the basis of their having putative disabilities. The implementation of these large-scale and long-lasting programmes of eugenic sterilization and euthanasia required the sustained activity of doctors and other medical professionals, offering diagnoses, performing surgeries, and supervising injections and gassings.

As mentioned in Section IV, the origin of at least American bioethics is sometimes described as a counter-reaction to revelations arising through the Nuremberg Trials about the role of racialized and ableist medicine during Nazi rule in Germany. Yet this view of the relationship between bioethics and eugenics ignores or even obscures how entrenched eugenic ideology was within the medical profession during the 1930s and 1940s.

In Nazi Germany, eugenic sterilization and euthanasia were motivated and justified by an appeal to medical ethics (Caplan 2005). Doctors, including paediatricians and psychiatrists, largely saw themselves as behaving morally in recommending sterilizations, undertaking non-consensual medical experimentation, and overseeing a programme of euthanization directed at people with disabilities. The view of bioethics as arising as an antidote within medical ethics to the dehumanizing medical treatment of people, including people with disabilities, is not only difficult to square with the actual roles that doctors played in Nazi Germany; it also ignores the widespread support that the Chicago-based physician Harry Haiselden's advocacy of the euthanization of what he called "defective infants" had within both medical and popular culture in the eugenic 1920s in the USA through his 1917 silent movie *The Black Stork* (Pernick 1996).

VI Newgenic Traits and Reproductive Technologies: The Expressivist Objection

Reproductive technologies—including contraception, prenatal screening, and in vitro fertilization—are generally viewed by able-bodied citizens and in public discourse as increasing parental autonomy and are portrayed within medical contexts as health-conducive. Such optimism is often viewed within disability bioethics, however, as naïve and ignorant about the realities of the eugenic past. In addition, for those with the traits that are seen as important to prevent the occurrence of in future generations—for example, Down syndrome, spina bifida, and blindness—the enthusiasm for the view that such technologies provide for the means of human improvement is often itself taken to be problematic.

Disability bioethics has thus identified a complacency in the endorsement of technologies that enhance reproductive choice and challenged some widespread practices in reproductive health. The best known of these challenges has focused on the routine practice of prenatal screening with selective abortion. Often called the *expressivist objection*, this is the claim that the practice of prenatal screening with selective abortion expresses a strongly negative view of people with those traits. This objection, originating in the work of Marsha Saxton (who has spina bifida) and Adrienne Asch (who was blind), reflects the standpoint of those with the newgenic traits targeted in prenatal screening.

The expressivist objection rests on three claims:

1   The practice of prenatal testing functions chiefly to detect foetuses that have a biological profile predictive of postnatal impairment.
2   The general expectation (but not requirement) in individual instances of this practice is that a foetus with such a profile will be terminated, rather than carried to term.
3   That expectation implies the judgement that such a foetus is not worth carrying to term to become, in turn, a baby, infant, child, then adult with that impairment.

Although one might challenge any of these claims, it is typically the inference from these to the expressivist conclusion—a conclusion not about the foetus terminated but more generally about people with these negatively valued traits—that has been challenged. For example, Bonnie Steinbock says that

> From the fact that a couple wants to avoid the birth of a child with a disability, it just does not follow that they value less the lives of existing people with disabilities, any more than taking folic acid to avoid spina bifida indicates a devaluing of the lives of people with spina bifida.
>
> (Steinbock 2000, 121)

What these claims about a practice that, in effect, aims to prevent the birth of a child with a given trait indicate, one might think, is simply that the corresponding *trait* is not value neutral but negative. And that need not express anything about those with the trait.

Steinbock's example of taking folic acid is developed in terms of the decisions of individuals, rather than in terms of an overall societal practice. Yet it is worth asking whether there is something distinctively devaluing of those with the trait that lies in the practice described by claims 1–3, i.e., something that is not present in other societal practices, such as taking folic acid, that aim to avoid or prevent those traits in future generations. One relevant difference that perhaps allows us to understand the attribution of devaluation is that claims 1–3 describe a practice of termination of an otherwise desired pregnancy, whereas the general practice of taking, recommending, or even prescribing folic acid does not. The first expresses a view of the trait that is so negatively valued that its presence provides a sufficient reason to terminate a process that would otherwise produce a child with that trait; the second only the view that it would be better, other things being equal, for that individual not to have that trait. That expression needs to be understood against the historical reality of the devaluation of the lives of people with disabilities. As Adrienne Asch writes,

For people with disabilities to work each day against the societally imposed hardships can be exhausting; learning that the world one lives in considers it better to 'solve' problems of disability by prenatal detection and abortion, rather than by expending those resources in improving society so that everyone—including those people who have disabilities— could participate more easily, is demoralizing. It invalidates the effort to lead a life in an inhospitable world.

(Asch 2000, 240)

The demoralization here is directly connected to the perception of devaluation, as it would be for parallel cases of sex selection or selection for lightness in colour of skin. In these cases, screening and selective termination arise against a background, respectively, of ableism, sexism, and racism. Traits such as Down syndrome, spina bifida, or blindness, unlike other less desired traits (such as having an elevated risk of high blood pressure or having haemophilia), are sufficiently devalued that individuals with them are better prevented from coming into existence than accommodated with the challenges they will face as people with those disabilities.

Like the eugenic traits of the past, such newgenic traits serve to identify individuals whose lives are not viewed as being as valuable as those without such traits. It is that connection to shared practices of non-inclusion or even outright elimination that makes the contemporary uses of reproductive technologies a site for a form of devaluation of people with a variety of disabilities, one perceived to be continuous with the eugenic past. From the standpoint of those with kindred disabilities, contemporary technologies deployed to prevent or eliminate disabilities in future generations raise more than the spectre of eugenic devaluation.

VII Disability Knowledge and Standpoint Eugenics

One theme arising in the preceding sections is that disability bioethics offers valuable perspectives on the relationship between disability and bioethics in part because of its distinctive embrace of the standpoint of those with lived experience of disability. In contrast with bioethics more generally, which often abstracts away from particular life experiences to arrive at general principles for decision-making and public policy, disability bioethics ascribes a central role to the perspectives of those marginalized by ableism and are often those whose lives are made worse by it.

Here the disability activism slogan *nothing about us, without us* should be understood more strongly than simply a call for inclusion. Instead, it needs to be seen as entailing the epistemic point that knowledge about disability fundamentally derives from knowing what it is like to have an embodied life with disability. One way in which this insight has been articulated is via standpoint theory or epistemology applied to the case of disability. Incorporating what has been called the *inversion thesis*—that those most deeply, negatively affected by a practice of oppression are sometimes in the best position to understand the nature of that oppression—standpoint theory provides general resources for articulating why reliance on the third-person epistemology dominant in bioethics is likely to limit thinking about how disability is best incorporated into bioethics. Part of the relevant standpoint is a commitment to recognize and change the conditions of disability oppression.

Recent recognition of the continuation of policies and practices of eugenic sterilization in the USA, Canada, and elsewhere beyond 1945 has created a space for what we might call a *standpoint eugenics*, eugenics from the standpoint of those who have survived it (Wilson 2018). The Canadian province of Alberta's eugenic sterilization programme, in place from 1928 until 1972, has been explored through video oral histories centred on the experience of eugenics survivors (www.eugenicsarchive.ca), which also structure the documentary film *Surviving Eugenics*. Such oral histories add further layers to the relationship between eugenics, disability, and bioethics. This is in part because many eugenics survivors in Alberta were wrongfully institutionalized and sterilized, and in part because self-advocates in the local disability community in Alberta, particularly those parenting with disability in some way, identify as survivors of the eugenic legacy in Western Canada. For them, eugenics is not a distant historical episode that ended in Germany in 1945 but part of the backyard in which they grew up.

References

Asch, Adrienne. 2000. "Why I Haven't Changed My Mind About Prenatal Diagnosis: Reflections and Reminders." In *Prenatal Testing and Disability Rights*, edited by Erik Parens and Adrienne Asch, pp. 234–258. Washington, DC: Georgetown University Press.

Barker, Matthew J. and Robert A. Wilson. 2019. "Well-Being, Disability, and Choosing Children." *Mind* 128 (April 2019): 305–328. https://doi.org/10.1093/mind/fzy039.

Caplan, Arthur J, ed. 1992. *When Medicine Went Mad: Bioethics and the Holocaust*. Totowa, NJ: Humana Press.

Caplan, Arthur J. 2005. "Too Hard to Face." *Journal of the American Academy of Psychiatry and the Law Online* 33: 394–400.

Dolmage, Jay. 2011. "Disabled Upon Arrival: The Rhetorical Construction of Disability and Race at Ellis Island." *Cultural Critique* 77: 24–69.

Galton, Francis. 1883. *Inquiries into Human Faculty and Its Improvement*. London: MacMillan.

Garland-Thomson, Rosemarie. 2012. "The Case for Conserving Disability." *Journal of Bioethical Inquiry* 9(3): 339–355.

Kittay, Eva Feder. 2017. "The Moral Significance of Being Human." *Proceedings and Addresses of the American Philosophical Association* 91: 22–42.

Laughlin, Harry H. 1922. "Model Sterilization Law." ch.XV of His Eugenical Sterilization in the United States. Chicago: Psychopathic Laboratory of the Municipal Court of Chicago.

Pernick, Martin S. 1996. *The Black Stork: Eugenics and the Death of 'Defective' Babies in American Medicine and Motion Pictures Since 1915*. New York: Oxford University Press.

Savulescu, Julian. 2001. "Procreative Beneficence: Why We Should Select the Best Children." *Bioethics* 15(5/6): 413–426.

Selgelid, Michael J. 2014. "Moderate Eugenics and Human Enhancement." *Medical Health Care and Philosophy* 17: 3–12.

Steinbock, Bonnie. 2000. "Disability, Prenatal Testing, and Selective Abortion." In *Prenatal Testing and Disability Rights*, edited by Erik Parens and Adrienne Asch, pp. 108–123. Washington, DC: Georgetown University Press.

Stramondo, Joseph. 2018. "Disabled by Design: Justifying and Limiting Parental Authority to Choose Future Children with Pre-Implantation Genetic Diagnosis." *Kennedy Institute of Ethics Journal* 27(4): 475–500.

Wilson, Robert A. 2018. *The Eugenic Mind Project*, Cambridge, MA: MIT Press.

———. 2019. "Eugenics Undefended." *Monash Bioethics Review* 37(1–2): 68–75.

Further Reading

Campbell, Stephen M., and Joseph A. Stramondo 2017. "The Complicated Relationship of Disability and Well-Being." *Kennedy Institute of Ethics Journal* 27(2): 151–184.

Garland-Thomson, Rosemarie. 2017. "Disability Bioethics: From Theory to Practice." *Kennedy Institute of Ethics Journal* 27(2): 323–339.

Miller, Jordan, Nicola Fairbrother, and Robert A. Wilson. 2014. *Surviving Eugenics*. Vancouver: Moving Images Distribution. *EugenicsArchive.ca/film*.

Scully, Jackie L. 2019. "The Responsibilities of the Engaged Bioethicist: Scholar, Advocate, Activist." *Bioethics* 33: 872–880.

Shakespeare, Tom. 2019. "When the Political Becomes Personal: Reflecting on Disability Bioethics." *Bioethics* 33: 914–921.

Shea, Matthew. 2020. "The Quality of Life is Not Strained: Disability, Human Nature, Well-Being, and Relationships." *Kennedy Institute of Ethics Journal* 29(4): 333–366.

Wilson, Robert A. 2018. "Eugenics Never Went Away." *Aeon Magazine.* https://aeon.co/essays/eugenics-today-where-eugenic-sterilisation-continues-now.

3

THEORIES OF DISABILITY

Joel Michael Reynolds

This chapter is adapted from chapter 3 of Reynolds, Joel Michael. *The Life Worth Living: Disability, Pain, and Morality* **(2022), University of Minnesota Press.**

Humans are creatures of categorization. We categorize for many reasons. For power. For research. For safety. We categorize for speed, ease, and even enjoyment. While the desires behind and effects of it are many, one thing is certain: categorization proves exceptionally useful for organisms like us. Safe/Dangerous. True/False. Near/Far. Working/Broken. We can't carry out the tasks of a day, much less a life, without the use of both binaries like these and also more complex, highly refined conceptual categories like those we use for colors, for time, for emotions, for labor, for friendship, for art, and, crucially, for all the many ways we categorize people. While categorization is unavoidable, sometimes we get things terribly wrong, and doing so can result in grave consequences.

In this chapter, I argue that disability is, historically, one of those "wow-we've-gotten-this-wrong" categories. Most people are taught as children that "disability" is a binary category: you are disabled or you are non-disabled/able-bodied.[1] But that is not how disability works. By canvassing a range of theories of disability, this chapter explains why thinking so is not just a result of lack of knowledge, but also a flint spark for deeply unjust ways of thinking about other humans and human societies more generally. My aim is not to advance one particular theory of disability over another, but instead to provide a brief introduction to the theoretical study of disability and to foster an appreciation of the complexity involved in careful reflection upon the meaning of disability and the many implications that follow.

One of these implications concerns history. More often than not, across historical time periods, cultures, and otherwise distinct social orders, one finds what I call the ableist conflation of disability with pain and suffering (Reynolds 2022). That is to say, one finds a pervasive habit of thought wherein experiences of pain and suffering are conflated with experiences of disability—experiences whose form, mode, matter, or style of living is considered categorically outside ableist norms. I offer *the ableist conflation* at the outset of this chapter on theories of disability as a concept to capture the underlying presuppositions that guide nearly all ableist discourses and practices in philosophy, ethics, politics, medicine, local, national, and international policy, and beyond.

DOI: 10.4324/9781003289487-5

Although it can take many forms, the ableist conflation involves some variation of at least the following four claims:

1   Disability necessarily involves a lack or deprivation of a natural good.
2   Deprivation of a natural good is a harm.
3   Harm causes or is itself a form of pain and suffering.
4   Given (1)–(3), disability comes along with or directly causes pain and suffering.

The ableist conflation functions in part by capitalizing upon the ambiguity of the array of terms it involves. "Disability," "harm," "pain," and "suffering" are all uncritically underdefined, as are the relations between them. A central goal of this chapter is to see how various theories of disability challenge the ableist conflation of disability with pain and suffering.

I Theorizing about Disability

Let's begin by appreciating the high stakes of theorizing about disability. What people think disability *is* impacts everything from the architecture and management practices of your local grocery store, to who gets a part in Hollywood blockbusters, to the tenets and obligations of international treaties. Claims about and understandings of what disability is impact everything: from whether your grandparents will receive state- or federal-level financial support for in-home care to whether you can be fired for "failing" to carry out certain tasks, even if you can't do so due to damages directly caused by your employer, to how technologies are developed and marketed at national and international levels, to whether or not your family thinks you are a "genius" or a "freak." In short, debates in disability theory touch upon and have the potential to impact nearly every aspect of human life and society. This means that debates over the meaning of disability are not debates merely about disabled people or "disability" as some abstract concept. On the contrary, these debates affect everyone and nearly everything.

Tellingly, it is commonplace for discussions of disability to begin by lamenting the difficulty of defining it. Disability studies scholar and theologian Nancy Eiesland demonstrates this well:

> The differences among persons with disabilities are often so profound that few areas of commonality exist. For instance, deafness, paralysis, multiple sclerosis, and [intellectual disability] may produce the same social problems of stigma, marginality, and discrimination, but they generate vastly different functional difficulties. Further, people with the same disability may differ significantly in the extent of their impairment…disabilities can be either static or progressive, congenital or acquired. The social experience of a person who becomes disabled as an adult may differ significantly from that of a person with a congenital disability. These dissimilarities make a broad definition of people with disabilities difficult, if not impossible.
>
> *(Eiesland 1994, 23–24)*

More provocatively, Fiona Kumari Campbell argues that "there is no literal referent for the concept" of disability. As soon as one starts to carefully reflect upon the concept of disability, "its meaning loses fixity and generality, and ultimately collapses" (Campbell 2001, 43). For Campbell, "disability" picks out nothing distinct because there is no coherent set of features of to which it refers—experiences of disability are simply too different to be grouped together. For other disability theorists, however, this claim gets things exactly backward. Disability rights have been won due to the fact that it is not the natural, but the social dimensions of disability that primarily define it for purposes of group identity, political solidarity, and the like (Barnes 2016). What unifies disability are experiences of being excluded on the basis of one's impairments. "Disability" is indeed "spoken in many ways," to invoke Aristotle's famous claim about being, but that does not relegate the concept to nonsense. On the contrary, the power of the concept of "disability" should instead be taken as an indication of the purchase that concept has on our lives.

Whether one sides with a deflationary approach ("the concept of disability refers to nothing"), a solidarity approach ("the concept of disability picks out shared experiences of oppression and discrimination"), or some other approach, most disability theorists agree on at least the following claim: appreciating the complexity of the meaning of disability involves an extremely wide range of considerations, including those relating to history, technology, religion, law, medicine, economics, gender, sexuality, race, ethnicity, indigeneity, class, citizenship, politics, science, culture, policy, architecture, and more (Nielsen 2012).

II Theories of Disability

During the first class of a typical "Disability 101" course, a distinction will be made between the medical model of disability and the social model of disability. As with all introductions to a complex issue, this is an oversimplification. In truth, there are multiple "medical models" of disability just as there are multiple "social models" of disability. And even if you could get disability theorists to agree on the various forms those models take, they are unlikely to agree on the precise content. Yet, perfect agreement is not necessary to appreciate the core distinctions and disagreements in disability theory. When all is said and done, there are three main types: individual theories of disability, social theories of disability, and historical theories of disability.

2.1 Individual Theories of Disability

The leading individual theory of disability is the medical theory of disability, often called the "medical model" of disability. This theory (or model, if you prefer—I will use these terms interchangeably) conceptualizes disability as an individual tragedy due to genetic or environmental misfortune. Accordingly, disability is understood in the medical model as a *pathology* in that term's etymological sense: it gives an account of a suffering one is undergoing. For example, in *Fundamentals of Nursing: Standards and Practice*, "disability" is mentioned nineteen times (De-Laune and Ladner 2011). However, the social dimensions and meanings of disability are *never* mentioned, even when referencing the ADA (197)! In the majority of cases, "disability" is listed alongside illness or disease (43, 62, 247, 312, 390, 490, 1305, etc.) Revealingly, "healing" is defined in the glossary as the "process of recovery from illness, accident, or disability" (1347). In a different medical textbook, this one on terminology, the word "disability" is used four times, once in a definition of Alzheimer's ("disease of structural changes in the brain resulting in an irreversible deterioration that progresses from forgetfulness and disorientation to loss of all intellectual functions, total disability, and death") and three times in the context of explaining a form of "developmental disability" (Willis 2006, 388, 413–16). Whether one looks to basic or clinical science, it is hard not to suspect that disability's meaning (namely, that the medical theory of disability *just is* what "disability" means) is assumed to be so obvious as to warrant no explanation.

The medical model of disability is also described as a *tragic* theory of disability. The cultural, social, political, and historical influence of the medical model and its tragic figuration of disability is hard to overstate. For example, it is only insofar as disability is seen as fundamentally tragic—and as individual problem in the medical model's sense—that the logics of inspiration porn and pity porn, of the supercrip and of Tiny Tim, can get off the ground. For those unfamiliar with these tropes, I'll discuss each in turn.

Inspiration porn is everywhere once you know how to see it. Consider the many TV ads, posters, infomercials, and stories that involve a person with a disability doing some activity—often one that is completely "normal"—and it being described as a "tale of inspiration." A person with Down syndrome goes to prom. A paraplegic becomes a successful attorney and finds romantic love. This framing only makes sense if one assumes that the person with a disability is fundamentally suffering or lacking to begin with. It then provides the (presumed) able-bodied viewer with an opportunity to feel good at seeing someone with misfortunes experience or achieve something one wouldn't expect them to. This narrative is structured in order to inspire an (able-bodied) person to try harder, believe in the impossible, and never give up (cf. Kafer 2013 Ch. 4). Insofar as disability is understood as a sign of something unfortunate, then in overcoming it, one is praised, even revered. The flip side of inspiration porn is pity porn. In failing to overcome disability, one is looked upon with compassion and mercy, such as in Jerry Lewis' infamously degrading MDA telethons (Longmore 2015). The supercrip is praised while the crip is pitied, yet both provide inspiration to the able-bodied observer who embalms the other's experience in the service of bolstering their sense of self. The medical model of disability assumes that the misfortune of disability is located in an individual and experienced as tragic. It then goes one step further in redirecting the psychosocial effects of that misfortune into one and just one possible response: the hope for rehabilitation, repair, cure, palliation, or, in certain cases, early death.

In conclusion, whether prompting pity and promoting cure for the crip or provoking praise and peddling progress for the supercrip, the medical theory of disability understands disability as a misfortune afflicting an individual. This misfortune is always assumed to have an etiology that could be discovered through symptomatology, whether via sophisticated diagnostic technologies or the careless gazes of hurried practitioners.

2.2 Social Theories of Disability

Social theories of disability are based upon a core conceptual distinction: impairment is different from disability. The classic example is as follows: while one might be *impaired* through paraplegia and use a wheelchair to get around, what *disables* such a person who confronts a flight of stairs is societal: a ramp could have instead been built and someone (or, more likely, many people) actively decided against that. Both with respect to its political and academic dimensions, the history of social theories of disability is complex and a proper analysis would far outstrip the aims at hand (Shakespeare 2014b). Paralleling the enormous practical and theoretical effects feminist theory achieved by demonstrating that neither gender nor sexuality are reducible to sexual differences and that inequalities resulting from these dimensions of human life are primarily shaped by mutable social norms as well as histories of oppression, disability activists made significant gains for people with disabilities by demonstrating that disability is not reducible to impairment and that disability inequality is primarily a product of mutable social norms and histories of ableist oppression. In the USA, the most obvious gain has been the landmark passage of the Americans with Disabilities Act in 1990, an act that created the largest legally protected minority identity in the country's history: people with disabilities.

The political and activist history of the social model of disability is often said to begin with the Union of the Physically Impaired Against Segregation (UPIAS) in the United Kingdom, established in 1972 by Paul Hunt, among others (Albrecht 2006). UPIAS's version of the social model is often called the strong social model, which picks out its Marxist framework and almost ideological emphasis on the social conditions of disablement. I can broach neither that fascinating history, nor the notable variations of the social model in other countries, including the way that the social model was deployed to conceive of disability as a "minority identity" on the heels of the civil rights movement in the USA. This variation drew strength from multiple smaller movements and activist projects, including the Independent Living Movement started at UC Berkeley through the work of Ed Roberts and others such as Judy Heumann (Nielsen 2012).

To the extent they can be separated, the history of the social model in academia is equally, if not more, labyrinthian than its activist history. For example, what could be called the "first wave" of criticism of the social model gained momentum in the 1990s. Susan Wendell argued in her seminal 1996 book *The Rejected Body* that the social model, especially the simpler version deployed by activists, is ill-suited to describe the experiences of people with chronic illness and chronic pain, especially with respect to their differential impact along lines of gender. Similar critiques honed in on how the social model both under-theorizes or simply ignores disabilities that involve chronic pain, chronic illness, and conditions involving constitutive or consuming pain. To take another example, Tom Shakespeare has argued that the distinction between disability and impairment is ultimately untenable. He calls his view a "critical-material" or "critical-realist" theory of disability.

> Any researcher who does qualitative research with disabled people immediately discovers that in everyday life it is very hard to distinguish clearly between the impact of impairment, and the impact of social barriers. In practice, it is the interaction of individual bodies and social environments which produces disability. For example, steps only become an obstacle if someone has a mobility impairment: each element is necessary but not sufficient for the individual to be disabled. If a person with multiple sclerosis is depressed, how easy is it to make a causal separation between the effect of the impairment itself; her reaction to having an impairment; her reaction to being oppressed and excluded on the basis of having an impairment; other, unrelated reasons for her to be depressed? In practice, social and individual aspects are almost inextricable in the complexity of the lived experience of disability.
>
> *(Shakespeare 2014b, 218–19)*

When placed in the larger context of theories of disability, I find it more accurate to term theories like that of Wendell and Shakespeare *biosocial*. They are of a kind with arguments from sociologists like Carol Thomas and Donna Reeve that focus on both personal and social dimensions of impairment, including its often complicated psychological effects (Thomas 2004, Reeve 2014). These theories bring the biological and psychological dimensions of impairment back into focus and demonstrate how impairments themselves, as it were, create effects, both personal and social, that can nevertheless prove resistant to social solutions for amelioration.

A perfectly just society will not solve neuropathic pain, for example, as Shakespeare often argues. Such a society could surely improve one's life along many important dimensions, but solving neuropathic pain, he contends, is less a question of justice and more a question of highly specific biomedical solutions. That is to say, a perfectly just society cannot directly solve the destructive grind of one's temporomandibular joints for one with TMJ or the slow breakdown of one's body due to congenital pain asymbolia, however much living in such a society diminishes the impact of "disability" as understood on the social model (and, to make the stakes even clearer, even an "ideal" society might not allocate the biomedical resources necessary to fix these particular problems in a sufficient way given larger concerns of equity and justice). Biosocial models of disability offer a crucial corrective to the bifurcation of material embodiment and social life as put forward by less-than-nuanced proponents of the social model. On biosocial models of disability, disability is a question not just of the socio-political environment in which a person lives, but also of how they experience their impairments and the extent to which various biomedical interventions—and the private and public research that supports them—can impact such experience.

In summary, the social model of disability marks the most significant shift in understanding disability in history. While there are notable intramural differences among the various types of social models, it is nevertheless the case that the social model, as Rannveig Traustadóttir puts it, "has provided the knowledge base which has informed the international legal development aimed at full participation and human rights of disabled people" (Traustadóttir 2009). On social models of disability, disability is a question of the socio-political environment in which a person lives. That is to say, on social models of disability, *disability is politicized*.

2.3 Historical Theories of Disability

Lennard J. Davis argues that the primary effect of social theories of disability was the creation of a political identity. Although this produced numerous positive effects, he finds the identity politics centered around disability both outdated and theoretically inept. It is especially so in light of increasing awareness of the need for intersectional concerns, whether with respect to the success of larger projects of social justice or just an individual's self-understanding. On the contrary, the phenomenon of disability should lead us to deconstruct prevalent ideas about embodiment and disability, leading to what he terms a "new ethics of the body."

If identities are products of social mediation—that is to say, if there is no such thing as identity categories without their social constitution—then the category of "disability" has no more absolute foundation than any other social phenomenon. Social kinds loop, to invoke Ian Hacking's and, before him, Michel Foucault's body of work. The hierarchies of disability within disability activism and studies demonstrate this point well. For example, those with physical disabilities are typically at the top of the so-called "disability food chain" in terms of political optics, representation, and more. It is hard to imagine the ADA emerging on the heels of activism by people who wear glasses. Or those with fibromyalgia. Or . . . One could go on and on with such counterfactual speculation.

The point is that any number of groups of people with disabilities appear ill-suited relative to the calculi of disability politics under late stage neoliberal capitalism. In short, social models of disability have created a political identity by carving prototypical forms of disability out of the cultural imaginary of the milieu of twentieth century. A cis, straight, white, male wheelchair user is an easier figure to leverage politically than a trans, queer, Black woman with invisible disabilities. Social models base their achievements in simplifications that actively ignore the fact that disability, Davis argues, is so diffuse as to be conceptually incoherent. Is the social model effective as a strategy to create and sustain political solidarity? Yes. Is it based in any defensible ontological solidity? No. Davis brings this set of problematics to a head by proclaiming:

> What we need now is a new ethics of the body [and] what I would like to propose is that this new ethics of the body begin with disability rather than end with it. To do so, I want to make clear that disability is itself an unstable category. I think it would be a major error for disability scholars and advocates to define the category in the by-now very problematic and depleted guise of one among many identities. . . [Disability] must not ignore the instability of its self-definitions but acknowledge that their instability allows disability to transcend the problems of identity politics. In setting up this model we must also acknowledge that not only is disability an unstable category but so is its doppelganger—impairment.
>
> *(2013, 271)*

In Davis' dispersive, historical theory of disability, disability functions to disorient everyone for whom it is a possibility—which is to say, everyone. Disability represents the flux and fluidity of being human, of the lived experience of human being-in-the-world in its many forms and figurations. Despite this flexibility, note that the "new ethics of the body" for which Davis calls nevertheless operates from a principle: corporeal variability. But because it is based upon a fundamentally unstable category, because based upon variability, such an ethics is without normative ground in any traditional sense. On dispersive models of disability, disability is an open question concerning the complex relationship between one's embodiment and the situation in which one finds oneself.

Some go farther than Davis, arguing that not just the modern concept of "disability," but also that of "impairment" are historical constructions. For example, scholars like Shelly Tremain argue that a Foucauldian, genealogical analysis demonstrates that even the concept of "impairment" does not refer to facts or states of affairs about individual bodies. On the contrary, "impairment" is a category arising out of governmental-juridical powers geared toward constituting the disabled subject. Beforehand, there was no such thing as "disability" in the sense we use it in English today. To think that "disability" picks out some fixed feature of the world is to fail to understand and to uncritically analyze how our concepts and epistemic frameworks have changed over the last few centuries and, more specifically, how these changes have correlated with the demands of modern modes of governance. As Tremain puts it:

> Notice that if the foundational (i.e., necessary) premise of the social model—impairment—is combined with the preceding claims according to which modern governmental practices produce—that is, form and deform—the subjects whom they subsequently come to represent by putting in place the limits of their possible conduct, then it becomes more evident that subjects are produced who "have" impairments because this identity meets certain requirements of contemporary social and political arrangements. Indeed, it would seem that the identity of the subject of the social model ("people with impairments") is actually formed in large measure by the political arrangements that the model was designed to contest.
>
> *(2015, 10)*

Tremain's overarching point is not merely that the concept of "impairment" is historically produced and contestable, but that it was specifically produced by governmental and juridical practices to control particular populations. Namely, those judged less valuable to the social and economic order and thus made fungible for enslavement, institutionalization, or other forms of state-sanctioned control. On Tremain's account, "impairment" actually functions to "legitimize the governmental practices that generated it in the first place" (2010, 11). That is to say, the very logic of differentiation at stake in the distinction between disability and impairment operates relative to a standard whose ultimate arbiter is modes of governance, i.e., power relations acting upon and constituting subjects of a nation-state or other such political order.

Recall that the ableist conflation of disability with pain and suffering undermines lived experiences of disability by conceptualizing disability in ways that flatten, hinder, or even eviscerate one's sense of self and purposivity. One of the ways it does so is by restricting the available social concepts—the hermeneutical resources—concerning disability, leaving public imaginaries to wallow in misguided personal, tragic theories of disability and in the ignorance of the actual varied lived experiences of disabled people. Although historical theories of disability are far more defensible than individual theories and are so on multiple counts, they also hinder or, at minimum, are neutral with respect to the conduct of a life. I thus find that there is more work to be done sussing out the normative and theoretical implications of historical theories of disability as they bear upon the lived experience of people with disabilities. This is not to say that these theories must be measured by their function or effect. It is only to say that such a limitation must be acknowledged and that such theories should be recognized as, at bottom, critical theoretical projects that leave much to be desired with respect to concrete projects of disability justice and broader social justice, as well as contributing to the flourishing of actual, existing disabled people.

III Conclusion

Multiple disability studies scholars and philosophers of disability have turned on its head the assumption that disability is ontologically problematic, that it necessarily presents a problem to be solved. As Titchkosky and Michalko aptly put it, "that disability is conceptualized as a problem is what we take to be our problem in need of theorizing" (Titchkosky and Michalko 2012, 127). I would argue that when analyzed as a whole, dominant theories of disability shed light on how disability has become a problematic of its own. Individual, social, and historical theories of disability all converge in one crucial respect: they assume that disability regulates the conduct of a life, and they seek to leverage its regulation toward various ends. Sentient life requires an enormous amount of sorting, classifying, and labeling. Disability is one of the ways in which we categorize humans, and how we deploy "disability" deserves far more attention.

Note

1 While the term "disability" is, of course, a historical product of modern English, the binarization at issue tellingly holds across an extremely wide range of cultures, historical periods, and more. The careful reader will wonder if my use of "categories" is meant to be different in kind from "concepts." For the aims at hand, I will treat these terms interchangeably.

References

Albrecht, Gary L. 2006. Encyclopedia of Disability. 5 vols. Thousand Oaks, Calif.: Sage.

Barnes, Elizabeth. 2016. *The Minority Body*. New York: Oxford University Press.

Campbell, Fiona Kumari. 2001. "Inciting Legal Fictions: 'Disability's' Date with Ontology and the Ableist Body of Law." *Griffith Law Review* 42: 42–62.

Davis, Lennard J. 2013. "The End of Identity Politics." In *The Disability Studies Reader*, edited by, pp.263–277 4th ed. New York: Routledge.

DeLaune, Sue C., and Patricia K. Ladner. 2011. *Fundamentals of Nursing: Standards and Practice*. Delmar: Cengage Learning.

Eiesland, Nancy L. 1994. *The Disabled God: Toward a Liberatory Theology of Disability*. Nashville, Tenn.: Abingdon

Hughes, Bill. 1999. "The Constitution of Impairment: Modernity and the Aesthetic of Oppression." *Disability and Society* 14 (2): 155–72. https://doi.org/10.1080/09687599926244.

Kafer, Alison. 2013. *Feminist, Queer, Crip*. Bloomington: Indiana University Press.

Longmore, Paul K. 2015. *Telethons: Spectacle, Disability, and the Business of Charity*. Oxford: Oxford University Press.

Nielsen, Kim E. 2012. *A Disability History of the United States*. Revisioning American History. Boston: Beacon Press.

Reeve, Donna. 2014. "Psycho-Emotional Disablism and Internalised Oppression." In *Disabling Barriers—Enabling Environments*, edited by John Swain, Sally French, Colin Barnes, and Carol Thomas, pp. 92–98. London: Sage.

Reynolds, Joel Michael. 2022. *The Life Worth Living: Disability, Pain, and Morality*. Minneapolis: University of Minnesota Press.

Shakespeare, Tom. 2014a. *Disability Rights and Wrongs Revisited*. 2nd ed. London: Routledge.

Shakespeare, Tom. 2014b. "The Social Model of Disability." In *The Disability Studies Reader*, edited by Lennard J. Davis, pp. 214–221. New York: Routledge.

Thomas, Carol. 2004. "Developing the Social Relational in the Social Model of Disability: A Theoretical Agenda." In *Implementing the Social Model of Disability: Theory and Research*, edited by. Colin Barnes and Geof Mercer, pp. 32–47. Leeds, U.K.: Disability Press.

Titchkosky, Tanya, and Rod Michalko. 2012. "The Body as the Problem of Individuality: A Phenomenological Disability Studies Approach." In *Disability and Social Theory: New Developments and Directions*, edited by Dan Goodley, Bill Hughes, and Lennard J. Davis, pp. 127–42. New York: Palgrave Macmillan.

Traustadóttir, Rannveig. 2009. "Disability Studies, The Social Model and Legal Developments," In The UN Convention on the Rights of Persons with Disabilities: European and Scandinavian Perspectives. Editors: Oddný Mjöll Arnardóttir and Gerard Quinn. International Studies in Human Rights. Vol. 100. Leiden, Netherlands: Martinus Nijhoff. pp. 1–16.

Tremain, Shelley, ed. 2015. *Foucault and the Government of Disability*. 2nd ed. Ann Arbor: University of Michigan Press.

Wendell, Susan. 1996. *The Rejected Body: Feminist Philosophical Reflections on Disability*. New York: Routledge.

Willis, Marjorie Canfield. 2006. *Medical Terminology: The Language of Health Care*. 2nd ed. Philadelphia: Lippincott Williams and Wilkins.

World Health Organization. 2001. International Classification of Functioning, Disability and Health. Geneva: World Health Organization.

Further Reading

Clare, Eli. 2015. *Exile and Pride: Disability, Queerness, and Liberation*. 2nd Edition. Durham, NC: Duke University Press.

Cureton, Adam, and David Wasserman. 2020. *The Oxford Handbook of Philosophy and Disability*. New York, NY: Oxford University Press.

Hall, Melinda C. 2019. "Critical Disability Theory", *The Stanford Encyclopedia of Philosophy*. edited by Edward N. Zalta. https://plato.stanford.edu/archives/win2019/entries/disability-critical/.

Garland-Thomson, Rosemarie. 1997. *Extraordinary Bodies: Figuring Physical Disability in American Culture and Literature*. New York: Columbia University Press.

McRuer, Robert. 2006. *Crip Theory: Cultural Signs of Queerness and Disability*. New York: New York University Press.

Siebers, Tobin. 2008. *Disability Theory*. Minneapolis: University of Minnesota Press.

PART II

Bioethics

Past and Present

4

A CRITICAL HISTORY OF BIOETHICS

John H. Evans

To further define "critical disability studies bioethics," it is important to be aware of what I will call mainstream bioethics – the bioethics that dominates the public sphere and that disability studies (DS) bioethics will need to engage with or perhaps draw boundaries against. The DS scholar needs to be aware of not only this field and its history, but also its various biases that will constrain or promote the use of discourse from DS. In this chapter I offer a history of mainstream bioethics along with a description of, and explanation for, its various biases. I conclude with a discussion of the trade-offs DS scholars need to be aware of when interacting with mainstream bioethics.

I begin with some disclaimers. First, I primarily write about the history of USA bioethics (for a history of UK bioethics, see Wilson 2014). That said, while every nation has their own particular bioethics, the influence of the USA on at least the Anglophone world has been large. Second, I do not have a disability and have also not previously engaged extensively with the DS field. I write as a historian of mainstream bioethics, and hope that my potentially outsider views will be productive for DS scholars. This text starts by summarizing the history of mainstream bioethics that I have more extensively described elsewhere (Evans 2002, 2012).

I The Jurisdictions in Contemporary Bioethics

To understand both the history of mainstream bioethics and particularly its intellectual biases that may or may not be compatible with DS, we have to think of bioethics as a "task" that various professions have been competing over. This is similar to how various professions have competed over the past 100 years for the task space or "jurisdiction" of controlling misbehaving children. The clergy once had jurisdiction, but now there is an ongoing struggle between psychology and psychiatry for this task. Similar stories can be had about competitions and jurisdictional settlements between nurses and doctors, lawyers and accountants, and many others. I consider DS bioethics to be a distinct competitor to mainstream bioethics. I will largely gloss over how the competition between various professions shaped what we now know as bioethics, and instead start with a description of the contemporary situation. When the DS scholar hears of "bioethics," this term is actually referring to one of four possible jurisdictions.

DOI: 10.4324/9781003289487-7

1.1 Healthcare Ethics Consultation

The first jurisdiction is healthcare ethics consultation (HCEC) which concerns the issues having to do with medical care within medical institutions like hospitals. The topics are extremely limited to those that would occur within the medical setting. The goal in this jurisdiction is to facilitate ethical agreement among individual "stakeholders" such as medical staff, patients, and the family of a patient. A typical debate would be whether to end the life support of someone in a permanent vegetative state. While DS scholars are interested in the topics in this space, mainstream bioethicists have iron-clad jurisdiction.

1.2 Research Bioethics

The second jurisdiction is research bioethics, which focuses on creating procedures for ethical research on humans. This activity occurs through Institutional Research Boards (IRBs). In the USA, every entity that receives government research money must have an IRB, although in practice essentially all human research is conducted with oversight from an IRB. This jurisdiction accounts for a large percent of all bioethical activity but, like the first jurisdiction, is constrained to a very narrow set of issues. A typical question would be whether a research protocol for a clinical trial to develop a treatment for Parkinson's disease using stem cells is conducted in an ethical manner. For this research to be considered ethical, researchers would have to demonstrate that the research accounts for the autonomous decision-making of individuals with informed consent; doing good for the subject and avoiding harm (technically called beneficence and non-maleficence) through risk-benefit analysis; and advancing "justice," which means not experimenting on disadvantaged groups like prisoners and orphans. Again, while DS scholars should be interested in this task, mainstream bioethics has an extremely strong jurisdiction, and DS scholars are largely not involved.

1.3 Public Policy Bioethics

The third jurisdiction is public policy bioethics which debates the ethics of technology and science affecting humans that can be incorporated into general policies that will be applied to all citizens. A recent example is the large group of recent commissions recommending policy concerning human gene editing. DS scholars occasionally participate in this jurisdiction – for example a DS scholar testified before the recent NAS panel on human gene editing. This activity is not just based on policy commissions, but any writing that ultimately is intending to influence law or policy. Jurisdiction is less settled in this task space, and since the topics in this jurisdiction are much broader than in the previous two, it will presumably be of more interest for DS scholars.

1.4 Cultural Bioethics

The final jurisdiction in bioethics is "cultural bioethics," which is the debate that tries to convince the public of the proper ethical course of action for – or proper understanding of – medical and scientific technology outside of the immediate framework of policy. For example, most philosophers are operating in this jurisdiction, debating questions like whether there is a moral obligation to genetically enhance one's children. My sense is that the majority of DS fits in this jurisdiction as DS scholars are debating topics such as what "ability" is, but with no direct connection to policy. Indeed, this very volume is largely in this task space. This is by far the most diffuse part of the debate both topically and, as I will focus on below, in terms of the forms of argumentation that are allowed.

II A Critical History of Bioethical Jurisdictions

The historical evolution of these jurisdictions shows their intellectual biases and points to the challenges for DS. For our purposes, the bioethical debate began in the 1960s, and originally there was only the cultural bioethics jurisdiction. During the 1960s, scientists were worried about the explosion of newfound technological abilities and had many meetings to discuss the emerging technologies of the time, such as birth control, genetic engineering, organ transplantation, and much else. Critically, the debate was about what our ends or goals should be: "Where are we taking ourselves with our new technological abilities?" was the central theme. Should our goal be the perfection of humanity? Obedience to God? The elimination of inequality? For example, Daniel Callahan, co-founder of the Hastings Center and one of the originators of contemporary bioethical debate, called for further debate over "some general, comprehensive, and universal norms for 'the human'" (Callahan 1972, 99). The point of one of the conferences was "not simply the question of the survival or the extinction of [humankind], but what kind of survival? A future of what nature?" (Jonsen 1998, 13).

This debate was also inherently social or cultural, and not about the relationship between particular individuals, as later debates would be. To foreshadow my later discussion, it is this social not individual level debate about what our ends should be that contemporary DS scholars would be most comfortable in.

This social or cultural debate of the scientists about setting the goals of humanity attracted the attention of another profession that arguably held jurisdiction over this task in this era: theology (with some participants from sociology, law, and philosophy). Theologians saw the scientists as infringing on their traditional professional tasks, and challenged the scientists' ability to set the ends or goals of society that should be pursued through biomedical technology.

The theologians and their allies were debating with the scientists what our goals or ends should be with technologies like human genetic engineering and the public began to pay attention, soon getting the attention of elected officials. Elected officials needed an ethics that could be executed by the bureaucratic state. The form of argumentation in the debate quickly changed when the bureaucratic state became the primary audience, and three new jurisdictions were soon formed that used this newfound form of argumentation.

2.1 The Emergence of the Public Policy Bioethics Jurisdiction

There had been congressional hearings as early as 1968 on creating a government commission to oversee research on emerging technologies such as behavior control, birth control, organ transplantation, and human genetic engineering. Congress seemed to be engaged in the same fundamental debate as the scientists and theologians. For example, one senator started the hearings by saying:

> Recent medical advances raise grave and fundamental ethical and legal questions for our society. Who shall live and who shall die? How long shall life be preserved and how should it be altered? Who will make decisions? How shall society be prepared?
>
> *(Jonsen 1998, 90–91)*

Scientists were fearful that Congress would try to directly determine which experiments could and could not be done. The scientists' attempt to avoid government involvement would not be fully successful as public concern increased. In 1972 it was discovered that the U.S. Public Health Service had been conducting a 40-year-long experiment on poor Black men in Tuskegee, Alabama by not treating their syphilis and waiting to see the results. The subjects of the Tuskegee Syphilis Study were under the impression that they were being treated. Additional revelations of similar experiments on unknowing subjects occurred in the same era. Historian David Rothman concludes that the public attention to these scandals provided the final impetus for government intervention into the ethics of researchers (Rothman 1991, 182–189). The state would develop bioethical policies and thus became the primary audience for ethical debate.

Congress could have, as the scientists' feared, directly decided what it thought were the goals or ends of the nation and banned certain technologies and certain practices – becoming the regulator of science and medicine. They did not. They first implicitly created the jurisdiction of public policy bioethics, and established the government as the audience, through the 1974 creation of the first commission whose task was to suggest bioethical policy to the government.

One of the mandated tasks of the Commission was to "conduct a comprehensive investigation and study to identify the basic ethical principles which should underlie the conduct of biomedical and behavioral research involving human subjects" and "develop guidelines which should be followed in such research to assure that it is conducted in accord with such principles" (Jonsen 1994, xiv). In other words, they were to create an ethical system that could be put into public law, and which could be used in a bureaucratic context.

These principles were reported in the Belmont Report, which made a transformation in ethical argument critical to the future of bioethical debate (National Commission for the Protection of Human Subjects of Biomedical and Behavioral Research 1978). The Commission did not decide what the ethics of the public should be regarding a public issue, but instead claimed to have discerned the existing values of all citizens in such a way as the values of the public can be used to create public policy – in this case, human research subjects policy of the executive branch. They were channeling what would later be called "the common morality."

The commission identified three primary principles that function like what a social scientist would call ends or values – and these principles were argued to be "among those generally accepted in our cultural tradition": respect for persons, beneficence, and justice. These three were satisfied through the practices of informed consent, risk-benefit analysis, and the selection of research subjects, respectively (National Commission for the Protection of Human Subjects of Biomedical and Behavioral Research 1978). These later were expanded and renamed as autonomy, beneficence, non-maleficence, and justice (Beauchamp and Childress 2009). To massively generalize, the ethical question in such an analysis is whether the means in question (human gene editing, cochlear implants, bone-lengthening surgery) maximize these four principles, ends, or values.

The emergence of this form of ethical argumentation, called "principlism," would come to dominate mainstream bioethics in at least the USA and the UK. Two British observers write that "by establishing itself as the state-sanctioned authority for converting discussions of good and bad in American medical science into a common language and concepts, the bioethics of principlism achieved the status of an ascendant political currency with global potential" (Salter and Salter 2007, 651). These few principles became established as the ends or purposes, which would inform future debate – and would not need to be questioned. The earlier debate about what our ends should be was replaced by a debate about whether various technological acts would maximize these four institutionalized and undebatable values.

2.2 Formation of the Other Jurisdictions

As described, the issue of the ethics of human experimentation was one of the first in bioethical debate, and the Belmont Report was concerned with this topic. The government strengthened its control over research bioethics by mandating that more and more research be overseen by what are now called Institutional Review Boards (IRBs) (Rothman 1991). The principlism as articulated by the Belmont Report came, through executive order, to be the ethical system to use in IRBs. This meant that this jurisdiction would not allow debate about what ends would be, but rather the task is whether the particular act of research maximizes the set ends of principlism. As we can imagine, this task is not of interest to theologians, who want to debate what the ends should be, so they left.

A similar story can be had in HCEC. Up until this time the profession of medicine had rock solid control over medical ethics and the ethics to be used in healthcare settings. However, according to historian David Rothman, between 1966 and 1976 "the new rules for the laboratory," by which he means the principlist ethical system used in the research bioethics jurisdiction, "permeated the examining room, circumscribing the discretionary authority of the individual physician" (Rothman 1991, 107).

Scholars agree that principlism is now equally dominant in HCEC as it is in research bioethics. For example, the *Handbook for Health Care Ethics Committees* states that the "core ethical principles that support the therapeutic relationship and give rise to clinical obligations include respecting patient autonomy ... beneficence ... nonmaleficence ... [and] distributive justice" (Post, Blustein and Dubler 2007, 15). Similarly, one of the influential textbooks for HCEC writes that "there is general agreement that modern medical ethics depends on a small group of moral principles: respect for the autonomy of patients, beneficence, nonmaleficence, and justice." The principles are so set that the book has a fourfold table with the principles, printed on card stock, so that it can be put in your pocket when consulting in the hospital (Jonsen, Siegler, and Winslade 2006, 2; 11).

III Why the Bureaucratic State Prefers Common Morality Principlism

Principlism is the utterly dominant form of argumentation in HCEC and research bioethics, and extremely strong in public policy bioethics. It is not the dominant form in cultural bioethics, where the consumer is not the bureaucratic state, but remains the general public and academia.

If DS scholars are going to engage the different jurisdictions in bioethical debate, it is important to understand the durability of principlism, because I will argue below that principlism is the main challenge DS scholars will have with interfacing with mainstream bioethics. Why does the bureaucratic state prefer principlism? The key is that the principles are portrayed as universally held values, goals, or ends.

The bureaucratic state prefers principlism and its purportedly universal values because the unelected employees cannot be seen as promoting their own values – they are simply promoting the universal goals of the American people. This makes their influence more democratically legitimate. Of course, this notion that bureaucrats cannot use discretion is very American, because in America, citizens do not trust the government. Historian Theodore Porter writes that in other countries government officials are "trusted to exercise judgment wisely and fairly. In the USA, they are expected to follow rules" (Porter 1995, 195). In this analogy, the principles are like rules, and the bureaucratic state does not appear to be using discretion. Imagine the outcry if the director of the National Institutes of Health were to say that they set a bioethical policy by reflecting upon their own personal moral beliefs.

Relatedly, the bureaucratic state likes a system with only a limited number of ends, values, or principles because it appears to be calculable and thus more transparent to the citizens. With only four non-debatable ends, all of a messy ethical discussion can be boiled down to four concepts. Moreover, since the principles can be set off against each other, we see a simple weighing or balancing decision, which also appears to be more transparent to those on whose behalf the decision is being made. (This is one of the allures of cost-benefit analysis.) As one of the participants in the Belmont Report later concluded, the principles "met the need of public-policy makers for a clear and simple statement of the ethical basis for regulation of research" (Jonsen 1994, xvi).

The particular ends that have been institutionalized also make this system very durable. The principles are not only portrayed as the common morality, but they are also part of the common morality because they are the basis of liberal democratic societies. That is, in a liberal democratic society everyone can pursue their own conception of what is beneficial (autonomy and beneficence), until the point they harm someone else's interests (non-maleficence). This makes principlism perfectly consistent with the nature of law, which clearly facilitates its use by the bureaucratic state.

Finally, principlism is durable not only because it is preferred by the bureaucratic state, but because the principles are also held by institutional science and medicine, which have an outsized influence on these debates. Despite the founding myth of bioethics that it is an oppositional force to medicine and science, most analysts would agree with historian Charles Rosenberg that bioethics has, as a condition of its acceptance:

> taken up residence in the belly of the medical whale; although thinking of itself as still autonomous, the bioethical enterprise has developed a complex and symbiotic relationship with this host organism. Bioethics is no longer (if it ever was) a free-floating, oppositional, and socially critical reform movement.
>
> *(Rosenberg 1999, 37–38)*

With the lack of independence in mind, we should consider that the principles are actually those that physicians largely held. For example, beneficence (doing good) and non-maleficence (avoiding harm) are the moral basis of medicine. While the early bioethics debate pushed physicians to consider autonomy of the patients more seriously, physicians and scientists have easily adapted, as they have also to "justice," which never weighs very heavily on bioethical discussions. The original debate between scientists and physicians and those promoting the principles was not over the content of the principles but more over who would have discretion in applying them.

So, to summarize so far, the dominant form of argumentation in three of the four bioethical jurisdictions is common morality principlism. Participants in research ethics, HCEC, and, to a slightly lesser extent, public policy bioethics are explicitly or implicitly limited to arguing whether the means in question (e.g. human gene editing) maximize four pre-set goals, ends, or values: autonomy, beneficence, non-maleficence, and justice. The cultural bioethics jurisdiction is still somewhat more diffuse. Principlism is very durable because it is strongly preferred by the bureaucratic state and by bureaucratic organizations like hospitals due to its pseudo-democratic qualities, its consistency with liberal democratic reason and law, the relative transparency of its reasoning to the citizens, and its consistency with the ethics of scientists and physicians. With this description and explanation of the bioethical jurisdictions, and the ethical system in use in each, the question is how critical DS scholars should engage with this field.

IV Disability Studies Engagement with Mainstream Bioethics

The first challenge that DS will have interfacing with mainstream bioethics is that principlism is resolutely individualist in orientation. For example, IRBs only examine whether the interests of an individual research subject in a trial for cochlear transplants are violated by the particular researcher – whether the research subject's autonomy is violated, whether the research will harm them, and whether they were selected to be in the trial because they have no social power.

On the other hand, DS is resolutely social in orientation. In the earliest days of DS the central claim was that disability is not a characteristic of an individual, but rather an orientation of society that constrained some individuals. In general, according to Goodley, critical DS "emphasizes the cultural, discursive and relational undergirding of the disability experience" (Goodley 2013, 634). So, imagine a DS scholar trying to argue that a cochlear transplant research trial should consider that they will teach the broader society to devalue bodily diversity. This social conception cannot be argued in this jurisdiction, and any broader concerns such as this can only be brought in if they are translated into the language of individualist principlism. That is, the concern about society would have to be re-described as individual harm. That will not be effective.

The second challenge is that DS is "critical," which obviously has many meanings, but at minimum means the questioning of institutionalized social assumptions from the perspective of the marginalized. How would we look at cochlear implants if we asked deaf people their view? Principlist bioethics is not critical but radically conservative in the technical meaning of conservative as aversion to change. Put bluntly, the ends or goals of the disabled are not universal, and thus this group must limit their concerns to the universal ends like everyone else. There is also not much room for critique if you must limit yourself to the four ends of principlism. DS scholars would have been more at home with the debate up until the mid-1970s where the point was not to maximize assumed ends, but how to convince others of what their ends should be. This debate is still alive in cultural bioethics, far from the three other jurisdictions of bioethics.

My third point is that DS actually has an advantage – at least compared to other challengers to mainstream bioethics – when interfacing with mainstream bioethics, which is that some of the concerns of DS are translatable to principlism. To contribute in a meaningful way to a jurisdiction that presumes principlism, one must translate your concerns into one of the four principles. To take an example from continental philosophers who challenge mainstream bioethicists, principalist bioethics does not accept deontological arguments that a technology is intrinsically wrong, and technologies can only be wrong if they violate one of the four principles. For example, if you want to argue that germline human gene editing is inherently wrong because humans should not have that power you will have to translate your claim into the idea that human gene editing somehow harms an individual. Indeed, in the history of debates over human genetic engineering, bioethicists struggled to identify any argument against germline editing that would fit with principlism, and the best they found beyond safety was that individuals who do not yet exist have not given their autonomous consent to be experimented upon (Evans 2002). This is an ineffective way to describe intergenerational responsibility – but intergenerational responsibility cannot be translated into principlism.

It is hard to imagine that DS could translate all of its concerns into principlism without losing part of the claim – particularly the "critical" part. However, I can imagine DS claims translated into the principle of "justice." Garland-Thomson writes that DS has been rooted in "an expansion of rights for people previously marginalized or excluded from full participation in exercising the obligations and benefits of equal citizenship" (Garland-Thomson 2017, 323). The idea that people with disabilities are not being treated equally, or are being harmed, is the easy interface with mainstream bioethics and explains why the DS perspective is actually invited into some mainstream bioethical discussions. Some of the concerns of DS can be translated.

V Conclusion

While there is undoubtedly little consensus on what DS bioethics should be, whatever it is it will have to account for the already established mainstream bioethical debate. In this chapter I have described the four jurisdictions of the field and how the dominant intellectual orientation of the field emerged. The dominant approach is common morality principlism, and by and large the concerns of critical DS may be hard to integrate. However, DS has the advantage of being concerned with justice, which is one of the principles of bioethics, so interface with the mainstream may well be possible, at least in part.

References

Beauchamp, Tom L. and James F. Childress. 2009. *Principles of Biomedical Ethics, 6th Edition*. New York, NY: Oxford University Press.

Callahan, Daniel. 1972. "New Beginnings in Life: A Philosopher's Response." Pp. 90–106 in *The New Genetics and the Future of Man*, edited by M. P. Hamilton. Grand Rapids, MI: Eerdmans Publishing Company.

Evans, John H. 2002. *Playing God? Human Genetic Engineering and the Rationalization of Public Bioethical Debate*. Chicago, IL: University of Chicago Press.

Evans, John H. 2012. *The History and Future of Bioethics: A Sociological View*. New York, NY: Oxford University Press.

Garland-Thomson, Rosemarie. 2017. "Disabilities Bioethics: From Theory to Practice." *Kennedy Institute of Ethics Journal* 27(2):323–339.

Goodley, Dan. 2013. "Dis/Entangling Critical Disabilities Studies." *Disability and Society* 28(5):631–644.

Jonsen, Albert R. 1994. "Foreword." Pp. ix–xvii in *A Matter of Principles? Ferment in U.S. Bioethics*, edited by E. R. DuBose, R. P. Hamel and L. J. O'Connell. Valley Forge, PA: Trinity Press International.

Jonsen, Albert R. 1998. *The Birth of Bioethics*. New York, NY: Oxford University Press.

Jonsen, Albert R., Mark Siegler, and William J. Winslade. 2006. *Clinical Ethics: A Practical Approach to Ethical Decisions in Clinical Medicine, 6th Edition*. New York, NY: McGraw Hill.

National Commission for the Protection of Human Subjects of Biomedical and Behavioral Research. 1978. *The Belmont Report: Ethical Principles and Guidelines for the Protection of Human Subjects of Research*. Washington, DC: GPO.

Porter, Theodore M. 1995. *Trust in Numbers: The Pursuit of Objectivity in Science and Public Life*. Princeton, NJ: Princeton University Press.

Post, Linda Farber, Jeffrey Blustein, and Nancy Neveloff Dubler. 2007. *Handbook of Health Care Ethics Committees*. Baltimore, MD: The Johns Hopkins University Press.

Rosenberg, Charles E. 1999. "Meanings, Policies, and Medicine: On the Bioethical Enterprise and History." *Daedalus* 128(4):27–46.

Rothman, David J. 1991. *Strangers by the Bedside: A History of How Law and Bioethics Transformed Medical Decision Making*. New York, NY: Basic Books.

Salter, Brian and Charlotte Salter. 2007. "Bioethics and the Global Moral Economy: The Cultural Politics of Human Embryonic Stem Cell Science." *Science, Technology and Human Values* 32(5):554–581.

Wilson, Duncan. 2014. *The Making of British Bioethics*. Manchester: Manchester University Press.

Further Reading

Fox, Renee C. and Judith P. Swazey. 2008. *Observing Bioethics*. New York, NY: Oxford University Press.

Hurlbut, J. Benjamin. 2017. *Experiments in Democracy: Human Embryo Research and the Politics of Bioethics*. New York, NY: Columbia University Press.

Stark, Laura. 2012. *Behind Closed Doors: IRBs and the Making of Ethical Research*. Chicago, IL: University of Chicago Press.

5

METHODS OF BIOETHICS

Alison Reiheld

Bioethics is an enormous field, spanning numerous ethical issues raised by health care as well as our treatment of plants, non-human animals, and even ecosystems. Some bioethics methods—skillful ways of analyzing issues—are useful across the breadth of bioethics, while others are particularly suited to the delivery of health care. Such a toolkit can help us to perceive when a situation is ethically tricky, to discern important features of these situations, to detect mistaken assumptions, and to avoid errors in reasoning that might cause us to miss the ethical mark. With these tools, we can do better than we might otherwise.[1]

I Casuistry

Casuistry is a method of case-based ethical reasoning. It is appealing in medical bioethics in part because cases are so heavily used in legal reasoning and in the practice of medicine and of nursing.

Casuistry as a method relies on finding paradigm cases that serve as a good example of a type of case. We then analyze these paradigm cases for the issues at stake and for unanswered questions we'd need to find the answer to in order to make a decision. This helps us learn what to look for in the future. If the paradigm cases are sufficiently richly described and our questions are answered, we can analyze what should and shouldn't be done in the paradigm case, by all the relevant moral agents. We then keep the paradigm case analyses in our ethical toolkit. When similar cases show up in our world, we have already analyzed a similar case. If the cases are sufficiently similar, we map our analyses and judgments from one on to the other. If they are relevantly different, the paradigm case has at least given us a running start so that we are not beginning from scratch in a situation that may be both urgent and important. We have to be careful, though, to occasionally revisit paradigm cases with new knowledge and tools.

For instance, the classic case of Dax Cowart, a young man who was badly burned in 1972, became blind, and lost mobility in his hands, is often taken as a paradigm case of a patient's right to refuse even life-sustaining treatment when informed about the benefits, alternatives, and risks—up to and including death. Dax tried multiple times to refuse painful burn treatments that prevented fatal infection. He repeatedly asked to be allowed to die because of both his pain level and what he believed to be his inevitably poor quality of life. In subsequent talks Dax

DOI: 10.4324/9781003289487-8

Cowart gave on the lecture circuit, he famously said that we have the right to act at our own peril. However, a disability lens might lead us to revisit Dax's case and ask whether one reason Dax did not want to live is because of ableism he had learned, and that he knew society would apply to him, resulting in likely social exclusion instead of programs of inclusion. In addition, we might revisit this case in light of the well-known "disability paradox" in which able-bodied people assess their likely quality of life with a serious disability as quite low, while disabled people who have had time to adjust to an acquired disability or one with which they were born often report that their quality of life is quite high.

Note that Dax was provided with minimal support after his treatment was complete, and it's not clear whether Dax ever met a blind person when deciding what quality of life he could live with. Dax was able to find a life as a lawyer and activist that he did not previously think possible. By revisiting paradigm cases, we can add to our understanding and develop an even richer toolkit over time. This allows us to take Dax's case as both a paradigm case of a patient's right to refuse even life-sustaining treatment, and the way that those preferences are not shaped in a vacuum but are inextricable from ableist attitudes and social structures. With this in our toolkit, we might be prepared to quickly analyze the case of Tim Bowers, who was badly hurt in a hunting accident in November of 2013. Bowers was told he would be paralyzed and could be on a ventilator for the rest of his life. Injured on a Saturday, he decided on a Sunday to have his breathing tube removed. Using the modified paradigm case of Dax Cowart, we can understand that Bowers has the right to choose to remove life-sustaining treatment, but might ask what kinds of things we'd need to investigate to be sure Bowers had all the information he needed for his refusal in order to be truly informed.

This is just a general outline of how the method of casuistry can be used. You might encounter some people who use the word "casuist" as an insult because they assume that case-based reasoning means that casuists are relativists who think that what is right and wrong differs entirely from case to case, and that there is no stable sense of right and wrong. But this is not what casuistry as an ethical method means: when done well using detailed case narratives, it grounds ethical reasoning by tethering our use of abstract ethical ideas to the concreteness of this actual world and gives us the tools we need to quickly reason through new cases by comparing them with paradigm cases.

II Narrative Bioethics

A close cousin of casuistry is the method of narrative bioethics. While casuistry relies on well-developed case narratives for its usefulness in decision-making, narrative bioethics considers how a story-telling mindset can lead us to perceive different aspects of a situation than we otherwise might. Sometimes, deliberately recasting the "story" of an ethically complex situation can open up new aspects that might have remained hidden, aspects essential to good ethical analysis. Who are the agents? Are we overlooking anyone? Whose concerns are being centered? If we shift the protagonist from the patient to the family, the physician, the nurse, the social worker, what new ethical issues are revealed? Is our narrative leaving some people out entirely who we know typically are involved in situations like this? Sometimes, if a situation has gone badly ethically wrong, we might even use narrative to tell a different story than the one that really happened in order to ask "How could this situation have unfolded differently?" (Montello 2014).

Narrative ethics can help us see, for instance, that stories about disability are often told in particular ways: the disabled person must overcome disability as an obstacle. An alternative narrative arc is not about overcoming a challenge, but of living as a whole person of whom the impairment is simply one part, a person with many capabilities as well as a disability. Or as someone for whom their disability identity in fact becomes a point of pride.

Narrative ethics is especially valuable for issues that affect members of vulnerable groups. Narratives can cast them as being inherently vulnerable or recast them as rendered vulnerable by the systems within which they live. Narratives can also help us to shift back and forth between individualistic framings of ethical problems and systematic framings of ethical problems (see next section).

The way we narrativize an ethically tricky situation will shape which tools in our toolkit we bring to bear as well as which solutions are apparent to us or remain hidden from view. Careless inattention to the narratives we cannot help but construct leads to fictional narratives (Chambers 1999); sometimes these can reveal what might have been, but sometimes these lead us astray. No narrative can fully and accurately convey all aspects of a situation. We inevitably make choices in the telling of tales, and those choices are ethically important. They, in turn, shape further ethical reasoning.

III Systems Analysis

All too often, bioethical analyses are distorted by reducing ethical issues to those of individual choice. This kind of reduction feels particularly comfortable and normal in societies that valorize individualism. However, the choices available to individuals are shaped by the systems in which we live. This notion of the systemic structuring of individual choices is captured in the idea of choice architecture. Careful ethical reflection on choice architecture can help us to see the difference between when people are "making good choices" and when people "have good choices," between when people are "making bad choices" and when people only "have bad choices" to choose from. This is one way in which individual choices are shaped by social conditions, and thus socially constructed. To think about this well, we need systems analysis.

Systems analysis can also help us to reflect on other ways that ethical issues are socially constructed. Consider the choice a wheelchair user in New York City makes to take an expensive ride-sharing or taxi trip rather than public transportation. This may seem wasteful. But such a judgment misses the fact that New York famously does not have reliable elevators down into subway stations; by missing this, we also might miss how more frugal choices are also shaped by the same issue, as wheelchair users may choose to stay home entirely because they cannot affordably get where they need to go. Instead of focusing on individual choice, we can turn our attention to aspects of the system. In these examples we must include wheelchair accessibility as part of what it means to think about public transportation. More generally, systems analysis includes attention to features such as:

- *the built environment* created by the way we choose to construct our structures and plan our cities
- *policies* supported by government and private organizations
- systems of *incentives and disincentives* that make it easier or harder to do the right thing
- the role of *social norms, stigma, and bias* in decisions by individuals to avoid certain settings or present themselves in certain ways; this includes considering the impact of racial discrimination, socioeconomic discrimination, gender discrimination, and more
- the *intersections* of membership in vulnerable and oppressed groups, the whole of which is greater than the sum of its parts; for example, a trans person who uses a wheelchair may have even more trouble with bathroom access than the average chair user

- *communication assumptions* that presume all individuals speak the same language or use oral speech to communicate, or have a particular level of general literacy or health literacy that would affect how they receive and process information
- *social determinants of health* such as income, access to health care in both cost and location, exposure to carcinogens and unsafe drinking water, risk of violence, and more
- *disabling features* of society that make a person with an impairment less able to navigate or function than they would otherwise be, such as a cultural preference for not using microphones, education that only allows one mode of presenting information, stairs at the entrances to buildings, and small desks in classrooms that only fit certain kinds of bodies
- *historical factors* that have led to current states of affairs, including the features listed above; ahistorical analyses often miss these important parts of systems analysis which might help us to see how to create a different future

Systems analysis is an important method to have in any bioethics toolkit. It is a skill, and getting good at systems analysis requires time and practice, until it becomes a habit of thought.

Until that happens, though, it's possible to insert a deliberate formal systems analysis check into decision-making processes: at some point, we must ask ourselves, have we considered the role that systems are playing here? Suppose we want to consider the case of an African-American woman who arrives at the hospital in labor with a high-risk pregnancy, and who has not received prenatal care. If we take an individualistic approach, we might judge her negatively for not seeking prenatal care when she realized she was pregnant. If we use systems analysis, we might consider her health insurance status, employment discrimination against Black folks by the kinds of employers that provide insurance and good wages, whether there were any prenatal care providers near her home, the cost and burden of transportation to prenatal care providers, whether she has the kind of job that allows people to take time off of work for preventive medical care, and even the history of housing segregation in the USA that has led to many Black folks living in racially homogenous areas with limited investment in health infrastructure and high levels of environmental pollutants such as lead in drinking water and industrial pollutants.

Once we engage in systems analysis, we are free to consider whether dealing with ethically tricky situations should focus on the immediate situation, reforming the system, or some combination of both. It can prevent us from making ethical errors that worsen existing systems of injustice, and help us see how we can create systems that help many more people than just the person before us. Those wishing to learn more about systems analysis can read about population health and public health by ethicists including Sean Valles and Rosemary Tong, and about the concepts of intersectionality and multiple axes of oppression as developed respectively by Kimberlé Crenshaw and Patricia Hill Collins.

IV Biomedical Principlism

Biomedical principlism is an ethical theory that lays out ethical principles specifically suited to making ethical decisions about an individual's health care. Classic biomedical principlism originated in the Belmont Report which sought to place limits on research with human participants.

It has since been developed into four principles of biomedical ethics (Beauchamp and Childress 2012) widely used to think about health care as well as research: autonomy, non-maleficence, beneficence, and justice. Let's start with autonomy.

4.1 Autonomy

Autonomy is about how we make decisions for ourselves, and how those decisions are supported by others. If we decide but are not allowed to follow through on it, autonomy is violated just as much as if we are rendered unable to make decisions by disease or impairment. Autonomy is the basis of informed consent and refusal. While some bioethicists only use the term "informed consent" I always include "refusal": it reminds us that without the possibility of meaningful refusal—the ability to say "no" and have it respected by others—there is no possibility of meaningful consent.

When our decisions are not respected under the guise that it is for our own good, we are treated paternalistically. Paternalism, also known as parentalism, captures the sense that an autonomous person whose choices are disregarded is treated like a child.[2] Disabled persons, whether their impairment is physical or cognitive, are often treated paternalistically. It is common for healthcare providers to speak to a disabled patient's support person instead of speaking to the disabled patient, and to assume that disabled persons are far more impaired than is the case. Indeed, such paternalism toward the patient is, itself, disabling to the patient.

There is a mechanism that I like to think of as the reverse of paternalism, in which the person's autonomy is scaffolded: they are supported in making their decisions, information is provided to them in the way that they need it in order to receive and process it, they are allowed to make all decisions they are able to make, others work to expand the array of decisions the person is able to make, and once they make decisions, others support them in making those choices real. So, for instance, a person with advancing (but not advanced) Alzheimer's is still able to make many decisions. If someone assists them in talking through and recording information that might pass in and out of working memory, they are able to make much more difficult decisions than if they are left in a room alone with a mountain of information; actually, this is true of anyone. Most people are not fully prepared and informed about the decisions they need to make when they first realize they need to make those decisions. Meeting patients and families where they are, and helping them get to where they need to be in order to engage in good decision-making, is a key component of respect for autonomy. Examples of scaffolding include videos that explain risks, benefits, and alternatives and show procedures, rather than just conveying the information in complex text as part of long informed consent form, as well as chances for patients or family members of a patient to talk to folks who have lived with the condition they are dealing with so that they are not deciding based on stereotypes or ignorance (Asch 2003). This kind of support is anti-paternalistic because it both preserves and enhances the person's autonomy.

4.2 Non-maleficence

Non-maleficence is principlism's version of "do no harm," and requires acting to prevent or reduce harm as well as refraining from actions that might cause harm. It is non-maleficence that drives palliative care, a specialty within medicine that seeks to relieve suffering such as physical pain—both temporary and chronic—and the grinding discomfort of thirst, nausea, "pins and needles," and other sensory issues caused by both medical conditions and their treatments. Medicine has long admitted that tests and treatments may both relieve suffering and cause it. The task of the ethical reasoner who uses non-maleficence is to judge which must be given priority: the prevention and relief of suffering, or the pain caused by the remedy. A classic example is vaccination by needle. It is painful in the moment and carries some risks, but also carries great benefits for both those vaccinated and others who live in society with them.

As important as it is to take non-maleficence into account during ethical reflection, it is equally important to reflect on what we consider to be harmful. An elegant example of this is the idea that the patient's own ranking of what is most harmful must guide considerations of autonomy, and that violation of autonomy imposes a particular kind of harm (Goldman 1980). This is a lovely little bit of re-thinking, for bioethics often uses the possibility that a patient's wishes and values would be harmful to them as a reason to override their autonomy. Such paternalism appeals to non-maleficence as a justification for violating autonomy. But the bit of re-thinking here allows us to consider violation of autonomy as also a violation of non-maleficence, and the patient's ranking of harms as paramount.

Careful reflection on what we consider to be harmful is especially important for thinking about disability. Medicine often considers core harms to be avoided as injury, pain, disability, and death; whether the differences associated with disability really are harms has an enormous impact on an ethical analysis that uses non-maleficence. To uncritically accept that these differences are inherently harmful, and to then apply non-maleficence, leads many providers and patients to conclude that a life with disability is worse than a life without it. However, a systems analysis can help us rethink this. We might consider whether the difference is harmful, or whether it is the way that society is built around the needs of people without that difference that is the true source of harm. We must also take care to consider how a disabled life can be a good life worth living that is full of joy and capabilities. Regardless of how we conceive "harm," taking it into account and attempting to avoid it are key elements of non-maleficence.

4.3 Beneficence

The principle of beneficence is oriented toward improving someone's life, and thus doing good for them. In medicine, beneficial actions traditionally attempt to restore lost function and improve health. It is the commitment to beneficence that leads many healthcare providers to feel ethically driven to provide care to those who cannot afford it, and to use systems analysis to reform health care as much as possible toward universal access to care. As with non-maleficence, it can be tempting to use beneficence to override autonomy. And as with non-maleficence, it is important to be able to reflect carefully on what is good and what is helpful.

Both individual clinical encounters and public health aim beneficently to improve health. But what does it mean to be healthy? And are there ways of aiding someone that in fact do harm? Should all of the conditions we consider unhealthy actually be called diseases? Prior to the 1970s, homosexuality was officially considered a mental illness. This justified imposing treatments such as institutionalization in mental hospitals or chemical castration of gay people. How we define health and disease, and the power beneficence gives them, need careful reflection. So, also, does an emphasis on "cure" or "restoring" someone to "normal." There are many ways to flourish as a human. Bioethics must not conceive of beneficence too narrowly. Someone whose differences could be erased by cure might not want them to be. Folks who have successfully incorporated their differences into their lives and identities and relationships may wish to retain those differences. Beneficence might best be used to help persons with these differences to live to their fullest. We then must ask whose judgment about what is helpful should be trusted. Again, we see autonomy appear in issues that might initially have seemed to be only about another principle—in this case, beneficence. Regardless of how we conceive what aid, help, and benefit mean, taking these into account and reflecting carefully on how we should provide them are key elements of beneficence.

4.4 Justice

The ethical principle of justice is concerned with fairness. It draws our attention to rights and corresponding obligations to fulfill those rights, the way that both labor and material goods and resources are acquired and distributed, and other aspects of organizing a fair society. Justice is essential for thinking through issues beyond individual cases and interpersonal encounters so that we can also reflect on the social context that shapes those cases and encounters. Systems analysis is built into justice.

When we think about rights, we are asking what the basic minimum is to which members of a society are entitled. This could be a basic minimum of safety such as a right not to be assaulted. It could be a basic minimum of resources such as housing and nutritional food. It could be a basic minimum of a political voice such as voting rights. And it could, indeed, be a basic minimum of health care.

Much ink has been spilled over whether there is a right to health or a right to health care. If there is a right to health or to health care, then there is a corresponding duty for someone to fulfill that right. Rights can be positive rights to receive the basic minimum and have it provided for the person with rights, or they can be negative rights to not have anyone stand in the way of our efforts to get the basic minimum. A positive right has a corresponding obligation for someone able to fulfill the right to positively act to do so. A negative right also has a corresponding obligation, but it is an obligation for anyone who might violate the right to refrain from acting in a way that does so.

These ideas show up in many ways in bioethical considerations. Let's consider reproductive justice. Reproductive rights could be negative rights that (1) prevent others from making reproductive technologies such as abortion, contraception, and fertility technologies illegal, and (2) prevent others from forcing or coercing sterilization and abortion and even unwanted reproduction upon us. However, reproductive rights could also be positive rights that entitle someone to have contraception, abortion, or fertility technologies provided to them by others. Reproductive justice is often concerned with reproductive rights, both positive and negative, and there is a long and painful history in the USA and elsewhere of reproductive injustice against persons with disability. This includes the famous case of Carrie Buck who was sterilized without her consent by those who claimed she was unfit to reproduce, a decision upheld in the Supreme Court case of *Buck v. Bell*.

Another way that rights show up in bioethical considerations is the allocation of scarce resources, from the expertise of personnel to ICU beds during a crisis. In the early days of bioethics in the mid-twentieth century, one of the great bioethical issues was how to fairly distribute access to new and rare kidney dialysis machines. One of the criteria that were sometimes used was the patient's value to society. Such a judgment is deeply fraught, and often shot through with prejudice: should dialysis go to a banker, or a mother? A young person or an old person? We see such judgments creeping into bioethics today, and the principle of justice is an important tool for seeing them before it's too late. The 2020 claim that disability should not play a role in whether patients needing respirators got them during the

US's early peaks of COVID-19 was an argument about justice as fairness and about positive rights to care. But rights and corresponding obligations aren't the only ways we can think about justice.

Justice requires us to consider fair allocation of resources even when there may be no moral rights under consideration. We often examine a combination of who needs the resource, and who deserves the resource, and we must take care to be sure our judgments of necessity and deservedness are not unjust discrimination. Of course, in one sense, to discriminate merely means to tell the difference between two things: if you can tell the difference between good wine and bad wine you have discriminating tastes. But to discriminate against something is to bring irrelevant features into consideration, to have wrongly considered features to be relevant that are in fact irrelevant. For instance, annual income is not a good measure of a person's worth to society. If it were, and we used it to determine who received scarce resources, parents and other caregivers who do not work outside the home would be considered largely useless and would not receive an allocation of resources. But of course, caregiving labor is labor, too, whether it is unpaid inside the home or paid outside the home. Furthermore, such labor is fundamental to the continued existence of human societies (Kittay 1999). Neither those who give it nor those who need it are any less valuable because they do so. Who among us has not received such labor during our lives, whether for medical or non-medical needs? In addition, most societies have been arranged such that it is primarily women who provide such labor. And thus, women are more likely to have their labor devalued. In addition, many disabled folks are likely to receive such labor, and somehow their need for it is perceived as both a reason to provide it and a reason that they are less worthy to receive resources. Fair allocation of resources—as a matter of necessity and deservedness—and the valuation of labor are matters of justice.

We've seen several aspects of justice, each of which has something to offer as a tool for thinking about a fair society and, for bioethics, fair provision of health care. Justice requires us to consider aspects of a situation that might be catastrophically overlooked by an ethical analysis that leaves it out.

V The Four Principles

Biomedical principlism is a valuable method because it directs our attention to ethically important elements of bioethical decision-making by providers, patients, and policy-makers. If you see a conflict between them, that doesn't necessarily mean you are doing it wrong; it may mean that you are simply using these principles to help you pick out relevant ethical features of the situation and principles are genuinely in conflict. When this happens, we must either pick the course of action that best balances the principles, being sure we've creatively considered many courses of action, or choose to violate some principles in order to fulfill others. We can use these principles when we do casuistry. We should also use systems analysis to ask ourselves what way of arranging policies, procedures, and people will reduce conflicts between the principles in the future. While these principles are valuable for ethical reasoning about life in general, they are particularly likely to be present in medical decision-making. This method draws our attention to them and demands that we look at all of them, not just the one(s) that support our desired conclusion.

As important as the method of biomedical principlism is, it only works well when we carefully reflect not only on how each principle affects the others but also on how we conceive of self-determination (autonomy), harm (non-maleficence), aid (beneficence), and fairness (justice).

This, too, must be part of the method if the method is to be done well. Readers wishing to learn more about classic biomedical principlism should consider the work of Tom Beauchamp and James Childress as well as Alan Goldman.

5.1 Reflective Equilibrium and the Theory-Practice Cycle

Whenever we use abstract concepts like ethical principles, socioeconomic classes, or health and disability, we use theory. Theory uninformed by this actual world is poorer for it, especially when theory must be implemented in this actual world. We see this when laws designed to be fair have unfair results because they are put into practice in a world that is already deeply unfair. Nonetheless, we do need theory to help us see patterns, to analyze this actual world, and to guide us across multiple cases and situations. Theory gives us tools. We put them into practice. By using our attempts to put theory into practice to then inform our revisions of theory, we engage in a theory-practice cycle (Figure 5.1).

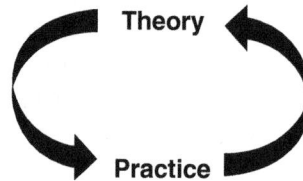

FIGURE 5.1   Theory-Practice Cycle.

We've already seen the core ethical concepts of biomedical principlism, and the way that thinking through their real-world implications can lead us to refine them and add new elements we should consider. The kinds of refinements we've seen of conceptions of disease, health, disability, benefit, harm, fairness, and self-determination arose from attempts to put theoretical concepts into practice. We would not have developed the ability to even examine these conceptions if we didn't use issues arising from practice—including the testimony of patients—to loop our reflections back onto the theoretical concepts themselves. Questioning core concepts using tools such as the theory-practice cycle is very important for making sure that we do not develop or sustain flawed foundations for bioethical thinking.

A concept related to the theory-practice cycle is reflective equilibrium. This is best thought of as the outcome of a successful theory-practice cycle, one that has produced a theory well informed by this actual world. Some ethicists think that once reflective equilibrium is done, we reach an endpoint and no further work needs to be done. However, this view of reflective equilibrium misunderstands the idea. As the world or our knowledge of it changes, this equilibrium may understandably—and correctly—be disrupted.

Disruption is bound to occur. New knowledge about what already existed in this actual world—say, because the reasoner now has a better sense of what life is like for members of certain groups rendered vulnerable by society—can be one source of disruption. So can social or other shifts that change the systems within which we live, reason, love, and make choices. Disruption of reflective equilibrium is not bad. Indeed, it may be wise to cultivate a healthy skepticism of long-settled views.

Just as it was necessary to revise our views of settled cases using casuistry when new tools become available, so we must welcome disruptions in reflective equilibrium and once again jumpstart the theory-practice cycle. The methods of the theory-practice cycle and reflective equilibrium remind us that we must keep thinking, keep learning, and keep revising. Indeed, the very word "revise" means to see again (from the Latin "visere" meaning "to see" and "re" meaning "to do it again"). These methods can keep us from becoming set in our ways when those ways are no longer good enough.

VI Conclusion

There are innumerable methods used across the interdisciplinary field of bioethics. The methods described here—casuistry, systems analysis, biomedical principlism, and the theory-practice cycle—contain components particularly well suited to helping us think through the kinds of issues that arise in disability bioethics.

Notes

1 Many philosophers who work in bioethics rely on the "Big Three" ethical theories: Kantian duty-based ethics, Aristotelian virtue ethics, and Utilitarianism. In addition, Care Ethics is often relied on to consider the ethics of human relationships and the value of care work. You can easily find information about these ethical approaches. I won't be discussing these because I'm not focusing here on all of the tools that could be useful, but specifically on bioethics methods that can help us avoid classic bioethics mistakes about disability.
2 Of course, perhaps we should not treat children this way. That's complicated, but I flag it for you here as something to think about.

References

Asch, Adrienne. 2003. In *Ethical Issues in Modern Medicine, 6th edition*, edited by Bonnie Steinbock, John Arras, Alex London, pp. 523–533. New York, NY: McGraw Hill.

Beauchamp, Tom and James Childress. 2012. *Principles of Biomedical Ethics, 7th edition*. London: Oxford University Press.

Chambers, Tod. 1999. *The Fiction of Bioethics: Cases as Literary Texts*. New York, NY: Routledge.

Crenshaw, Kimberlé. 1991. "Mapping the Margins: Intersectionality, Identity Politics, and Violence against Women of Color." *Stanford Law Review* 42(6): 1241–1299.

Goldman, Alan. 1980. *The Moral Foundations of Professional Ethics*. New York, NY: Rowman & Littlefield.

Kittay, Eva. 1999. *Love's Labor: Essays on Women, Equality, and Dependency*. New York, NY: Routledge.

Lindemann Nelson, Hilde, ed. 1997. *Stories and Their Limits: Narrative Approaches to Bioethics*. New York, NY: Routledge.

Montello, Martha. 2014. "Narrative Ethics." *Hastings Center Report* 44(s1): S2–S6.

Schalk, Sami. 2018. *Bodyminds Reimagined: (Dis)ability, Race, and Gender in Black Women's Speculative Fiction*. Durham, NC: Duke University Press.

Scuro, Jennifer. 2018. "Intersectionality—A Dialogy with Devonya N. Havis and Lydia X.Z. Brown." In *Addressing Ableism: Philosophical Questions via Disability Studies*, pp. 41–94. Lanham, MD: Lexington Books.

Further Reading

Asch, Adrienne. 2001. "Disability, Bioethics, and Human Rights." In *Handbook of Disability Studies*, edited by Gary Albrecht, Katherine Seelman, and Michael Bury, pp. 297–326. New York: Sage.

Garland-Thomson, Rosemarie. 2017. "Disability Bioethics: From Theory to Practice." *Kennedy Institute of Ethics Journal* 27(2): 323–339.

Kafer, Alison. 2013. *Feminist, Queer, Crip.* Bloomington: Indiana University Press.

Wolbring, Gregor. 2003. "Disability Rights Approach toward Bioethics?" *Journal of Disability Policy Studies* 14(3): 174–180.

6

DISABILITY BIOETHICS

From Theory to Practice

Rosemarie Garland-Thomson

This chapter is excerpted from "Disability Bioethics: From Theory to Practice" by Rosemarie Garland-Thomson, in *Kennedy Institute of Ethics Journal* 27:2 (2017) pp. 323–329.

© 2017 Johns Hopkins University Press. Reprinted with permission of Johns Hopkins University Press and Rosemarie Garland-Thomson.

What has come to be called critical disability studies is an emergent field of academic research, teaching, theory building, public scholarship, and something I'll call "educational advocacy." The critical part of critical disability studies suggests its alignment with areas of intellectual inquiry, sometimes awkwardly called identity studies, rooted in the political and social transformations of the mid-20th century brought forward by the broad civil and human rights movement. These movements pressed both the law and the social order toward an expansion of rights for people previously marginalized or excluded from full participation in exercising the obligations and benefits of equal citizenship. The ideas of equality and equal access for all that propelled the broad USA civil rights movement led to the legal desegregation of schools in the mid-20th century and changed the composition of the learning environment; with that came changes in what counted as knowledge in educational settings. In other words, when people excluded from the educational environment were included, knowledge about who we are as a community expanded along with that. Beginning, then, in the USA in the early 1970s, new knowledge perspectives and bodies of knowledge began to emerge, first perhaps as women's studies, African-American studies, then as critical race theory, feminist theory, queer theory, and more recently, critical disability studies. So while critical disability studies is a sister to women's and gender studies or critical race studies, it is distinctive in several ways. First, it grew out of a civil rights movement in the USA that was stealth in comparison to the women's movement or the Black civil rights movement.[1] The social justice that the disability rights movement achieved moved forward largely through desegregation laws and policies carried out through changes in the built environment. For people with disabilities to be integrated into the educational system required not just opening previously closed doors, but retrofitting schools with the technologies that people with disabilities needed to be present and to learn. To be integrated into public transportation, cultural institutions, spaces of

DOI: 10.4324/9781003289487-9

citizen practice, and the marketplace required building and rebuilding sidewalks, buses, train cars, voting booths, paths, businesses, restaurants, not only to ramp public and private space but to develop technology—from curb cuts to software, prosthetics, lifts, automated devices, to signage. Indeed all built and designed material aspects of the world we share and use together were transformed so that people with disabilities entered into places and institutions from which we had been excluded not only through discriminatory attitudes but through the very way that we built that shared world. As with all integration initiatives in modern liberal democracies, when excluded populations enter into previously segregated spaces and institutions, everything changes.

The work of critical disability studies, carried out largely as a research- and knowledge-building enterprise in higher education, has been to document that transformation of the social order and communal consciousness through the varied lenses of knowing that are our academic disciplines. The human variations that we call disabilities have always been the target of research and analysis, but until interdisciplinary critical disability studies arose as I have described it above, these ways of being in the world, the people who bear them, and the culture they make have been the objects of narrow focus in medical science and health studies.[2] Critical disability studies has in one sense been a corrective to this limited understanding of the enduring human experience of what we think of as disability. By aiming the perspectives and knowledge tools of the humanities and social sciences toward disability in its most pervasive manifestations—from concept to history, data, culture, human experience, narrative, theory, and aesthetic expression—the academic world broadly defined has illuminated disability and in doing so made it new for all of us who have encountered the perspectives and knowledge that is interdisciplinary critical disability studies.[3]

[...]

Even though disability rights legislation and the changes in attitudes and opportunities it has made available for people with disabilities have helped us both become and understand ourselves as political subjects, entering into disability and living as people with disabilities are still largely a medicalized affair. Indeed, all subjects in modern, liberal, technologized, and consumerist social orders are medical subjects—or perhaps more precisely medical consumers. Access to healthcare, medical treatment, and an environment that sustains our biological selves is a fundamental right in liberal democracies. Medicine, like all other institutions in modern capitalistic liberal orders, has become an industry. I do not mean this as some simplified condemnation of what has been called in disability studies "the medical model." Rather, my claim is that all of us—including people with the particularities we think of as disabilities—are overdetermined in our medical subjectivity. Medical science understands us and treats us all according to its logic and practices. This is the appropriate role for medical science. What critical disability studies can do is enlarge our shared understanding of what it means to live with disabilities and be counted as disabled. To do this, the insights and knowledge of critical disability studies need to be applied—to be brought into—medical science as a knowledge base and to its practitioners.

My proposition here, then, is that the field of bioethics is an appropriate arena of knowledge building and practice into which critical disability studies can be brought to bear. I offer here a speculative proposal for developing a practice I call disability cultural competence that can be developed as a component of the emergent field of disability bioethics. In other words, I explore how and why interdisciplinary critical disability studies can be applied to both the knowledge and practice of biomedicine and healthcare as disability cultural competence.

[...]

I A Brief Review of Disability Bioethics

Before I elaborate and suggestively offer the term disability cultural competence, I need to review the term disability bioethics, a larger concept within which disability cultural competence can be placed. By briefly reviewing the history and purposes of bioethics as a recent interdisciplinary academic field of practice, I hope to show why bioethics as an applied field is an appropriate discourse through which to engage the questions, insights, and knowledge offered by critical disability studies. While much bioethics work focuses on disability, not very much of it names itself explicitly as disability bioethics. Part of what I call for here is to claim the term disability bioethics as an enterprise in deep conversation with interdisciplinary critical disability studies as it is practiced in the humanities and social sciences. Articulating a named disability bioethics can signal connections between critical disability studies in educational environments and the applied fields of healthcare, medical research and education, biomedical policy, or commercial biomedicine in general.

Bioethics is generally defined broadly as "the study of ethical, social, and legal issues that arise in biomedicine and biomedical research" (National Institutes of Health 2016). Different definitional sources name varying subfields such as medical, animal, environmental, and public health ethics, all of which suggest the applied aspect of bioethics. Other lists of bioethics subfields organize the field by theoretical approach to include feminist bioethics, virtue ethics, deontological approaches, utilitarianism, principlism, and practical bioethics. Some definitions emphasize the theoretical while others highlight the applied aspects. The Center for Ethics and Humanities in the Life Sciences, for example, stresses practice by saying that bioethics "is an activity; it is a shared, reflective examination of ethical issues in health care, health science, and health policy." It goes on to claim that bioethics "has brought about significant changes in standards for the treatment of the sick and for the conduct of research. Every health care professional now understands that patients have a right to know what is being done to them, and to refuse. Every researcher now understands that participants in their studies have the same rights, and review boards to evaluate proposed research on those grounds are almost universal" (Michigan State University Center for Ethics and Humanities in the Life Sciences 2016).

[…]

What we think of as disability begins in human variation and the inherent dynamism of enfleshment. Because the human body is made from flesh, its movement through time and space in the process we call life constantly transforms us. So even though human development follows a genetic script, the human variations we think of as disability emerge as we develop within a standard script of human form, function, behavior, or perception which medical science calls "normal." What we think of as disability is the transformation of flesh as it encounters world, as our body's response to its environment. This call and response between flesh and world makes disability. The discrepancy between body and world, between that which is expected and that which is, produces disability as a way of being in an environment. So disability is certainly an index of capability in context but it is also a witness to our inherent receptivity to being shaped by the singular journey through the world that we call our life. Although our modern collective cultural consciousness denies vulnerability, contingency, and mortality, disability insists that our bodies are dynamic. We evolve into disability. Our bodies need care and assistance to live. Disability is the essential characteristic of being human.[4]

Even though the terms disability and bioethics rest uneasily with one another in a wide range of definitions of both terms, two important books do join the words disability and bioethics fruitfully. Drawing from the logic of equal rights, law professor Alicia Ouellette offers in Bioethics and Disability what she calls a "disability-conscious bioethics" drawn from the principles of the United Nations Convention of Rights of Persons with Disabilities (2011). These "less-attended principles" of equal rights implements generally are: non-discrimination; full and effective participation of people with disabilities in society; respect for difference; and accessibility. To emphasize the quality and participation for people with disabilities suggests a socially conscious disability bioethics that attends to what Valerie Fletcher, Director of the Institute for Human Centered Design, calls "social sustainability" (2012).

In *Disability Bioethics*, the scientist and bioethicist Jackie Leach Scully argues persuasively that a distinctive moral knowledge can arise from the experience of living in a disabled body (2008). Scully refers to what psychologists call "embodied cognition" to suggest that people draw on their bodily experiences not only to think and know but also to construct our social reality.[5] In other words, our bodily form, function, comportment, perceptual apprehension, and way of mind shape how we understand our world.

[…]

The emphasis on what bioethics does in the world—its impact—arises in part from the origin of bioethics as an institutional enterprise. Bioethics is generally understood to have begun as a response to the medical and scientific immorality of the Holocaust. Medical and scientific practice and practitioners repeatedly commit moral errors, often egregiously harming people in their efforts to treat and create new knowledge. But the widespread, state sponsored aspects of the Holocaust's unethical scientific and medical procedures and experiments goes beyond individual or even single institution unethical practice, standing as exceptional and suggesting the need for communal and state responsibility in monitoring unethical medicine and science. Structural implements from authoritative organizations that provided ethical guidelines emerged, such as the Nuremberg Code of 1947, the Declaration of Helsinki in 1964, and the Belmont Report of 1979, which responded to the Tuskegee Syphilis Study. In this bioethics origin story, the focus of unethical practice seems to be the scientific experiments carried out by Nazi medicine on individual prisoner patients rather than the larger enterprise of mass murder through eugenic euthanasia of groups understood as socially inferior. A specific disability bioethics can amend that narrative by following, as I do, the tradition of several historians of the Holocaust—such as Henry Friedlander, Michael Burleigh, and Robert Proctor—who link mass gassing with medical treatment, understanding the relationship between social judgment, political policy, and what might be called eugenic medicine—the use of life-ending procedures based on judgments of individual or group worth. Disability bioethics understands that Nazi medicine and Nazi social and political policy merged traditional disability and illness categories with ethnic categories, using extreme eugenic measures of extermination to address what the Nazi regime framed as social otherness and biological inferiority.[6]

The appropriate goal, I assert, of disability bioethics is to strengthen the cultural, political, institutional, and material environment in which people with disabilities can most effectively flourish. The principle of democratic equality and inclusion that seeks to integrate people with disabilities into the civic world by creating an accessible, barrier-free material environment thus informs disability bioethics. Disability bioethics frames disability as valued social diversity and supports the civil and human rights-based understanding of disability encoded in legislation such as the Americans with Disabilities Act of 1990 and 2009 and broader initiatives such as the United Nations Treaty on the Rights of People with Disabilities, which aim to integrate people with disabilities as full citizens. This definition of disability bioethics moves both disability and bioethics out of a primarily medical or healthcare context to expand the domain of disability bioethics into material environments, civic institutions, cultural structures, and interpersonal interactions. The disability bioethics I am defining here is theoretical and descriptive. To put theory into practice, however, requires implementation. The implement I offer here to operationalize disability bioethics is, as I suggested above, disability cultural competence.

II What Is Disability Cultural Competence?

Disability cultural competence expands the scope and content in rehabilitation programs, clinical treatment, medical humanities, and disability services as we know them. It supports my disability bioethics aim of strengthening the cultural, political, and institutional climate in which people with disabilities can most effectively flourish as they are. Disability cultural competence is a form of what Jonathan Metzl and Helena Hansen call "structural competency" in that it focuses attention on how social and cultural structures influence health outcomes and shape personhood at levels above individual interactions (2014). Informed by Ouellette's "disability principlism" and Scully's "disability epistemology," disability cultural competence goes beyond sensitivity to language and adjustments in activities of daily living; it is developing competencies for using the world effectively, maintaining our dignity, exercising self-determination, cultivating resilience, recognizing and requesting accommodations, using accessible technology, finding community, maintaining successful relationships—all as persons living with disabilities. Thus, disability cultural competence is for people with disabilities and about living with a disability at the same time. It brings disability culture to people currently identified as disabled and their families and caregivers as well as people who may in the future identify as disabled. With its competencies grounded in the disability principles of nondiscrimination, full and effective participation of people with disabilities in society, respect for difference, accessibility, moral knowledge, and disability epistemology, disability cultural competence extends beyond the usually understood contexts of healthcare environments to include social institutions and structures in which people with disabilities act and are acted upon, such as the workplace, marketplace, domestic spaces, public spaces, and cultural spaces.[7]

[…]

III Why We Need Disability Cultural Competence as a Practice of Disability Bioethics

One fundamental premise supporting the need for disability cultural competence is what I call our shared disability illiteracy. By this I mean that most people don't know how to talk about disability or how to be disabled. Yet disability is fundamental to being human. The human lifecycle and our encounters with the environment as we move through life transform our bodies and minds in ways we think of as disabilities. The business of medicine is making and unmaking people with disabilities. So the human variations we think of as disabilities are a part of every family and will enter into every life sooner or later. Indeed, people with disabilities are the largest minority group in the USA and a growing constituency as the American population ages and new disability categories such as neurodiversity, psychiatric disabilities, and learning disabilities emerge and grow.

In spite of this reality, Americans remain unprepared for disability. We get little information about living with disabilities and few opportunities to practice being disabled. But more important, what we learn about life with a disability is relentlessly grim. Medicine and health care focus on normalizing us and eliminating disability. But the limitation of medicine's view of disability is that it cannot provide a context outside of medical treatment about living with disabilities. All people need to learn how to be disabled—how to achieve well-being and good life quality as disabled people. For us to live effectively in a world not yet fully built for disabled people, our health care organizations as well as all of the institutional structures in which we participate need to develop a robust disability cultural competence.

[…]

IV A Suggestive Plan for Implementing Disability Cultural Competence

Disability cultural competence, as I offer it, is the promotion and development of bioethical, cultural, technological, and legal supports for people living with disabilities as they are. Its primary aim is to identify disability cultural competencies in disability culture, history, technology, and law, along with the tools needed to develop those competencies in order to augment the medical and rehabilitation environments that now address disability. This disability cultural competence for all would begin in healthcare environments, where the concept of cultural competence seems now most fully developed, but it would extend into workplace, commercial, government, cultural, and private organizations and structures as well. It should be a part of leadership training and workplace development. In other words, disability cultural competence is a skill set or a toolkit everyone will need to navigate life and to implement the promises and obligations of egalitarian democratic societies.

To operationalize disability cultural competence requires translating disability theory into usable disability bioethics by identifying competencies and developing tools to foster competencies and evaluate success. It also requires identifying institutional settings and audiences appropriate to disability cultural competence training and evaluation. This involves expanding the reach of disability bioethics beyond academic environments to a wide and diverse array of audiences, civic institutions, and corporate organizations and then tailoring disability cultural competence to organizations' aims, products, and operations where it is implemented.

The tools of disability cultural competence in healthcare training environments would be curricula, training, presentations, expert patients, certification, speakers, exhibits, media products. Primary leadership in disability cultural competence development and implementation would come from expert communities in disability bioethics and from subject experts in disability culture. Leaders and tool developers would be people with a high degree of disability cultural proficiency. A disability cultural competence initiative would produce research, policy papers, events, education, curation, and support for disability cultural competence implementation. Addressing an implementation plan in detail is beyond the scope of this paper or my expertise. I can nonetheless address here in more detail the elements of disability cultural competence.

Disability cultural competence involves five interconnected elements: 1) biomedical decision-making, 2) disability culture and history, 3) accessible technology and design, 4) disability legislation and social justice, and 5) disability cultural competence research. Biomedical decision-making includes practices, policies, laws, education, and attitudes in relation to disability. Crucial to disability cultural competence is awareness about, support for, development of, and promotion of disability history, culture, material culture, and arts as cultural competence. User-based design, development, and promotion of accessible technology assure high quality of life for people with disabilities. We need to know our legal rights, obligations, and protections due to us when we identify as people with disabilities.

Moreover, patients—which includes all of us potentially—need to develop disability cultural competence as well. Unlike other cultural or ethnic groups for which health care providers learn cultural competence, disability identity and culture are foreign territories for most patients and health care institutions. Disability is an experience, identity, and culture that people generally enter unexpectedly and unwillingly. Entering into disability requires more than medical treatment and rehabilitation; becoming disabled requires adjusting to new functioning, appearance, and social status. It requires learning how to flourish as a person with disabilities, not just living as a disabled person trying to become nondisabled. Awareness about disability rights and identity, such as the ADA and GINA, help people understand the rights, protections, and benefits of identifying as disabled. Research on and promotion of quality of life, dignity maintenance, access, self-determination, and cultural proficiency are necessary for people with disabilities.

The aspirational constituency, then, for disability cultural competence education extends from health care providers to patients and to all people. This is what distinguishes disability cultural competence from other cultural competence constituencies—it is capacious in both content and user base. All people are either active patients or patients in waiting. Developing disability cultural competence will prepare all of us to navigate that status and experience effectively.

Notes

1 For histories of the USA disability rights movement and disability rights legislation, see Scotch (1984) and Shapiro (1993).
2 See Linton (1998) for an account of the development of critical disability studies.
3 A review of key canonical works of critical interdisciplinary disability studies from the 1980s to 2010s follows: Eiesland (1994); Davis (1996); Wendell (1996); Garland-Thomson (1997); Linton (1998); Kleege (1999); Mitchell and Snyder (2000); Padden and Humphries (2005); McRuer (2006); Davidson (2008); Siebers (2008); Garland-Thomson (2009); Schweik (2009); Carlson (2010); Straus (2011); Hall (2011); Price (2011); Chen (2012); Brune and Wilson (2013); Kafer (2013); Baynton (2016); Hamraie (2017).
4 A version of this definition of disability can be found in Garland-Thomson (2012).
5 I would characterize what Scully does here as using phenomenology to explore the relationship between ontology and epistemology. Scully stresses the idea of moral knowledge, a crucial term in bioethics.
6 See Friedlander (1995); Burleigh (1997); Proctor (1988); and also Kevles (1985).
7 Rebecca Garden and others have suggested that the concept of cultural competence implies arrogance on the part of healthcare workers— particularly doctors—because it is based on the premise that information and training can provide mastery and effective relational skills. Garden has argued for the concept of cultural humility as the appropriate affect of an informational approach that health-care workers might use in treating patients. In the case of disability as experience, social status, and cultural group, *competence* rather than *humility* is the goal. The difference between disability cultural competence and cultural humility is that no firm border exists between the healthcare worker attaining certain competencies in disability and the immediate patient in the social and medical category of disability. The goal of disability cultural competence would be to build an affect of pride and positive identity in people experiencing disability and in patients in waiting. Humility is an antidote to arrogance, overconfidence, and privilege. Yet the competencies of pride, confidence, and status development are the social capital that disability status often attenuates; what people with disabilities need is not humility, but awareness of the tools for flourishing and high quality of life that they can access. This is what disability cultural competence provides. For explications of cultural humility, see Tervalon and Murray-Garcia (1998) and Garden (2019).

References

Baynton, Douglas C. 2016. *Defectives in the Land: Disability and Immigration in the Age of Eugenics*. Chicago: University of Chicago Press.
Brune, Jeffrey A., and Daniel J. Wilson, eds. 2013. *Disability and Passing: Blurring the Lines of Identity*. Philadelphia: Temple University Press.
Burleigh, Michael. 1997. *Ethics and Extermination: Reflections on Nazi Genocide*. Cambridge, UK: Cambridge University Press.
Carlson, Licia. 2010. *The Faces of Intellectual Disability: Philosophical Reflections*. Bloomington, IN: Indiana University Press.
Chen, Mel Y. 2012. *Animacies: Biopolitics, Racial Mattering, and Queer Affect*. Durham, NC: Duke University Press.

Crossley, Mary. 2015. "Disability Cultural Competence in the Medical Profession." *Legal Studies Research Paper Series Working Paper* No. 2015–30, August 24, University of Pittsburgh School of Law. http://ssrn.com/abstract=2650092.

Davidson, Michael. 2008. *Concerto for Left Hand: Disability and the Defamiliar Body.* Ann Arbor: University of Michigan Press.

Davis, Lennard J. 1996. *Enforcing Normalcy: Disability, Deafness, and the Body.* London: Verso.

Donchin, Anne, and Jackie Scully. 2015. "Feminist Bioethics." *The Stanford Encyclopedia of Philosophy.* http://plato.stanford.edu/archives/win2015/entries/feminist-bioethics/.

Eiesland, Nancy. 1994. *The Disabled God: Toward a Liberatory Theology of Disability.* Nashville, TN: Abingdon.

Fletcher, Valerie. 2012. "Socially Sustainable Design: Making the Case for Design that Includes." *Architecture Boston Expo*, Boston, MA, November 15. http://www.humancentereddesign.org/sites/default/files/ABX2012/MakingtheCaseforSociallySustainableDesign11-15-12.pdf

Friedlander, Henry. 1995. *Origins of Nazi Genocide from Euthanasia to the Final Solution.* Chapel Hill: The University of North Carolina Press.

Garden, Rebecca. 2019. "Who Is Teaching Whom? Deaf and Disability Approaches to the Health Humanities." In *Teaching Health Humanities*, edited by Olivia Banner, Nathan Carlin, and Thomas Cole. Oxford University Press.

Garland-Thomson, Rosemarie. 1997. *Extraordinary Bodies: Figuring Physical Disability in American Culture and Literature.* New York: Columbia University Press.

———. 2009. *Staring: How We Look.* New York: Oxford University Press.

———. 2012. "The Case for Conserving Disability." *Journal of Bioethical Inquiry* 9 (3): 339–55.

Hall, Kim Q., ed. 2011. *Feminist Disability Studies.* Bloomington, IN: Indiana University Press.

Hamraie, Aimi. 2017. *Building Access: Universal Design and the Politics of Disability.* Minneapolis: University of Minnesota Press.

Kafer, Alison. 2013. *Feminist, Queer, Crip.* Bloomington, IN: Indiana University Press.

Kevles, Daniel. 1985. *In the Name of Eugenics: Genetics and the Uses of Human Heredity.* New York: Knopf.

Kleege, Georgina. 1999. *Sight Unseen.* New Haven, CT: Yale University Press.

Linton, Simi. 1998. *Claiming Disability: Knowledge and Identity.* New York: New York University Press.

McRuer, Robert. 2006. *Crip Theory: Cultural Signs of Queerness and Disability.* New York: New York University Press.

Metzl, Jonathan M., and Helena Hansen. 2014. "Structural Competency: Theorizing a New Medical Engagement with Stigma and Inequality." *Social Science and Medicine* 103: 126–33.

Michigan State University Center for Ethics and Humanities in the Life Sciences. 2016. "What is Bioethics?" Accessed October 21, 2016. http://www.bioethics.msu.edu/what-is-bioethics.

Mitchell, David T., and Sharon L. Snyder. 2000. *Narrative Prosthesis: Disability and the Dependencies of Discourse.* Ann Arbor: University of Michigan Press.

National Institutes of Health. 2016. "What Is Bioethics?" *National Institute of Environmental Health Sciences.* https://www.niehs.nih.gov/research/resources/bioethics/what_is_bioethics/index.cfm.

Ouellette, Alicia. 2011. *Bioethics and Disability: Toward a Disability-Conscious Bioethics.* Cambridge, UK: Cambridge University Press.

Padden, Carol, and Tom Humphries, eds.. 2005. *Inside Deaf Culture.* Cambridge, MA: Harvard University Press.

Price, Margaret. 2011. *Mad at School: Rhetorics of Mental Disability and Academic Life.* Ann Arbor, MI: University of Michigan Press.

Proctor, Robert. 1988. *Racial Hygiene: Medicine Under the Nazis.* Cambridge, MA: Harvard University Press.

Robey, Kenneth L., Paula Minihan, Linda Long-Bellil, Joane Earl Hahn, John Rice, and Gary Eddey. 2013. "Teaching Health Care Students about Disability within a Cultural Competency Context." *Disability and Health Journal* 6: 271–79.

Sandel, Michael. 2007. *The Case Against Perfection.* Cambridge, MA: Belknap Press of Harvard University Press.

Schweik, Susan. 2009. *The Ugly Laws: Disability in Public.* New York: New York University Press.

Scotch, Richard. 1984. *From Good Will to Civil Rights: Transforming Federal Disability Policy.* Philadelphia: Temple University Press.

Scully, Jackie Leach. 2008, *Disability Bioethics: Moral Bodies, Moral Difference*. Lanham: Rowman and Littlefield Publishers.

Siebers, Tobin. 2008. *Disability Theory*. Ann Arbor: University of Michigan Press.

Shapiro, Joseph. 1993. *No Pity: People with Disabilities Forging a New Civil Rights Movement*. New York: Times Books and Lennard Davis Enabling Acts.

Straus, Joseph N. 2011. *Extraordinary Measures: Disability in Music*. New York: Oxford University Press.

Tervalon, Melanie, and Jann Murray-Garcia. 1998. *Journal of Health Care for the Poor and Underserved* 9 (2): 117–25.

Wasserman, David, Adrienne Asch, Jeffrey Blustein, and Daniel Putnam. 2016. "Disability: Definitions, Models, Experience." *The Stanford Encyclopedia of Philosophy.* https://plato.stanford.edu/archives/sum2016/entries/disability/.

Wendell, Susan. 1996. *Rejected Bodies: Feminist Philosophical Reflections on Disability*. New York: Routledge.

PART III

Philosophy of Medicine and Phenomenology

7

DISABILITY AND THE DEFINITION OF HEALTH

Sean Aas

This chapter considers a philosophical question that has run throughout critiques of the medicalization of disability. It asks whether and to what extent disabilities are or imply health conditions: that is, states of the body inconsistent with the kind of 'full' or 'perfect' health that medicine tries to achieve. Along the way, reasons will emerge why this question is difficult, if not impossible, to avoid, as we try to interpret, reconstruct, and vindicate central ethical and political claims of the disability movement.

Now: on one common sense way of thinking about disability, it might seem obvious that disabled people cannot be fully healthy. Medicine aims to prevent and cure conditions that cause pain and premature death. But that is not all that it does, or should do – health care can and should aim at preventing even painless and non-fatal loss of function. A non-life threatening condition that caused blindness, but no pain would still be a disease. Doesn't it follow that disabilities like blindness are inconsistent with full or perfect health? Can't we infer that disability, generally, is a deficit in health?

This reasoning goes too fast, in more ways than one. A first set of points concerns the definition of disability. Many scholars sympathetic to a social model of disability would say that the actual inability to see is an impairment, rather than a disability. Disability is the social response to the body, not the state of the body itself. If we accept this distinction, then a disability itself is not a health problem, even if it sometimes involves or implies an impairment that is. Health problems are problems with the body, disabilities are problems with society.

On the other hand, critics of the social model argue that our definition of disability ought to foreground features central to the experience of disabled people as a group, including bodily limits on function (Shakespeare 2006). If functional differences are central, in this way, then we should be open to saying that conditions of limited function are, literally, disabilities. This is consistent with recognizing the main point of the social model – that many of the disadvantages disabled people experience arise from the social response to bodily difference, rather than the bodily difference itself. It's just that we should reserve the word 'disability', rather than 'impairment', for the actual bodily differences.

This definitional debate is important for framing discussion about disability and health. We need to make sure we are not talking past each other here. But adherents of both definitions – disability as social response and disability as functional difference – can agree on the importance

DOI: 10.4324/9781003289487-11

of two questions, which I will use to frame this discussion of disability and health. First, whether or not one thinks disability is a functionally significant bodily difference, one can ask whether disability implies or entails the existence of a difference like that between those who (for example) can see and those who cannot. Second, one can ask whether any bodily differences implied by disability are (always or typically) problems with health.

Most of this entry will focus on that second question, about which bodily differences are inconsistent with health. But first it's worth saying something about whether disability always entails any bodily difference whatsoever. Here we have a real difference between proponents of most social models of disabilities and their opponents, regarding disability and health. Those who think that what disabled people have in common is not a bodily state but a social response to a bodily state will tend to think that some people are disabled in virtue of a mistaken social response – because society treats them the way society treats people perceived to have impairments, even though they don't actually have one. Perhaps, for instance, societies that excluded women from productive activity based on beliefs about their frailty or emotional instability thereby disabled these women – this, despite the fact that the underlying beliefs about functional difference were entirely false. Or, to take a more obviously contemporary example, it might be that some of the differences in social and emotional processing that tend to produce a diagnosis of 'autism spectrum disorder' are just that, differences, a matter of the mind-body complex working differently, rather than working less well.

For this reason, many proponents of social models will think that it is reasonably obvious that disability is consistent with full health, since you can be disabled without having any bodily differences at all; or by having differences that actually have nothing to do with health.

Still, even the most strident social model ought to be open to questions about the relationship between the conditions we typically think of as disabilities, and health. After all, many disabled people do have functional differences, and understanding how they are socially positioned may require understanding how these differences fit into our medical thinking. And of course, the social model might be wrong, in whole or in part; and knowing whether it is wrong seems to require knowing something about whether and to what extent disability is essentially or characteristically a matter of medical concern.

I Disability and the Definition of Health: Factualist Theories

So, what is health, and what does it have to do with disabilities/impairments like blindness, or achondroplasia, or autism? For the purposes of debates about disability, it will be helpful to divide views of health into two broad categories: those who make health a flatly factual matter, the proper subject matter of some science or other, and those who think of it as involving evaluative judgments. Among the factualist views, there are two main kinds: naturalists, who think that health is a matter for biology, to do with whether the human organism is functioning in a 'normal' or 'natural' way; and constructivists, who think that health judgments are judgments of social theory, about how bodily functioning compares to standards set by social norms, not by nature.

1.1 Social Constructivism

There are various constructivisms about health and disease (see Engelhardt 1974, for one influential example). What each has in common is a conviction that social practices somehow determine whether someone is actually healthy. Notice this isn't just the uninteresting claim that whether we call someone 'healthy' depends on facts about social norms; everyone would agree to that. Rather, it's the more provocative claim that being healthy is more like being a citizen of the USA, than like being a member of the species *Homo sapiens*. On this view, whether someone is healthy is a fact about society, not a fact about the nature or functioning of their body.

This perspective interacts with disability in a number of interesting ways, depending on just what we think disability is. If, for instance, we think that to be disabled is, also, to occupy a certain social role, a question of social theory arises: can someone play the 'disabled' role and the 'healthy' role at the same time? It seems unlikely, at least on the most obvious social constructionist interpretation of these terms: disability on this sort of view will tend to involve social representations to the effect that someone is duly excluded from significant aspects of social life, 'on grounds' of what society regards as an impairment. But a perceived impairment involves a problem with health, as the prevailing society understands it. So, the vast majority of people who are 'disabled' as social modelers use that term will be at least somewhat unhealthy, in the social constructivist sense of 'health'.

Still, per above: both social modelers and their opponents will also want to ask about the relation of specific bodily differences to health on this view. The answer will just depend on a society's prevailing view about health, disease, and normality. So, for instance, schizophrenia would be inconsistent with health now, since we classify it as disease. But other societies might see it instead as a metaphysical matter – as a sign of revelations from the divine or demonic possession, perhaps. In those societies, people with schizophrenia might count as healthy or at least not unhealthy. In that sense, this view can allow that what we, here and now, call 'impairment' (or 'disability', in the non-social sense) is not always or necessarily actually a health problem.

This brings us to problems arising from disability for these constructivist views. A central concern in much disability scholarship and activism has been about the appropriateness of medicalization. Some conditions – alcoholism, schizophrenia – have historically been under medicalized, regarded as signs of evil or weakness of will rather than unchosen afflictions. More recently, there has been more concern about overmedicalization – say, of differences in social and emotional functioning as 'Autism Spectrum Disorder', or of reliance on signs instead of sounds to communicate as 'hearing loss'. Social constructivists about health can certainly say that what is now medicalized should not be, and vice versa. But they cannot make sense of the most basic sort of move in these debates: the claim that, in our own culture, a condition like alcoholism should be medicalized because it is in fact a health problem; or that a condition like deafness should not be medicalized because it is not in fact a health problem. If something is medicalized, on these views, it is a health problem; otherwise, it is not. Constructivist views, ironically, give us less critical purchase on health talk than we might want.

1.2 Naturalistic Views

This brings us to naturalistic views. Though various versions have been developed, by far the most influential view is Boorse's (Boorse 1977). On Boorse's view, being a healthy human being is just like being a healthy cow or healthy paramecium. Each of these organisms has an evolved 'plan' – a way in which parts work together to allow the organism as a whole to fulfill basic biological goals, like survival and reproduction. An organism, including a human organism, is healthy to the extent that all of its parts are working according to the relevant plan. These plans are not made up by social institutions (even the institutions of science): facts about health and disease are discovered by biomedical science, not invented by processes of medicalization.

On a strictly social definition of disability, it will be entirely contingent whether and to what extent disabilities, as social responses to perceived impairment, are consistent with health in this naturalistic sense. On this view, society can get it wrong about health – say, failing to recognize real dysfunction in the brains of alcoholics, or misunderstanding social and emotional differences that are actually within the normal species range as dysfunction. Thus, the fact that a society excludes someone on the basis of a perception that they have a health problem will not entail that they do, in fact, have any health problem, on the naturalistic view of health and disease.

That said, the naturalist view will probably give less congenial answers about the health implications of many actual, paradigm bodily differences that go with disability. Boorseans might recognize that capital D Deafness is a culture, rather than a disease; but little d deafness, the actual inability to hear, is clearly a failure of some part of the body to make the sort of contribution to survival and reproduction that it typically makes in human organisms. So, even if Boorseans can recognize that some people who are disabled are fully healthy, many of the people we might have thought as healthy disabled people won't be, on this view of health.

To this we can add another problem: it can seem that Boorse's austerely descriptive view blocks intuitively appealing arguments regarding disability and health. For don't we want to say that healthy disabled people are so precisely because they are faring well; more specifically, because their bodies do not seem to be getting in the way of their flourishing any more than anyone else's? That seems to be a common, and plausible, form of argument in these debates. This observation about actual well-being will be neither here nor there, on a naturalistic theory of health: functioning in a species typical way might conduce to happiness on average, but no such view will hold that evidence of happiness is evidence of health.

II Disability and the Definition of Health: Normative Theories

Other theories of health foreground the evaluative considerations that a scientific theory like Boorse's leaves out. These may seem promising for disability bioethics, where arguments about health often seem to involve implied premises regarding well-being.

2.1 The WHO: Health as Well-Being

In perhaps the broadest possible view of this evaluative kind, the 1948 Constitution of the World Health Organization famously defined health as a 'state of complete physical, mental, and social well-being'. This idea can seem congenial for disability. Health care on this view should be focused broadly on what makes a life go well, rather than narrowly on the presence or absence of biological 'abnormalities' which may or may not make any difference to disabled lives. And, assuming disabled people can possess 'complete well-being', the WHO view does seem to allow for healthy disability.

This view, however, ultimately obscures matters that the disability movement has long been interested in clarifying. It does not, for instance, allow us to discuss the extent to which disadvantages due to disability are social problems, rather than health problems – for, on this view, any problem, anything that detracts from well-being, is ipso facto a problem with health. Thus, it seems to undermine the central argument associated with the social model of disability. For this reason and others, few bioethicists accept this as the correct definition of health.

2.2 Phenomenological Views

There are a number of other, more specific proposals that try to link health to well-being. Most focus on ability or capacity, following naturalistic theories in understanding health and disease in terms of what we or our bodies can do. A minority strain, however, focuses instead on the relevance of our body in our experience (Svenaeus 2000). These phenomenological views tend to understand a healthy body as one that remains in the background of our experience.

Someone who is sick notices their body in a way a healthy person does not – as something that sticks out as disruptive, opaque, and unhomelike.

Unlike the WHO view, this phenomenological view maintains the distinction between health problems and problems of other kinds. There are many challenges in life which do not manifest in an unusual awareness of our own body. And it certainly does seem to allow the possibility of healthy disability. Someone with an impairment to which they are fully adapted and accustomed may not be any more aware of their body and its limitations than anyone else.

It is not clear, however, that the phenomenological view does much better than the totalizing WHO view in drawing a line between health problems and problems of other kinds. As proponents admit, conditions external to us can make our bodies opaque to us, as well. Someone who is stigmatized and excluded from social participation because of false perceptions of serious impairment is likely to be painfully aware of their bodily differences, as a result. Someone who is fired from their job because their employer believes that their distinctive way of speaking ('speech impediment') indicates intellectual limitations might, as a result, be much more aware of how they produce speech than they were before. Thus, it is not clear that phenomenology per se is sufficient to make the classic distinction – accepted by almost all models of disability – between those disadvantages of disability that are due to society, and those properly attributable to the state of the body instead.

2.3 Capability Views

A different broadly value-based approach to health focuses not on how healthy people feel but on what they can do. A healthy body is one that facilitates certain sorts of agency. Now, some versions of this account may suffer from their own problems about the cause of a lack of capacity. Sometimes, people cannot do, or be, what they need to do or be to be happy simply because of the state of their society. Thus, if health is simply being able to do or be certain things, then again it seems difficult to distinguish social from medical problems on grounds that the latter but not the former are problems with health. But if following Lennart Nordenfeldt, we distinguish opportunity, or the external conditions for action, from ability proper, or internal conditions for it, this move to ability may avoid problems concerning the cause of challenges to well-being (Nordenfeldt 1995). Many disabled people might have the ability to thrive, but lack the opportunity to do so, thanks to social barriers.[1] These people could, easily, be healthy, on a view which defines health in terms of intrinsic ability.

Different versions of this view say different things about what we have to be able to do or be to count as fully healthy. Nordenfeldt's version appeals to the idea of vital goals – those things we have to be able to accomplish, to achieve a minimal or adequate level of well-being or happiness. For Nordenfeldt, what we have to do to be happy, and hence what we have to be able to do, to be healthy, depends on what we want, and how much we want it. That raises at least two concerns from the perspective of disability.

On the one hand, some people can have vital goals that are highly demanding. Consider some cases of acquired disability, where people find themselves unable to accept that they can still lead rich lives even if they can no longer do all the specific things they used to be able to do. Some such people we may want to count among the healthy disabled: to counsel that the problem is not with their health, or the state of their body, but with how they take their body into account in deciding what is most important to them. This kind of advice is hard to make out on Nordenfeldt's view. Health for him is a match between body and vital goals, addressable symmetrically by adjusting either the body or the goals.

Conversely, Nordenfeldt's account has always faced serious concerns regarding people with unusually or problematically low ambition. This, again, is particularly concerning in some contexts of disability. Disabled people have, to be sure, tended to resist the claim that achieving their goals somehow counts for less because their preferences are 'adaptive', modified in light of what is possible given the realities of impairment. In that way, some may welcome the thought that their health as well as their happiness depends on what they actually want, not on what others think they should want. But matters seem different in cases where reduced ambition is a result of the proverbial 'soft tyranny of low expectations'. Nordenfeldt's account may, in those cases, overgenerate 'healthy disability', implying implausibly that people who cannot leave their homes as a result of treatable but under-treated impairments are 'healthy' because social barriers have lowered their ambitions to the point that leaving the house is no longer part of any 'vital goal' of theirs.

Now, it may seem that the main problem here isn't the appeal to ability or welfare in defining health, but rather the idea that what we need to be able to do, to be healthy, depends so much on what we happen to want.

Some have therefore proposed, instead, a conception of health on which it means being able to do (or be) those things that one must do or be to be minimally happy, by some objective standard of happiness, a standard uninfluenced by unrealistic ambition or problematically adaptive preference (Ram-Tiktin 2011). Whether this is plausible depends on what we think objective well-being is. Any such view, however, will face serious challenges. If the narratives of disabled people about the value of their own lives teach us anything, it is that full human flourishing is consistent with many different forms of embodiment, including some that seem to involve serious diseases obviously inconsistent with health. Here again, then, an evaluative theory of health seems to conflict with what we should want to say, if we want to take seriously the personal testimony and political claims of disabled people.

2.4 Harm-Based Views

This kind of concern, that well-being-based views of health are overdemanding, suggests that health is less about what it takes to thrive and more about avoiding certain, specifically bodily, sorts of infirmities or harms.

Perhaps the most prominent harm-based view of health is due to psychologist Jeremy Wakefield, who understands health as the absence of harmful dysfunction (Wakefield 1992). This view has two parts. One is not so different from Boorse's: you have health so long as everything in your body is working (at least) normally. But you don't have to have a normally functioning body to be healthy, on Wakefield's view. For a biological difference to be a health problem, it has to also be harmful; harmless 'dysfunction' is just variability, difference without disease.

This view thus has obvious advantages, in thinking about disability and health. Thriving disabled people, people whose impairments are no impediment to living good lives, can be fully healthy on this view. That looks like an improvement on Boorse's more austerely factual naturalism. And we might think it improves on Nordenfeldt's subjectivist evaluative conception of health as well, if we think that people are not really harmed when they can't accomplish unrealistically ambitious goals, but can be harmed by conditions that cause them to revise their ambitions radically downward.

Still, Wakefield's view inherits some problems of social constructionist approaches. Though Wakefield thinks that whether you have a dysfunction is an objective matter, determined by how your body parts fit into the evolutionary 'plan' for the human organism, whether you are harmed is for him a social question, concerning what standards of harm are operative in your society. This, however, raises real problems regarding health judgments in ableist societies.

Suppose most members of our society believe that being unable to hear is profoundly harmful, wrecking the lives of Deaf people even in the presence of the best sorts of accommodation. Given that the inability to hear invariably involves some organic dysfunction, this would entail that deafness is really a health problem. This might seem wrong to many of us, who think that Deaf people have a better conception of what harms them than the one prevailing in the broader society in which they live.

As above, this sort of problem could probably be avoided by moving to a more objectivist notion of 'harm', on which you can be harmed whether your society thinks so or not. Another problem, however, goes deeper. Suppose that those who think that deafness or other disabilities are harmful are right to think so – not because society constitutes these conditions as harmful, in relativist fashion, but because social conditions cause harms to disabled people, in familiar and unjust ways. Wakefield's view seems to imply another objectionable kind of relativity here, identifying an impairment as a health problem even if it only causes difficulty because of stigma, prejudice, or other sorts of unjust responses. But if that's true, it's hard to see how disabled people can use the concept of health to resist medicalization: how they can say that they deserve social accommodation rather than medical treatment, because their problems are not health problems. So, Wakefield's view, like many of the views above, doesn't do a very good job accounting for the argumentative role of claims about health in debates around disability policy.

Responding to these sorts of problems for Wakefield's version of the harm-based approach, David Wasserman and I have proposed an alternative (Aas and Wasserman 2016). On the view we suggest for consideration, health problems or diseases are those states of the body that dispose their bearers to experience certain specific and objective bodily harms: primarily pain, suffering, and untimely death.

This view easily allows disabled people to argue that they are healthy – even if their impairments cause difficulties for them. If those difficulties are not intrinsic to the impairment, if they only arise from social conditions, then they do not dispose their bearers to harm, any more than setting a glass on a narrow ledge makes it more fragile or disposed to break. Moreover, if the nature of the harms in question is not itself bodily – if the harm is, only, in being insulted or stigmatized, or not fitting dominant assumptions about embodiment – then again, this view can and will say that the problem is not a problem with 'health'. Indeed, this account is tailor-made to address debates about which conditions deserve health resources intrinsically – only those that are, really, bodily problems, in that they both arise from and primarily affect the body.

That said, Aas and Wasserman's account faces significant challenges as well – particularly, in making sense of arguments for access to health resources in order to avoid prejudice and stigma, or to function better in social institutions. Consider the cochlear implant. Deafness per se does not seem to be a health problem on Aas and Wasserman's account, since, in and of itself, it does not shorten life, cause suffering, or automatically eliminate the possibility of participating in social interaction. Yet, some people who cannot hear well might reasonably want medical treatment for hearing loss. If their bodily differences are not health problems, it is not clear how they could press these claims – how, for instance, they or other people with static non-painful impairments could argue that care needed to function well is 'medically necessary', as they would need to in order to have this care paid for in most existing medical systems.

2.5 The Institutional View

This brings us to our final account of health, also tailor-made to address the concerns of the disability movement. This institutional conception of health, proposed by Quill Kukla (writing as Rebecca Kukla), is a kind of hybrid between social constructionist and more value-based views (Kukla 2015). On Kukla's view, health is, basically, the absence of problems that ought to be medicalized, given existing knowledge and resource constraints. This view is, in one way, certain to give the right answers on medicalization: there will by definition be no case in which a health problem ought not to be medicalized, or a non-health problem ought to be. And it improves on normative views like Nordenfeldt's in explaining why excessively lofty goals, or depressively low ones, do not affect our health claims – on the one hand, others don't have to give me everything I want; on the other, if I want little or nothing, that can be a moral problem even I don't see it as one. And clearly the account can accept the possibility of healthy disability, and even 'healthy impairment': anyone whose bodily differences (even, functionally significant differences) should not be medicalized, has no Kuklain health problem, however different their embodiment might be from that of others.

Still: the way Kukla's institutional view accounts for the morality of medicalization appears to remove a major weapon from the arsenal of disabled people, in their arguments about justice, health, and accommodation. For Kukla, the claim that I have a health problem is pretty much the same as the claim that requires that this sort of problem be addressed by medical institutions. Conversely, to say that I don't have any health problems is to say that I don't have any bodily conditions that should be medicalized. But that means that we can't argue, as the social model often seems to, from the claim that something is not a problem with health, to the conclusion that it shouldn't be medicalized – repetition, after all, is not argument. So, while Kukla's institutional view will, by stipulation, never get the wrong answer about whether a health condition ought to be medicalized, by deflating claims about health into claims about the ethics of medicalization, it may in the process make it hard to see why medicalization is or is not appropriate, when it is or is not appropriate.

★★★

Where does this all leave us? With a dilemma, regarding how we understand the role of health talk in the critiques of over- and under-medicalization that have been so central to disability bioethics. Traditional views of health seem to tend in the direction of an overly medical model of disability, implying that many or most of the bodily differences that distinguish people with disabilities from others are health problems. These views make it hard to understand apparently cogent arguments that static impairments like blindness or the inability to hear ought not be medicalized because they are not health problems. But attempts by pro-disability philosophers of medicine to avoid this problem, by restricting health problems to conditions that cause specifically bodily harms, raise concerns in the other direction, making it hard to understand why people sometimes have a valid claim to medical resources to address problems that are about how their body functions in its actual contingent social setting, rather than the state of their body itself.

It may be that the right answer, here, is (with Kukla) to move away from arguments that require judgments about health to justify judgments about medicalization. But that leaves important questions, of roughly the same kind: such a view needs to tell us what claims we have, to have normative bodies as defined by our societies. Such claims might amount to having a body suited for existing social infrastructure or a body that avoids certain kinds of stigmatization. It remains to be seen whether this can be done, without reference to health. This, then, is where the action ought to be in debates about disability, medicalization, and health: less about who gets to be called 'healthy' and more about what it will take to understand which bodily states are appropriate objects of medical intervention.

Note

1 That said, it is less clear that this move, or Nordenfeldt's ability-focused account in general, can capture the importance to health of *beings* rather than doings – some people seem to have health problems solely in virtue of being in a state of suffering or pain, even if this does not rise to the level of impairing their activities. See here Wendell (2001).

References

Aas, Sean, and David Wasserman. 2016. "Disability, Disease, Health Sufficiency." In *What Is Enough? Sufficiency, Justice, and Health*, eds. Carina Fourier and Annette Rid. Oxford: Oxford University Press, 164–184.

Boorse, Christopher. 1977. "Health as a Theoretical Concept." *Philosophy of Science* 44(4): 542–573.

Engelhardt, H Tristram. 1974. "The Disease of Masturbation: Values and the Concept of Disease." *Bulletin of the History of Medicine* 48(2): 234–248.

Kukla, Rebecca. 2015. "Medicalization, 'Normal Function', and the Definition of Health." In *The Routledge Companion to Bioethics*, eds. Jonathan Arras, Elizabeth Fenton, and Rebecca Kukla. New York: Routledge, 515–530. http://dx.doi.org/10.4324/9780203804971.

Nordenfeldt, Lennart. 1995. *On the Nature of Health*. Dordrecht: Kluwer Academic Publishers.

Ram-Tiktin, Efrat. 2011. "The Right to Health Care as a Right to Basic Human Functional Capabilities." *Ethical Theory and Moral Practice* 15(3): 337–351.

Shakespeare, Tom. 2006. *Disability Rights and Wrongs*. London: Routledge.

Svenaeus, Fredrik. 2000. *The Hermeneutics of Medicine and the Phenomenology of Health: Steps towards a Philosophy of Medical Practice*. Dordrecht: Kluwer Academic Publishers.

Wakefield, J C. 1992. "The Concept of Mental Disorder. On the Boundary between Biological Facts and Social Values." *The American Psychologist* 47(3): 373–388. https://doi.apa.org/doiLanding?doi=10.1037%2F0003-066X.47.3.373

Wendell, Susan. 2001. "Unhealthy Disabled: Treating Chronic Illnesses as Disabilities." *Hypatia* 16(4): 17–33. https://doi.org/10.1111/j.1527-2001.2001.tb00751.x (June 22, 2015).

Further Reading

Aas, Sean. 2016. "Disabled – Therefore, Unhealthy?" *Ethical Theory and Moral Practice* 19(5): 1259–1274.

Amundson, Ron. 2000. "Against Normal Function." *Studies in History and Philosophy of Biological and Biomedical Sciences* 31(1): 33–53.

Bickenbach, Jerome. 2013. "Disability, 'Being Unhealthy,' and Rights to Health." *Journal of Law, Medicine & Ethics* 41(4): 821–828.

Shakespeare, Tom. 2006. *Disability Rights and Wrongs*. London: Routledge.

8

THE LIVED EXPERIENCES OF ILLNESS AND DISABILITY

Havi Carel

I Introduction

Disability can be caused by, and can involve, illness. For example, respiratory illness can cause restricted mobility; a mobility disability can contribute to further illness, for example, cardio-vascular disease. The two concepts overlap to an extent, but they are also profoundly different and have been discussed in largely separate literatures. In this chapter, I offer an overview of a particular approach – phenomenology – and how it sheds light on illness, followed by some thoughts on how disability may be analyzed phenomenologically. I write as a philosopher of medicine, specializing in the phenomenology of illness (Carel 2016, 2018) and primarily focus on the experience of illness. Building on that I then ask what this analysis might tell us about disability.

Three questions are examined in this chapter. First, what is phenomenology and how does it contrast with other approaches to illness? Second, what areas of overlap and shared interest may there be between disability and illness? Third, can the existing phenomenology of illness serve as a model for a phenomenology of disability? In this chapter I discuss these questions in the hope that they open further avenues of communication between philosophical discussions of illness and the largely separate area of disability studies.

II What Is Phenomenology and How Does It Contrast with Other Approaches to Illness?

Phenomenology is a philosophical approach aimed at examining and articulating experiences. As a philosophical tradition it dates to the early years of the twentieth century, with figures such as Edmund Husserl, Martin Heidegger, Maurice Merleau-Ponty, and Simone de Beauvoir, among others. Phenomenology focuses on phenomena (what appears to us) rather than on the reality of pragmata, of things (what there is). It focuses on how we perceive and encounter the world: how phenomena appear to consciousness (Moran 2000). In other words, phenomenology studies the encounter between consciousness and world, viewing the latter as inherently human-dependent. It is primarily a descriptive philosophical method (as opposed to, say, an explanatory method), providing a philosophical analysis of consciousness and its engagement

DOI: 10.4324/9781003289487-12

with the world, examining perception, subjectivity, intersubjectivity, and meaning making. As such, phenomenology aims to be a practice rather than a system (ibid.).

As a philosophical practice, we can use phenomenology to focus on the experience of illness: an experience that until the late twentieth century has not been the topic of much academic study. Like disability, illness was generally considered shameful and tragic, and like other areas of life that reveal our profound and ineliminable vulnerability, it was largely excluded from philosophical discussion until recently (Barnes 2016; Carel 2018; MacIntyre 1999). Illness is also often painful and destabilizing, and thus hard to articulate, and therefore requires sophisticated descriptive tools that can capture its nuance, diversity, and idiosyncratic nature. Perhaps most centrally, the body is the prime site in which illness, including, to a degree, mental illness, plays itself out. This makes phenomenology, with its acute awareness of the embodied, situated, and enacted nature of consciousness, an ideal theoretical approach for the study of illness. It is also much more than that; in Section IV I suggest that phenomenology is a tool of resistance and social justice. As Virginia Woolf writes:

> [the soul] cannot separate off from the body like the sheath of a knife or the pod of a pea for a single instant; it must go through the whole unending procession of changes, heat and cold, comfort and discomfort, hunger and satisfaction, health and illness, until there comes the inevitable catastrophe [...]
>
> *(Woolf 1974, 33)*

This essential lock of body and mind, indeed, their being only a single, unified "body-subject", as Merleau-Ponty (2012) claims, gives illness (and disability) a distinctive sense of ineluctability. Illness is unrelenting, unforgiving: it is a permanent state that allows little respite. One cannot choose to take a break from one's illness or disability. Although some conditions may be episodic in nature, and some chronic conditions may have periods of relative stability that allows adaptation, the impairment or illness remains integral to one's embodiment.

While other approaches within philosophy of medicine aim to provide an objective definition of disease (see for example Boorse 1977; Cooper 2002), phenomenology's focus is different. Instead of attempting to provide a philosophical definition of the term, phenomenology of illness focuses on what it is like to be ill. It thus articulates the richness and importance of the first (and second) person description and lived experience of illness, and its project is to develop such accounts into a systematic philosophical framework that can illuminate illness experiences in their full distinctiveness, richness, and hermeneutical force (Carel 2016).

Phenomenology does this by utilizing concepts, some of which are general phenomenological concepts (e.g. Heidegger's conception of humans as "being-in-the-world"), while others specifically address the lived experience of health and illness (e.g. Sartre's notion of the transparency of health) (Heidegger 1962; Sartre 2003). These are concepts that can be used to study the experience of illness not empirically (a task better left to qualitative health research and medical anthropology) but philosophically.

What does such a philosophical study consist of? Consider S. Kay Toombs' account of illness as a series of five losses. She suggests that certain features of illness are manifest regardless of the particular disease affecting the ill person. These, claims Toombs, are the typical characteristics of illness that are integral to the experience of being ill regardless of varying empirical features (Toombs 1987). These characteristics, she writes, "transcend the peculiarities and particularities of different disease states and constitute the meaning of illness-as-lived" (ibid., 229). The five characteristics are loss of wholeness, loss of certainty, loss of control, loss of freedom to act, and loss of the familiar world. Cumulatively, they represent the impact of the illness on the ill person's being-in-the-world (ibid.). These losses represent the lived experience of illness in its qualitative immediacy and are ones that any ill person, in whatever disease state, will experience. This is part of what makes phenomenological inquiry unique: it can uncover the general structures of experiences like illness.

Toombs' account provides no mention of a particular disease or even class of diseases; there is no causal enquiry about the origins of disease; and there is no attempt to define disease as a somatic dysfunction, like disease definitions in philosophy of medicine often do. Instead, phenomenology asks: what has happened to the world of the ill person? How have their experiences, abilities, and situation changed? What happens to the ill person's experience of space and time? How has the social architecture of their world been recast? What have they lost through their illness? This type of questioning can reveal what the experience of illness is like.

While the ill body requires attention, care, and concern, the healthy body is often characterized as transparent, as Sartre and Leder suggest (Leder 1990; Sartre 2003). I will first explain this idea before offering some caveats on its unqualified use. Sartre says: "consciousness of the body is lateral and retrospective; the body is the neglected, the 'passed by in silence'" (2003, 354). And Leder writes: "while in one sense the body is the most abiding and inescapable presence in our lives, it is also essentially characterized by absence. That is, one's own body is rarely the thematic object of experience" (1990, 1). On this view, the healthy body is transparent, taken for granted. This transparency is the hallmark of health and normal function, and is experienced on the axis of health and disease, although there are obvious aberrations of this transparency within health too (Carel 2016). We do not often stop to consider the healthy body because if everything is going smoothly, the body remains in the background, the vehicle through which we experience our "medium for having a world" but not the thematic focus of experience.

This does not mean that we have no experience of the body but, rather, that many of the sensations it constantly provides are neutral and tacit, enabling it to stay in the background of our experience. An example is the sensation of clothes against our skin. This sensation is only noticed when we draw our attention to it or when we undress (Ratcliffe 2008, 303). We do not feel our clothes against our skin throughout the day, unless we turn our attention to this sensation and such tacit sensations do not interfere with the focus of our attention.

This transparency of the healthy body is somewhat idealized in philosophical descriptions of health, since it is often pierced by experiences in which the body comes to the fore, sometimes in negative ways. We do, of course, have many moments of explicit attention to our body, and in particular to its wellness: for example, when a headache goes away or while exercising, we have a positive thematic experience of our wellness. We also have many negative experiences, still within the context of health: for example, we can experience bodily failure, if we trip up awkwardly, fail to pick up a dance step, or find a parcel too heavy to carry. Such bodily failure is – importantly – experienced within the norm of health and does not disrupt this norm. I therefore contrast it with ill health, in which the norm of health itself is replaced by a state of ill health, disease, or impairment.

There are also other cases of healthy, non-impaired bodies that nonetheless draw negative explicit attention; for example, we can feel shame or self-consciousness when social attention is drawn to our body as a sexual object or as a gendered or racialized object (Wieseler 2019). These are social experiences of one's body as it is perceived or objectified by others. Sartre's famous analysis of the gaze as annihilating my subjectivity and objectifying my body, which becomes an object in the other's field of vision, recognizes the tension between the naïve unthematized body and the social body (2003, 276ff).

So, our "healthy" bodies are not free of experiences in which the body is explicitly thematized, and these can be both positive and negative, or sometimes neutral. Christine Wieseler deploys Rosemarie Garland Thomson's concept of "misfitting" to describe experiences in which one's body does not fit, or conform, to the environment one is in, which is a useful way to describe such experiences (ibid.).

However, I would maintain a distinction between misfitting that arises from social stigma, prejudice, and bias (for example, stigma arising from one's race or gender) and between misfitting that arises from impairment. Gendered and racialized bodies are not impaired (although they can be), so the misfitting is the result of social prejudice and injustice. Misfitting that arises from impairment is often also accompanied by misfitting arising from disability. But the two should be held separately as they are underpinned by different sets of norms: impairment is an aberration of a physiological norm while disability arises from social norms.

What cases of disruption of transparency – whatever its cause – have in common is in such situations the body moves from the background to the foreground of our attention. And returning to the transparency of health, we can now say that when our bodies function normally (in the minimal biological sense: when we digest our food or breathe normally) attention is deflected away from our body and toward our intentional goal or action. While my stomach digests the lunch I have just eaten, it does not get in the way of my writing this paragraph. That is what is meant by transparency: the body lets you get on with things, whatever the "things" might be. The body is, as Leder says, "absent" – not because it is not there but because our experience of it remains in the background while the object of our focus is in the foreground. "The body is in no way apprehended for itself; it is a point of view and a point of departure" (Sartre 2003, 355).

In contrast, when we become ill our attention is drawn to the malfunction, which becomes the focus of attention. Leder contrasts the healthy, "absent" body, when the body is simply there in the background, with illness and other situations when the body becomes an explicit object of negative attention and appears as a "dys" (function). "In contrast to the 'disappearances' that characterize ordinary functioning, I will term this the principle of dys-appearance. That is, the body appears as the thematic focus, but precisely as in a dys state […]" (Leder 1990, 84). The body can dys-appear as ill, impaired, aesthetically flawed, or socially awkward, objectified, or sexualized, or as attracting negative attention from others, as in the experience of shame discussed by Sartre (2003).

A third and final theoretical scaffold will complete this brief account of the phenomenology of illness. This is the distinction between the objective body and the body as lived. The objective body is the physical body, the object of medicine: it is what becomes diseased. Sartre calls this body the "body of Others": it is the body as viewed by others, not as experienced by me (Sartre 2003). The body as lived is the first personal experience of this objective body, the body as experienced by me: the person whose body it is. It is on this level that illness, as opposed to disease, appears.

This distinction is fundamental to any attempt to understand the experience of illness: the ill person is only and ever the one who experiences the illness from within. Only the ill person can definitively say if they feel pain or what a medical procedure or a particular symptom feels like. Of course, we can surmise much from another people's behavior; sometimes we can know better than them. For example, we can tell from a toddler's grumpy behavior that she is tired, whereas she may not be able to formulate this to herself. However, in most cases at least when it comes to conscious adults, each person is the ultimate authority on their own sensations, feelings, and experiences. Because of this first-person authority the experience of illness contains a measure of incommunicability that should be acknowledged (Carel 2018). Or, as Sartre put it more strongly, "the existed body is ineffable" (Sartre 2003, 377).

Disease, however, is a process in the objective body that may be observed by any other person or third personally via, e.g., test results, and may yield information that is not available through first-person reports. For example, one may have elevated cholesterol or blood pressure, or an early stage of renal disease, while having no experience of these. Often such knowledge comes from medical tests that yield objective facts that have no experiential correlates. For example, hypertension may not be experienced at all. It is only once it is revealed via a blood pressure test that it begins to feature in the diseased person's experience. A person can be diseased without being ill.

Armed with these three theoretical accounts of illness, we can move to questions two and three, and ask: how do these analyses fair when applied to disability, and are they as useful in the case of disability as they are in illness?

III What Areas of Shared Interest Are There between Disability and Illness?

So far, I have characterized illness as a series of losses (Toombs), as a loss of bodily transparency (Sartre and Leder), and as generating the useful distinction between disease and illness. Let us now see how these might work in disability. I propose that disability is a loose and open-ended set of somatic and mental states that range widely in terms of their impairment, social barriers, and self-definition. Some such states are recognized as disabilities by some bodies, but not others (e.g. color blindness). Other conditions are recognized as difference rather than disability (e.g. autism, Deafness). And yet others are problematically and inconsistently operationalized (e.g. borderline personality disorder). I do not suggest that there is a singular definition that will unproblematically capture all of these states, but instead I offer an open-ended invitation to consider individual cases as disabilities and their claims to social justice. Importantly, there is some, but by no means total, overlap between illness and disability. As Reynolds notes, "to be disabled is not automatically or necessarily to suffer or be in pain or to have an illness or disease. Many people with disabilities do not experience pain and suffering, and many are not ill or diseased" (2018, 1183–1184). However, illness often causes impairment (and disability) so that there is considerable overlap in such cases. That illness causes impairment is undeniable; what is more complex is the relationship between that type of impairment and other types that do not involve illness and indeed enable a person to be both disabled and healthy (ibid.). I suggest that one way to approach this question is by drawing the distinction between congenital and acquired disability, to which I now turn.

Much of what I say about illness in Section II does not apply to congenital disability, but may apply to a disability acquired for example via a road traffic accident. I suggest that it is the transition from one state – the state of health transparency – to a state of occluded concern for the body that characterizes the movement into illness. A previously healthy person who becomes paraplegic through an accident may be a case of disability which fits the analysis offered above. They may experience Toombs' five losses, the Sartrian loss of transparency of the body, and have a tumultuous transition from health into disability.

Such a case will strongly contrast with a person born with a missing limb, for example, who experience their bodies as whole and able and have a strong sense of integrity, completeness, agency, and ability, untouched by what others may see as disability. This, to me, is a key difference: whether there is an event that marks the beginning of a transition from one state to another. Is there a fundamental contrast between "before" and "after"? Has the impaired person lost something because of becoming impaired? Does their experience fulfill the sense of lost transparency and "dys-appearance" Sartre and Leder describe? A positive answer to these questions would put the case as firmly analogous to illness; a negative answer would not.

Another area of shared interest between illness and disability is what I call "being unable to be" or "dis-ability". That analysis calls into question Heidegger's description of the human being as "being-able-to-be" (1962). Human existence is characterized by its openness, potential, ability to become this or that thing. This underpins a powerful picture of human agency: one can become what one wants. If I want to be a polar explorer, I will have to train, build up my strength, learn to navigate, and so on. Eventually, I would join a polar expedition and fulfill my plan, achieve my goal.

This definition of the human being is best understood by Heidegger's notion of projection. Projection means throwing oneself into a project, through which a human being's character and identity are enacted. If my project is being a teacher, I project myself accordingly by training to be a teacher, applying for teaching positions, and so on. This, Heidegger claims, is the essence of human existence: the ability to be this or another thing, to assume a role as a teacher, a polar explorer, and so on (ibid.). This view of the human being as becoming, as able to achieve her aims, as constantly changing according to the project she pursues, is appealing in many ways. It credits us with the freedom – and the responsibility – to become what we want, to shape ourselves and our lives in ways we find fulfilling: to transcend our present self with a future self that is more developed, more able.

But what happens when we become less able or unable to do things, what we might term, contra Heidegger, "unable to be"? What about decline and insufficiency? And what of illness and impairment? Does Heidegger's definition exclude this important aspect of life, that of decline, inability, failure to be?

When we fall ill or become impaired, we become unable to do some things, perform particular roles, and engage in certain activities. This poses a problem for Heidegger's definition because it reveals that his definition of human existence excludes important human states. In some illnesses, especially mental and chronic illness, a person's ability to be, to exist, is radically changed and sometimes altogether curtailed. Certain projects must be discarded and sometimes, as in severe psychosis, the possibility of having a project at all becomes impossible. Could Heidegger's account allow radically differing abilities to count as forms of human existence?

I suggest that Heidegger's definition needs to be modified in two ways. First, the notion of "being-able-to-be" must be broadened to include radically differing abilities. Second, varied types of "inability to be" ought to be recognized as ways of being. Heidegger's definition can be a useful launch pad for thinking about being unable to be – or dis-ability – as a form of existence that is worthwhile, challenging, and, most importantly, unavoidable.

The opposite of "being-able-to-be" is not just "being unable to be"; this presupposes that the two concepts form a binary. But we can replace this binary with a spectrum of abilities to be. There are other modes of being able to be that are excluded from this binary. Being partially able to be, learning to be able to be, and rehabilitating an ability to be are a few examples. The "ability to be" that characterizes human existence is territory to be experientially explored and developed, rather than delimited through an opposition between "able" and "unable".

We should interpret the notion of "being-able-to-be" as broadly as possible. It should include cases in which the smooth operation of the body, its assistance in carrying out plans and projects, is no longer there. Current projects may have to be abandoned and new projects created. These new projects must be thought of in light of new limitations and therefore arise within a restricted horizon. But radically differing abilities count as abilities to be. Take a person who uses a wheelchair, someone with stage IV cancer, or a person with Down syndrome – all of these are ways of being that differ profoundly from the virile Heideggerian "being-able-to-be". But they are nonetheless human ways of being. As Iris Marion Young writes, for any lived body, the world appears as the system of possibilities that are correlative to its intentions. For any lived body, moreover, the world also appears to be populated with opacities and resistances correlative to its own limits and frustrations.

(Young 2005, 37)

This interplay between possibility and resistance has been overlooked by Heidegger, who emphasizes the former but underplays the latter. Applying Heidegger's notion of "being-able-to-be" to cases of illness and disability can open the door to acknowledging the diverse ways in which it is possible to be and the ways in which human beings differ from one another.

We can also think of processes such as rehabilitation from drug use or stroke; learning to be able to enjoy life after severe depression; being only partially able to walk, hear, see, or talk; and so on. None of these conform to Heidegger's definition, but if we understand "ability to be" more flexibly, we can have a more inclusive definition that can encompass such cases. Furthermore, in many cases of aging, illness, or disability we need to acknowledge an inability to be as a way of being.

One way of thinking about aging and illness and of becoming impaired in these ways is as processes of coming to terms with "being unable (or less able) to be": as coming to think of one's existence as more reliant and less independent, more interlinked and less autonomous. The inability (or the altered ability) to be and do is the flipside of Heidegger's account. For some individuals it is there throughout life, as in cases of chronic illness or disability. For all of us it is there as a stage in life, the stage of aging and decline. Inability and limitation are part of human life, just as ability and freedom are. By introducing the notion of "being unable to be" as an integral part of human life we can move from seeing ability as the exclusively positive and desirable state, to seeing it as part of a broader, more varied flux of life.

Being unable to fly or being unable to breathe under water are not examples of "being unable to be". Otherwise, the concept would be so broad as to be meaningless and we would be more unable than able to be. It is a lost ability or an ability that is never achieved viewed against a background in which ability is expected. "Being unable to be" is therefore mutually implicated by an "ability to be", and vice versa. "Being-able-to-be" is not unlimited. It is a way of existence that is granted temporarily, for a number of years, and is never guaranteed, never certain. It is a fragile, transient gift.

IV Can This Serve as a Model for a Phenomenology of Disability?

We can now ask whether the model laid out so far in this chapter can serve as a model for a phenomenology of disability? I propose – tentatively – that the answer is "yes". The phenomenological approach is descriptive, non-prescriptive, nonreductive, respectful of lived experience, and offers a theoretical framework within which the richness and diversity of disability experiences can be accommodated, appreciated, and communicated (Carel 2016). The phenomenological approach as a whole looks, therefore, like a fitting approach to the philosophical study of disability, within which many different disability experiences can be appreciated, studied, and better understood. These are positive features of a phenomenology of disability.

However, phenomenology is much more than simply an apt tool for such work. The act of describing, articulating, and sharing illness and disability experiences is also an act of resistance: it counters the eschewing of bodily vulnerability, whether rightly or wrongly judged to be so, as an inappropriate philosophical topic. It also combats what Fricker has termed "epistemic injustice":

in this case, the passing over, ignoring, or even silencing of voices calling to include minority voices, minority bodies, and minority experiences in the philosophical discourse (Barnes 2016; Fricker 2007). So, a phenomenology of illness – and by extension, I propose, a phenomenology of disability – is a contribution to the struggle for social justice and for the epistemic inclusion, within and beyond philosophy, of the perspectives of illness and disability.

The claim that phenomenology is quietist or non-political can, I suggest, be firmly rebutted. Providing ill or disabled persons with the tools and space to describe and share their experiences is a social justice-promoting positive step toward the amelioration of discrimination, ignorance, stereotyping, and prejudice towards ill and disabled persons. An example of such a tool is my "phenomenological patient toolkit", developed in the form of workshops for ill persons and health professionals (Carel 2016). In the workshop, ill persons are given space and non-prescriptive tools to describe their illness experiences in their idiosyncratic and non-reducible fullness and from their own perspective. This kind of tool gives ill persons a space in which they can reflect on and describe their illness experience, while avoiding the Scylla of social scripts (e.g. the monolithic view of illness as a private tragedy) and the Charybdis of medical discourse (reducing the lived experience of illness to a set of medical symptoms).

Released from those kinds of external restrictions, and from the need to conform to a pre-given understanding of their illness, ill persons are then free to forge their own interpretation and point of view on their illness. The creative and personal nature of this process is mirrored by a comment Virginia Woolf makes in her essay, "On being ill":

> The merest schoolgirl, when she falls in love, has Shakespeare, Donne, Keats to speak her mind for her; but let a sufferer try to describe a pain in his head to a doctor and language at once runs dry. There is nothing ready made for him. He is forced to coin words himself, and, taking his pain in one hand, and a lump of pure sound in the other […] so to crush them together that a brand new word in the end drops out.
>
> *(1974, 34)*

This process, freeing the ill person from external meanings and social scripts, can support them and promote the acquisition of important epistemic skills and tools: a clear understanding of their life situation; confidence to articulate and share it; the support of others who listen without prejudice or judgment; the desire of others (in particular health professionals) to hear such accounts; support to coin their own terms and words to talk about their experiences; a sense of the value of their own testimonies and interpretations. Together these amount to significant social and epistemic scaffolding which can enable unique interpretations of one's own illness experience to emerge. Although the toolkit has not been shared with disability groups, there is good reason to think that it may prove a useful tool there too.

These epistemic skills can, in turn, provide interpretative tools needed to ameliorate and begin to address the hermeneutical injustice to which ill and disabled persons are especially vulnerable (Kidd and Carel 2018). Kidd and Carel define what they term "pathocentric hermeneutical injustice" as "interactive and/or institutional constraints that limit the capacity of a person or group to create or share the meanings of some of their social experiences in appropriately intelligible ways" (4). Such constraints arise from an absence of appropriate labels, categories, terms, or concepts for recognizing, understanding, and appreciating forms of social meaning.

But hermeneutical injustice can also be generated when a subject fails to perform, epistemically and socially, in legitimated ways. Perhaps the styles of expression or forms of communicative performance are unfairly regarded as unintelligible, or, at least, as less intelligible than others, as when ill or disabled persons use highly subjective, narrative, autobiographical, or anecdotal styles in their efforts to describe their lived experiences. Such testimonies are often derogated as imprecise, vague, irrelevant, lacking relevant detail, or communicatively and epistemically deficient in other ways. Such derogations ignore the fact that ill and disabled persons determine the content and styles to best suit their expressive capacities and hermeneutical needs. Thwarting that agency prevents the ill or disabled person from performing hermeneutically; it obstructs their efforts to make or share a sense of one's bodily, social, and existential experiences as ill or disabled, making them especially vulnerable to hermeneutical injustice (ibid.).

Ill and disabled persons' ability to express and interpret their experiences can be supported, enhanced, and developed using phenomenological tools such as the patient toolkit. Therefore, I conclude, a phenomenological framework is useful for understanding and legitimizing divergent and poorly understood experiences such as the experiences of illness and disability. It is useful in providing a space in which ill and disabled persons' testimonies and interpretations can be sought, reflected upon, and used for decision making in healthcare, for example. It is also useful in cultivating epistemic confidence and hermeneutical skills that can play an ameliorative role in our efforts to overcome pathocentric and ableist epistemic injustice.

References

Barnes, Elizabeth. 2016. *The Minority Body*. Oxford: Oxford University Press.

Boorse, Christopher. 1977. Health as a Theoretical Concept. *Philosophy of Science* 44(4): 542–573.

Carel, Havi. 2016. *Phenomenology of Illness*. Oxford: Oxford University Press.

Cooper, Rachel. 2002. Disease. *Studies in History and Philosophy of Science Part C: Studies in History and Philosophy of Biological and Biomedical Sciences* 33(2): 263–282.

Fricker, Miranda. 2007. *Epistemic Injustice: Power and the Ethics of Knowing*. Oxford: Oxford University Press.

Heidegger, Martin. 1962 [1927]. *Being and Time*. London: Blackwell.

Kidd, Ian and Havi Carel. 2018. "Pathocentric Epistemic Injustice and Conceptions of Health." In *Overcoming Epistemic Injustice: Social and Psychological Perspectives*, edited by Benjamin R. Sherman and Stacey Goguen, 153–168. New York: Rowman and Littlefield.

Leder, Drew. 1990. *The Absent Body*. Chicago: University of Chicago Press.

MacIntyre, Alasdair. 1999. *Dependent Rational Animals*. London: Duckworth.

Merleau-Ponty, Maurice. 2012 [1945]. *Phenomenology of Perception*. New York: Routledge.

Moran, Dermot. 2000. *Introduction to Phenomenology*. London: Routledge.

Ratcliffe, Matthew. 2008. Touch and Situatedness. *International Journal of Philosophical Studies* 16(3): 99–322.

Reynolds, Joel. 2018. Three Things Clinicians Should Know About Disability. *AMA Journal of Ethics* 20(12): E1181–E1187.

Sartre, Jean-Paul. 2003 [1943]. *Being and Nothingness*. London & New York: Routledge.

Toombs, S. Kay. 1987. The Meaning of Illness: A Phenomenological Approach to the Patient–Physician Relationship. *Journal of Medicine and Philosophy* 12: 219–240.

Wieseler, Christine. 2019. "Challenging Conceptions of the 'Normal' Subject in Phenomenology." In *Race as Phenomena: Between Phenomenology and Philosophy of Race*, edited by Emily S. Lee, 69–85. London: Rowman & Littlefield.

Woolf, Virginia. 1974 [1930]. "On Being Ill." In *The Moment and Other Essays*, pp. 32–45. New York: Mariner Books.

Young, Iris Marion. 2005. "Throwing Like a Girl: A Phenomenology of Feminine Body Comportment, Motility, and Spatiality." In *On Female Body Experience: "Throwing Like a Girl" and other Essays*, pp. 27–45. Oxford: Oxford University Press.

Further Reading

Carel, Havi. 2018. *Illness: The Cry of the Flesh*. London: Routledge.
Kidd, Ian James. 2012. Can illness be Edifying? *Inquiry* 55(5): 496–520.
Toombs, S. Kay. 1988. Illness and the Paradigm of Lived Body. *Theoretical Medicine* 9: 201–226.

PART IV

Prenatal Testing and Abortion

9

ABORTION, DISABILITY RIGHTS, AND REPRODUCTIVE JUSTICE

Elizabeth Dietz

I Introduction

Prenatal genetic testing is substantially—one could argue mostly—about disability. On some accounts, it is about how to live with disability: a tool for identifying conditions that could be treated prenatally, to make neonatal intervention possible, and to allow families to prepare in advance for the advocacy work that disability in an ableist world demands. But it is also fundamentally about the management of disability: a tool for allowing parents (and doctors, financial systems, cultural pressures—there are no isolated individual choices in a world that acts on genomic information) to decide what disabling conditions to bring into the world and which to prevent. This management can occur through moments of discrete choice, such as a termination decision following a prenatal diagnosis of trisomy 21.[1] But the process by which an embryo that could have become a child with a disability does not become a child is neither necessarily one of active choosing against disability, nor is it necessarily the result of anti-disability sentiments held by individuals intending to bring about the elimination of disability. Such a framing would be reductive and miss the complex, structurally ableist (but also racist, sexist, and economically unequal) forest for the individual decision-making trees.

A 2020 cover story for the *Atlantic Magazine* by Sarah Zhang offers an account of the relationship between prenatal testing and Down syndrome that sits with this complexity, and in so doing raises questions about the nature of harm that is part and parcel of the existence and availability of prenatal testing in an ableist world (Zhang 2020). Zhang's piece focuses on people with Down syndrome in Denmark, which is significant in part because Denmark provides a version of the kinds of care, and economic and social support, that disability advocates call for as a precondition for flourishing with disabilities (and without them). Denmark also makes both prenatal genetic testing and access to abortion widely available. And this story puts those two facts in stark relief, for in Denmark, virtually no babies are today born with Down syndrome. At the very least, this means that we need to interrogate arguments for disability justice rooted purely in changing the acceptance and support conditions of the outside world: what does it mean that more social and parenting support, coupled with access to genetic information and the ability to act on it, has not prevented the elimination of people with Down syndrome?

DOI: 10.4324/9781003289487-14

This case offers a window into potential tensions between commitments to reproductive and disability justice, and it demands thoughtful reflection upon the issue of how to build social worlds in which both are allowed to exist. The question of how to achieve both reproductive and disability justice—or to put it differently, to conceive of disability as reproductive justice (and vice versa)—is fundamental to any effort aimed at reckoning ideas of justice with those of liberation for the people most forcefully disciplined by our biomedicalized world. Through modern biomedicine, pregnancy is mediated by a plethora of forms of testing, monitoring, and obligatory points of choosing that are supposed to depend on biomedical information. Many of these choices—choices which one must make—are framed to pregnant people as necessary for the preservation of their autonomy. Prenatal information is presented as the right basis for making trustworthy, informed, independent decisions in accordance with one's values. But autonomy is complicated, and an individual framing of autonomy gets the facts wrong, for individual choices are never really, simply individual (Fineman 2004; Ho 2008). When individual autonomy is maximized, collective notions of interdependence and the way that individual decisions are inseparable from their social contexts fall away.

This chapter examines two harms that can emerge as a result of the use of prenatal genetic testing for disabling conditions. First, I consider the harms of elimination: of potential individual disabled people and of entire categories of disabled people. This occurs when, as in Denmark, the availability of prenatal testing combines with cultural pressures and the choices presented to prospective parents such that people with particular disabilities simply stop being born.[2] Second, I turn to the harms that result from making complicated things simple. More specifically, from socio-legal systems that require the translation of complicated, interdependent, collective issues into problems adjudicated by legal systems with individual perpetrators and individual victims. In both cases, people are required to make choices that presuppose the existence of free, independent, autonomous decision-making, and whose very structure means they cannot be so.

II Prenatal Testing and the Harm of Elimination

Prenatal genetic testing is often framed as a necessary and quotidian part of reproduction under biomedicine: a process with such clear and obvious benefits that the ethical questions fall away in favor of a sense of requirement or inevitability. This naturalization of reproductive medicine should give us pause.

Prenatal genetic testing provides genomic information, which is translated to prospective parents with the help of genetic counselors into presumptive knowledge about disability. But this knowledge is highly uncertain. This is the case at the level of modern scientific knowledge, where the meaning of a given variant may be highly probabilistic, and at the level of the meaning of disability, where category boundaries are fuzzy and socially as well as historically contingent. Some variants lead with a high enough probability to a particular phenotypic deviation from what medicine considers to be typical that they can be communicated as certain. At the other end of the spectrum, those known as "variants of unknown significance" offer uncertain meanings that might change over time (Johnston 2017). In other words, in some cases there is agreement by scientific and medical experts that X variant will lead to Y condition, in some cases there is contestation, and in some cases genomic information is little more than a shot in the dark.

This process is also complicated by uncertainty about what counts as an impairment. Some forms of body-mind variation are diagnosable as pathology until social forces catalyze their recategorization into forms of mere difference. For example, homosexuality was a psychiatric diagnostic category until activism lead to its 1973 removal from the DSM (Drescher 2015). This illustrates one way that cultural understandings of disability are expressed through clinical determinations. But it also points us to how interventions at the level of knowledge, of presenting better information to combat bad information, are insufficient for the task of reckoning with the profound societal consequences of widespread genetic testing. Simply knowing more about the likelihood of flourishing with a particular disability (which, to be clear, is very likely—an entire body of literature points to the real potential for disability to be mere difference, rather than bad difference[3]) won't correct for the structural ways that flourishing is hindered.

Part of the issue here is fundamental uncertainty concerning the meaning of disability, an uncertainty about which the meaning of genetic testing helps produce. Genetic counseling tries—valiantly, to be sure—to step into this breach, clarifying what is known and formulating what is uncertain into matters of individual reproductive choice. Prenatal testing with genetic counseling is understood as a pathway to making informed, and thus legitimate and trustworthy, choices (Hodgson and Spriggs 2005). Crucially, this route is imagined to be a way for individuals, rather than government policies or paternalistic doctors, to make reproductive decisions. Testing and counseling are implicated in the biomedicalization of pregnancy, and call into being a kind of autonomy that regards people as autonomous if and only if they are capable of making informed choices.

But behind the choices made possible by the knowledge produced through prenatal testing stand a multitude of hopes and dreams that people have for their offspring. People imagine things for their future children: what kinds of activities they will like, the sort of people they will be, the kind of pride they will take in their accomplishments, and the care they will offer their parents. Prenatal testing gears into that process of imagining. But, as David Wasserman notes, in contrast to the slow process of a person coming into their own, wherein they both resist and follow paths imagined for them, prenatal testing enters with the presumptive certainty of scientific knowledge. It offers concrete, actionable, and transformative knowledge about the kind of person that will soon emerge (Zhang 2020). This is the sin of synecdoche: one part (disability) being made to stand in for the whole (a person) (Asch and Wasserman 2005).[4] Prenatal testing facilitates this error because it provides a kind of certainty about a fetus that is not otherwise available. For all the hopes and dreams of parents preparing for their children, they are just that: imaginings of what could be, hopes for what might be. Prenatal testing fundamentally changes that imaginative process when it introduces not only (supposed) certainty about the person that will be but also (supposed) certainty that the thing that is now known about that person is bad.

The disability critique of prenatal testing is a powerful rejoinder to the purported necessity of prenatal testing (cf. Parens and Asch 2003). This critique goes beyond the mere assertion that life with disability is not tragedy (though that is worth stipulating at the outset). The argument, in brief, is that disabling traits are overdetermined in projections of how parents will shape their children, that termination on the basis of disability may project harmful attitudes about disability into society (a claim often discussed as the "expressivist objection"), and that prenatal testing is based in misinformation about what disability is. Parens and Asch conclude that more robust informed consent is needed to combat the elimination that can result from prenatal testing. While necessary, this is insufficient: it is not enough to say that the information that we get out of prenatal testing needs to be augmented with additional information about quality of life, or that our assumptions about life with disability are flawed; it is not enough to presume that correcting these misperceptions will lead us on a path toward justice. Rather, we need to acknowledge that the very ways we build scientific systems are not neutral. Consequently, those systems are the places upon which we need to intervene if we care about justice.

A good place to start is to see the work done by the systems that take pains to assert their neutrality—in this case, the way that prenatal testing translates the complexity of genomic knowledge and social prejudice into what it considers to be an individual decision about what to do with a pregnancy. This points us to see how prenatal testing collapses a complex world into an individual decision: where prenatal testing renders these societal-scale outcomes matters of individual choice.

III The "Perpetrator Perspective" and the Danger of Making Systemic Problems Individual

Modern pregnancy involves forms of monitoring, diet management, and genetic technologies that are routinely used to ensure healthy pregnancies and babies. Consequently, pregnant people are faced with individual care choices that bear on their own individual lives and which are often undertaken under conditions of profoundly unequal and under-resourced social worlds. It is possible for a person to, on the one hand, feel abject horror at the systematic elimination of disabled people whose lives they sincerely value and to, on the other hand, live in communities with scarce disability resources, underfunded public schools, inadequate healthcare, and without the networks of companionship and formal and informal kinds of care that might (they don't know for sure—this person recognizes the unknowability of the future needs a child might have) be required in order for their child to flourish.

Care is a material thing that demands time and energy and resources. Care is unequally distributed, falling predominantly on women in the home, and the paid caring professions are disproportionately staffed by people (again, predominantly women) of color (Hartmann et al. 2018). It is important to pay attention to the value and labor of care because, particularly in the USA, individuals and families are left to take it on themselves. If we take on board the idea that aspiration to autonomous individualism is unrealizable not only for disabled people but for us all, we then have to confront head on the structures, informal and formal, paid and unpaid, that bind us together and make possible the appearance of independence for some and presumption of dependence for others (Crosby and Jakobsen 2020).

Care has costs; the availability of supportive care environments is not a peripheral factor in reproductive decision-making—it is central.[5] I offer this framing not as a defense of termination decisions on the basis of disability, but rather as an entry point into an analysis of some of the harms built into the tools available for addressing social failures to provide care resources. The forms of prenatal testing outlined above have the consequence of transforming reproductive outcomes from matters of chance to matters of choice, and US law and culture treat matters of choice as demanding individual responsibility, for which there must also be individual blame.

Dorothy Roberts' work calls for a deeper understanding of what she terms "reproductive dystopia." Roberts notes that

> In the neoliberal future, the state may rely on the expectation that all pregnant women will undergo genetic testing to legitimize not only its refusal to support the care of disabled children but also its denial of broader claims for public provision of health care.
>
> *(Roberts 2009, 798)*

Here she makes an explicit connection between choice and responsibility: insofar as prenatal testing turns matters of reproductive chance into elected features of offspring and insofar as certain offspring are not in the economic interests of the state, the state may use the fact of choice as grounds to not financially support them.

There will always be people born with disabilities. But care continues to be costly (in all senses of the word, for all parties involved) and if the state fails to take responsibility for its provision by continuing to underfund or further reducing disability services, care must nevertheless come from somewhere. And despite the effectiveness of genetic testing programs in places like Iceland and Denmark in eliminating certain kinds of disabilities, disability does and will still exist. Everybody—those with disabilities and those without—needs care. Some kinds of care, such as that delivered through educational institutions or employer-funded healthcare or taxpayer-funded road maintenance, already have methods in place to, however inefficiently, take care of it. Some forms of disability, of course, require only average amounts of care, and others require a bit more. And some require that significant human, technological, and environmental resources come together to make it possible for people to go about their lives. We in the USA do not yet live in a world that recognizes those forms of care as needed without strings attached, or even as possible to receive without significant (and often impossible) individual financial outlay.

Enter the wrongful birth lawsuit.[6] These are undertaken on behalf of parents against medical providers such as genetic counselors or testing companies when pregnancy results in the birth of a child that would not have been, if the parents had had different information about the future child's disability. They quantify the dollar value of the care a person with significant impairments will need, and assert that parents should not have to pay it: not only are the parents blameless, but someone else is to blame. Such lawsuits, then, require that parents assert that there is something wrong with their children—so wrong with them, in fact, that if their disability had been known, they would have opted to terminate.[7] Wrongful birth suits thus codify disability as an economic problem with economic redress. And so, in some senses, wrongful birth suits make an argument consonant with that made by people who wish to expand the welfare state, who argue that it ought to be the state's responsibility to pay for things like healthcare, education, and family leave time. Both groups are arguing that care is valuable, that it has material costs, and cannot be regarded only as a labor of love. But where advocacy for universal programs sees care as something given and required by everyone (albeit in different ways and as the responsibility of the state), wrongful birth suits do so by finding individual parties liable for its provision, and responsible for redress.

My aim here is neither to defend nor to recuperate wrongful birth suits as responses to birth with disability, but instead to point to how they exemplify important patterns of thought that are encoded into US legal, political, and social life. In transforming what could be expansive political claims about the scope of welfare programs or the responsibilities of governments to their citizens into matters of individual harm and individual redress, legal actions like wrongful birth suits adopt what Dean Spade and others call the "perpetrator perspective" (Spade 2015). This is a critique of the idea that in the event of complex harm such as systemic racism or ableism, an individual culpable party (such as a doctor or genetic counselor) can be found and held responsible.

Instead of orienting us toward the complex social structures that systemically sustain ableism (and its relationship to racism, sexism, transphobia, and other forms of systemic marginalization), the perpetrator perspective sees individual actors as enacting the harms of ableism. It frames the birth of a disabled child as something for which there ought to be blame. In adopting the perpetrator perspective, wrongful birth suits translate the complexity of all the reasons that a person might make a reproductive decision (lack of social supports, inadequate information, fear, worry about suffering, individual bias, etc.) into regretted choices with singular causes for that regret. Individuals associated with the supposedly faulty genetic tests represent a concrete point of intervention; assigning responsibility to a single part is more practical and legible than changing entire biopolitical regimes that make lives more or less livable. And so wrongful birth suits are complicated: in a society that fails to provision resources at scale for all people to thrive, it makes an unfortunate sense that those advocating for people they love, who have arrived into the world with heightened care needs, would adopt whatever means necessary to get them. Even if it involves telling a simpler story than the one they experience. Nevertheless, if we collectively notice the frameworks for simplification and responsibility allocation built into social systems as policy inventions that are the results of human decisions, we can begin to ask if there are other paths to justice that entail fewer harms.

IV Conclusion: Reproductive Justice Is Disability Justice

The analytic framework and goals laid out by the reproductive justice movement (RJM) offer some guidance as to how to navigate the paradox of choice. It takes a systems approach to understanding how choices are made and made possible, and more specifically to understanding the conditions that would make having, not having, and parenting children all realizable in practice. The collection of reproductive advocates, scholars, practitioners, and community members who fall under the banner of the RJM conceive of commitment to bodily autonomy as an interdependent, rather than atomistic, proposition. An RJM perspective helps us see that in our biomedicalized world that renders pregnant people what Rayna Rapp terms "moral pioneers," there is no position free of choice, no path free from the potential for harm (Rapp 1999).

The harm of elimination and the harm of the perpetrator perspective both orient us toward processes through which in prenatal testing, complexity and uncertainty are transformed into matters of individual decision-making and individual redress. As a starting point, this logic needs to be noticed. Where does it simplify that which ought to linger in complexity? Where does it make people accountable for things that are not within their sphere of control? Scholars and activists working within the RJM have long seen disability justice as within the remit of their thinking, as something that is part and parcel of the landscape of the claims they make about rights to social and political support in all aspects of family building (Ross 2017). They are invested in meaningful choice not as a project of more informedness or optimizing autonomy, but rather because it doesn't make sense to think about choice until you have created the conditions under which it could be possible: the conditions under which the decision to terminate a pregnancy would be supported through abortion protections and the decision to give birth would be supported by the availability of education, shelter, and care. Without provisioning the conditions of possibility for choice to be meaningful, translating choice into a project of optimizing the availability of prenatal information is little more than a project of shifting responsibility from social systems onto individuals and calling it freedom.

Prenatal testing entails complex harms, in tension with one another. But a robust account of harm ought not engage in thinking it an individual thing; rather, it needs to see "harm" as something that is brought forth systemically, and does not buy into the individuation that our system of rights and redress attempts to instantiate. A strong commitment to autonomy, to reproductive rights, to the value of notions of interdependence may orient a person faced with the need to make decisions about prenatal testing toward particular outcomes. But even such commitments do not make the question of what to do (skip prenatal testing? Refuse termination? Make choices while stipulating the need for ignorance about sex, disabling traits?) obvious. This chapter, accordingly, has not pretended to offer easy answers as to what to do. Instead, it offered an account of one reason why reckoning with those harms is so complicated: because the translation of things that are complicated and collective into the assertion that they are concrete and individual doesn't actually make them so.

Notes

1 The extra copy of chromosome 21 associated with Down syndrome.
2 This articulation of harm resembles the framing of the non-identity problem, wherein harm is done because people who might have come into being are prevented from doing so and instead other,

non-identical people are allowed to. In this chapter, I do not attempt to answer the question of whether harm can occur to potential but ultimately non-existent people, and I instead point to the effects of aggregate decisions about disability on the lives, cultures, and societies of people who do.

3 For an overview of the "mere difference" account of disability, see C2 in Barnes (2016). For an account of why we shouldn't discount disabled people's self-reports of their positive quality of life, see Stramondo (2021).

4 Also Chapter 10 in this volume.

5 Some important critiques of formulations of care ethics note the harms of framing care as a burden, as something that a more powerful party gives to a less powerful one, and how that tends to center caregivers as ethical subjects while marginalizing disabled people and those receiving care. I focus on the costs of care here even while taking those criticisms as fundamentally important: my intent here is not to frame disability as burden or caregiving as saintly, but rather to see how it is fundamentally important to demand that social systems take up the need to support care in order to transform its demands from individual sites of valorization and marginalization into collective understandings of our mutual dependency and obligation.

6 See Valentine, this volume (Chapter 12).

7 In this way, wrongful birth lawsuits are a key pathway through which the expressivist objection, that termination on the basis of disability may project harmful attitudes about disability into society, is expressed, because such suits establish a clear relationship between the presumptive badness of disability and claims about why it ought not exist.

References

Asch, Adrienne, and Wasserman, David. 2005. "Where Is the Sin in Synecdoche? Prenatal Testing and the Parent-Child Relationship." In *Quality of Life and Human Difference*, 1st ed., edited by David Wasserman, Jerome Bickenbach, and Robert Wachbroit, pp. 172–216. Cambridge; New York: Cambridge University Press. https://doi.org/10.1017/CBO9780511614590.008.

Barnes, Elizabeth. 2016. *The Minority Body: A Theory of Disability*, 1st ed. Studies in Feminist Philosophy. Oxford: Oxford University Press.

Crosby, Christina, and Janet R. Jakobsen. 2020. "Disability, Debility, and Caring Queerly." *Social Text*, *38*(4): 77–103. https://doi.org/10.1215/01642472-8680454.

Drescher, Jack. 2015. "Out of DSM: Depathologizing Homosexuality." *Behavioral Sciences*, *5*(4): 565–575. https://doi.org/10.3390/bs5040565.

Fineman, Martha. 2004. *The Autonomy Myth: A Theory of Dependency*. New Press.

Hartmann, Heidi, Jeff Hayes, Rebecca Huber, Kelly Rolfes-Haase, and Jooyeoun Suh. 2018. "The Shifting Supply and Demand of Care Work: The Growing Role of People of Color and Immigrants." *IWPR 2020* (June). https://iwpr.org/iwpr-issues/race-ethnicity-gender-and-economy/the-shifting-supply-and-demand-of-care-work-the-growing-role-of-people-of-color-and-immigrants/.

Ho, Anita. 2008. "The Individualist Model of Autonomy and the Challenge of Disability." *Journal of Bioethical Inquiry*, *5*(2–3): 193–207. https://doi.org/10.1007/s11673-007-9075-0.

Hodgson, Jan and Merle Spriggs. 2005. "A Practical Account of Autonomy: Why Genetic Counseling Is Especially Well Suited to the Facilitation of Informed Autonomous Decision Making." *Journal of Genetic Counseling*, *14*: 89–97. https://doi.org/10.1007/s10897-005-4067-x.

Johnston, Josephine, Ruth Farrell, and Erik Parens. 2017. "Supporting Women's Autonomy in Prenatal Testing." *New England Journal of Medicine*, *3*: 505–507.

Parens, Erik, and Asch, Adrienne. 2003. "Disability Rights Critique of Prenatal Genetic Testing: Reflections and Recommendations." *Mental Retardation and Developmental Disabilities Research Reviews*, *9*: 40–47. https://doi.org/10.1002/mrdd.10056.

Rapp, Rayna. 1999. *Testing Women, Testing the Fetus: The Social Impact of Amniocentesis in America*. London; New York: Routledge.

Roberts, D. E. 2009. "Race, Gender, and Genetic Technologies: A New Reproductive Dystopia?" *Signs: Journal of Women in Culture and Society*, *34*(4):783–804. https://doi.org/10/b77nv4.

Ross, Loretta, Lynn Roberts, Erika Derkas, Whitney Peoples, and Pamela Bridgewater (eds.). 2017. *Radical Reproductive Justice: Foundation, Theory, Practice, Critique*. The Feminist Press at CUNY.

Spade, Dean. 2015. *Normal life: Administrative Violence, Critical Trans Politics, and the Limits of Law* (Revised and expanded edition). Duke University Press.

Stramondo, Joseph A. 2021 "Bioethics, Adaptive Preferences, and Judging the Quality of a Life with Disability." *Social Theory and Practice*, 47(1): 199–220. https://doi.org/10.5840/soctheorpract202121117.

Zhang, Sarah. 2020. "The Last Children of Down Syndrome." *The Atlantic* (December). https://www.theatlantic.com/magazine/archive/2020/12/the-last-children-of-down-syndrome/616928/.

Further Reading

Jarman, Michelle. 2015. "Relations of Abortion: Crip Approaches to Reproductive Justice." *Feminist Formations 27*: 46–66. https://doi.org/10.1353/ff.2015.0008.

Kaposy, Chris. 2018. *Choosing Down Syndrome: Ethics and New Prenatal Testing Technologies*. Basic Bioethics. Cambridge, MA: The MIT Press.

Roberts, Dorothy E. 2015. "Reproductive Justice, Not Just Rights." *Dissent Magazine*. https://www.dissentmagazine.org/article/reproductive-justice-not-just-rights.

10

A FATAL ATTRACTION TO NORMALIZING

Treating Disabilities as Deviations from "Species-Typical" Functioning

Anita Silvers

This chapter is a significantly shortened reprint. The original chapter appears in *Enhancing Human Traits: Ethical and Social Implications*, edited by Erik Parens, pp. 95–123. Washington, DC: Georgetown University Press, 2000. © Georgetown University Press Reprinted with the permission of Georgetown University Press and Erik Parens.

Health Care as a Social Good

[…]

> None of us deserves the advantages conferred by accidents of birth... It is... important to use resources to counter the natural disadvantages introduced by disease... This does not mean we are committed to the futile goal of eliminating or 'leveling' all natural differences between people... [But] health care has normal functioning as its goal: it concentrates on a specific class of obvious disadvantages and tries to eliminate them.
>
> *Norman Daniels (1987, 312)*

The notion of "leveling" that Daniels introduces here is a traditional theme in American political morality. Dissenting religious groups like the Quakers urged that society be arranged to show more respect for the commonalities of human nature, our essential humanity, than for artificial distinctions of class, caste, or role. Those committed to "leveling" were motivated by the conviction that all souls were equally valuable to God; therefore, all souls should have an equal voice in the community. The accidents of wealth and birth ought not to disadvantage people by limiting their opportunities for social participation.

Traditional leveling theories did not propose that all souls were identical, of course, but only that they have equal opportunity for community involvement and influence. Nor did these theories propose to eliminate the natural differences among people, only those accidental disadvantageous differences attendant on wealth and birth. Indeed, the argument for diminishing the importance assigned to wealth and birth (that is, to inherited rank) was that these socially constructed characteristics should not be allowed to obscure or impede the expression of natural talents and traits, the properties that naturally differentiate one individual from another.

DOI: 10.4324/9781003289487-15

In this tradition, Daniels's policy proposes that health care should eliminate, to the degree possible, the disadvantageous adventitious differences that occur when poor health impairs physical, sensory, or cognitive functioning. People should not be leveled in every way, for they naturally differ in skills and talents: only artificially disadvantageous differences should be eliminated. One of Daniels's important contributions to the traditional discussion is to suppose us to have become so proficient in the practice of medicine that the disadvantageous differences attributable to unrepaired poor health are artificial as when people remain in ill health due to an unjust system of distributing effective medical interventions. If the disadvantages associated with poor health are thus as much a social as a natural product, our "leveling" tradition urges that deficiencies in people's functioning that result from poor health should be remedied so as not to diminish the opportunities their skills and talent would otherwise secure for them. Daniels writes:

> We are obliged to help others achieve normal functioning, but we do not "owe" each other whatever it takes to make us more beautiful, strong, or completely happy. [...] The uses of health care that most of us believe we are obliged to make available to other are uses that maintain or restore normal functioning, not simply any use that enhances our welfare. . .. This distinction between the treatment of disease and disability and the enhancement of otherwise normal appearance or capabilities is reflected in the health care benefit package of nearly every national health insurance system, whether public or mixed, around the world.
>
> *(Daniels et al. 1996, 21)*

[...]

On this (bio)ethical emendation to political morality, medical treatment has a public value because it is an instrument of the state's commitment to protect all citizens equally against arbitrary disadvantage. Interventions that merely enhance the welfare or well-being of individuals in respects in which they are not disadvantaged do not have a similar public value.

> The central function of health care services is to keep us functioning as close to normal as possible. Since maintaining normal functioning protects the range of opportunities open to people, by providing an appropriate set of health care services, we make a significant contribution to preserving equality of opportunity.
>
> *(Daniels et al. 1996, 41)*

Treatments, then, are those interventions that are used to reduce or remedy whatever disadvantage is occasioned by abnormal functioning that is associated with ill health. Because treatments are so defined, they are necessarily equalizing, in the sense that to be treatments they must be aimed at preventing or rectifying disadvantageous functioning and, consequently, at reducing or eliminating a specific kind of disadvantage the patient has in comparison with normally functioning individuals. Treatments can be prospective as well as retrospective on this view. For example, as the purpose of vaccination programs is to prevent some individuals from becoming disadvantaged by the sequelae of disease, to vaccinate children against polio or measles is to treat them.

Treatments are processes, however, and a process that has a definitive objective may not always succeed in reaching it. For example, it can be accurate to describe what we do in relation to our students as "educating them" even if some of the students are not educated. Similarly, treating someone may not always succeed in restoring that individual to the desired mode and/or level of functioning. But the key to a medical process's being a treatment is the plausibility of our casting it as a procedure to eliminate a disadvantage by restoring functioning.

For example, breast reductions often count as treatments now that a convincing case for the disadvantageousness of very large breasts is made; for example, she can't buy clothes that fit, she can't run because of their weight, they make her an object of derision in the workplace. But we can imagine social contexts in which it is much harder to make this case. If women custom-made their own clothes, rarely ran (because society insisted it isn't lady-like to run) and never, never pursued careers in the work place (because fathers and husbands did not want women to work outside the home), it would be harder to argue that the breast reduction procedure remedies disadvantages rather than merely increase a woman's comfort. For in that context women would not normally engage in the performances the procedure rehabilitates or restores. But this does not totally resolve the issue, for some women may desire to transform the roles females are permitted to adopt and so may argue that breast reductions remove one of the barriers to women's assuming such roles. They might argue that in their case breast reduction is not merely a means of enhancing the welfare of large breasted women with unfashionable preferences for comfort over sexual attractiveness; rather, it responds to a legitimate need to eliminate a social disadvantage.

Considerations such as these raised by the breast reduction procedure led to questions about the neutrality of appealing to normal fashions of functioning. Sometimes, people who function in the normal fashion are, for that very reason, confined to roles that are disadvantageous and detract from their flourishing. This restriction has surely been the case for women in societies in which women have been assigned to disadvantageous roles on the ground that their normal fashion of functioning prohibited their achieving in more highly valued roles. Since the goal of treatment is to remove disadvantage, but normal fashions of functioning can be disadvantageous, why does Daniels believe that (maintaining or restoring) functioning in a normal fashion is the standard for determining whether an intervention is a treatment?

Normalizing

Whether or not I am an individual whose disadvantage is reduced because I receive treatment, social arrangements providing for the reduction of undeserved disadvantage occasioned by physical, sensory, or cognitive dysfunction are for the public good, Daniels says, and thus for my good insofar as I am a community member. Daniels comments:

> I abstract from the special effects that derive from an individual's conception of the good. This level of abstraction seems appropriate given our search for a measure of the social importance, for claims of justice, of impairments of health. My conclusion is that we should use impairment of the normal... as a measure of the relative importance of health care needs.
>
> *(Daniels 1987, 306)*

Because treatment is a public good, the condition which occasions or invites it should be objective and independent of transitory social accidents, Daniels believes (Ibid, 300). What he takes to be the natural difference between normal functioning and functioning corrupted by illness or accident suggests to him a fixed and objective, and therefore an appropriately public, standard for ascertaining the occasions when treatment should occur. "Where we can take as fixed, primarily by nature, a generally uncontroversial baseline of species-typical functioning," we can show, he thinks, "which principles of justice are relevant to distributing health care services" (Ibid, 303). Daniels thinks that the way the species typically functions constitutes a natural and therefore a neutral standard to which the public can assent.

First, all "people have a fundamental interest in protecting their share of the normal range of opportunities" (Ibid, 306–07). Second, maintaining "normal species functioning" is necessary to protect this high[er]-order interest persons have in maintaining a normal range of opportunities: "Life plans we are otherwise suited for and have a reasonable expectation of finding satisfying or happiness-producing are rendered unreasonable by impairments of normal functioning" (Ibid, 301). Third, and crucial to Daniels's argument, is his assumption that normal functioning is natural and thereby neutral in that the criteria for determining what functioning is normal are biological rather than social. "The basic idea is that health is the absence of disease, and diseases (I here include deformities and disabilities that result from trauma) are deviations from the natural functional organization of a typical member of a species" (Ibid, 302). This step of the argument is critical, for it is here that we are told why not functioning as people typically do is disadvantageous. When disease is the reason individuals do not function in typical fashion, their resulting performances must be inferior to those that issue from individuals whose natural functional organization has not been corrupted by disease.

Of course, this argument leaves open what counts as being diseased. Daniels thinks that the line between disease and its absence generally is noncontroversial and publicly ascertainable through the methods of the biomedical sciences (Ibid, 303). Others argue to the contrary, of course. For example, Susan Sherwin points out that some elements of women's lives for instance, menstruation, pregnancy, menopause, body size and feminine behavior-have been medicalized and treated as diseases because they have been viewed as disruptive of normal functioning (Sherwin 1992, 179). In the same vein, genetic conditions that result in what we think is inferior functioning are equated with disease. These examples suggest that not disease but functioning in the normal fashion is the controlling notion here.

[...]

On the view that being a well-functioning individual is critical to performing the social responsibilities of citizens, normalizing is seen as qualifying functionally defective individuals for citizenship by repairing them so they can execute the usual social interactions and sustain common social responsibilities. To do so they must conduct themselves normally and be able to comply with other people's natural expectations of them. For whoever cannot perform competently as a cooperating and contributing and, therefore, an equal, social partner is fully neither citizen nor person.

[...]

Yet when we probe more deeply into Daniels's account of just health care, we find that it is a social rather than a biological value that informs and validates reparative interventions. Indeed, it is not even a sociobiological value, for rather than surrendering individual benefit to the species' collective evolutionary good, the value to which Daniels appeals is simply an extension of the liberal sociopolitical commitment to preserving equal access to opportunity for the individual.

Daniels's standard is biological, but the principle that implements it clearly is not. For biology tends to eliminate truly dysfunctional individuals, not repair them. The principle that advises us to restore people's normal functioning through health care is also not an expression of an impersonal sociobiological drive of individuals to maintain their species. At most it is an intersubjective principle with the potential to unify the personal interests that individuals have in maintaining a competitive position. It suggests that we should be suspicious of claims that there is a biological mandate that accredits policies of normalizing people by restoring them to typical or familiar modes and levels of functioning.

"Normalizing" has a passionate component, of course, namely, our tribal preferences to congregate with individuals like ourselves. But our attraction to the company of our counterparts, which can be intense, is not usually thought to justify a public commitment to allocate resources to repair those who do not measure up. Simply avoiding or excluding those who fall away from the common standard is the usual concomitant of our passion for congregating with those who most resemble us. The difficulty with thinking of a policy of normalizing as a component of democratic political morality becomes even more evident when we notice that normalizing is sometimes privileging rather than equalizing. For instance, interventions that help some individuals more closely approach species typical functioning may deprive, disadvantage, or otherwise reduce opportunities for individuals who function normally already. And, as we saw when we considered the value of breast reductions, making normalizing our policy can also be unfair if it worsens the position, or otherwise oppresses, the very individuals whose functioning it purports to repair by, for instance, depriving them of anomalous but effectively adaptive alternative modes of functioning.

We should be wary of policies that cloak privileging certain fashions of functioning in the mantle of the "normal." Normalizing then is not the self-evidently right thing to do. Nor can we justly allocate health care without careful attention to the circumstances of whoever is normalized.

What Is Being Normal?

How useful is the concept of normalizing in warranting medical interventions considered case by case? Whether, in fact, there is much clarity about what normalizing is raises a second kind of concern about its compatibility with egalitarian ends, for there is a tendency to equivocate to a dangerous degree as to the meaning of this admittedly circumstantial standard. In "The Meaning of Normal," Phillip Davis, and John Bradley comment:

> Medicine uses the word normal to express . . . various meanings. . . . In medicine, normal can refer to a "defined standard," such as normal blood pressure: a "naturally occuring state," such as normal immunity: . . . "free from disease," as in a normal pap smear . . . "balanced" as in a normal diet, "acceptable" as in normal behavior, or it can be used to describe a stable physical state. In all these meanings . . . normal is used to describe an "ordinary finding" or an "expected state." But medicine allows another meaning . . . that differs significantly from the ordinary. [M]edicine has come to understand normal as a "description of the ideal." . . . Defining the norm as an ideal leads to significant problems. . . . Disease and ill health are a normal part of the human condition. The constant pursuit of health . . . leads easily to blaming those who bear the burden of illness. . . . More important . . . are the problems that result from defining variation from the ideal as "abnormal.". . . Accepting the ideal as the norm begs the question of how uncommon something must be to be considered abnormal.
>
> *(Davis and Bradley 1996, 69–70)*

Current practice assigns pathological conditions the role of being signifiers of unhealth. But because it is not cost effective to intervene wherever pathology occurs—that is, to conduct an all-out campaign to normalize all parts of all people—current practice takes a further conceptual step by identifying some departures from the norm as incapacitating, while others are tolerated as benign. Davis and Bradley observe: "When the ideal is taken as the norm, variation becomes defined as disease—an especially peculiar circumstance insofar as much variation has no particular clinical significance or biological consequence" (Davis and Bradley 1996, 70).

[. . .]

A Social Instrument for Normalizing

Accordingly, we need to clarify the nature and consequences of a public policy that gives expression to a mandate to normalize, an inquiry that I propose to pursue by exploring an analogous sphere, the domain of education. At least two reasons compel us to consider what we can learn from comparing analogous practices in education and health care. First, our current public policy intersects the two domains by legislating entitlements to preventative health education and rehabilitative special education within the public educational system. Indeed, a surprisingly large amount of what Daniels describes as basic health care needs are served by the public education system. Given that it is an egalitarian value, equalizing opportunity, that ultimately justifies the allocation of health care on Daniels's view, he is relatively expansive in delineating what is needed to maintain, restore, or provide functional equivalents (when possible) to normal species functioning. These include adequate nutrition and shelter, safe and unpolluted living and working conditions, preventative and rehabilitative personal medical services, and nonmedical personal (and social) support services (Daniels 1987, 304).

Using public education to deploy preventative, curative, and rehabilitative health care services are, of course, a very familiar practice. During the past century, day and residential schools in this country have, to give some examples, enforced preventative vaccination policies; promoted safe, sanitary, unpolluted, and tobacco-free living; ensured that children are given the basic principles of nutrition; instructed students about maintaining the health of their reproductive systems; identified and referred children in need of reparative care for vision, hearing, and other impairments; and offered rehabilitative speech, psychological, and other therapies, or adaptive education for blind children or deaf children, on their premises. Second, Daniels himself not only finds the analogy between educational and medical benefit apt but relies on it to argue for social support for (universal) health care:

> There is an important analogy between health care and education. Both are strategically important contributors to fair equality of opportunity. Both address needs that are not equally distributed among individuals. Various social factors . . . may produce special learning needs; so too may natural factors, such as the broad class of . . . disabilities. . . . Educational needs, like health care needs, differ from other basic needs . . . Both at the national level and in many states, legislation to meet special educational needs is justified by reference to the opportunities it protects.
>
> *(Daniels, 1985, 46)*

Daniels construes education as a reparative technology. The difference between their views lies in whether, as Daniels thinks, our educational priority should be to teach all children the knowledge and skills normally required to be participating, contributing citizens, or whether, as Brock has it, our educational priority should be to advance the widest array of children's cognitive abilities so as to nourish their different talents and give each appropriate personal opportunities for flourishing. In education as in health care, then, there is a question about the priority to assign to engendering normal functioning.

The Ascendancy of the Normal

In this regard, the battles that raged for nearly a hundred years about how best to educate deaf children provide an opportunity to assess the benefits of giving priority to normalizing. In the late nineteenth century, educators of the deaf bifurcated into two camps, one that passionately supported and one that vehemently opposed educating deaf children in the language of manual signs. Each charged the other with protracting the dysfunctionality of deaf people and consequently with unfairly constricting their opportunity. At the heart of this debate lay another divide, namely, an unbridgeable chasm that separated two very different beliefs about the relationship between biology and opportunity.

As Douglas Baynton writes in *Forbidden Signs: American Culture and the Campaign Against Sign Language*:

> The real battle (over sign language) was fought on a . . . rarefied plane, encompassing such questions as the larger purposes of education in a democratic and industrializing society . . . and the locus and character of cultural authority in America. Indeed, occupying a central place in the fight was a late-nineteenth century debate over the nature of nature itself.
>
> *(Baynton 1996, 107)*

That deaf individuals talk by means of manual signs has been recognized since antiquity. Plato refers to this mode of communicating in the Cratylus (422e). In the early part of the nineteenth century manual sign language schools were established to equalize deaf people's access to the Word, understood to be the conveyance of that moral and religious knowledge which is the goal of human imagination, intelligence, and understanding. Because manual signing was thought to rely on natural symbols that were self-interpreting, Sign was believed to engage the intelligence directly and lucidly, and to stimulate the moral sense. Sign therefore was the instrument to repair deaf people's dysfunction and permit them to develop to greatest perfection. America followed Europe in becoming fascinated by signing. Teachers were imported to systematize and disseminate the gestural communication used by the deaf in this country.

Manualists considered deaf people to be a singular class, distanced from the transient fashions of speech and therefore less corruptible, out of the ordinary, remarkable, unique. Citing such eighteenth-century sources as Daniel Defoe and Denis Diderot, Lennard Davis writes in *Enforcing Normalcy: Disability, Deafness and the Body* that, for different reasons and in different respects, both the blind and the deaf were often thought to exhibit certain heightened and purer sensibilities than the ordinary person (Davis 1995, 53, 57–59). Far from being roleless, deaf people were assigned a special place in the eighteenth- and early nineteenth-century imagination.

But, as Baynton observes, "by the late nineteenth century, naturalness as an ideal was being challenged and eventually was not merely defeated but colonized by the competing ideal of 'normality' (1996, 110). For one thing, naturalness had lost its status as a trait independent of and superior to the artifice, convention, and craft characteristic of social organization. "This intellectual and indeed moral shift in American culture was crucial to the reversal in attitudes toward sign language and the deaf community."

An 1884 speech by oralist Alexander Graham Bell shows how naturalizing the preferred behaviors of the dominant class propelled a program of normalizing, where this meant conforming the behaviors of deaf children to those of the hearing majority:

> I think we should aim to be as natural as we can. I think we should get accustomed to treat our deaf children as if they could hear. . . . We should try ourselves to forget that they are deaf. We should teach them to forget that they are deaf. We should . . . avoid anything that would mark them out as different from others.
>
> *(Baynton 1996, 136)*

By 1899 we find the President of Amherst College, John Tyler, folding the mantle of science around these normalizing practices. Tyler assured a convention of oralists that America would "never have a scientific system of education until we have one based on . . . the grand foundation of biological history...[T]he search for the goal of education compels us to study man's origin and development" (Baynton 1996, 36). Such scientifically-based or biologized education would maintain the functional strategies that seemed to place speech higher on the scale of evolutionary development than the expressive gestures of lower primates. The political morality of the time made it a moral and social obligation to increase the opportunities of both deaf signers and gesticulating foreigners by repairing them, a duty which a scientific system of educating them in the language and communication behaviors of the dominant class could help discharge.

It is important to understand that the fundamental division here is between competing ideas of what organizes a well-ordered society. The eighteenth-century's ideal of individualized moral perfection had given way to an ideal of communal or social participation by people who functioned dialogically in common in the public sphere. Where once language's highest function had been to engage individuals with ideas understood as transcendent sources of right belief and right conduct, now its most important use was to engage people with one another in productive commercial and civic interaction.

Arguing against the idea that deaf people could flourish with a language of their own, an oralist insisted: "To go through life as one of a peculiar class is the sum of human misery. No other misfortune is comparable to this" (Baynton 1996, 145). This thought typifies the shift of priorities from personal to social improvement and the correlated elevation of the importance of collective over idiosyncratic individualistic identities.

This urge to create fair opportunity by leveling the players rather than the playing field is a theme which has come more and more to dominate American egalitarianism over the past hundred years. What is striking is that systematically "normalizing" how deaf people communicate (and many other rehabilitation strategies) may amplify anomalous individuals' opportunities by making them more fit to pursue these [opportunities], but concomitantly may make them less able to perform [them] alternatively or adaptively. By devaluing alternative or adaptive modes of functioning, the policy transgresses liberal political theory's requirement that the state remain neutral between different citizens' ideas of the good life. Oralism's defense of its violation of this dictum was that, until deaf individuals communicated and consequently contributed in the normal mode, they could not be qualified for the protection due citizens.

Normalizing is played out in both medical and educational programs that intervene to repair or restore or revise members of nondominant groups so they qualify as citizens. In education, normalizing has been expressed as a mandate to assimilate the children of immigrant families to the dominant culture, and to impose the practices and preferences of males upon females. In being presented with arguments for normalizing deaf children, the public was invited to decide whether the management of "deaf schools" should be awarded to hearing people who promised to assimilate deaf children. By allocating resources to educational techniques intended to normalize deaf people, public policy imposed a conception of the good under which they did not flourish. We need to ask now whether there are some people who may not flourish, or whose well-being will be compromised, if normalizing similarly warrants and consequently guides the allocation of resources that go to health care.

A Cost/Benefit Assessment of Normalizing

Normalizing has costs. If maintaining or restoring normal function is of such public significance that a system of benefits is made available for this purpose, it is hard to resist supposing that those whose functioning is anomalous ought to acknowledge the system by assigning the same priority to being restored. Baynton reminds us: "Oralism meant that many deaf people had access only to limited or simplified language during the crucial early years of language development" (Baynton 1996, 1). For fear they would fall back to communicating in a more convenient but "abnormal" or "unnatural" way, deaf children were often not taught to write unless they had mastered intelligible speaking. This practice left a legacy of reduced literacy among deaf people.

Interventions that reduce rather than expand already limited functionality surely extract too high a price, but such is the history of oralism in the education of the deaf. "Oralism failed," Baynton concludes, "and sign language survived, because deaf people themselves chose not to relinquish the autonomous cultural space that their community and language made possible." That is, for many people who do not hear, the opportunity of communicating fully within a limited group appears to be more satisfying, more equalizing, and more meaningful than the opportunity of communicating in a limited way with the larger community. This is not to say that all deaf and hard of hearing people make this choice, but merely to point out that the alternative to normalizing often is not a limitation of full functioning but merely a limitation in the expanse of environment in which one functions successfully.

Tribalism, our partiality for interacting with those most like us, undoubtedly influences us to assign preeminence to (the appearance of) normalcy. But to the degree it corrupts the positive balance of benefits over personal and public costs, serious questions about whether the policy of normalizing compromises fair opportunity rather than promotes it must be addressed.

The Canadian Health Care system's intervention in the cases of children born with missing or shortened limbs because their pregnant mothers took thalidomide illustrates this last point. In their treatment, appearing more normal was the priority, so much so that large public sums were expended to design dysfunctional painful prostheses which actually decreased their dexterity and mobility. They could walk with these, but only painfully and slowly. Reminiscent of the oralist ban on signing, they were forbidden to roll or crawl, although these modes offered much more functionality, at least within their home environments.

The direction of resources to fund artificial limb design and manufacture rather than wheel chair design was influenced by the supposition that walking makes people more socially accept able than wheeling does. As the children became independent adults, less vulnerable to the aggressive elements of institutionalized health care, they discarded the dysfunctional prosthetics in favor of wheelchairs, some made to their own designs. Here is another case (among many such examples I could adduce) in which the tyranny of the normal cost anomalous individuals to sacrifice an effective level of functioning at the altar of social preference for a particular mode of functioning, and in so doing compromised rather than equalized their opportunities.

We should not underestimate the coercive potential of policies that validate a particular mode of functioning by directing resources to efforts to restore that mode. When oralism dominated in schools for the deaf, deaf children could either try to lip read and speak, or have no education at all. For the Canadian children with no usable lower limbs, mechanical limbs were the only mobility option offered because policy directed the resources to institutions that designed and engineered limb like prostheses, not wheelchairs. More generally, then, to commit public policy to restoring individuals to species-typical modes of functioning diminishes public recognition of, and consequently resources for, alternative modes of functioning.

So far, we have seen that normalizing equalizes opportunity primarily for those who can be maintained in or restored to the image of the dominant group. But no natural biological mandate nor evolutionary triumph assures that the functional routines of this group are optimally efficient or effective. Rather, the members of this group have the good fortune to find themselves in a social situation that suits them.

For others, there is the choice of limited functionality in an ordinary environment, or ordinary functionality in a limited environment. How much opportunity need be absent from the former alternative, or sacrificed by the latter, depends upon how expansive the nonhostile environment can be made. Replacing staircases with spiral ramps for wheelchair users, adding captions to televised programs for deaf viewers, alt-tags to computer icons so that the screen readers used by people who are blind can identify them, all these make the constructed environment less hostile to and more inclusive of people who function in anomalous, alternative, or adaptive modes.

The main ingredient of being (perceived as) normal lies in finding or creating social situations that suit one. Contrary to Daniels's claim, normalizing is no self-warranting process that deserves the allocation of resources because it furthers democratic values. For individuals with disabilities, for example, such values are better advanced by developing social environments accustomed to people like one's self. The record of their history does not support assuming that broad social or moral benefits accrue to normalizing interventions. The attractiveness of warranting health care interventions that maintain/ restore normal functioning on the ground that they are instruments of justice therefore appears to be much dimmer than the initial enthusiasm of macro-(bio) medical ethics for normalizing suggests.

Disability, Self-Respect, and Lowered Quality of Life

Nothing said so far should be interpreted to mean that interventions to maintain or restore familiar modes of functioning never enhance individuals' welfare. But as we have seen, no clear difference in social benefit, or strict difference in obligation, separates these interventions, ones Daniels would call treatments, from interventions that enhance already average functionality. Then why is functioning normally of such value that maintaining or restoring this level becomes a decisive standard?

A critical component of a good quality of life, Dan Brock says, depends on each of us measuring our capacities and capabilities favorably against the standard of normal human functioning (1994, 31). That is, whether we function normally or not influences how we rate ourselves in comparison to others and consequently affects our confidence and self-esteem. This observation suggests that the paramount benefit of being normal is to maintain the psychosocial well-being of tribalism.

Brock may well be correctly describing a self-reflective process our current cultural standards promote. But this is not sufficient to defend the process as reliable or otherwise reasonable. We have seen that normal functioning is hardly a firm and reliable mark of the quality of our performance. Indeed, it is so fragile a standard that Brock worries about how easily a program of genetic intervention might shatter it. Brock, and others, are alarmed by the potential genetic intervention has for disrupting our confidence in the standard of normalcy.

First, if we manipulate genes to raise the level of performances that typify our species, any pretense that typical functioning is a natural rather than manipulated standard vanishes. Heretofore, a social structure that privileges some people to control communications and construction to suit themselves has determined what modes of functioning are considered normal. Henceforth, a social structure that privileges some people to influence genetic research and the allocation of genetic interventions might determine what levels of functioning become normal. What is feared is that our current confidence in a firm and impersonal, because natural, standard of normality will be undermined by a new and widespread recognition of seemingly normal functioning as being merely the artifactual expression of the interests of whichever members of our species are positioned to deploy technology.

Will such an eventuality constrict rather than enlarge opportunity? Applying genetic technology that increases disparities of access to opportunity initially appears to be inconsistent with a democratically informed health care system, regardless of how much personal welfare the applications might bestow. Consequently, justice appears to advise constraining, or even prohibiting, these important broad applications of genetic technology.

For instance, genetic intervention could result in improving how some people function, so that someone who performs at a level that was comfortable for his species-average parents might find that the naturally good genes he inherited from them are surpassed by great genes installed in his competitors as a compensatory or even as a privileging measure. Constraining applications of this technology so that no individual can acquire an abnormally large number of desirable characteristics may seem advisable. But it is hardly an implementable policy, for the desirability of many of our characteristics is itself provisional and dependent on environment. Whether, and how, adding specific characteristics benefits the recipient—whether it privileges, equalizes, or just makes one more comfortable—must be decided with regard to the context in which the patient will function.

Brock is also concerned about whether, as we come to understand genetic structures accurately enough to identify the potentially anomalous functioning consequent on every species' member's inheritance, some members of the species will find themselves devalued by their own futures. Although performing splendidly at the time, they will be labeled, and consequently marginalized, as being at greater risk than others of deteriorating function. So, for instance, those at risk of Alzheimer's would be rejected as mates by whoever wanted the services of a spousal caretaker, while employers desiring to keep medical insurance costs down would not hire individuals genetically disposed to developing various kinds of cancer.

With the widespread use of genetic testing, he worries:

> [P]eople who feel healthy and who as yet suffer no functional impairment will increasingly be labeled as unhealthy or diseased. . . . For many people, this labeling will undermine their sense of themselves as healthy, well-functioning individuals and will have serious adverse effects both on their conceptions of themselves and the quality of their lives.
>
> *(Brock 1994, 29)*

Notice that, at this point, the idea, rather than the reality, of nonnormal functioning has become the signifier of whether someone is equally well off, or is advantaged or disadvantaged in comparison to others. This observation suggests that it is one's psychosocial rather than physical functioning that is most vulnerable to variations from accustomed states or normal prognoses—that is, to deviations from what is typical of our species. Brock describes how this occurs: "Generally it is when we have noticed an adverse effect or change in our normal functional capacity that we contact health care professionals and begin the process which can result in our being labeled as sick or diseased" (idem). An adverse outcome of a genetic test could trigger this same process, though deterioration in physical functionality has not been and perhaps never will be manifested. Here being labeled as likely to become nonnormal initiates psychosocial processes that themselves are dysfunctional. The perception of being disadvantaged thus precedes and causes, rather than follows upon, dysfunction.

So, the standard constructed to identify who is disadvantaged itself becomes the facilitator of disadvantage. Applying genetic technology then is merely the occasion, not the cause, of an unjust constriction of opportunity. It need not be categorically constrained for fear it will do so. For notice now how the disconnect between actually functioning differently and being disadvantaged has opened even wider. In the case about which Brock worries, individuals are functioning normally but are disadvantaged by having a significant potential, perhaps never to be actualized, for anomalous functioning. Here the social convention of the sick role, rather than the realities of effective performance, determines what modes and levels of functioning are advantageous, indifferent to advantage, or disadvantageous.

The prospect of increasing the power of the standard of species-typical functioning to consign individuals to the sick role undoubtedly is alarming. However, it is not the standard, but the science that could extend its applications, that is typically attacked. So, Adrienne Asch and Gail Geller express their concern that "the Human Genome Initiative could turn out to make 'species-typical functioning' a guide to joining or remaining part of the human community" (Asch and Geller 1996, 330).

To counter these worries about genetic research, we should turn from policies of normalizing to approaches that make us more receptive to alternative ways of functioning. The strategy of protecting against discrimination those who function differently, are genetically disposed to function differently in future, or are perceived as functioning differently is of great help here. The 1990 Americans with Disabilities Act offers one strategic example; another is the recent stream of legislation protecting patients against disclosure of the results of genetic testing.

[...]

Normalizing, What Priority?

[...]

Because there may be no firm answer as to whether a health care intervention does or does not level social advantage, we often cannot deduce from principles of justice whether a medical intervention effects treatment or enhancement. Consequently, even in a just system—indeed, especially in a just system—this distinction is unlikely to guide us in determining what should be provided for particular patients.

Of course, health care's primary mission is to keep us functioning . . . But what kind of intervention is most just remains an issue. To justly liberate group members' many talents, its members could be altered to better satisfy the expectations that pervade their social environment. On the other hand, altering the environment to better support their flourishing may correct their disadvantage equally well.

[...]

This last consideration remains too much neglected by prominent strategists of health-care justice. As we have seen, our normal modes and levels of functioning are, to an extent that often goes unrecognized, socially relative constructions rather than independent biological facts. Adjusting the environment so anomalous individuals can better flourish can be as compensatory as leveling them. Moreover, enhancing individuals or their groups by magnifying their exemplary performance in some domains can, under some circumstances, sometimes compensate for barriers to their performance in other domains of functioning. Wherever strategies that equalize the amount of opportunity individuals have available rather than homogenize the kinds of opportunities they can access are feasible, there is even less reason to suppose that restoring anomalous individuals to normal modes of functioning is a better instrument of justice than enhancing the effectiveness of their anomalous modes.

In positing justice as the regulatory ideal of health care, macro-(bio)medical ethics initially proposed a deductive model on which principles of justice would inform our picking out and prioritizing those medical interventions that further equality. Interventions that qualify as treatments because they aim effectively at restoring normal function were, on this model, to take precedence in the allocation of resources. As we have seen, however, endorsing maintenance or restoration of normal functioning as the standard for allocation can itself, all too readily, prolong disadvantage. Macro-(bio) medical ethics must therefore overcome its fatal attraction to normalizing in order to open itself to other strategies for advancing justice.

References

Asch, A. & Geller, G. (1996). Feminism, bioethics and genetics. In S. M. Wolf (Ed.), Feminism and Bioethics (pp. 318–350). Oxford: Oxford University Press.

Baymon, D. (1996). Forbidden Signs: American culture and the campaign against sign language. Chicago, IL: University of Chicago Press.

Boorse, C. (1987). Concepts of health. In D. Vandeveer and T. Regan (Eds.), Health Care Ethics: An Introduction (pp. 359–393). Philadelphia, PA: Temple University Press.

Brock, D. (1993). Life and Death. New York: Cambridge University Press.

Brock, D. (1994). The Human Genome Project and human identity. In R. Weir, S. Lawrence and E. Fales (Eds.), Genes and Human Self-Knowledge: Historical and philosophical reflections on modern genetics (pp. 18–33). Ames, IA: University of Iowa Press.

Daniels, N. (1985) Just Health Care. Cambridge: Cambridge University Press.

Daniels, N. (1987). Justice and health care. In D. Vandeveer and T. Regan (Eds.), Health Care Ethics: An introduction (pp. 290–325). Philadelphia, PA: Temple University Press.

Daniels, N., Light, D., & Caplan, R. (1996). Benchmarks if Fairness for Health Care Reform. Oxford: Oxford University Press.

Davis, L. (1995) Enforcing Normalcy: Disability, deafness and the body. London: Verso.

Davis, P., & Bradley, J. (1996). The meaning of normal. Perspectives in Biology and Medicine, 40 (1): 68–77.

Rawls, J. (1985). Justice as fairness: Political not metaphysical. Philosophy and Public Affairs, 14: 223–251.

Sherwin, S. (1992). No Longer Patient. Philadelphia, PA: Temple University Press.

11

BEING DISABLED AND CONTEMPLATING DISABLED CHILDREN

Jackie Leach Scully

I Introduction

In 2007, the 1990 United Kingdom Human Fertilization and Embryology Act was revised in the light of new developments in assisted reproductive medicine and research. The revision inserted a clause, section 14(4)(9), stating that

> Persons or embryos that are known to have a gene, chromosome or mitochondrion abnormality involving a significant risk that a person with the abnormality will have or develop—(a) a serious physical or mental disability, (b) a serious illness, or (c) any other serious medical condition, must not be preferred to those that are not known to have such an abnormality.

In other words, gametes or embryos that are likely to produce a child with a known disability may not be used in any form of assisted reproduction if alternatives are available.

To many people this was an uncontroversial and frankly pointless change: what parent would choose to have a child with an avoidable genetic disability? Yet the clause was (and is) controversial. In this chapter I will discuss the ethics of what is often referred to as 'choosing disability': disabled people expressing a preference to have a child also with a disability.[1] As we'll see, there is more nuance to such preferences than bioethical analyses tend to reflect.

This chapter focuses specifically on the use of selective reproductive technologies to 'choose disability'. I will not be discussing broader issues of the right of people with disability to be parents, although I want to acknowledge in passing that parents with disability still experience extensive discrimination, despite explicit legal protection of disabled people's right to a family in Article 23 of the UN Convention on the Rights of Persons with Disability.

II Technologies of Reproductive Identification and Selection

There is now an array of means through which potential parents can not only identify, but often choose, the characteristics of their future child, at least in the sense of choosing between different options. Technologies of prenatal diagnosis (PND) range from ultrasound, through biochemical or genetic analysis of fetal tissue obtained via amniocentesis or chorionic villus sampling (CVS),

DOI: 10.4324/9781003289487-16

to non-invasive prenatal testing (NIPT), which tests fetal tissue from a sample of the mother's blood. Coupled with in vitro fertilization, preimplantation genetic diagnosis (PGD) makes it possible to genetically profile in vitro-produced embryos before deciding which one(s) to transfer for pregnancy. Most recently, attention has turned to the future and more speculative possibilities of genome editing: a different mode of genetic reproductive selection that I discuss later on.

The prenatal identification of characteristics doesn't necessarily lead to selection, but it is still true that these technologies were developed to avoid the birth of infants with anomalies. The underlying assumption here is that parents, given the opportunity, will use prenatal information to select against disability. There is considerable empirical evidence that in the majority of cases this is what happens, although exact figures vary between countries and according to condition: termination rates of over 90% following identification of Down syndrome have been reported from numerous countries (Hill et al. 2017).

Many bioethicists have argued that the use of selective reproductive technology (SRT) to avoid having a child with disability is ethically justified by a broad principle of reproductive or procreative liberty or, as some put it, reproductive autonomy. This principle involves the idea that control over the number, timing, and kind of children is an important, if not necessary, feature of individual autonomy. Increasing the number of options open to prospective parents enables them to exercise their reproductive autonomy, thereby maximizing personal autonomy overall. In modern liberal democracies, this is generally seen as a good in itself.[2]

Some bioethicists have gone further. The principle of procreative beneficence as originally proposed (Savulescu 2001) suggested that potential parents are not just permitted, but ought to select the child that would, out of the available options, go on to have the best life, however construed (for a more in depth discussion, see Amundson, this volume).[3] In the wide literature on procreative beneficence it is not always made clear whether the principle is to be understood as providing good moral reasons for exercising choice or as articulating a genuine moral duty, and commentators are often careless about distinguishing between 'the best possible life' and 'the best possible child'. Procreative beneficence has been criticized on several grounds, including the difficulty of providing objective criteria for the best possible life across all countries, cultures, and individual biographies (Parker 2007).

Other than critiques of the principle of procreative beneficence, standard bioethical responses to prenatal selective technologies generally do not challenge the permissibility of the overall aim. In contrast, some disability scholars and activists suggest that selecting to avoid disability is straightforwardly eugenic (e.g. Hansen et al., 2008; and many others) in both motivation and practical consequences. Others hold that prenatal selective policies might be discriminatory but not fundamentally eugenic, since the ostensible motivation is presumably about increasing parental autonomy and nothing to do with improving the genetic composition of the population. Against that, critics have argued that exercising reproductive autonomy through controlling the number and timing of pregnancies is substantially different from determining the kind of children that you have.

Feminist bioethicists find the decontextualization of the debate especially troubling. For arguments about reproductive autonomy to be credible, women must be genuinely able to choose whether or not to undertake prenatal screening, and between termination and continuation of an affected pregnancy. A contextually critical approach asks to what extent freedom of choice exists in any society where there is discrimination against people with disability, generally abysmal provision of social supports for disabled people and their families, and a growing routinization of prenatal screening. In this context it is difficult to see a simple linear relationship between increasing the number of choices available to parents and increasing parental autonomy (Seavilleklein 2009).

III Changing Concepts of Disability

Over the last 50 years the work of disability activists and researchers has made it possible to challenge our fundamental concepts of disability and attitudes toward disabled people. Questions include whether disablement is due to a body that deviates from a biological standard, or due to social parameters that disable, or the interaction of both; whether being disabled can be an individual or collective identity; and if so, whether it is possible for a disabled identity to flourish. Thinking of disability as ontological rather than pathological (at least in some respects) has begun to shift attitudes but also, crucially, increased public receptiveness to disabled people's own accounts of their lives and experiences. And some of these accounts disrupt easy assumptions about parental preferences toward disability.

Much of the data on choosing disability are anecdotal or come from surveys about intentions and preferences, rather than actual choices. The evidence confirms that the majority of people without disability would prefer to have children who are not disabled. But there are also indications that shifting attitudes toward disability may go along with greater openness to continuing a pregnancy with a known impairment. This might be called 'accepting' rather than 'choosing disability'. While many, perhaps most, people with disability would also prefer to avoid having a disabled child, the personal encounter with disability introduces new variables, meaning that any sweeping generalizations about what disabled parents want, or will do, are misleading. Some disabled people will think that an anomaly isn't an impairment but simply part of the wide range of possible, quite livable, human embodiments; while some disabled people nevertheless want to avoid their child having to cope with the additional difficulties they themselves face, others feel able to go ahead and deal with whatever happens. This is also 'accepting disability': not avoiding it, but not choosing it either.

IV Choosing Disability

The step beyond acceptance is active preference. The paradigmatic examples of disabled parents expressing, and occasionally acting on, a strong preference for a similarly disabled child are essentially provided by two disabilities: deafness, and achondroplasia or related forms of restricted growth. It is an interesting question why this preference is most common in these two groups. It is undoubtedly relevant that both are associated, in different ways, with distinct community identities that have a relatively strong political voice, and that neither is routinely associated with reduced lifespan or physical pain.[4] Since most of the available evidence comes from deafness, I focus on this community for the rest of the discussion.

Audiologically deaf people who are members of the signing Deaf community, and who consider themselves to be a distinct sociolinguistic group rather than disabled, have claimed the right to use selective reproductive technology to maximize their chances of having an audiologically deaf child. By far the most well-known case is that of Sharon Duchesneau and Candace McCullough, a lesbian couple from Washington, USA, both with genetic deafness, whose reproductive story was widely reported in 2002 (Camporesi 2010; Scully 2011). Like any lesbian couple wanting to have children genetically related to one of them, the couple used donor sperm; this couple was unusual in that they actively sought out a donor with the 'right' kind of genetic deafness to raise the odds that their resulting children would be deaf. It was the publicity (and widespread condemnation) of this case that launched a minor industry of bioethical comment and led indirectly to the insertion of clause 14(4)(9) into the UK's 2008 Act I described earlier.

There are few solid data on how widespread a preference for a deaf child is among deaf people. Some surveys, now dated, suggest that up to a quarter of Deaf people would prefer deaf over hearing children. But data on attitudes and responses to hypotheticals do not tell us much about active choices. Empirically based bioethics has tended to assume that preferences lead to actions, ignoring the myriad political, sociological, economic, and emotional factors that profoundly influence how a preference is worked out in real life. I may want to use PGD to have an audiologically deaf child, but not want it enough to undergo medical intervention; or I may just not be able to afford it. Real people have a diversity of preferences that sometimes come into conflict and have to be weighed against each other: I might want a child of a certain sex, for example, but also want to live in a society that bars people from enacting sexism in their reproductive choices, and so decide not to operationalize my preference. It would be quite possible for a Deaf couple to hold a very strong preference for a deaf child but at the same time feel it would be wrong to do anything about it, if they believe that their preference is outweighed by their parental obligation to accept a child unconditionally (Scully et al. 2006).

V The Preference Not to Have a Preference

The implication that parents have a dichotomous 'preference for' either a deaf or a hearing child is also misleading. First, deaf (or indeed hearing) parents may genuinely have no preference insofar as they see deafness as having little or no relevance to their child having a good life. Less obviously, they may not have any preference for a particular kind of child but a strong preference against expressing any preference. This is a key reason why the 2008 UK legislation is so problematic. Consider a situation where PGD has produced four embryos, two carrying a mutation in the connexin 26 gene that is associated with deafness, and two without. Under UK law, parents who prefer a deaf child and would want to use one of the two 'deaf' embryos would be unable to do so: they would be obliged to use one of the others. But in addition, anyone who actively wants to show no preference is similarly blocked from doing so. Although both the professional and public discussions have concentrated on parents who want either deaf or hearing offspring, the real difficulty is for those who would rather just leave it up to chance – which is, after all, what happens in most cases when two people with genetic deafness have a child together. The law forces them to enact a preference that they do not hold, that directly counters what might be a strongly held principle of equality, and which they might also feel entails a rejection of their own identity and of the traditions of their community. All of which, it can be argued, means the UK law (and other analogous regulation) significantly constrains their reproductive autonomy.

VI Is Disability a Harm?

These ethical discussions hinge on the question of whether an embodied difference like deafness or achondroplasia is a harm. Does choosing disability constitute a harm to the child, one severe enough to outweigh respect for reproductive autonomy (remembering that appeals to autonomy are only frictionless up to the point that someone's autonomous actions start to harm others). Many deaf people consider audiological deafness to be a neutral difference, or near enough neutral in developed societies to be treated as such; the responses to stories like that of Duchesneau and McCullough indicate that to the majority of hearing people this is patently untrue.

Bioethical responses to the question of what kind of harm we're talking about here fall into two main categories: *harms to well-being* and *harms to future autonomy*. To address the first, we need to know whether a disability like deafness or restricted growth inevitably leads to a reduction in individual well-being. The sheer diversity of impairment and disability means there is no simple answer to this. The skeletal dysplasias that cause short stature are also often associated with pain, for example, but that may neither be inevitable nor unmanageable, and many other disabilities like deafness don't involve physical pain or shortened lifespans. With any disability, there may be indirect suffering caused through the stigmatization and discrimination endemic in any disablist society; arguably, though, this is better resolved by tackling these societal responses than by treating the subjects of those responses as the source of the problem.

The second group of concerns about harm attend to the impact of disability on future life possibilities, famously described as a child's 'right to an open future' (Millum 2014). How does it affect the opportunities that would otherwise be available, and the child's capacity to make use of them (Schroeder 2018)? It seems obvious that even if well-being is unaffected, disabilities like deafness and achondroplasia effectively close off certain life roles and activities. Selecting for, or failing to select against, those disabilities would be wrong in view of the limitations they impose on a child's autonomy: the child's future would be less 'open'.

However, no one, disabled or not, is born into a truly open environment, in the sense of the availability of all possible opportunities. Real lives are constrained in some ways because of the time, place, social system, families, and bodies into which people are born; moreover, what are considered 'normal' or 'constrained' lives varies across cultures. The proponents of 'right to an open future' arguments generally fail to provide compelling evidence that the restrictions imposed by disability are consistently and meaningfully different in kind or magnitude from the restrictions imposed by class, gender, or the contingencies of a global pandemic. Many disabilities are linked to incapacities, but it is at this point that too many bioethicists reveal their gross unfamiliarity with disability: Burke (2011) quotes a 2004 article stating, as proof of reduced opportunity, that deaf people can't drive, play sports, join the armed forces, or enjoy music, which would be news to the many deaf people who do all of those activities. Judging whether a disability still offers a 'good enough' range of opportunities requires factual evidence, and looking beyond the impairment itself to consider the impact that accommodation and acceptance by the community has on the openness of disabled people's futures (Stramondo 2020).

These discussions, judgments, and evaluations can reach an impasse of mutual incomprehension, in which people outside an experience are just unable to make sense of the perspectives of those within it. To many (though not all) Deaf people, deafness is just not a disadvantage, or at least not one big enough to make this parental choice ethically wrong. Parents are not choosing deafness or short stature because they want their child to suffer or be disadvantaged, but precisely because they believe that being Deaf or short will lead to a life at least as good, or even better, for their child and family. This is why many people with disabilities find it baffling that the discussion is always couched in terms of harm to well-being or constraints on freedom, rather than care or benefit. Parents who want to select for disability, or not select against, may think not only that they can provide a more disability-friendly environment for their child than nondisabled parents, but that they can actually be better parents to children with 'their' disability, and prepare them better for life. For example, Deaf parents worry, with some justification, that a hearing child will inevitably be treated as an unpaid interpreter for their deaf parents; achondroplastic parents might be anxious about caring properly for a child who rapidly outgrows them. In both cases the concern is about maintaining a family environment within which a child is best able to develop in an unconstrained and affirming way.

VII Identity

Understanding disability as constitutive of identity provides valuable insight into the significance of a preference for disabled children. These preferences are typically oversimplified as being about the desire of potential parents for a child like themselves. For most people, building a family has something to do with who you are, what you want in your life, and how you embody continuity between your past and the future. In choosing disability a potential parent may be identifying with their future child, but they are also affirming their own identity as a disabled person. Furthermore, the 'wanting a child like themselves' argument focuses exclusively on the dynamics of parent and child rather than the wider network of family and community relationships; yet it is significant that both the exemplar disabilities have a clear and distinctive collective identity. This is particularly strong for deaf parents who identify with the signing Deaf community, believe that Deaf culture is valuable, and want to share that with their child and enable the culture to continue. The fact that vertical transmission of Deaf culture and language from parent to child is relatively rare means that deaf children of deaf parents are particularly valued for their continuation of an ongoing, vibrant Deaf world. The question is whether sustaining the existence of a precarious community provides enough moral justification for Deaf parents actively choosing to have deaf children – or even a moral obligation to do so. The continued existence of the Deaf community may be a compelling argument for deaf people, but is unlikely to do the trick for those in the hearing world.

VIII Burdens and Benefits

Although the bioethical discussion has primarily focused on the well-being or future autonomy of the child, among the other issues sometimes raised is the potential burden or cost to society of the disabled children that might be born as a result of parental choice. It is unreasonable, so the argument goes, to expect society simply to absorb the additional costs of disability (taxpayer-funded healthcare and social care, specialized education, and so on) through the lifetime of a disabled person whose birth could have been avoided in the first place.

Counterarguments include the lack of evidence and problematic assumptions about the size and nature of these costs; the bias of focusing on the (economic) costs of disability, and not the benefits that people with disability bring; and the risk of oversimplified cost/benefit arguments being a backdoor to eugenics. In practice, given the low numbers of people involved, the actual economic and societal impact of any form of choosing disability is likely to be minuscule, indicating again that the concerns reflect ignorance about and fear of disability rather than any empirical reality.

IX Expressivism

Another issue that has yet to be addressed in detail is the role of expressivism: a surprising neglect, since the 'expressivist objection' is one of the best-known and well-established elements of the disability critique of selective reproductive technologies. As originally formulated by Adrienne Asch and Erik Parens, the expressivist objection says that measures to prevent the birth of children with disabilities are ethically objectionable, because they 'send out a message' – specifically a hostile one – to people with disabilities as well as to nondisabled people (Edwards 2004). These messages are psychologically damaging because they attack disabled people's sense of self-worth; but they are also socially damaging because they reinforce persisting hostile attitudes to disability.

The original version of expressivism focused on the impact of individual reproductive choices. This focus has since been challenged on the grounds that individuals making reproductive decisions probably have no intention of sending any kind of message to anyone, and in practice individual decisions are too personal and ambiguous to communicate effectively. Expressive power is much more effectively channeled through official policy and regulation, both of which actually aim to influence behavior. Policy on selective reproductive technologies is likely to have a profound influence on public thinking and decision making about what is 'normal' for bodies, families, and citizens. A policy that blocks prospective parents either from actively 'choosing disability' or from leaving things to chance will reinforce the collective normative belief that disability is always accompanied by such intolerable suffering or disadvantage that it is better not to be born: responsible parents will always do their utmost to avoid inflicting this on their child (and society). Individual reproductive choices that align with the collective normative moral framework and its policy go unnoticed, while choices that run against the norm are disproportionately visible and carry clearer meaning. The anomalous decisions of individuals to prefer or be neutral toward children with disability have an expressive power that decisions against disability never can.

X What Is Different about Genomic Editing?

Appearing on the horizon of feasibility is the technology of genome editing and its implications for the ethics of reproductive choice. So far, the discussion has been about the ethics of selecting for/not selecting against existing embryos or fetuses likely to be born with disability. Genome editing is similarly presented as a way of avoiding disability, this time by directly changing (editing) gene sequences associated with genetic anomalies – for instance, replacing the mutated connexin 26 gene with a version that is not associated with deafness. Hypothetically, those people who would use prenatal diagnosis or PGD to act on a preference for a disabled child might want to use genomic editing to achieve the same goal: in the absence of the gene(s) associated with short stature, for example, parents might ask for these sequences to be edited in.

The goal may be the same, but the means are significantly different. Selective reproductive technologies enable parents to choose between options that nature or technology has made available, discarding some and using others. In discussing the potential harms of selective technologies, mainstream bioethical literature often draws on the so-called non-identity claim. The philosopher Derek Parfit (1984) argues that no child can be made worse off by their parents failing to prevent them being born with a disability, because the only alternative for that child would be not to exist at all. So long as a person's life is worth living on balance, it cannot be said that their life has been harmed by being born with an impairment, because the only way such harm could have been avoided would have been to preclude their existence. But this is not true of embryonic genome editing which aims to affect the kind of person who will come into being by manipulating the genes of an existing embryo. Unlike PGD, which is about selecting between existing options, somatic genome editing is about changing the nature of the embryo into something that didn't exist before the editing.

Like the distinction between letting something happen and actively ensuring it does happen, which is crucial to numerous healthcare ethics dilemmas, many people share the intuition that increasing the chances of a child being born with a particular disability is morally distinct from actively intervening to change a disability status. Even people who would be sympathetic to parents preferring an embryo that already carries a gene associated with deafness might balk at the idea of deliberately inserting a mutated connexin 26 gene into a 'hearing' embryo. If and when genome editing rolls out, it will be interesting to see whether in fact potential parents do show any desire to use it to have children with deafness or restricted growth, or whether they too find this to be an ethical step too far.

XI Conclusion

In the case of deafness Rob Sparrow concludes robustly, 'we have no legitimate grounds to deny Deaf parents the right to use these technologies in order to have deaf babies, if they wish to do so' (Sparrow 2002, 16). It's true that deafness has some unique features, but the same may plausibly be said of at least some other disabilities. Yet it is clear that the idea of 'choosing disability' is troubling to many people. In a context of widespread lack of acceptance of disability equality, increased routinization of reproductive selection, and where citizens are 'responsibilized' to ensure their own and their children's bodies align with implicit economic and societal demands, these anomalous decisions are likely to remain controversial.

The idea of choosing disability is still of bioethical interest not so much because of its practical impact (in no country have prenatal clinics been overwhelmed by demand for disabled children), but for what it reveals about the ethical and societal framing of disability. Engaging with it is an opportunity to ask important and profoundly contextual questions about normality and abnormality, reproductive autonomy, being a parent, the forces that shape our moral landscape, and the responsibilities of legislators and policy makers.

Notes

1 As this chapter shows, 'choosing disability' is a major misunderstanding of what is going on; but for ease of reading I will continue to refer to it as that, without quote marks.
2 Even though the view that increasing choice is an obvious good is often challenged by research showing public ambivalence about it, especially in reproductive contexts.
3 Chapter 13.
4 Some skeletal dysplasias do involve physical pain and other complications, which influences how parents feel about their children having the same disability.

References

Camporesi, Silvia. 2010. "Choosing Deafness with Preimplantation Genetic Diagnosis: An Ethical Way to Carry on a Cultural Bloodline?" *Cambridge Quarterly of Healthcare Ethics* 19(1): 86–96. https://doi.org/10.1017/S0963180109990272.

Edwards, S.D. 2004. "Disability, Identity, and the 'Expressivist Objection'." *Journal of Medical Ethics* 30(4): 418–420.

Hansen, Nancy E., Heidi J. Lanz, and Dick J. Sobey. 2008. "21st Century Eugenics?" *The Lancet Supplement* 372, (December): 104–107. https://doi.org/10.1016/S0140-6736(08)61889-9.

Hill, Melissa, Angela Barrett, Mahesh Choolani, Celine Lewis, Jane Fisher, and Lyn S. Chitty. 2017. "Has Noninvasive Prenatal Testing Impacted Termination of Pregnancy and Live Birth Rates of Infants with Down Syndrome?" *Prenatal Diagnosis*, 37: 1281–1290.

Millum, Joseph. 2014. "The Foundation of the Child's Right to an Open Future." *Journal of Social Philosophy* 45(4): 522–538. https://doi.org/10.1111/josp.12076.

Parfit, Derek. 1984. *Reasons and Persons*. New York: Oxford University Press.

Parker, Michael. 2007. "The Best Possible Child" *Journal of Medical Ethics* 33, 279–283.

Savulescu, Julian. 2001. "Procreative Beneficence: Why We Should Select the Best Children." *Bioethics* 15: 413–426. https://doi.org/10.1111/1467-8519.00251.

Scully, Jackie Leach. 2011. "'Choosing Disability', Symbolic Law, and the Media." *Medical Law International* 11(3): 197–212. https://doi.org/10.1177/096853321101100304.

Scully, Jackie Leach, Tom Shakespeare, and Sarah Banks. 2006. "Gift not Commodity; Lay People Deliberating Social Sex Selection." *Sociology of Health and Illness* 28: 749–767.

Seavilleklein, Victoria. 2009. "Challenging the Rhetoric of Choice in Prenatal Screening." *Bioethics* 23(1): 68–77. https://doi.org/10.1111/j.1467-8519.2008.00674.x.

Sparrow, Robert. 2002. "Better off Deaf." *Res Publica* 11(1): 11–16.

Stramondo, Joseph. 2020. "Disability and the Damaging Master Narrative of an Open Future." *The Hastings Center Report*. https://doi.org/10.1002/hast.1153.

Further Reading

Burke, Teresa Blankmeyer. 2011. "Quest for a Deaf Child: Ethics and Genetics." PhD diss. University of New Mexico. http://digitalrepository.unm.edu/phil_etds/14.

Kaposy, Chris. 2018. *Choosing Down syndrome: Ethics and the New Prenatal Testing Technologies.* Cambridge, MA: MIT Press.

Schroeder, S. Andrew. 2018. "Well-being, Opportunity, and Selecting for Disability." *Journal of Ethics and Social Philosophy* 14(1): 1–27.

Scully, Jackie Leach. 2008. "Different by Choice?" In *Disability Bioethics: Moral Bodies, Moral Difference,* 59–81. Lanham, MD: Rowman & Littlefield.

12

THE WRONGS OF 'WRONGFUL BIRTH'

Disability, Race, and Reproductive Justice

Desiree Valentine

The legal claim of wrongful birth is a specialized type of tort claim. Tort claims seek to recognize civil 'wrongs' that have occurred due to negligence that can be traced to the defendant's action or inaction. Typically, wrongful birth cases involve negligence of a physician to identify and inform parents of the existence of—or probability of —birth anomalies, disabilities, or other such factors thought to be medically pressing issues. Plaintiffs often sue for emotional and economic burdens associated with raising a child with a disability. In these cases, the burden of proof is on the plaintiffs to show that a medical provider "owed a duty to the parents, breached that duty, and the duty was a factual and proximate cause of the parent's injury" (Yakren 2018, 587). Wrongful birth cases inextricably link parents' injury to the child's existence, making it a particularly controversial kind of tort claim.

In this chapter, I provide a brief historical and conceptual overview of wrongful birth. I then discuss disability studies critiques, followed by a broader examination of contemporary issues with wrongful birth in the context of race. I argue that we ought to understand the problematic nature of wrongful birth through a history of intersecting and co-constituting ideologies of marginalization. I show how race and disability ought not be understood as disparate phenomena, but as a nexus. By seeing them in this manner, we can reconceive opportunities for developing liberatory responses to the problems wrongful birth raises.

I Wrongful Birth: An Overview

Understanding the history of wrongful birth and its common usage today requires an understanding of related tort claims such as wrongful conception, wrongful pregnancy, and wrongful life. Wrongful conception and wrongful pregnancy emerge from medical malpractice torts. The legal 'wrong' in question is the actual conception or pregnancy. Such cases typically involve a faulty contraceptive leading to the plaintiffs relying on incorrect medical information. Examples include a sterilization procedure that did not result in sterilization or incorrect medical information about one's infertility that produces an unplanned pregnancy.

Wrongful life is more similar to wrongful birth in that the injury described is related, either directly or indirectly, to the existence of the child. Wrongful life cases directly label the life of the child as the injury. In these cases, it is the child who has direct standing to bring a case.

DOI: 10.4324/9781003289487-17

Wrongful birth cases indirectly note the existence of the child as the injury by claiming a lack of choice to abort given an absence of or inaccurate medical information. In these cases, it is the parent(s) of the child that can bring a suit.

In wrongful life cases, since the legally cognizable injury is the life of the child, the burden of proof must show that non-existence would have been preferred over the life now had by the claimant.[1] Many courts have refused wrongful life claims on account of the metaphysical conundrum they produce, arguing that existence is always preferred over non-existence, so 'life' cannot be expressed as a particular 'harm' itself. Though courts have generally looked unfavorably on wrongful life claims, they have been far more approving of wrongful birth cases. As Wendy Hensel (2005) notes, this is likely because "it is more palatable to identify lost parental choice as the injury than to answer the metaphysical question of whether non-existence is ever preferable to life" (143). In wrongful birth cases, parents must argue that (1) there was a "deprivation of choice" regarding the ability to make fully informed decisions about a pregnancy given a physician's misdiagnosis or failure to disclose medical information and (2) had they been informed sufficiently, they would have terminated the pregnancy (Rinaldi 2009, 5).

Historically, this bundle of tort claims—wrongful conception, pregnancy, life, and birth—have all tended to be treated disapprovingly by the courts when the action brought was related to a nondisabled child. Yet in the context of disability, wrongful birth has had considerable success. This is undoubtedly linked to the proliferation of prenatal testing, an increasing sense of personal responsibility for procreation, and biased assumptions about disability. Additionally, since wrongful birth claims have gained specificity as a claim against a physician for the deprivation of reproductive choice and autonomy, these tort claims would be unintelligible without the legal decision of *Roe v. Wade*.

For example, in 1967, the first wrongful birth case based on 'impairment' as an injury in the USA occurred in the New Jersey Supreme Court. In Gleitman v. Cosgrove, the plaintiffs sued because their doctor failed to inform them that rubella, which Mrs Gleitman had contracted early on in her pregnancy, would likely lead to birth defects. They sued for emotional distress and costs for caring for their child. The court rejected these claims, stating that the difficulty of determining damages given the "intangible, unmeasurable, and complex human benefits of motherhood and fatherhood" (Gleitman) weighed against the alleged emotional and financial injuries. As a pre-*Roe v. Wade* case, the court concluded that the prohibition against abortion disallowed damages to be recouped on account of a missed opportunity to abort a fetus. Following Roe, however, many courts have rejected Gleitman's reasoning. In 1975, the first successful wrongful birth case in the context of disability was brought in the Supreme Court of Texas. Jacobs v. Theimer involved a mother who contracted rubella during pregnancy and was neither properly diagnosed by her doctor, nor informed of the risks associated with rubella during pregnancy. The court recognized compensation due for the costs of care and treatment of the child, though not for the emotional distress of the parents.

Since this time, wrongful birth has largely been associated with disability-related claims. While some USA states specifically prohibit wrongful birth, more than half recognize the claim, and around the world many courts permit such claims (Frati 2017, 344). Wrongful birth remains contentious, however. In what follows, I outline the disability studies critique of wrongful birth.

II The Disability Studies Critique of Wrongful Birth

Disability studies critiques illuminate the harmful impacts of wrongful birth claims. These critiques identify wrongful birth's problematic deployment of disability as a problem of individual lack rather than a largely structural issue, as an expression of able(ist)normativity and ableism, and as resulting from eugenic logics.

Wrongful birth claims rely on a medical model of disability that assumes disability as an objective, biological determination that necessarily has significant and negative impacts on life. This assumes a standardized 'norm' of human functioning. However, many disability studies scholars have shown that this norm is informed by certain patterns of valuing, specifically those presuming a masculinist picture of health as stable, fixed, and conducive to productivity according to (neo)liberal capitalist logics. Opposed to this medical model, scholars have offered models of disability that attend to the ways in which disability is socially and politically produced.[2] That is, rather than an objective, scientific 'fact,' disability is understood as generated within the encounter between bodies, environments, and the norms and values circulating within such contexts. From this perspective, the disadvantage and injury identified within wrongful birth torts are only identifiable as such within an ableist society that lacks support for a variety of ways of life and, moreover, actively stigmatizes and pathologizes those forms of living that differ from the ableist norm. It makes disability an individual lack rather than a structural problem to which states and governments should respond. Currently, wrongful birth links disability with harm and that harm as an individual problem for which individuals can seek compensation, but as Rinaldi (2009) argues "the responsibility for improving the lives of disabled people should be borne by society and taken up as a human rights issue" (1).

Scholars also critique wrongful birth torts with the charge that they contribute to the discrimination of disabled individuals by expressing a harmful message to people currently living with disabilities—namely, that one's life is 'less' valuable than able-bodied lives or perhaps even not valuable at all. Such devaluing of disabled life and living occurs in the context of implicit "hegemonic ablenormativity," described by Lydia X. Z. Brown (2018) as "the cultural dominance of a pathology paradigm that assigns the values of health, normality, worth, and functionality to normatively abled bodies while simultaneously marginalizing and medicalizing deviant bodies" (4). Critical disability studies scholars have specifically identified wrongful birth's contribution to sustaining ableism through a devaluation of disabled life. This is evident in wrongful birth claims since at base the plaintiff must show that had they been aware of existent disabilities or risks for disabilities in utero, they would have aborted the fetus. This forces parents to employ the language of 'burden' and 'harm' when discussing their children in court; otherwise, they are much less likely to qualify for compensation. In the case of Arndt v. Smith, for example, the court's decision rested on the question of whether Arndt would have terminated her pregnancy had she been made aware of testing that would have identified genetic impairment. The case was lost because the judge determined Arndt would not have aborted the fetus had she known about its genetic impairment (Rinaldi 2009, 5).

Finally, many disability studies scholars describe the eugenic logic and impact of wrongful birth claims. We tend to think of the term 'eugenics' as something of the past, particularly associated with Nazi Europe. However, the term originated in the late nineteenth century by British Naturalist Francis Galton. Combining the Greek 'eu' and 'gen-' the term means "'good in stock, hereditarily endowed with noble qualities'" (Stein 2009, 1138). Supporters of eugenics historically supported selective breeding processes, both negative and positive, in an attempt to gain control over reproduction at the level of populations.[3] The USA and Britain have a long and ongoing history related to eugenics, which actually pre-dates and inspired the Nazi regime. At the turn of the twentieth century, eugenics in the USA was practiced through multiple means, including the forced sterilization of the 'feeble-minded,' 'criminals,' and other so-called 'defectives.'[4] Additionally, control of immigration and marriage practices contributed to eugenic practices. These practices continue today in forms that often evade detection as such.[5] The intersection of law and reproductive technologies is one such site of a 'new' eugenics. Wrongful birth claims are legally cognizable only via assumptions that align with and promote eugenic logics. Consider what it takes for wrongful birth statutes to be intelligible. First it must be assumed that the birth of a disabled child is a harm or injury. Second, because of this so-called injury, it is assumed that it would be reasonable (and perhaps even responsible) for parents, if informed of the possibility of the birth of a disabled child, to choose to terminate the pregnancy. This, in turn, enables healthcare providers to suggest termination, lest they be held legally culpable for damages resulting from the birth of an 'injurious' child. As Jill T. Stein argues, the government participates in this "new eugenic movement" judicially and legislatively by imposing such a liability on physicians and recognizing an injury to parents in the case of a child born with a disability but not in the case of a child born without a disability (1138). Together, rhetoric and actions like this produce the "cumulative effect of eliminating the genetically impaired" (1138). We see evidence of this given the rates of termination of pregnancies with fetuses identified as having Down syndrome or cystic fibrosis, for example, which reach up to 85–95% (Lenon and Peers 2017, 153).

Though many scholars hesitate to embrace wrongful birth for the aforementioned reasons, some argue that such claims should be available as a meaningful form of response and redress that bolsters reproductive autonomy. The idea here is that such claims provide a legal mechanism for strengthening the right to informed decision making in matters related to pregnancy and childbirth by working to deter healthcare professionals from withholding or offering false information to pregnant individuals. But when situated in a context that assumes disability is an inherent harm and that it is reasonable (and even expected) to abort in the case of disability, choice is not what is at issue, but the lost opportunity to have (what are perceived to be) better reasons to abort. Instead of prenatal tests and diagnoses strengthening reproductive autonomy, they become conflated with the narrow option to abort (Campbell 2009, 157).

Disability studies critiques helpfully articulate problematic assumptions associated with wrongful birth and provide a necessary reassessment of the stakes of the matter. My goal in the remainder of this chapter is to understand 'wrongful birth' in the broader social, political, and biomedical nexus of contemporary life. I examine the case of Cramblett v. Midwest Sperm Bank and further explore the problematic intersecting ideological elements which make wrongful birth socially and legally cognizable.

III The Curious Case of Cramblett v. Midwest Sperm Bank

In 2014, Jennifer Cramblett, a white lesbian woman from Uniontown, Ohio, sued for wrongful birth and breach of warranty against Midwest Sperm Bank. As the complaint alleged, Cramblett had chosen the sperm of a white, blue-eyed man and instead received the sperm of a Black man. It was only after a successful insemination that Cramblett was made aware of the 'mix-up.' At this time, she was four to five months pregnant. She carried the pregnancy to term and in 2012 gave birth to a healthy, black/mixed-race girl. The wrongful birth complaint suggested that this child's existence caused damages—both economic and emotional—for which Cramblett ought to be compensated. Such 'damages' included how the town Cramblett and her partner moved to is all-white and "too racially intolerant" and far from any area where Cramblett can take her child to get a "decent haircut." Additionally, Cramblett expressed worries about her "unconsciously insensitive family members," and noted her "limited cultural competency relative to African Americans, and steep learning curve" given the fact that she "did not know African Americans until her college days" (Cramblett).

Originally heard in an Illinois Circuit Court, the case was dismissed in 2015 with the judge's suggestion that Cramblett refile as a negligence claim. Newspapers reported that "[her] claim of wrongful birth 'could not be legally sustained in a case where a healthy child was born' given that wrongful birth case law largely revolves around the birth of a child deemed to have a significant impairment" (Lenon and Peers 2017, 142). Cramblett appealed and the case is currently pending as of this writing.

Cramblett garnered much media attention. Unsurprisingly, commentators reacted with outrage, disbelief, and dismay. Many responded by saying that the parents should simply be happy that they had a healthy child and that it was racist for them to claim specific harms had been leveraged against them because they had a nonwhite child. Some commentators identified the problematic features of a lawsuit that presents parenting of nonwhite children as a particular difficulty. As columnist Aya De Leon (2014) writes, "it is painful and difficult to witness the journey of parenting brown children posited as a legal liability and a quantifiable set of damages," and as another commentator noted sarcastically, "I hope this lawsuit wins. If it does, I am suing somebody" (Dwyer 2014), describing in jest the potential of a class action lawsuit on behalf of all parents of kids of African descent because of the emotional distress caused by parenting Black children in an anti-Black, racist environment.

While there was much media alarm surrounding the racial dynamics of this case, there was very little surrounding the disability-related aspects. As commentator Ki'tay Davidson (2014) questions, if "wrongful birth lawsuits have justified the devaluing of and violence against disabled lives for decades, [why] are all my friends just getting upset about the concept now that it's based on race?" Not only was disability dismissed as a locus of concern in this media matrix, but when disability was recognized as a part of the case, it was positioned as heightening the racial offense. Take for example Gayle King's comments in response to a legal correspondent noting wrongful birth's "grounding in disability"; King replies, "That makes it even worse," signaling a sort of "repulsion" to the idea of "saddling race next to disability" (Kearl 2018).

There is historical precedent for well-founded repulsion to such a comparison given intertwined histories of racism and ableism. For, as Eli Clare (2017) reminds us, "Black people kidnapped from Africa and enslaved in the Americas were declared defective as a way to justify and strengthen the institution of slavery" and "Immigrants at Ellis Island were declared defective and refused entry to the United States" (23). Clearly, such comparisons ought to be contested and rejected. Yet, King's revulsion signals something else: there is a particular ease with which 'disability' is assumed to be inherently harmful. And other marginalized social positions, such as racial blackness, are often leveraged to make this assumption. In these instances, 'race' also becomes solidified as a singular, separate marker, depoliticizing and dehistoricizing both race and disability in the process.

Cramblett's analogization of race and disability is cause for us to deepen our analysis of wrongful birth and identify further its historical inheritances and present-day sociopolitical conditioning. In what follows, I suggest that we ought to reject a discourse of comparison that upholds race and disability as discrete social categories and instead embrace one that foregrounds their intersectional relationship and co-constitution. In so doing, we illuminate deeper 'wrongs' associated with wrongful birth and work to refigure the conceptual and sociopolitical landscape currently making some births legally cognizable as 'wrongful.'

IV Race and Disability: Cramblett in Broader Context

We tend to think of race and disability as separate 'traits' that 'attach' to individuals; someone is white or someone is disabled. As such, when we leverage social positions like race and disability against one another, we assume they are separate statuses with individual, though perhaps parallel, histories. However, this assumption is conditioned by and masks the historical and social processes naturalizing our present, dominant ways of understanding 'race' and 'disability.'

Both standard disability-related wrongful birth cases and Cramblett use wrongful birth to naturalize disability and race, respectively. What I mean by 'naturalize' is that they pose disability and race as fixed, biological, and inherent to the individual; they generate the appearance of a 'natural' or 'neutral' occurrence that aims to cover over histories of the co-constitution of race and disability via social and political mechanisms of power. Examples of the sociopolitical factors that construct disability particularly in the case of pregnancy and reproduction include the following: epigenetic effects of racism,[6] destruction of base resources such as food, water, clothing, shelter, contamination (think Flint, MI, and lead pipes or the Dakota Access oil pipeline and its impact on the environment and Indigenous nations), high risk working conditions, the violence of wars, terrorism, spread of disease, gun violence, rape, and any other lack of supports in one's environment, which often coincides with marginalized racial communities.[7] Thinking back to the original cases of disability-related wrongful birth claims, we might ask about the sociopolitical, economic, and public health situations leading to the pregnant claimant contracting rubella.

As disability studies critiques have pointed out and as I discussed above, wrongful birth claims assume disability to be an objective, fixed, biological fact and therefore makes of it a personal responsibility rather than a matter of social welfare and human rights. The fact that one is requesting individual compensation for having to care for a disabled child disallows broader conversations about what is owed socially to those living with disabilities in an ableist world.

Likewise, Cramblett emerges in an age of what I would call the re-biologiziation and naturalization of race through the use of assisted reproductive technologies and prenatal and genetic ancestry testing, together resulting in reprogenetics, or the use of genetic technologies to generate or prevent the inheritance of particular genes (Russell 2015, 609). Reprogenetics, coupled with the strong social expectation of 'racial matching' in assisted reproduction, construes "the transmission of race from parent to child… as natural and inevitable" (Russell 2018, 158). Cramblett appeals to a loss of "shared genetic traits" which she presents as an implicit—or 'natural'—desire. Her claim describes her daughter as "obviously mixed-race" and herself as "obviously different in appearance" (emphasis mine), followed by a list of difficulties this presents. Cramblett describes a personal loss generated by structural problems (all-white neighborhood, insensitive white family members, psychological impacts of racism on their daughter) all the while refusing to vilify those structures as problems in themselves. For example, it apparently was not a problem for Cramblett to raise a white child in a racially intolerant neighborhood or among racially insensitive relatives. In effect, Cramblett sued for 'wrongful racism'; she did not receive the whiteness bargained for and so sued under terms suggesting she is due compensation for the fact—not that there is racism—but that she now has to deal with it personally.

Cramblett showcases the emergence of what Camisha Russell calls "liberal or neo-liberal eugenics" (609) and how this can be weaponized in reproductive matters related to race. This era is characterized in part by "the sense that procreation carries with it a personal responsibility, such that one must try to avoid bearing any children with genetic 'flaws' who might prove to be a burden on society" (609). This shifts responsibility away from the state for improving social conditions of society and instead demands of parents a heightened responsibility for procreation and parenting. Cramblett's case is made intelligible by the context of neo-liberal eugenics within a racist social world.

While criticisms of wrongful birth tend to focus on its disability-related eugenic inheritance, eugenics is historically informed by biological racism and continues to be deeply racialized, as evidenced by Cramblett. Biological racism involves a set of pseudo-scientific theories of evolution that claimed superior evolution for some racialized sub-groups of people ("light-skinned, upper class, Western European, heterosexual men") (Lenon and Peers 2017, 151). This thinking "informed how eugenicists understood and treated a series of human variations, many of which we now categorize as disabilities, mental illnesses, and some of which we now consider sexualities and gender identities" (152). Such pseudo-scientific discourses affected social norms and politics as well, generating instances of what we might call the "disable-ization of race" and, in tandem, the "racialization of disability."

Calvin Warren's (2018) discussion of the 1840 Census, the first of which to enumerate the 'mentally defective,' is instructive here. What emerged in statistical analysis was a large difference in free states vs. slave states of Black people with mental illness—1 in 144 in free states, 1 in 1558 in slave states. This large difference was an ideological move by pro-slavery advocates to claim that slavery was good and necessary for the Black, lest they suffer "lunacy under the burden of freedom," as described by Albert Deutsch (1944), quoting John Calhoun, ex-Vice President of the USA and political leader of slave states (473). Wildly inaccurate and riddled with internal inconsistencies regarding the enumeration of Black people and the 'insane' in the north, these statistics reflected towns as (1) having 'insane negroes' when there weren't any negro residents; (2) having more 'insane negroes' than they had negroes; (3) counting white 'insane' patients at state hospitals as 'insane negroes.'

While we could read this as a statistical 'error' or clear case of deception, Warren suggests we account for these 'errors' philosophically. How did a confluence of discourses 'materialize' 'Black insanity' where there were no Black people or generate white people who were deemed Black by way of mental defect? Blackness and disability were (and I would argue, still are) more like floating signifiers that can 'attach' to serve the needs of oppressive power structures. They are co-constituted by processes that 'make appear' or naturalize bodies as discretely raced and disabled; race and disability appear as such under the conditions of the oppressive power structure of white supremacist ableism.

The work of liberatory social justice, however, demands that we uncover the social and political mechanisms of power at work in such instances. It requires first and foremost that we seek to illuminate and actively contest those structural conditions producing patterns of oppression operating through idea(l)s of normalcy/ability, whiteness, and naturalized race and (dis)ability positions. Therefore, to close, I will identify some frameworks for disability bioethicists to think through resistant responses to wrongful birth which take into account the intersecting and co-constituting nature of race and disability as well as a host of other social categorizations.

V Reproductive Justice, Disability Justice

In this chapter, I've discussed the legal claim of wrongful birth and the issues raised by disability studies scholars concerning its eugenic inheritance and problematic reliance on an individual, medical model of disability and able-bodiedness. In this final section, I build on the previous discussion of Cramblett to propose an intersectional disability bioethics response to wrongful birth.

In the context of reproductive politics, the framework of reproductive justice is central to understanding what a radical and inclusive critique of wrongful birth might look like. Reproductive justice emerged in response to a narrow reproductive rights platform focused on abortion. This limited focus on expanding abortion rights did very little to take into account the populations experiencing reproductive oppression, or the "control and exploitation of women, girls, and individuals through…bodies, sexuality, labor, and reproduction" (SisterSong 2006, 2). Black, Indigenous, and other women of color often face state and social control of reproduction, which exceeds the narrow issue of abortion rights. Therefore, advocates of reproductive justice fight for the right to have or not to have a child, the right to control birthing options, and the right to raise children in safe environments that contribute to flourishing. Additionally, reproductive justice works to build "the enabling conditions to realize these rights," (2) which are sought via a variety of avenues not originally thought to be part of a "reproductive rights" platform: economic justice, quality education, accessible public services such as healthcare and affordable childcare, and environmental safety. A reproductive justice critique of wrongful birth would shift our attention from the so-called 'injurious' child to the lack of education, social welfare programs, and legislative and judicial decrees truly allowing for reproductive autonomy, or for pregnant individuals to make reproductive decisions under conditions of flourishing. Part of building these conditions necessarily includes working to end ableism.

Typically, ableism is described as the stigmatization and exclusion of disabled individuals in society, but we ought to expand such a definition. Take for example the recent working definition of ableism by Talila "TL" Lewis (2020) "in conversation with Disabled Black and other negatively racialized folk, especially Dustin Gibson" that describes ableism as something that can be experienced by a variety of folks: "You do not have to be disabled to experience ableism" because "ableism is a system that places value on people's bodies and minds based on societally constructed ideas of normalcy, intelligence, excellence, and productivity. These constructed ideas are deeply rooted in anti-blackness, eugenics, colonialism, and capitalism." The framework of disability justice is central to this (re)definition.

Launched in 2005 by queer women of color working within progressive and radical movements fighting ableism, disability justice commits to anti-capitalist, intersectional solidarity-building and includes a "commitment to cross-movement organizing" and "cross-disability solidarity" (Sins Invalid 2015). As the authors of *10 Principles of Disability Justice* write,

> We cannot comprehend ableism without grasping its interrelations with heteropatriarchy, white supremacy, colonialism, and capitalism, each system co-creating an ideal bodymind built upon the exclusion and elimination of a subjugated 'other' from whom profits and status are extracted. Disability justice understands all bodies as "unique and essential" and "[having] strengths and needs that must be met."
>
> *Sins Invalid 2015*

Having unique strengths and needs is not a problem to be overcome, but the position from which we must generate new ways of relating and belonging. We need to shift the 'wrongs' associated with wrongful birth from the individual child to the social and political world in which they were born.

Race and disability are shaped by social power. The development of reproductive technologies heightening 'parental responsibility' alongside our reliance on naturalized notions of race and disability masks the ways in which social inequities are reproduced by these very technologies and the social norms and values we attach to them. The true 'wrongs' of wrongful birth arise within this nexus and the responsibility for rectifying such wrongs lies with us all.

Notes

1 Much hangs on who the claimant is. When the claimant is the parent(s) of some child rather than the child themself, whether non-existence would have been preferable might be a claim about what would have been better for the parents, rather than a claim about the quality of the child's life.

2 See Susan Wendell, *The Rejected Body* (1996); Alison Kafer, *Feminist, Queer, Crip* (2013); Shelley L. Tremain, *Foucault and Feminist Philosophy of Disability* (2017); Nirmalla Erevelles, *Disability and Difference in Global Contexts: Enabling a Transformative Body Politic* (2011); Sami Schalk, *Bodyminds Reimagined: (Dis)ability, Race, and Gender in Black Women's Speculative Fiction* (2018).

3 The term 'positive eugenics' is used to refer to the former and can describe any sort of encouragement of those with 'desirable' characteristics to reproduce while the term 'negative eugenics' refers to the latter, or the disincentivizing or disallowing of those with 'undesirable' characteristics to reproduce.

4 See *Buck v. Bell*, the Supreme Court case which approved state's right to sterilize people with intellectual disabilities.

5 Consider the forced or highly coerced sterilization of incarcerated people, those with substance abuse issues, and others deemed 'unfit.'

6 See Shannon Sullivan (2013) "Inheriting Racist Disparities: Epigenetics and the Transgenerational Effects of White Racism."

7 Many of these examples emerge from Susan Wendell's "The Social Construction of Disability" in *The Rejected Body* (1996).

References

Brown, Lydia X.Z. 2018. "Legal Ableism, Interrupted: Developing Tort Law & Policy Alternatives to Wrongful Birth & Wrongful Life Claims." *Disability Studies Quarterly*, 38(2).

Campbell, Fiona. K. 2009. *Contours of Ableism: The Production of Disability and Abledness*. New York: Palgrave Macmillan.

Clare, Eli. 2017. *Brilliant Imperfection: Grappling with Cure*. Durham: Duke University Press.

Cramblett v. Midwest Sperm Bank, LLC, No. 1:2016cv04553- Document 29 (N.D. Ill. 2017)

Davidson, Ki'tay. 2014. "Angry About the White Lesbians Suing for Having a Black Child? You're Missing Something." *BGD* (blog): October 6. http://www.bgdblog.org/2014/10/angry-white-lesbians-suing-black-child-youre-missing-something/.

de Leon, Aya. 2014. "We Wish the White Mom with Black Sperm Said This." *Denene Millner's: My Brown Baby* (blog): October 24. http://mybrownbaby.com/2014/10/ white-mom-black-sperm-donor/.

Deutsch, Albert. 1944. "The First U.S. Census of the Insane (1840) and Its Use as Pro-Slavery Propaganda." *Bulletin of the History of Medicine*, 15(5): 469–482.

Dwyer, Liz. 2014. "If Those White Lesbian Moms Win Their Biracial Baby Suit, I'm Totally Suing America for Racial Pain and Suffering." *Los Angelista* (blog): October 4. http://www.losangelista.com/2014/10/if-those-white-lesbian-moms-win- their.html?m=1.

Erevelles, Nirmala. 2011. *Disability And Difference In Global Contexts: Enabling A Transformative Body Politic*. New York: Palgrave Macmillan.

Frati, Paola, Vittorio Fineschi, Mariantonia Di Sanzo, Raffaele La Russa, Matteo Scopetti, Filberto M. Severi, and Emanuela Turillazzi. 2017. "Preimplantation and Prenatal Diagnosis, Wrongful Birth and Wrongful Life: A Global View of Bioethical and Legal Controversies." *Human Reproduction Update*, 23(3): 338–357.

Gleitman v. Cosgrove, 227 A.2d 689, 693 (N.J. 1967)

Hensel, Wendy F. 2005. "The Disabling Impact of Wrongful Birth and Wrongful Life Actions." *Harvard Civil Rights-Civil Liberties Law Review*, 40: 141–195.

Jacobs v. Theimer, 519 S.W. 2d 846, 850 (Tex. 1975)

Kafer, Alison. 2013. *Feminist, Queer, Crip*. Indiana University Press.

Kearl, Michelle K. 2018. "The Stolen Property of Whiteness: A Case Study in Critical Intersectional Rhetorics of Race and Disability." *Rhetoric Review*, 37(3), 300–313.

Lenon, Suzanne and Danielle Peers. 2017. "'Wrongful' Inheritance: Race, Disability and Sexuality in Cramblett v. Midwest Sperm Bank." *Feminist Legal Studies*, 25(2), 141–163.

Lewis, Talia. 2020. Ableism 2020: An Updated Definition. January 25. https://www.talilalewis.com/blog.

Rinaldi, Jennifer A. 2009. "Wrongful Life and Wrongful Birth: The Devaluation of Life with Disability." *Journal of Public Policy, Administration and Law*, 1: 1–7.

Russell, Camisha A. 2018. *The Assisted Reproduction of Race*. Bloomington: Indiana University Press.

Russell, Camisha A. 2015. "The Race Idea in Reproductive Technologies: Beyond Epistemic Scientism and Technological Mastery." *Bioethical Inquiry*, 12: 601–612.

Schalk, Sami. 2018. *Bodyminds Reimagined: (Dis)Ability, Race, and Gender in Black Women's Speculative Fiction*. Duke University Press.

Sins Invalid. 2015. *10 Principles of Disability Justice*. September 17. https://www.sinsinvalid.org/blog/10-principles-of-disability-justice.

Sistersong: Women of Color Reproductive Health Collective. 2006. Understanding Reproductive Justice.

Stein, Jillian T. 2009. "Backdoor Eugenics: The Troubling Implications of Certain Damages Awards in Wrongful Birth and Wrongful Life Claims." *Seton Hall Law Review*, 40(1):1117–1168.

Sullivan, Shannon. 2013. "Inheriting Racist Disparities in Health: Epigenetics and the Transgenerational Effects of White Racism," *Critical Philosophy of Race*, 1(2): 190–218, https://doi.org/10.5325/critphilrace.1.2.0190.

Tremain, Shelley. 2017. *Foucault and Feminist Philosophy of Disability*. Ann Arbor: University of Michigan Press.

Warren, Calvin. 2018. *Ontological Terror: Blackness, Nihilism, and Emancipation*. Durham: Duke University Press.

Wendell, Susan. 1996. *The Rejected Body: Feminist Philosophical Reflections on Disability*. New York: Routledge.

Yakren, Sofia. 2018. "'Wrongful Birth' Claims and the Paradox of Parenting a Child with a Disability." *Fordham Law Review*, 87(2): 583–628.

Further Reading

Erevelles, Nirmala. 2011. *Disability and Difference in Global Contexts: Enabling a Transformative Body Politic*. New York: Palgrave Macmillan.

Kafer, Alison. 2013. *Feminist, Queer, Crip*. Indiana University Press.

Paul-Emile, Kimani. 2018. "Blackness as Disability." *Georgetown Law Journal* 106: 293–364.

Roberts, Dorothy. 2011. *Fatal Invention: How Science, Politics, and Big Business Re-Create Race in the Twenty-First Century*. New York: The New Press.

PART V

Disability, the Life Course, and Well-Being

13

DISABILITY, IDEOLOGY, AND QUALITY OF LIFE

A Bias in Biomedical Ethics

Ron Amundson

The central philosophical concepts regarding disability were constructed not by philosophers but by disability rights activists. Only recently have these concepts received attention in the philosophical literature. This chapter will argue that an important discussion in biomedical ethics is biased against the civil rights interests of people with disabilities because of the failure of philosophers to come to terms with the disability rights movement. Quality of life is conceived in a way that directly conflicts with the Social Model of disability, and the conflict is deeply rooted in biomedical ethical discussion. One particular application will be discussed: the reduction of health care for disabled people because of their allegedly low quality of life.

I Two Models

A defining characteristic of the disability rights movement is a particular explanation of the disadvantages experienced by disabled people. Disadvantages are explained as effects not of biomedical conditions of individuals but of the socially created environment that is shared by disabled and nondisabled people. This environment (it is said) is so constructed that nondisabled people are privileged and disabled people penalized. Disability is a social problem that involves the discriminatory barriers that bar some people but not others from the goods that society has to offer. For this reason, the view is often called the Social Model of disability. It contrasts with the traditional view, sometimes termed the Medical Model, according to which disability is a problem of individuals whose biomedical conditions disadvantage them. On the Medical Model, disadvantages are natural and inevitable outcomes of simple biomedical facts. Reductions of these natural disadvantages can be accomplished only by individual cures (changing the biomedical facts) or by charitable donations intended to compensate the victims of disability

DOI: 10.4324/9781003289487-19

for their inevitable and pitiable conditions. The Social Model depicts disability as a problem experienced by a class of people, a problem that is caused by social organization and that can be remedied by social change. The Medical Model is an individualistic rather than a social theory. Disabilities are properties of individuals, and remedies (e.g., cures, rehabilitations, charitable donations) are meted out one individual at a time.

[...]

This chapter will examine one of the manifestations of the Medical Model in contemporary biomedical ethics. It will do so from the perspective of the Social Model. The concept of quality of life (QOL) is entwined in many issues of biomedical ethics. I will try to show that prominent discussions of the QOL of disabled people reveal an unjustified and unexamined commitment to the Medical Model of disability, and for that reason a bias against the interests of disabled people. This is not a simple factual error, of course. The treatment of QOL is embedded within a well-articulated framework of literature involving prominent ethicists and other philosophers. All of these people display a sophistication in philosophical ethics that far exceeds my own. I am a philosopher and historian of biology, a disabled person, and a disability activist. I come to this discussion as an activist, representing a perspective that I consider poorly understood among biomedical ethicists.

[...]

The Medical and Social Models of disability are ideological, as explanations of social disadvantage often are.[1] My claim that they are ideological amounts to the following. Each model presents an account of the causal relations that hold between disability and other phenomena. The causal accounts at first look like other causal explanations – like the gravitational explanation of the tides, for example. The causal accounts involve or entail the identification of various phenomena as *natural* or *unnatural,* and as *inevitable* or *contingent* and *changeable.* On closer inspection, it can be discovered that the contrasting causal accounts of the same phenomenon (here disability) serve or harm the interests of different groups of people. A causal account that depicts a social phenomenon as natural and inevitable (or changeable only at great cost) works to the advantage of the people who benefit from the phenomenon, and to the harm of the people who are hurt by the phenomenon. When the same phenomenon is depicted as artificial and changeable, the reformist interests of those harmed by the phenomenon are served. My goal in this chapter is not to argue for the correctness of the Social Model. Rather, I will show that important bioethical discussions presuppose its falseness, and assume the correctness of the Medical Model. This unexamined assumption produces a bias against the goals of the disability rights movement within the practice of biomedical ethics.

II Disability and QOL: The Standard View

The Standard View is that disabilities have very strong negative impacts on the quality of life of the individuals who have them. This view is widely held by nondisabled people, both in popular and in academic culture. The Standard View is confronted by a fact that I will term its Anomaly: when asked about the quality of their own lives, disabled people report a quality only slightly lower than that reported by nondisabled people, and much higher than that projected by nondisabled people. Both the Standard View and its Anomaly have been robustly demonstrated in a number of studies. Disabled and nondisabled people have very different assessments of the quality of disabled people's lives.

Many factors are commonly assumed to lower one's QOL. Poverty, the loss or the lack of loving relationships, thwarted ambitions, and frustrated hopes are all assumed to reduce QOL. Some of these are contingently related to disability, as they are to other causes. But disability itself holds a privileged position in the catalog of QOL reducers. The life badness that is assumed to follow from disability goes beyond the badness that comes from the partial and contingent associations that disability has with loss of love, loss of income, and so on. In other words, disability is conceived to have a *surplus badness*, over and above its specific and identifiable affects. It is assumed to be categorically bad, bad beyond its contingent effects, bad to the bone, butt-ugly bad. I will argue that the surplus badness attributed to disability comes not from a rational appreciation of the consequences of disability itself, but from the stigma that disability carries, both in popular and in academic culture. In other words, the Standard View is an *expression* of the stigma of disability. It is not (as it presents itself to be) an estimate of the objective consequences of impairments.

I am interested in the Standard View for three reasons, two of them specific and one generic. The first reason is that the Standard View generally devalues the lives of disabled people. It enforces the "pity" aspects of the Medical Model that disability rights advocates find so objectionable, and thereby obscures the civil rights basis of the disability rights movement. The second reason is that under certain theories of health care rationing, the supposed low QOL of disabled people can imply that they have less claim on health care than nondisabled people do. This is a concrete example of the harm that the Standard View can cause. The third, generic reason is that the Standard View is well integrated within the literature of biomedical ethics. Its flaws are reflected in a wide range of philosophical discussions of disability.

[...]

III The Links: Normality, Opportunity, and QOL

Within current biomedical thinking, the Standard View of QOL and disability is tied into a well-articulated set of views involving the notions of biological normality and the importance of a wide opportunity range for quality of life. The area includes important work by Christopher Boorse, Norman Daniels, and Dan Brock. The work is founded on Boorse's concept of what he calls "species typical function." Despite the statistical-sounding term "typical," Boorse's concept is of the *normal* functioning of members of a species (not just the most common or usual functioning). Boorse claims that the distinction between normal and abnormal function is an empirically grounded implication of biomedical science, not a prejudice of human observers. In effect, normal and abnormal function are distinct natural kinds. Impairments (the biological aspects of disabilities) are objectively and scientifically defined as species-abnormal functioning (Boorse 1977).

Boorse's claim is empirical, not normative. Nevertheless, it is very widely cited in the normative literature. His apparently moderate claim about biology supports normative consequences in the work of Norman Daniels, Dan Brock, and others. Daniels uses Boorsian normality to explain the moral importance of health care. On Daniels's view, the importance of health care is to maintain normal function – that is, to avoid disease and impairment. Normality is important because of its essential connection to opportunity. Species-abnormal functioning reduces the "normal" opportunity range, while health care maintains and restores species-normal functioning, thereby protecting opportunity (Daniels 1981;1985). Brock then argues that a "normal" wide opportunity range is a necessary condition for a high quality of life (Brock 1993; 1995; 2000).

The objective normality of biological function was inferred from biomedical science. The linkage from Boorse to Daniels to Brock completes an argument, apparently founded on biological fact, that people with impairments will (must?) have a low quality of life. High QOL depends on wide opportunity range, which is dependent on biological normality, which is an objective fact of the natural world.

I believe that this chain of reasoning is flawed at every step. Boorse's contribution misrepresents biomedical science. Daniels's step embodies the prejudices of the Medical Model of disability, and so its shortcomings. Brock's contribution shares Daniels's commitment to the Medical Model, and in addition assumes an epistemologically privileged knowledge of others' lives that is unjustified by the facts.

[...]

IV The Standard View and "Objective" Quality of Life (Brock)

Dan Brock discusses the quality of life in the context of the work of Boorse and Daniels. In a series of papers beginning in the early 1990s, he acknowledges the Anomaly but does not consider it an important challenge to the Standard View. He sometimes presents the Standard View as a definitional truth. "Serious disabilities or handicaps will, by definition, typically reduce a person's quality of life" (Brock 1995, 179). "[S]ince disabilities by definition under the ADA substantially limit one or more major life activities, they will reduce an individual's health related quality of life" (Brock 2000, 226–7). According to the ADA definition, disabilities limit one or more major life activities. Why must we assume that unlimited major life activities are *by definition* required for a high QOL? Walking and seeing are often given as examples of major life activities. It might be argued that it is a matter of empirical fact that limitations in walking and seeing are associated with lower QOL. This would require empirical data about the correlation. Brock offers no data of this sort. The language of the ADA makes a semantic link only between disability and limitations, not between limitations and reduced QOL. Brock is assuming some additional conceptual connection. I suggest that this conceptual connection is the Boorse/Daniels linkage between biological normality and "normal" opportunity discussed earlier. With the right kind of philosophical account, empirical evidence can appear irrelevant.[2]

In order to explain the Anomaly (and by the way protect the Standard View from refutation), Brock distinguishes between subjective and objective QOL. Subjective QOL is how happy or satisfied one is with one's life. Objective QOL is how well one's life is *really* going. "To be satisfied or happy with getting much less from life, because one has come to expect much less, is still to get less from life or to have a less good life" (Brock 1993, 309). Reports of high QOL from disabled people are merely subjective.

What are the objective aspects of QOL? In the 1993 paper, Brock discusses with approval an instrument called the Health Status Index (HSI). One of the scales on the HSI is said to judge the "mobility" of the individual. The highest ranking is 5, for those who are able to use public transportation alone. Someone who requires assistance to use public transportation is scored 4, and someone who needs assistance to go outside scores 3. Brock refers to this sort of scale as measuring "functions of the 'whole person'" (Brock 1993, 298). The mobility measurement is conceived as a biomedical attribute of the individual being measured. A low mobility score constitutes a part of the objective basis on which the person's QOL is judged as objectively low, whatever the person's subjective opinion about that life.

Is "mobility" as thus measuring a biomedical attribute of the individual? The slightest acquaintance with the Social Model of disability will show that the answer is no. Imagine a set of identical triplets, sisters with paraplegia who use wheelchairs and who live in different cities. The first lives in an inaccessible building and needs assistance to go outside. The second lives in an accessible building, but in a city in which public transportation is not wheelchair-accessible. The third lives in an accessible building in a city with accessible public transportation. These three scores 3, 4, and 5 respectively on the "mobility" measure, even though they are biomedically identical. It is utter confusion to attribute their scores to their biomedical condition. The least mobile (call her Sister 3) is biomedically identical to the most mobile (Sister 5). The difference in mobility scores is caused by differences in environmental barriers. Blinkered by the

Medical Model of disability, Brock and the HSI present "mobility" as a measure of biomedical traits of individuals. It is not.

The same example illustrates the flaw in Daniels's identification of biological normality with opportunity range. The three sisters are biomedically identical, and so equally "abnormal" by Boorse's criteria. But Sister 5's opportunities are immensely broader than Sister 3's. This demonstrates that individual normality does not determine opportunity. The HSI "mobility" scores do measure something, but what they measure is not a biomedical characteristic of the sisters. As the Social Model shows us, it is the accessibility of their respective environments.

In 1993, Brock explained the Anomaly as a mere effect of the lowered expectations of disabled people, and in 1995 he presented the Standard View as a definitional truth. In later papers, he has recognized that disabled people might accommodate their goals to their opportunities without necessarily lowering their expectations. But even with this less-belittling possibility, Brock remains committed to the Standard View and dismisses the Anomaly as reflecting subjective rather than objective QOL. Recent endorsements no longer appeal to specific definitional truths (i.e., disability defined as QOL-reducing). His intention is more generally to articulate our concept of a good life. This concept, according to Brock, contains both subjective and objective factors, and the objective factors include the absence of significant disabilities.

I am not at all sure what kind of evaluation standards are appropriate to an analysis of "our concept" of QOL. Social critics, after all, claim that some of society's dominant concepts are flawed and objectionable. A correct account of a flawed concept has great charms to those who do not recognize the flaw. Is such an account good because it is true to the concept, or bad because the concept is flawed? A correct account of "our concept of race" to white Americans in the 1830s would surely have included innate variations in intellectual and moral capabilities among races. So, I am tempted simply to grant Brock the correctness of his account of the Standard View as "our concept of QOL" and go on to critique that concept itself. The fact that it's "our concept" doesn't make it right.

[...]

I am willing to accept the coherence of the distinction between objective and subjective QOL. But the epistemic status of this kind of judgment about other people's lives is far from sturdy. How are we to separate judgments that are objectively grounded from those that merely express one's prejudices? Separating the wheat from the chaff requires more than logic. It requires epistemology.

[...]

Take, for example, the Anomaly. The disabled subjects in studies know full well their own impairments. They are able to describe in detail the day-to-day difficulties that they can cause, and they realize (sometimes with amusement) that nondisabled people assume them to have a low QOL (Albrecht and Devlieger 1999). Many acquired their impairments late in life, and so have lived both with and without impairments.

By contrast, most members of the nondisabled public have never experienced life with a significant impairment, and know about impairment only through its social stigma. Who is judging from ignorance? On what grounds does Brock favor the opinions of nondisabled over disabled people, when the issue is itself the QOL of disabled people? What ignorance burdens disabled people, and why is it so much greater than the ignorance of the nondisabled majority? Does the advocate of the Standard View really want to claim that nondisabled people *know better* than disabled people what the different lives are like?

Come to think of it, the answer is obvious. *Of course*, the advocate of the Standard View wants to claim this. The testimony of disabled people about their lives has been dismissed in favor of that of nondisabled "experts" for a very long time, as historians have documented (Longmore and Umanski 2001). The real question is this: when nondisabled people claim to know better than disabled people what the different lives are like, why should they be believed? Why should the opinions of nondisabled people be epistemologically privileged over those of disabled people?

Judging the lives of others is epistemologically hazardous, even if irresistible. The life of a pious person, or an urbanite, or a parent of a large family is bewildering to me. I gossip and joke about it when I'm with like-minded people. I would never voluntarily exchange my situation for that of a person in any of those categories. Even so, it would be presumptuous of me to claim that the quality of one of those lives is merely subjectively high, and that on objective grounds (known to me and my like-minded friends) it is low.

[...]

The Standard View is very widespread. Is that because of the superior knowledge of nondisabled people about the lives of disabled people? Or is it merely a reflection of the stigma of disability? The fact that we *can* trump subjective QOL judgments with judgments *that we believe* are objective does not mean that we are correct when we do so. When our "objective" judgments happen to match our own social prejudices, that coincidence alone should make us wary about our own objectivity.

[...]

V Penalizing Health Care Priorities by QOL

This section will discuss one application of the Standard View of disability and QOL to health care policy. The particular application is not one that the philosophers so far discussed have endorsed. In fact, Dan Brock specifically rejects it. But the case does show that the Standard View can be a direct and tangible threat to disabled people. The notion of *quality-adjusted life years* (QALYs) is common in discussions of health care rationing and prioritization. If health care funds are prioritized to maximize QALYs, then funding for people with low QOL should receive a lower priority for the same kinds of care than funding for people with high QOL. On this approach, the Standard View would entail lower levels of health care for disabled people than for nondisabled people.

[...]

Let us use the term "discrimination" in a morally neutral sense, so that some cases of discrimination are just and some are unjust. Unjust cases of discrimination are those in which it is directed against a member of a "protected class," a class of people against whom discrimination has been practiced for social reasons unrelated to the legitimate goals of some activity. If one is hiring a carpenter, discrimination against people who are unskilled carpenters is not unjust. This is true even if the unskilled carpenter happens to be a woman, a member of a protected class. It is unjust to refuse to hire the woman only if her sex rather than her carpentry skill is the basis of the decision not to hire. An employer could defend against a complaint of unjust hiring practices by showing that other women with higher carpentry skills were hired, and that men who had skills comparable to the complainant were not hired. This would demonstrate that discrimination was based on skill level rather than protected class membership, and so was not unjust.

Health care discrimination might be defended in the same way. Suppose a member of an ethnic minority were refused a heart transplant on the grounds that the operation had a low probability of success. Suppose the decision were challenged as unfair discrimination against an ethnic minority. The decision could be defended by showing that other members of the minority whose probability of success was higher did receive the procedure, and that nonminority patients whose probability of success was similar to the patients were also refused the procedure.

Consider disability in place of minority status. If some feature X that happens to correlate with disability reduces a patient's ranking in health care prioritization, but the priority judgments are made on the basis of feature X and not based on the disability status per se, then it could be argued that the discrimination is just (if discrimination on the basis of X is otherwise just). Evidence that the priority rankings were based on X rather than on disability could come from the fact that disabled people who do not have trait X are not priority penalized, and that nondisabled people who do have trait X are priority penalized.

[...]

VI Just and Unjust QOL Reduction

One problematic feature of QALY criteria for prioritization is that they are insensitive to the causes of reduced QOL. Intuitively, one might think that unjust reductions in life quality might receive a different treatment than just ones. Perhaps a naïve assumption is being made that all of the influences on QOL are matters of fortune, undeserved by the recipient but not unjust. But suppose that reduced QOL arises from social oppression. If we use opportunity range as the operationalization of QOL, it would be easy to argue that women and African Americans have significantly lower opportunity ranges in the USA with respect to both employment and freedom from harassment. Could these very real opportunity losses be used to justify reduction in health care priority? I seriously doubt it – at least not without strong resistance from the affected groups.

Why haven't QOL reductions been discussed as a factor in health care prioritization for *nondisabled* people? I suspect that it is because opportunity losses due to sex or race are seen as unrelated to health care, while the same reductions are seen as a health care issue when they coincide with disability. This is simply another begging of the question in favor of the Medical Model and against the Social Model of disability. According to the Social Model, the opportunities lost to disabled people are taken away by unjust and discriminatory social barriers, not by biomedical conditions. Similar social barriers disadvantage women and racial minorities. If we are unwilling to penalize women and racial minorities for the QOL consequences of the discrimination they experience, it is unjust to penalize disabled people for the same consequences.

The concept of *health-related* quality of life (HRQOL) is sometimes used in this context. This use of HRQOL as a substitute for QOL is simply the gerrymandering of social problems into medical ones. If disability is defined as a health-related problem, then the QALY advocates can use health care priorities as a stick to beat disabled people. The same treatment would not be tolerated with respect to sex or race. Consider again the three sisters. They are biomedically indistinguishable, but they differ immensely in their HSI-defined "mobility." The opportunity restrictions experienced by Sister 3 (the least mobile) are caused not by her biomedical condition but by her inaccessible surroundings. This fact is merely disguised by referring to her "health-related quality of life." Sister 3 is already penalized by her inaccessible environment. To compound the penalty by cutting her health care because of her inaccessible environment would surely be unjust.

I submit that a policy of priority penalization for people whose low QOL stems from social oppression could not be socially negotiated. (Surely the wretched medical treatment of slaves in the American South prior to emancipation is in no way excused because they had a low QOL anyway.) I further submit that a policy of priority penalization based on "objective" factors such as reduced opportunity range could not be negotiated so long as it was applied without discrimination. For these reasons, I conclude that the application of these penalties to disabled people is based *only on their stigmatized status*, and not on their alleged low QOL.

[...]

We need not rely on intuitive conceptual connections between disability and low QOL – we can study the patterns of correlation between impairments and the objective correlates of QOL.

If the factors that relate to high and low QOL are really the same for disabled and nondisabled people, then one way of making an objective assessment of the QOL of disabled people is to measure those correlates. Here we can actually find evidence of a lower QOL for disabled people. Consider QOL-lowering factors such as unemployment, isolation, and being a crime victim. Disabled people score significantly higher than nondisabled people on these factors. These are demographic facts, not philosophical intuitions or implications of "our concepts." Wouldn't these empirically measurable facts serve the biomedical ethicists better than conceptual analysis in proving the inherent superiority of the normal?

In fact, I believe they would not. When we get down to actual causes of disadvantage, and we study them in a way that allows unbiased empirical comparison between disabled and nondisabled people, the social causes become more apparent. Each of the demographic QOL lowering factors that applies to disabled people at a higher rate than to nondisabled people *does so for social reasons*. The impact of the Social Model is much clearer when we attend to specifics than when we abstractly think of reified *abnormality* as a person-type. For example, consider crime victimization. Disabled people are no more responsible for the crimes committed against them than are the victims of rape; victimhood is no more essentially tied to disability than it is to womanhood. Consider unemployment and isolation. An important cause of unemployment and isolation is the lack of suitable transportation. This fact is true for disabled and nondisabled people alike. A wheelchair user in a town with wheelchair-inaccessible transportation is in a position very similar to that of a nondisabled person in a location that has no transportation. Neither can hold down a job, and each has limited social contacts.

Certain customary ways of talking disguise the fact that disabled and nondisabled people share the problem of transportation. Our bioethicists (and others) often label transportation that is accessible to disabled people as "special transportation." This label is merely one more way of stigmatizing disability, by falsely making it appear that "abnormal" people have different needs than "normal" people. Everyone needs transportation. No one needs *special* transportation! (Are racially integrated lunch counters special lunch counters?) When we look at the details, we see shared social problems.

[...]

VII Conclusion

The Medical Model of disability and the Standard View of the low QOL of disabled people are shared by popular and academic culture. Biomedical discourse assumes that the disadvantages of disability are intrinsic to the disabled state itself, and that abnormality is penalized by nature itself.

[...]

The stigma of disability is embedded in those biases. Even a familiarity with the literature of the disability rights movement does not change that bias very much. The recent book *From Chance to Choice: Genetics and Justice* is co-authored by two of the bioethicists discussed here and two others – Allen Buchanan, Dan W. Brock, Norman Daniels, and Daniel Wikler (Buchanan et al. 2001). The book contains a great deal of discussion of disability activists' critiques of the uses of genetic technology. The authors give their liberal endorsement to the general goals of the disability rights movement, while rejecting almost every specific argument of its advocates. My interest is not in the authors' rejection of the disability rights arguments. It is rather in the fact that the authors show only a verbal understanding of the Social Model. Even after reporting, reasonably accurately, on the perspective itself, the authors immediately refer to the biomedical conditions of impairment and disability as the *direct causes* of disadvantage. Two examples: "We devalue disabilities because we value the opportunities and welfare of the people who have them. And it is because we value people, all people, that we care about limitations on their welfare and opportunities. We also know that disabilities as such diminish opportunities and welfare. ... People with disabilities have more to gain from these [genetic] techniques than others do since their deficits, real and imagined, serve to marginalize and exclude them." (Buchanan et al., 278, 332)

Disabilities (all by themselves) limit welfare and opportunities. Deficits (all by themselves) marginalize and exclude people who have them. Elsewhere in the book, the authors acknowledge that social arrangements contribute to the disadvantage of people with impairments. But when they find themselves pledging their respect for people with impairments, the social causes of disadvantage are forgotten. This is the power of the Medical Model over biomedical ethics. No matter how sincere the authors' present, the social causes of disadvantages are the first things to slip from their minds. From the perspective of the Social Model, the problem of disability has been whitewashed. The ethicists' explanation of exclusion as due to impairment makes no more sense than if they were to explain racial segregation as caused by race itself, as if the social phenomenon of racism played no part in the matter.

I confess that I have given the reader very little reason to actually accept the Social Model, to think of the range of impairments in much the same way that we now think of race and gender. That argument must be given elsewhere. But I will end with two observations that I consider relevant to the question. First, less than a century ago race and sex were themselves considered by the scientific community to be *literally* disabling. It was not a simple scientific discovery but a social change that gave rise to modern egalitarianism regarding sex and race. Disability activists envision a similar social change with respect to disability itself. It will require a change in "our concept of a good life," but a change no greater than those that have already happened regarding race and sex.

Finally, many bioethicists express a widespread but utopian hope that medical advances can wipe out or drastically reduce impairments ("... we are committed to the judgment that in the future the world should not include so many disabilities... " [Buchanan et al., 278]). This vision is misplaced. Medical science does more than repair and prevent impairments. It also allows people to survive while living with impairments. A simple example is the fact that a person newly quadriplegic from a spinal cord injury had a life expectancy of less than a year prior to World War II. Today, the same person's life expectancy approaches that of an unimpaired person. The demographic consequence is that quadriplegics are a larger proportion of the population today than they were fifty years ago, and the same applies to many other impairments. Greater numbers of increasingly "abnormal" people are living among us, and the trend will continue. This spectacular achievement goes unnoticed by the biomedical advocates of normality. Despite the utopian rhetoric one sometimes hears from some enthusiasts of the Human Genome Project, tomorrow's world will contain a greater proportion of people with impairments than today's. The social movement for the civil rights of disabled people will certainly continue. It will not be rendered moot by idealistic dreams of biological perfection.

Notes

1 Notice that I am already speaking as an activist. The Medical Model was given its name by advocates of the Social Model. Advocates of the Medical Model typically do not see it as a model at all, but rather as the simple truth.
2 The correlation between extremely wide opportunity range and QOL asserted by Daniels and Brock deserves further discussion. It does not seem to me to describe the expectations of average people. I consider it a kind of American cultural ideal that is satisfied only in a very few, very privileged people. Most people expect to make good lives within the limitations they encounter, and to make the best of what they have. Many things in life have very large effects on one's opportunity range. Few of these carry the stigma that disability carries.

References

Albrecht, Gary, and Patrick Devlieger. 1999. "The Disability Paradox: High Quality of Life Against All Odds." *Social Science and Medicine* 48: 977–88.

Amundson, Ron. 2000. "Against Normal Function." *Studies in the History and Philosophy of Biological and Biomedical Sciences* 31C: 33–53.

Bach, John R., and Margaret C Tilton. 1994. "Life Satisfaction and Well-being Measures in Ventilator Assisted Individuals with Traumatic Tetraplegia." *Archives of Physical Medicine and Rehabilitation* 75: 626–34.

Boorse, Christopher. 1977. "Health as a Theoretical Concept." *Philosophy of Science* 44: 542–73.

Brock, Dan W. 1993. *Life and Death.* Cambridge: Cambridge University Press.

———. 1995. "Justice and the ADA: Does Prioritizing and Rationing Health Care Discriminate against the Disabled?" *Social Philosophy and Policy* 12: 159–85.

———. 2000. "Health Care Resource Prioritization and Discrimination against Persons with Disabilities." In *Americans with Disabilities: Exploring Implications of the Law for Individuals and Institutions,* edited by Leslie Pickering Francis and Anita Silvers, pp. 223–35. New York: Routledge.

Buchanan, Allen E., Dan Brock, Norman Daniels, and Daniel Wikler. 2000. *From Chance to Choice: Genetics and Justice.* Cambridge: Cambridge University Press.

CBS *Morning Show.* Interview with Jerry Lewis. May 20, 2001.

Daniels, Norman. 1981. "Health-Care Needs and Distributive Justice." *Philosophy & Public Affairs* 10: 146–79.

———. 1985. *Just Health Care.* Cambridge: Cambridge University Press.

Fuhrer, Marcus J., Diana H. Rintala, Karen A. Hart, Rebecca Clearman, and Mary Ellen Young. 1992. "Relationship of Life Satisfaction to Impairment, Disability, and Handicap among Persons with Spinal Cord Injury Living in the Community." *Archives of Physical Medicine and Rehabilitation* 73: 552–7.

Longmore, Paul K., and Lauri Umanski (eds.). 2001. *The New Disability History: American Perspectives.* New York: New York University Press.

Menzel, Paul T. 1992. "Oregon's Denial: Disabilities and Quality of Life." *Hastings Center Report* 22(6): 21–5.

Nosek, Margaret, Marcus Fuhrer, and Carol Potter. 1995. "Life Satisfaction of People with Physical Disabilities: Relationship to Personal Assistance, Disability Status, and Handicap." *Rehabilitation Psychology* 40: 191–202.

Silvers, Anita. 1998. "A Fatal Attraction to Normalizing." In *Enhancing Human Traits: Ethical and Social Implications,* edited by Erik Parens, pp. 95–123. Washington, DC: Georgetown University Press.

14

THE CASE OF CHRONIC PAIN

Emma Sheppard

Chronic pain, as distinct from acute pain, is defined in medical terms as pain lasting for a minimum of 12 weeks (Merskey and Bogduk 1994; British Pain Society 2014a); chronically pained philosopher Susan Wendell defines chronic pain as having lasted for months or years, with no probable end in sight (1996, 2001). This chapter will explore how pain is social, "emerging only at the intersections of bodies, minds and cultures" (Morris 1991, 1). I will pay special attention to how pain's sociality impacts the lived experience of chronically pained people, particularly those who live with pain over many years or even a lifetime. I argue that while chronic pain experiences are distinct from short-lived, acute pain, they are nonetheless shaped by an underlying social assumption that pain is a fundamentally singular experience.

The incidence of chronic pain in the UK is approximately one in ten – with the British Pain Society's National Pain Audit stating that around 11% of the UK's adult (over 18 years of age) population is chronically pained (C. Price et al. 2012). Other estimates vary a little because medical studies tend to group chronic pain by a diagnosis or diagnostic category, rather than in terms of the presence of chronic pain itself. A meta-study by Fayaz et al. (2016) stated a slightly higher prevalence of moderate to severe disabling chronic pain, between 10.4% and 14.3%, in the adult population in the UK; figures for the USA and Australia are similar.

This chapter will explore how narratives and expectations shape not just what – and who – is regarded as disabled, but also expectations of normal and ideal physiology, functions, capacities, and emotions. In taking this approach, the chapter will focus on the stigma of chronic pain, and the resultant drive for and expectation of cure, as well as how people live with chronic pain.

I Pain as Social

Taking pain as social is to expand on the idea that "pain" encompasses a wide range of somatic sensations, experiences, and feelings. How we talk about and describe those sensations, experiences, and feelings – and how we respond to others' situational talk, everyday talk, and expression of pain – is shaped by the social contexts in which we live, the language we use, and the relationships of power between individuals and social groups.

Dominant discourses frame pain as isolating, overwhelming, unliveable, and, above all, individual; doing so ignores – and even hides the role of social structures in making pain isolating

DOI: 10.4324/9781003289487-20

and unliveable (Patsavas 2014; M. Price 2015; Sheppard 2018). These social structures might range from gender, race, and age as well as geographical location, income, and the training given to health care providers or the specialist support available. In effect, while pain is experienced by an individual, their experience and responses to their pain are shaped by a large array of social forces (some of which they may not be aware of), which to many may not be directly apparent – or which one might prefer to pretend to have no impact on one's pain and its treatment.

II The Stigma of Chronic Pain

Like pain, stigma is social. Living with chronic pain is a stigmatized identity. Stigma is known to track inequality, in that higher rates of stigma are experienced by those who are poor, disabled, from a racialized group, LGBTQ+, and have lower educational attainment, among other groups. Stigmatized identities intersect and overlap, and stigma is shaped by social discourses – those same discourses which shape experiences of chronic pain. Stigma directly impacts interpersonal relationships between chronically pained people and their medics, caregivers, family, friends, and peers; being stigmatized means a lower quality of care, and more limited access to care overall.

Goldberg (2017b, 2018) argues that the stigma of chronic pain is directly tied to its invisibility. Medicine, Goldberg argues, depends on visibility – on lesion, including symptoms, test results, and a diagnosable root cause; in effect, whether or not the pain can be seen and proved to exist, or at least have a reason for existing. For many chronically pained people, however, there is no lesion, no sign of injury, and no test result. If there ever was a direct injury, that injury has often healed and is no longer visible. In this sense, some forms of chronic pain seem to be a meaningless repetition (Conrad and Muñoz 2010; Sheppard 2018) – a sign and experience of an injury repeated long after the injury has healed, if it was ever even present in the first place.

This positioning of chronic pain as meaningless reflects an inadequate explicit differentiation in medicine between acute pain – as sign of injury – and chronic pain (Goldberg 2018). In effect, acute pain is visible pain with a known cause, and pain that is made reasonable through its visibility and etiology. Chronic pain is assumed to be the same experience – to feel the same – but to be unreasonable, in that it is either the result of an invisible (and thus non-existent) injury, or is a repetition of a long-healed injury. In effect, chronic pain is unreasonable because it does not fit within expectations of how bodies experience pain, or within medical understandings of why pain is experienced.

Both Goldberg (2017b) and Mollow (2014b) explain that chronic pain does not always fit neatly within Western medical knowledge, which also increases the stigma experienced by those whose chronic pain is unrecognized and/or dismissed, as well as those with pain as a part of illnesses and conditions that are not necessarily consistently recognized within medical care. These conditions include fibromyalgia and chronic fatigue syndrome/ME (CFS/ME). For those experiencing chronic pain without lesion (without identifiable cause), the stigma rooted in this lack of a causal diagnosis often results in their pain being labeled as psychosomatic – that it is "all in their head" – and thus deemed to be the result of that individual's psychological (rather than physical) ill-health or weakness. In this way, the experience of these chronically pained people is dismissed as "not real" (in that the cause is not identifiable), positioned as a more stigmatized mental illness. For many, this epistemological invalidation – this dismissal of their self-knowledge as "not really happening" – combines with the stigma of mental illness; for those experiencing disbelief around their chronic pain without lesion this can be profoundly unsettling, an experience which calls into question their selfhood (Wendell 1996). In addition, there is an implication that the pained individual is either seeking attention, and/or hysterical, and not in need of medical care. As Jones writes on endometriosis, "rewriting pain as psychological suggests that those with endo are hysterical, denies them necessary medical intervention, and reduces social support" (Jones 2016, 557). This implication of hysteria is a highly gendered label, associated with very young women as well as menopausal women. Stigma-bearing diagnoses associated with hysteria and psychosomatic illness are particularly given to ciswomen, especially as those living with chronic pain conditions such as fibromyalgia and CFS/ME (Wolfe et al. 1995; Goldenberg 1999; Lim et al. 2020) – although there are questions that remain as to whether this is the result of selection and/or confirmation bias during diagnosis (Wolfe et al. 2018).

I wish to note at this point there are very little data on the incidence of chronic pain which explicitly includes trans, non-binary, and gender fluid identities; while some chronic illnesses may be tied to reproductive organs (endometriosis, for example), this does not exhaust concerns over the gendering of people with chronic pain. In effect, the label of hysterical requires a social gendering, an understanding that the person in pain is in some way feminine.

At the same time, we do see that there is a higher incidence of chronic pain among stigmatized people beyond considerations merely of gender; there is some research which indicates a higher incidence of chronic illness, including chronic pain, in older lesbian women, gay men, and bisexual people (over 50 years of age) in the USA (Fredriksen-Goldsen et al. 2017). That same research indicated that adult LGB people of all ages experience disparities in access to and quality of care; it does not seem a stretch to conclude that LBG people's lower access to and poorer quality of care leads to higher incidence of chronic illness in later life, just as we see higher rates of chronic illness among people who have worked a lifetime in manual jobs, and/or those who live in poverty. In effect, stigmatized people's lesser access to and lower quality of care as children and adults – combined with further issues such as limited access to "healthy" food,[1] low-pollution living or working environments, and so on – means they are more likely to experience chronic illnesses as older adults. However, it can be much harder to place chronically pained younger people – children and working-age adults – within this, especially when there is no obvious cause of pain, or their diagnosis is poorly understood or dismissed entirely; thus, we return to the label of the hysteric.

The label of psychosomatic or hysterical, particularly when applied to women and those experiencing no visible lesion, depends on the sociocultural positioning of the medical professional as knowing and the patient as known; consider the way in which "consulting doctor Google" is discussed. Within this positioning, the doctor assumes (or is assigned) the heroic role (Wendell 1996), with a mission to cure the illness of their patients, and also to be the ultimate arbiter of what is and is not, what is known, what is true, and what is to be done. Seeking information online (and bringing it to the doctor) goes against this heroic role; it challenges the discourse of medical knowledge as distinct from everyday knowledge, and challenges the doctor's position – and their ability to carry out their mission. It is no coincidence that ignorance and epistemological disablement is cast as the lot of women, while knowledge and heroic medicine is cast as the lot of men.

Mollow, identifying the gendered, ableist roots of the narrative of hysteria, further identifies a need to challenge the "epistemological disablement of people with undocumented impairments… [and to] undermine commonplace cultural figurations [of them] as deficient in self-knowledge" (2014b, 187). For those seeking information and support online – or through in-person support groups – there is a validation of their experience, in finding others with the same or similar experiences of invalidation, stigma, and self-doubt (Sheppard 2018; Newhouse, Atherton, and Ziebland 2018; Main 2020; Smith 2020). Much of this work in challenging epistemological disablement is done through sharing narratives online.

The positioning of the doctor as an all-knowing hero is in turn bound up with an assumption that pain is always experienced as suffering – that a chronically pained person exists in a state of constant suffering. This assumes that the experience of chronic pain is the same as the experience of acute pain, that pain is felt the same on day one as it is on day one thousand. The assumption of pain as suffering is exposed by taking a crip approach and exploring discourses of living with chronic pain. If instead we position pain as value-neutral, neither good nor bad,[2] we quickly see that being pain-free is a part of compulsory ableism; the ideal-normal person is able-bodied, mentally well, white, young, cis, and pain-free.

This assumption of pain as suffering is behind the ethical imperative of ending pain, positioned particularly in terms of a need to end pain by whatever means necessary. This discourse has arguably shifted somewhat in favor of a narrative in which the need for long-term and lifelong pain relief is, in certain circumstances, outweighed by the fear that opiate painkiller use will result in addiction. These circumstances rely on the individual judgment of medics and caregivers, but inevitably disadvantage people of color, women, trans and non-binary people, and poor people, because the closer one is to the ideal-normal, the easier it becomes to access adequate pain relief over long periods (Wilbers 2015; K. M. Hoffman et al. 2016; Goldberg 2017a; D. E. Hoffman and Tarzian 2001).

At the same time as this drive to end pain (with or without the risk of addiction), there is an assumption that a life with chronic pain is vastly inferior to a life without chronic pain, to the point where a life with chronic pain is assumed to be not worth living at all, filled as it is with overwhelming suffering. This ableist assumption is, I argue, as important in the stigma of chronic pain as its invisibility. However, it is important to understand the discourses that this assumption draws upon, as part of the broader stigmatizing of chronic pain.

This assumption of living with chronic pain as not worth living is rooted in a broader discourse which takes pain as profane (Boddice 2017), pain as disgusting, as matter out of place. Within this discourse pain requires elimination – i.e. cure – in order to restore the person in pain to cleanliness and wholeness, as well as containment until cure is achieved. Pain as matter out of place – which triggers a response of visceral disgust and horror – means we can understand pain as a form of bodily fluid, one which can be considered contagious, or perhaps the carrier of contagion, much like other bodily fluids (Sheppard 2019), hence the need for containment. Pain needs to be controlled and contained, hidden from polite public view, in that the person experiencing pain must either demonstrate appropriate self-control (in much the same way as there is a broad expectation that people control their bladder and bowels (see Liddiard and Slater 2017), or their emotions). If they are unable to control their bodyminds, there is an expectation that they either (or more frequently, both) have their bodyminds contained and controlled for them through appropriate care, or they are removed from polite public view. In this way, it is not just their chronic pain which is disgusting, but the chronically pained person who becomes out of place, who is rendered the object of disgust.

However, this is not straightforward. The discourse of pain as profane also impacts expressions of pain – whether non-verbal or spoken – as expressing pain makes it visible to the observer, makes it a contagious fluid – as though hearing of another's pain makes it somehow possible for the hearer to become pained themselves (Sheppard 2018, 2019). This contagion is instead the non-pained confronting the possibility of their own painful futures – in the same way that disability confronts the non-disabled with their own vulnerability. While first expressions of pain – in other words, when pain is acute, especially when the cause of pain is visible – may be heard with sympathy, repeated expressions of chronic pain are responded to with much less, if any, sympathy, especially if the responder has heard these expressions before. When those living with chronic pain make expressions of pain or complain of pain, this can result in accusations of whining, exaggerating, or looking for sympathy, especially when that person has been living in chronic pain for a long time.

For those hearing of another's pain, the disgust they feel on discussing pain (or hearing about it) is tied to both pain as bodily fluid and pain as stigmatized when invisible. Their disgust contains within it a drive to effectively contain expressions of pain, and to only discuss pain within settings deemed appropriate (while this might at first appear to be medical settings, experiences of stigma within those settings make it apparent the only appropriate setting is "never"). For those living with chronic pain, this stigma means that they feel unable not just to complain, but unable to discuss their pain or ask for help.

The disgust and stigma that chronic pain is met with is not experienced uniformly, however; experiences differ depending on the sociocultural context in which pain is experienced. In North American and Western European contexts, and reflecting broader structural inequalities, disgust is stronger and sympathy harder to come by for people of color, reflecting racist assumptions about bodily capacities to feel and withstand pain, how easily a person vocalizes pain, and also whether or not an individual seeking medical care is engaging in so-called "drug-seeking" behavior – in other words, whether or not they are assumed to have an addiction (K. M. Hoffman et al. 2016).

Women – whether women of color or white women – also have a harder time accessing appropriate care (D. E. Hoffman and Tarzian 2001; Wilbers 2015; Jones 2016); their pain is dismissed more easily and over a longer time period. Disgust goes along with the epistemological invalidation of chronic pain when experienced by particularly gendered and/or racialized bodyminds; not only are they disbelieved but they are also responded to with disgust, acquiring a stigmatized identity.

III Wellness and Cure

Within this stigmatized experience, people living with chronic pain – with any chronic illness or disability – are faced with a dominant narrative of cure (Goldingay 2018). Tied to the previously mentioned narratives of the medic as specialist knower, this narrative builds an expectation that the chronically pained person will be cured of their pain – that they can be cured, and that they want to be cured (using the contemporary tools of Western medicine). Cure is regarded as desirable within an ableist-eugenic discourse wherein disability (including chronic pain) is regarded as tragic and unwanted for both the individual and society; rehabilitation and cure narratives position a desire to be cured as the only reasonable response to disability (Kafer 2013; Patsavas 2014; Sheppard 2018). Furthermore, this positioning is underpinned by a discourse in which bad things only happen to bad people, in effect – that acquired disability is a proxy for moral worth (Edwards and Imrie 2003; Hughes 2012). In effect, this means that disabled adults, positioned as dependent and economically unproductive (and thus morally inferior to "independent," economically active non-disabled people) regardless of their actual interdependencies, productivities, and loving relationships, are thus read as being morally inferior, of lesser worth. Cure thus acts as affirmation – if they have not been cured, it is down to their personal failure; the only way to escape failure is to be seen to seek out a cure.

Being cured requires the chronically pained person to subject themselves to further "expert" examination – with the stigma it entails – and there is the implication that if they are not cured then they have failed either to find the right expert, or to obey the right instructions with sufficient dedication and desire for to be free of pain. This narrative of cure ties with the narrative of life with chronic pain as not worth living; it drives the assumption that cure is always better. However, this has been resisted by some chronically pained people, explaining that while they may want their personal pain to go away, they have found moving outside of "cure" into a space of "management" and "acceptance" to improve their quality of life (Patsavas 2014; Sheppard 2020).

The narrative of cure runs alongside discourses of responsibilization and wellness (see: Basas 2014). Responsibilization positions a person's health and/or wellness as their individual responsibility, regardless of socioeconomic factors – such as race, class, location – which impact their health or ability to access health care, food, or exercise. It works to increase the stigma experienced by sick, disabled, and fat people by positioning their experience as a choice while ignoring systemic factors – or positioning those factors as further choices. Responsibilization means that when a person continues to be chronically pained – when a cure fails, is not sought, or is out of reach – they experience stigma from being seen as making poor choices, or for being lazy.

With the term "wellness," I am referring to the practice of healthy behaviors – but also the performance of healthy behaviors, in that these behaviors must be seen or recorded, and thus recognized as healthy by others (in effect – is a gym workout really a workout if it is not documented on social media?). Wellness also ignores systemic factors impacting a person's ability to engage in healthy behaviors – and to have their behaviors recognized as healthy – and rewards the performance. For chronically pained people, wellness involves not just engaging with medical care, but being seen to engage in performative healthy behaviors – a particular diet or an exercise routine, and so on – in ways that observers are able to recognize as attempts to regain their lost unpained selves. For chronically pained people, wellness and responsibilization act to further stigmatize them; they continue to be in pain, and this is regarded as the result of their own failure – cure thus becomes compulsory, while being presented as a choice.

Responsibilization and wellness also act to position pain as something to be feared and avoided – and something which can be entirely prevented, rather than a part of human experience. In this way, the prospect of pain, and the prospect of pain continuing, becomes much greater – and the pressure to end pain increases as a life with pain becomes seen as not just one of suffering, but as not liveable. Avoiding disability and illness – including pain – becomes a central concern of engaging in wellness. This concern can lead to pain being regarded as catastrophic – but also underpins the response to pain labeled as "pain catastrophizing."

IV Conclusions

Living with chronic pain is assumed to be hellish, unwanted, and unsustainable. Sometimes it is. However, it also becomes boring, unremarkable – and perhaps most surprising of all – forgettable. For chronically pained people who have lived with pain for years, chronic pain becomes a part of their sense of who they are, of their everyday self (Sheppard 2018, 2020). In effect, being in pain becomes an everyday experience.

However, this does not lessen the stigma or its impact – or the assumption from non-pained people that their lives are awful. Living with chronic pain results in one taking on a stigmatized identity. Within medicine, the lack of visible, measurable cause for chronic pain results in disbelief, in stigma – which in turn impacts the care and support chronically pained people receive. Beyond medicine, as well as within it, chronically pained people are positioned as responsible for their pain, and for their pain continuing; they are stigmatized as having failed.

The dual stigmatizing of pain – and the underlying discourse in which living with pain is assumed to be impossible – contribute to people's fear of living in pain. These stigmas also negatively impact the care and support chronically pained people receive from medics and caregivers – and their relationships with family, friends, and employers.

Notes

1 I refer here to food which is both nutritious and tasty – in the broadest sense – and an overall diet rich in the required nutrients, rather than rich in food which is high in calories and low in nutrients.
2 Pain as value-neutral is about viewing pain as an experience that can be good/positive, or bad/negative, or both, depending on the situation; pain does not in and of itself hold value. There are situations and times when pain can be welcomed, and even sought out, from tattooing to BDSM, and for people living with chronic pain, the presence of an 'everyday' amount of pain can be regarded as unremarkable (Sheppard 2018).

References

Basas, C. G. 2014. "What's Bad about Wellness? What the Disability Rights Perspective Offers about the Limitations of Wellness." *Journal of Health Politics, Policy and Law* 39 (5): 1035–1066. https://doi.org/10.1215/03616878-2813695.

Boddice, Rob. 2017. *Pain: A Very Short Introduction.* Oxford: Oxford University Press.

Conrad, P., and V. L. Muñoz. 2010. "The Medicalization of Chronic Pain." *Tidsskrift for Forskning: Sygdom Og Samfund* 13: 13–25. https://doi.org/10.7146/tfss.v7i13.4147.

Edwards, Claire, and Rob Imrie. 2003. "Disability and Bodies as Bearers of Value." *Sociology* 37 (2): 239–256.

Fayaz, A., P. Croft, R. M. Langford, L. J. Donaldson, and G. T. Jones. 2016. "Prevalence of Chronic Pain in the UK: A Systematic Review and Meta-Analysis of Population Studies." *BMJ Open* 6 (e010364). https://doi.org/10.1136/bmjopen-2015-010364.

Fredriksen-Goldsen, Karen I., Hyun-Jun Kim, Chengshi Shui, and Amanda E.B. Bryan. 2017. "Chronic Health Conditions and Key Health Indicators among Lesbian, Gay, and Bisexual Older US Adults, 2013–2014." *American Journal of Public Health* 107 (8): 1332–1338. https://doi.org/10.2105/AJPH.2017.303922.

Goldberg, Daniel S. 2017a. "On Stigma & Health." *The Journal of Law, Medicine and Ethics* 45 (4): 475–483.

———. 2017b. "Pain, Objectivity and History: Understanding Pain Stigma." *Medical Humanities* 43: 238–243. https://doi.org/10.1136/medhum-2016-011133.

———. 2018. "Pain, Stigma, and Neuroimaging: History, Ethics and Policy." In *Developments in Neuroethics and Bioethics,* edited by Daniel Z Buchman and Karen D. Davis, 1:85–103. Elsevier. https://doi.org/10.1016/bs.dnb.2018.08.005.

Goldenberg, Don L. 1999. "Fibromyalgia Syndrome a Decade Later: What Have We Learned?" *Archives of Internal Medicine* 159 (8): 777. https://doi.org/10.1001/archinte.159.8.777.

Goldingay, Sarah. 2018. "Act Like It Hurts: Questions of Role and Authenticity in the Communication of Chronic Pain." In *Painscapes,* edited by E. J. Gonzalez-Polledo and Jen Tarr, 61–81. London: Palgrave Macmillan. https://doi.org/10.1057/978-1-349-95272-4_4.

Hoffman, Diane E., and Anita J. Tarzian. 2001. "The Girl Who Cried Pain: A Bias Against Women in the Treatment of Pain." *Journal of Law, Medicine and Ethics* 29 (1): 13–27. https://doi.org/10.1111/j.1748-720X.2001.tb00037.x.

Hoffman, Kelly M., Sophie Trawalter, Jordan R. Axt, and M. Norman Oliver. 2016. "Racial Bias in Pain Assessment and Treatment Recommendations, and False Beliefs about Biological Differences between Blacks and Whites." *Proceedings of the National Academy of Sciences of the United States of America* 113 (16): 4296–4301. https://doi.org/10.1073/pnas.1516047113.

Hughes, Bill. 2012. "Civilising Modernity and the Ontological Invalidation of Disabled People." In *Disability and Social Theory: New Developments and Directions,* edited by Dan Goodley, Bill Hughes, and Lennard Davis, 17–32. Abingdon: Palgrave MacMillan.

Jones, Cara E. 2016. "The Pain of Endo Existence: Toward a Feminist Disability Studies Reading of Endometriosis." *Hypatia* 31 (3): 554–571. https://doi.org/10.1111/hypa.12248.

Kafer, Alison. 2013. *Feminist, Queer, Crip*. Bloomington and Indianapolis: Indiana University Press.

Liddiard, Kirsty, and Jen Slater. 2017. "'Like, Pissing Yourself Is Not the Most Attractive Quality, Let's Be Honest': Learning to Contain through Youth, Adulthood, Disability and Sexuality." *Sexualities* 21 (3): 319–333. https://doi.org/10.1177/1363460716688674.

Lim, Eun-Jin, Yo-Chan Ahn, Eun-Su Jang, Si-Woo Lee, Su-Hwa Lee, and Chang-Gue Son. 2020. "Systematic Review and Meta-Analysis of the Prevalence of Chronic Fatigue Syndrome/Myalgic Encephalomyelitis (CFS/ME)." *Journal of Translational Medicine* 18 (1): 100. https://doi.org/10.1186/s12967-020-02269-0.

Main, Susanne. 2020. *Exhibiting Pain: Using Creativity to Express Chronic Pain*. Milton Keynes: Open University.

Merskey, H., and N. Bogduk, eds. 1994. "Part III: Pain Terms, A Current List with Definitions and Notes on Usage." In *Classification of Chronic Pain*, 2nd Edition, edited by Harold Merskey and Nikolai Bogduk, 209–214. Seattle: IASP Press.

Mollow, Anna. British Pain Society. 2014a. "Useful Definitions and Glossary." *British Pain Society*. 2014. https://www.britishpainsociety.org/people-with-pain/useful-definitions-and-glossary/.

———. 2014b. "Criphystemologies: What Disability Theory Needs to Know about Hysteria." *Journal of Literary & Cultural Disability Studies* 8 (2): 185–201. https://doi.org/10.3828/jlcds.2014.15.

Morris, David B. 1991. *The Culture of Pain*. Berkeley: University of California Press.

Newhouse, Nikki, Helen Atherton, and Sue Ziebland. 2018. "Pain and the Internet: Transforming the Experience?" In *Painscapes: Communicating Pain*, edited by E. J. Gonzalez-Polledo and Jen Tarr, e-book, 129–155. London: Palgrave MacMillan. https://search.ebscohost.com/login.aspx?direct=true&scope=site&db=nlebk&db=nlabk&AN=1615541.

Patsavas, Alyson. 2014. "Recovering a Cripistemology of Pain: Leaky Bodies, Connective Tissue, and Feeling Discourse." *Journal of Literary & Cultural Disability Studies* 8 (2): 203–218. https://doi.org/10.3828/jlcds.2014.16.

Price, Cathy, Hoggart, Ola Olukoga, Amanda C de C Williams, and Alex Bottle. 2012. "National Pain Audit, Final Report 2010–2012." HQIP, The British Pain Society and Dr Foster. http://www.nationalpainaudit.org/media/files/NationalPainAudit-2012.pdf.

Price, Margaret. 2015. "The Bodymind Problem and the Possibilities of Pain." *Hypatia* 30 (1): 268–284. https://doi.org/10.1111/hypa.12127.

Sheppard, Emma. 2018. "Using Pain, Living with Pain." *Feminist Review* 120 (1): 54–69. https://doi.org/10.1057/s41305-018-0142-7.

———. 2019. "Chronic Pain as Fluid, BDSM as Control." *Disability Studies Quarterly* 39 (2). https://doi.org/10.18061/dsq.v39i2.6353.

———. 2020. "Performing Normal but Becoming Crip: Living with Chronic Pain." *Scandinavian Journal of Disability Research* 22 (1): 39–47. https://doi.org/10.16993/sjdr.619.

smith, s.e. 2020. "The Beauty of Spaces Created for and by Disabled People." In *Disability Visibility: Twenty-First Century Disabled Voices*, edited by Alice Wong, First Vintage Books edition, pp. 271–275. New York: Vintage Books.

Wendell, Susan. 1996. *The Rejected Body: Feminist Philosophical Reflections on Disability*. New York: Routledge.

———. 2001. "Unhealthy Disabled: Treating Chronic Illness as Disabilities." *Hypatia* 16 (4): 17–33.

Wilbers, Loren E. 2015. "She Has a Pain Problem, Not a Pill Problem: Chronic Pain Management, Stigma, and the Family—An Autoethnography." *Humanity & Society* 39 (1): 86–111. https://doi.org/10.1177/0160597614555979.

Wolfe, Frederick, Kathryn Ross, Janice Anderson, I. Jon Russell, and Liesi Hebert. 1995. "The Prevalence and Characteristics of Fibromyalgia in the General Population: Fibromyalgia Prevalence and Characteristics." *Arthritis & Rheumatism* 38 (1): 19–28. https://doi.org/10.1002/art.1780380104.

Wolfe, Frederick, Brian Walitt, Serge Perrot, Johannes J. Rasker, and Winfried Häuser. 2018. "Fibromyalgia Diagnosis and Biased Assessment: Sex, Prevalence and Bias." Edited by Claudia Sommer. *PLOS ONE* 13 (9): e0203755. https://doi.org/10.1371/journal.pone.0203755.

Further Reading

Boddice, Rob. 2017. *Pain: A Very Short Introduction*. Oxford: Oxford University Press.

Goldberg, Daniel S. 2017. "Pain, Objectivity and History: Understanding Pain Stigma." *Medical Humanities* 43: 238–243. https://doi.org/10.1136/medhum-2016-011133.

Patsavas, Alyson. 2014. "Recovering a Cripistemology of Pain: Leaky Bodies, Connective Tissue, and Feeling Discourse." *Journal of Literary & Cultural Disability Studies* 8 (2): 203–218. https://doi.org/10.3828/jlcds.2014.16.

Price, Margaret. 2015. "The Bodymind Problem and the Possibilities of Pain." *Hypatia* 30 (1): 268–284. https://doi.org/10.1111/hypa.12127.

15

CHRONIC ILLNESS, WELL-BEING, AND SOCIAL VALUES

Lydia Nunez Landry

I Introduction

I am writing this chapter amid a devastating viral pandemic. When the outbreak began, governmental officials and media assured the public that the majority of people infected recover with few complications, most with mild or no symptoms. To further ease public concerns, they stressed that the virus threatened serious illness and death only for older and chronically ill people.

Soon the rising number of people seeking aid overwhelmed hospitals, and dwindling resources hindered hospital staffs' ability to adequately treat all patients. In such circumstances, hospitals turn to institutional or state crisis care plans to help them determine who receives life-preserving treatment. These plans shift focus from the needs of individual people to what best benefits society overall.[1] This utilitarian focus on how resource allocation benefits society partially displaces biomedical facts and triage principles with social values, including socially constituted biases—we should save as many lives as possible, but the right and worthwhile lives. To attain this "greater good," many crisis plans favored younger and nondisabled people and categorically excluded from life-preserving treatments people with certain chronic illnesses, cognitive and physical disabilities, and some who simply needed assistance with activities of daily living (Mello, Persad, and White 2020).

We have the option to shift to practices that purportedly benefit society overall only because most of us accept that some predetermined categories of people may have to be sacrificed in circumstances deemed dire. The groups from which our society chooses those who will be sacrificed reflect beliefs so deeply ingrained that they have taken on the guise of an obvious, unquestionable truth: when people must die, society benefits when the dead are chronically ill, disabled, and older people.

In this chapter, I explore reasons why our society devalues chronically ill people and why people fear chronic illness, even though chronically ill and impaired people live happy, fulfilling, and meaningful lives. I further argue that ableist[2] attitudes and responses to illness and impairment represent a far greater threat to the well-being of impaired people than do their conditions.

DOI: 10.4324/9781003289487-21

II Chronic Illness, Objectified and Lived

We are, throughout our lives, liable to injury and disease. In 2016, 88% of deaths in wealthy countries involved noncommunicable conditions. Nearly half of the US population has at least one chronic condition, and in countries like the USA and the United Kingdom, most disability results from chronic conditions (Scambler 2020).

A wide range of conditions are classified as chronic, including hypertension, lupus, multiple sclerosis, bipolar disorder, rheumatoid arthritis, diabetes, fibromyalgia, ulcerative colitis, and depression. Equal in variety are the experienced effects of chronic conditions, including weakness, lethargy, pain, nausea, and difficulty breathing. This diversity impedes consensus on how best to define chronic illness. We could point to a few presumptive commonalities, but in short chronic illnesses are lumped together by what they are not: relatively brief and medically resolvable.

Disease classification will, anyhow, be of little use to us. For when we consider questions relevant to this chapter—questions about the experiences of chronically ill people, about their well-being or the quality of their lives—we soon discover that theories of disease and biomedical descriptions and explanations of chronic conditions contribute remarkably little insight. Modern medicine was fashioned in imitation of experimental laboratory sciences, adopting over the first half of the twentieth century the philosophical commitments and methodological values and practice of science (Krieger 2011). These commitments and values—objectivity, reductionism, value-neutral evaluation of facts, quantifiable evidence, emotional detachment—while perhaps well-suited (if also idealistic) to the laboratory, are particularly ill-suited for what matters in evaluations of quality of life (QoL), as they severe the objects of medical research and practice from the concerns of human meaning and experience.

The body-as-machine metaphor is a founding axiom of medicine (Lee 2012). This objectified body has no biography, no identity, no personality. Like a clock, it functions either normally (is intact) or abnormally (is broken). Our feelings, values, and personal experiences are epistemologically suspect, for the objects of medical science are "free of values and qualities… [and] divorced from the context and time in which they naturally occur" (Cassell 2004, 18).

QoL considerations, however, are necessarily value-laden—it is only against the backdrop of our individual conceptions of a good human life that we can make the qualitative distinctions these judgments involve. They are not objective in any scientific sense. Indeed, researchers developed QoL instruments to fill the gaps left by objective medical judgments (about health states) with experiential evaluations from patients themselves. The idea was that we could better assess medical interventions if we looked beyond the biochemical level and to whole persons (with their values, preferences, and emotions) interacting with others within social environments.[3]

Healthcare professionals (HCPs) nevertheless routinely talk of patients' "quality of life," though often citing an "intuitive and spontaneous assessment" carried out by physicians themselves and without standardized measures (Caillault et al. 2020, 1899). Physicians sometimes make these intuitive judgments not to assess treatment outcomes, but to rationalize denials of treatment.[4] When patients and their physicians both complete QoL instruments, physicians routinely underestimate patients' QoL compared to patients' own assessments (Srikrishna et al. 2009; Barata et al. 2017).[5] As Slevin and colleagues (1988) explained, patients focus on the psychological and social aspects of QoL, while physicians focus on health states.[6] Whether intentional or not, physicians who attempt to be their patients' QoL proxies restore the epistemic imbalances that QoL assessments were created to correct. This practice reasserts both the insignificance of patients' lived experiences and the preeminence of biomedicine's narrow and reductive preoccupations—it reasserts the physician's authority to define what matters.

III The Lived Body

We will therefore explore the evidence and practices of those people best positioned to know what it is like to live with chronic illnesses, those who live those lives. S. Kay Toombs (1992) surveys what she considers the typical characteristics of our lived experience of illness, irrespective of biomedical classification. She depicts these five characteristics as "losses": the losses of (the perception of) wholeness, certainty, control, freedom to act, and the familiar world.

A stable sense of being in the world, the grounds for familiarity, wholeness, and certainty, depends on the (kinds of) possibilities that structures our bodily expectations—patterns of anticipation that our everyday actions will or will not produce given outcomes. Usually, possibilities are realized as habitual certainties. We rely on a multitude of habitual actions to anchor the more complex routines we need for daily activities. Our deeply embedded confidence in their integrity allows them to operate in the background. We can reckon the feasibility of actions or projects without being overwhelmed and perhaps immobilized by concerns about whether these innumerable sub-tasks will succeed or fail. A person who wants a cup of coffee may wonder whether any is left in the pot, but usually they need not contemplate lifting an arm, straightening it, grasping the cup handle, scooting feet closer to the chair, applying downward force with their feet to rise, and so on. In illness, that confidence may be lost and the background springs to the fore; bodily certainty is replaced, to some degree, by "bodily doubt" (Carel 2016).

Toombs's loss of wholeness is linked to this bodily doubt. What worked without supervision now requires constant attention, endurance, and physical and mental exertion. Uncertainty of a condition's cause or the course a condition will take (when or whether symptoms will flare up, abate, change, or worsen) introduces precariousness into every action, plan, and project. Diffuse unpredictability and uncertainty undermine the meaningfulness and coherence of the decisions we make. And the hopes we fostered when we sought medical counsel likely expired when we received diagnoses but very little information or solace. We feel powerless, overwhelmed, and unsure how to proceed.

The natures of some symptoms, of course, contribute more directly to a feeling of alienation. Symptoms such as incontinence and involuntary bodily movements particularly disturb us; we fear humiliation—harms to dignity—if others witness these symptoms. Humiliation and dignity implicate the involvement of others in our experiences of illness, as indeed all of our experiences are mediated through diverse social representations and relationships. Others, then, contribute to the character of our experiences of illness, and the quality of social support looms large in how our new lives and self-conceptions are reconstituted. Without stable social support, despair may replace anxiety.

However stable or supportive our intimate relationships, wider social forces shape our emotions and behaviors. We wonder whether we will be able to work or pay bills. We worry what others will think of us; will we be believed, and if so, will we be blamed for our illnesses—and many of us will blame ourselves. We feel guilt when we cannot fulfill the obligations our social roles demand or a piercing shame at the thought of becoming burdens to our loved ones. We feel like frauds when we have "good days" and can do some of the things we used to do. In public, we may comport our bodily movements with nondisabled people's expectations so as not to appear malingerers; for example, if we can stand or walk, however precariously, we remain seated in our wheelchairs when shopping rather than stand to reach items on the higher shelves. The experience of chronic illness is chiefly the fear and experience of stigmatization (suspicion, discrimination, and social exclusion).[7]

As Carel (2016) points out, Toombs's characteristics are not experienced by all chronically ill people.[8] Different people experience the same condition differently. The feelings of loss will be experienced as more or less significant by different people. More broadly, the sense of loss may rise or fall according to one's social location or a society's dominant values or understandings of health and illness. For example, a society uncommitted to foundational individualism would not view vulnerability and dependency as the lamentable remains of unrealized ideals; and a society wherein ill people face little or no stigmatization would utterly alter the long-term experience of chronic illness.

IV Transformative Experiences and Adaptation

Notwithstanding the difficulties outlined above, chronically ill people (as disabled people generally) live happy, fulfilling, and meaningful lives. In fact, many appraise their well-being at levels comparable to the self-appraisals of people without chronic conditions (Angner et al. 2009). Some conditions (rapidly progressing conditions, for example) present unique and continuing issues, and generally the age of onset can increase the time required to adjust, but people typically adapt to illness within a year or two of diagnosis or serious symptoms (Carel 2016).

But, what do I mean by "adapting" to illness or impairment? Many people writing on this topic connect adaptation to a return to previous levels of well-being. We cannot, however, identify adjustment with a return to former or increased well-being. A person may adjust to a condition in the manner I propose but be dissatisfied with their life for any number of other reasons but particularly the effects of ableism.

Adapting to chronic illness is not overcoming disease or returning to a vestige of one's preill life; no more than becoming a parent involves overcoming the conditions of childrearing to get back lost freedoms, abilities, and opportunities. While our values are apt to change, what I refer to is not centrally about adapting "preferences" or settling for less by lowering one's expectations, nor any other purely psychological process whereby we "come to terms" with our "misfortunes." Nor is adaptation merely learning new ways of doing things. Since it is based on the distinctive character of "what it is like" to live with chronic illness with others in an ableist world, how one adapts is always deeply social. When we experience the sense of loss described by Toombs, our ruptured relationship to the world renders considerations of "preserving well-being" absurd. First, we must reestablish a sense of coherence and meaning through which such considerations might make sense.

Chronic illness, like parenthood, can be a transformative experience—that is, a circumstance that significantly alters our relationships to the world, our sense of who we are as persons. Transformational experiences bring unavoidable uncertainty, and thus the opportunity for discovery. These circumstances transform us in two ways, epistemically and personally. Experiencing the world in a way we hadn't before, in a way we couldn't anticipate or imagine, transforms us epistemically. We learn new things that allow us to evaluate the experiences in ways we couldn't beforehand; we understand the world and our belonging in it differently; and we can then imagine further ways the world and ourselves might be (Paul 2014). Understanding ourselves differently implies a transformation of our conceptions of who we are as persons, of our values and preferences, or, we might say, what we care about. Though we may have loved drinking with our co-workers every night, becoming parents replaces those preferences with new ones. And it isn't by the pull of obligation, guilt, or bitter resignation that we desert that old lifestyle for an all-consuming babyworld. Nor has our esteem for our friends suddenly plummeted. It is simpler than that: many activities we reveled in as the persons we were no longer hold sway for our newly forming selves. Adapting to these kinds of circumstances entails reorienting ourselves to the world with reconstituting self-conceptions.

To reorient ourselves to the world with chronic illness we must take our bodily experiences seriously. This includes both our experiences of illness and the emotions, such as anxiety, evoked by uncertainty and uncontrollability. We experience bodily uncertainty, Toombs says, as a threat to the self. The body or mind, then, for a time becomes the focal point in our lives.

Given the magnitude of the consequences, the cognitive resources we need to lessen uncertainty and unpredictability necessarily limit the number of other things to which we can attend. Anxiety both provides the focus and summons those cognitive resources.

We tend to judge anxiety as a negative and harmful emotion; it conjures images of people in states of panic, havoc, or extreme and irrational worry. Clinical and social psychologists, however, have shown that typical anxiety benefits us; it prepares us for future negative events (Barlow 2002).[9] In fact, we anticipate potentially negative outcomes throughout each day, though, to be sure, with varying degrees of severity and significance. Whether preparing for a hurricane, a job interview, a math exam, or a novel social interaction, anxiety drives us. It figures prominently, as well, in personal growth, creativity, and moral decision making and agency. Since any goal may be undone by countless contingencies, we can, with Barlow (2002), safely assert that "without anxiety, little would be accomplished" (9).

Like the closely related affective states of pain and fear, anxiety impels us to immediately address a threat; and like a blaring smoke detector, the subjectively unpleasant aspect of each emotion is essential to its function. As part of a defensive-motivational system, anxiety targets (is directed at) the uncertainty surrounding a threat and its relation to our need for control.[10] We experience anxiety when a possible negative outcome threatens to disrupt our lives, and we are uncertain what will happen, when it will happen, what we should do about it, or whether we have the power to act against it. Barlow (2002, 64) characterizes this state as a sense of "helplessness," a condition that anxiety's hypervigilance counteracts. Anxiety then is a forward-looking emotion. It forces one to attend to potential future recurrences of negative events.

To achieve this end, anxiety motivates two types of behaviors, avoidance behaviors (those that minimize risks) and particular cognitive behaviors (e.g., information-gathering, reflection, deliberation).[11] Anxiety during the COVID-19 pandemic, for example, spurred people to enact behaviors (social distancing, hand-washing, mask-wearing) that lowered risks and to seek information (cable news viewership substantially increased).

There are no equivalent experts to advise chronically ill people, but we do possess privileged access to our experiences of illness, and anxiety incites a vigilant awareness of our bodies. Initially, information pours in unfiltered and unguided.[12] Every twinge of pain, tightness in the chest, feeling of pressure, weakness, or shortness of breath, which anyone may occasionally experience, we assiduously register as potentially meaningful. We integrate (at least provisionally) bits of information into our understanding of illness as we reflectively answer questions such as, Did I feel that before severe symptoms first appeared? Did I eat the same kind of food before I last had that symptom? These observations and questions start what has been variously called "monitoring," "reading," or "listening" to one's body, a habit of interpreting our experiences of illness, looking for signs, and measuring tolerances and responses. This, itself, begins to bring meaning and structural coherence to our lives. Taking bodily insights seriously also begins the longer process (whether consciously implemented or not) of hypotheses creation and experimentation. As we spot patterns, we test for correlations or causal connections, and we gather and incorporate the knowledge others with our conditions have discovered. Over time we learn to recognize signs of coming episodes, and, in some instances, we learn how to mitigate them with preparation or other discovered countermeasures.

Meyer (2002) describes how women who experienced unpredictable and debilitating migraine headaches developed their own health regiments by closely monitoring their individual experiences of illness. Through this practice they reduced recurrences by identifying triggers and early warnings of imminent attacks.[13] Meyer calls this behavior the "art of watching out," a "perpetual vigilance" (1221). Better would be to refer to this vigilance by its instigating emotion—that is, a continual anxiety, one that sustains extended watchfulness.[14] The language used by Meyer's informants suggests a systematic investigation, as they frequently used phrases such as

"I would notice that…," "figuring it out," "trial-and-error learning," "making associations," and "putting it together." These processes, however, needn't be so highly structured or experienced in ways we can clearly express. An ineffable sensation might be ascertained that consistently alerts them of impending migraines. When nothing beyond "a feeling" could be articulated, they nevertheless ascribed meaning to the feeling as a basis for prediction and control.

V Self-efficacy, Control, and Noncompliance

After they had worked to make their migraines more predictable, the women in Meyer's study occasionally chose not to avoid triggers or take medicine to prevent impending migraines. This behavior may seem counterintuitive, since ultimately, many will assume, they strove to restore their health, to be migraine-free. While it is likely true that if offered a (side-effects-free) treatment to end their migraines, they would accept it, they clearly did not strive to be migraine-free as an end itself. This they showed when they readily accepted the pain and nausea of a migraine rather than disappoint people they cared about. Predictability and controllability establish a relatively stable set of possibilities from which to choose, not so we can determine which courses of action are most healthful or least painful per se, but which best permit and sustain (hinder least) the relationships and activities we most care about—sometimes they coincide with illness management, sometimes they do not.

This is not unusual. Parents, for example, routinely forgo or relinquish things that matter to them for the sake of their children, but because those relationships constitute essential parts of who they are, they may never consider those losses harms to their well-being. In relationships of all kinds, intimate and distant, we together create common goods and goals and shared meaning (relational interests), and we cannot often conceive of our loved ones' well-being or interests as independent from our own.[15] The contribution any bodily "health state" makes to well-being is indeterminate and contingent. When we make decisions about which desires or goals to pursue and how to pursue them, we rarely consider them singularly, isolated from other interests. We instead consider whether they mutually fit with (at least some of) our other goals, desires, and values. In all cases we go beyond an overly individualistic view of well-being by incorporating relational interests or accounting for relational well-being—that is, "the same old notion of what is good for someone, except where we acknowledge that certain relationships are built into who that someone is" (Carbonell 2018, 349). In our lives as we live them through and with others, health and wellness spring not from the bodily states of individuals, but the caring relationships constitutive of who we are.

In part for these reasons, "noncompliance" with HCP directives is common in the self-management of chronic illness, for compliance requires subordinating one's own intricate structures of values and desires to the one-dimensional biomedical focus on disease treatment.[16] Considering the feelings of powerlessness and hopelessness initially experienced by many chronically ill and impaired people, one can hardly overstate the remedial feelings of self-efficacy, control, and agency we experience from weighing the consequences of options and choosing whether, when, and how to act on information about our conditions in light of our own values.

Chronic illness forces us to listen intently to our bodies. In doing so, we acquire knowledge that makes our lives more predictable and controllable. During this time we also learn new skills and create novel ways to complete the tasks impairment made unreliable, thereby forging new habitual certainties, new structures of anticipation and fulfillment, and a stable relationship with the world. Some of us learn to use devices—such as wheelchairs and canes—that begin as clumsy and cumbersome objects and become integrated parts of lived bodies.

Even where some unpredictability and uncertainty remain, as it must for us all, we learn what to expect, how to prepare, and what we must accept. The experience of illness itself grows ever more predictable and familiar. We grow accustomed to its vicissitudes, understanding that difficult periods wane, or we become so habituated to symptoms that they become (at least for a time) inconspicuous. Illness can, as Carel (2016) writes, "recede into the background in a way unimaginable to the healthy outsider" (132).

VI Social Stigma and Personal Transformation

The preceding largely deals with the epistemic transformation associated with adapting to chronic illness. But we still have to consider the second aspect of transformative experience, personal transformation.

Many of us develop, perhaps for the first time in our lives, an intimate familiarity and deep understanding of our bodies and minds. Not infrequently, chronic illness foists a profound recognition of the vulnerability, contingency, and dependency of the human condition, a recognition previously confounded by social demands for self-reliance and independence. J. H. van den Berg (1966) suggests that such insights take hold viscerally, and not as mere objects of intellectual dalliance, only in conflicts with certainty. Illness motivates reflection; it draws our attention to what should be treasured rather than despised, our interdependence and universal needfulness, and that "the really healthy person possesses a vulnerable body and … is aware of this vulnerability" (74).

The point here is not to make a virtue of suffering or to trivialize the distressing experiences of illness. What we have seen is that "health states" and a happy, meaningful life have no firm association, and a life transformed through chronic illness may be experienced as better for the person living it. But, again, we cannot predict how or whether we will be personally changed through a given experience. One may be transformed epistemically and yet believe one's life robbed of meaning and happiness. To understand why, we must account for the significant influence of others in how we conceive ourselves, for neither our pre-ill nor ill selves are self-made. To end the chapter, we examine social attitudes and meanings of chronic illness and impairment and how they influence the courses of our changing self-conceptions.

VII The Social Meanings of Chronic Illness

While most people agree that managed hypertension and diabetes impose few limits on a person's ability to live a good life, the vast majority of people believe that untreatable chronic illness to be grave misfortunes that necessarily reduce or prohibit the ability to lead meaningful and happy lives. Dozens of ways of living with chronic illness and disability have been deemed "worse than death" by people with no experience living those lives (Dolan 1997; Rubin, Buehler, and Halpern 2016). HCPs tend to appraise chronically ill and disabled people's expected QoL with even greater pessimism (Iezzoni 2021).[17] As noted, people living those lives demur, and often rate the quality of their lives substantially higher than the dreary imaginings of nondisabled people.

This discrepancy between chronically ill people's QoL reports and nondisabled people beliefs is sometimes explained by general difficulties imagining the lives of others. Transformative experience compounds the difficulties. Yet, while adaptation as transformative experience may clarify one side of the discrepancy, it cannot wholly explain why these reports are met with incredulity or hostility. If a man devastated by the death of his beloved partner of 30 years remarries five years after his loss, few will say that he does not truly love his new partner or he has irrationally adapted his preferences. While childless people may never know what it is like to be a parent, they seldom doubt the sincerity of parents who profess values and preferences altered by parenthood. No social ideals or ideological commitments require our rejecting the belief that the man loves his new partner or that parenthood alters our values. Disabled people's testimonies are rejected not only because of general difficulties imagining the lives of others, but also because happy, competent, smart, sexual, caregiving, and productive disabled people leading meaningful and fulfilling lives contradict some of our society's fundamental ideals and ideological values. More to the point, the supposed "deviance" of stigmatized[18] groups defines through opposition how dominate social groups view themselves and model citizenship and personhood. Given this, we obviously face considerable obstacles when we attempt to imagine ourselves members of a social group we devalue, reject, and exclude from society; nor will we accept their claims that lives we imagine worse than death stand equal to our own. The thought of joining the ranks of that stigmatized group certainly fills non-members with fear and dread few other horrors can match.

Despite the vast web of veiled interdependencies that everyone relies on to do the things they want to do (Lindemann 2003), our culture remains transfixed by an implausible conception of persons as atomistic and self-obsessed: a person is independent, autonomous, self-reliant, self-sufficient in pursuing her self-interests. Each of these ideals links to the human capital conception of human worth; as market agents, our worth is entirely wrapped up in our contributions to economic growth. Economic productivity has become a touchstone of personhood and self-worth and hence central to our self-conceptions. And although many ill and disabled people work—and many more would but for discriminatory hiring practices—we are globally stereotyped as incompetent and incapable of productive work.[19] From this stereotype flows the stigmatizing rhetoric of right-wing politicians and ordinary people. The dehumanizing associations mount quickly from dependency to social burden, incapacity to uselessness, and economic unproductivity to parasites and drains on productive taxpayers (Burch 2018).

We cannot discount other sources of ableist fears and exclusionary practices. Encounters with illness and impairment may summon existential fears or disgust because of life's fragility or norms of bodily aesthetics. Primarily, these fears concern not the fragility of biological life, but the threat of social death or diminishment. The supposed traits of a person—independence, self-reliance, productivity, and self-sufficiency—bear fearful complements that underlie both threats to personhood (vulnerability, impairment, immorality) *deficiencies of personhood itself* (dependency, needfulness, burdensome, uselessness). Dehumanization is built into the social conception of impairment.

VIII Stigmatization and Self-Constitution

Our self-conceptions (how we view ourselves as persons) are ongoing projects, projects formed, maintained, and altered through and by our relationships with others in our communities. Being socially created, our identities and agency are constricted by a vast array of social norms, representations, attitudes, practices, institutions, conventions, and expectations, many of them tailored to specific social groups. How others view who we are or ought to be necessarily influences how we view ourselves.[20]

These third-person and first-person viewpoints rarely integrate completely, and this can lead to conflicts that make our identities less stable. We resolve conflicts of this kind in two broad ways, by deferring to others' accounts of who we are or by resisting those accounts.[21] For example, many women feel shame or believe themselves unattractive or failures as women because their lived bodily experiences misalign with internalized cultural ideals of fitness, body shape, and beauty. To allay the dissonance threatening their identities, many attempt to nudge their bodies closer to those ideals with habitual malnourishment, exercise regimens, and cosmetic surgery. Mackenzie (2009) notes that these "experiences of alienation from one's bodily perspective are clearly bound up with oppressive social and cultural practices," and so "that some women … experience difficulty integrating their subjectivity with their lived embodiment … is hardly surprising" (121).

Our relationships can also be ameliorative. Processes of stigmatization imbue human difference with arbitrary negative associations and transform them into "common sense." Individuals who stand against these narratives are presumed wildly eccentric or provocateurs, at best. Without at least minimal uptake from others, our counternarratives succumb to social forces; they succeed through affiliative communities (Lindemann Nelson 2001). People dissatisfied with bodies that fall short (as all bodies do) of cultural ideals may experience a salutary change (identity stabilization), were they to interact with feminists who critique the dominant narratives and offer alternative representations of gender, health, and sexuality or express solidarity and acceptance of counternarratives. So, how identities are constituted and how we resolve differences between our self-understandings and the perspectives of others (from intimates to society at large) raise significant normative issues; balancing competing perspectives on our identities in unjust conditions could be good or bad for us (Baylis 2012).

Chronic illness can globally destabilize self-conceptions. It alters our experience of the world, partly as a reflection of how the world experiences us. Friends and others we encounter begin to treat us differently, and society assigns us a new diminished status. And so, given the above, the quality of our reconstituted self-conceptions and our well-being will largely depend on the extent to which we accept negative social representations of ourselves. And this crucially depends on our relationships and experiences of the affiliative emotions of support, validation, trust, and concern—that is, whether others supply or affirm counternarratives or affirm or amplify harmful social representations.

The more thoroughly we internalize the dominant narratives of our existence as tragic, of our worthlessness, untrustworthiness, and shamefulness, the less likely we will "challenge structural forms of discrimination that block opportunities [we] desire" (Link and Phelan 2001, 375). Thinking oneself a useless drain on society can altogether block previous desires and push from one's thought available opportunities one now considers undeserved.[22] This diffuse loss of potentially meaningful and fulfilling engagement with the world, this change in preferences, cannot be laid on the "misfortune" of illness or impairment. Rather, it results from self-conceptions shaped by, what Thomas (1998) calls, the hostile misfortunes[23] of ableist oppression—selves constituted "so as not to see themselves as full and equal members of society" (365).

People who reject affirmative disability identities follow the biomedical script and individualize their responses to stigmatization. Many adapt to their conditions under the aegis of professionals who unreservedly embrace a biomedical understanding of impairment. These professionals convey to their clients the normalizing imperatives of their services and the cultural devaluation of lives with impairments. Without opposing narratives, even the unspoken promise of normality can powerfully influence the reconstitution of the self. From that perspective, people endorse "overcoming" narratives; they hope only to be fixed with cures, treatments, or new technology. When the promise fades, they resign themselves to stigmatization, thinking it justified, inevitable, and impervious to social activism. Consequently, they recognize ableist injustice and engage in collective political action less often than those with positive disability identities (Nario-Redmond and Oleson 2016). When possible, they conceal their impairments to minimize job discrimination and other consequences of ableism. People who adapt to these orientations report lower self-esteem and life satisfaction and higher levels of anxiety and depression than people with positive disability identities (Nario-Redmond, Noel, and Fern 2013; Bogart 2014, 2015). Associating disability with shame and deviance, they limit their social interactions, further diminishing well-being. Quite often they avoid or outright repudiate association with disability communities. In doing so, they deprive themselves of the affiliative emotions and meliorative counternarratives these communities could offer.

In social environments where ableist ideology is not so stridently affirmed, space opens to nurture resistance to harmful social perspectives and to reconstitute a sense of self with values, beliefs, and preferences that express a positive disability identity. Positive disability identity correlates with high self-esteem, self-worth, and sense of belonging, each considered an essential aspect of a global sense of well-being (Nario-Redmond et al. 2013; Silverman et al. 2015; Bryson and Bogart 2020). Finally, and significantly, for many of us the relationships built and the fights won and lost as members of disability liberatory communities vest our lives with a meaning and purpose matched only by our children.

IX Conclusion

In this chapter I sought to show that chronically ill and disabled people live happy, fulfilling, and meaningful lives, and that the communal creation of meaning through our relationships with others primarily determines those outcomes. Like everyone, we experience obstacles and misfortunes, but the "hostile misfortunes" of ableist oppression pose significantly greater and unending threats.

Many chronic illnesses and impairments occur suddenly; they can dramatically alter our lives and our sense of self. How we, with those around us, reconstruct our identities relies delicately and directly on how those others respond to us in our altered circumstances. This period is critical, for HCPs and family and friends easily reinforce the idea that a person's "true" identity matures to some fixed core of "authentic" inclinations, traits, and desires, and so, though someone lives, "who they really were" is irretrievable.

Given their influence on those recently impaired, HCP must be educated on the value of a positive disability identity, how it promotes a sense of self-worth and self-efficacy and feelings of belonging in the world without self-hatred.

Notes

1 When questions about rationing first arose, a *Washington Post* article quoted George L. Anesi, a critical care specialist at the University of Pennsylvania, as saying "in a public health emergency, you shift from a focus on individual patients to how society as a whole benefits, and that's a big change from usual care." Ariana Eunjung Cha, "Spiking U.S. Coronavirus Cases Could Force Rationing Decisions Similar to Those Made in Italy, China," Washington Post, March 15, 2020, Health, http://www.washingtonpost.com/health/2020/03/15/coronavirus-rationing-us.

2 By "ableist," I mean of or relating to a "network of beliefs, processes and practices that produce a particular kind of self and body (the corporeal standard) that is projected as the perfect, species-typical and therefore essential and fully human. Disability, then, is cast as a diminished state of being human" (Campbell 2001, 44n5).

3 No one, after all, experiences chronic illness at the biochemical level of a "disease state"; we experience it at a higher interpretative and social level.

4 For a couple of recent examples, see Ariana Eunjung Cha, "Quadriplegic Man's Death from COVID-19 Spotlights Questions of Disability, Race, and Family," *Washington Post*, July 5, 2020, https://www.washingtonpost.com/health/2020/07/05/coronavirus-disability-death; and Joseph Shapiro, "As Hospitals Fear Being Overwhelmed by COVID-19, Do the Disabled Get the Same Access," *National Public Radio*, December 14, 2020, https://www.npr.org/2020/12/14/945056176/as-hospitals-fear-being-overwhelmed-by-covid-19-do-the-disabled-get-the-same-acc.

5 I am not supposing that QoL instruments actually capture and quantify any robust understanding of well-being or the worth or quality of a life. What matters are the consistent disparities seen in the assessments.

6 Slevin et al. (1988) conclude that physicians simply "could not adequately measure patients' [QoL]," and that if one needs such measures, then "the patients themselves, and not their doctors or nurses," should supply them (110).

7 In asserting that the primary experience of chronic illness and disability is the experience of ableism and stigmatization, I do not mean to suggest that these experiences are unitary. Ableism is deeply interconnected with racism, misogyny, and the stigmatization of most other socially marginalized groups.

8 Note that Toombs's typical characteristics describe the experiences of illness of any kind and not exclusively chronic or serious illness. She packs, then, rich and variegated experiences with wide-ranging intensities into these abstract features. Acute illness is far less likely to lead to (the need for) adaptation or integration into one's self-conception simply because acute illness resolves relatively quickly and everyone expects such resolution. Consequently, the experiences of illness Toombs characterized may remain the entire course of an acute illness while never indicating one of the principal experiences of chronic illness, social stigmatization.

9 As with any emotion, anxiety can appropriately fit a circumstance or not. When it goes wrong, its benefits invert: instead of focusing us on what matters, anxiety diverts us from it and toward something relatively unimportant; instead of helping us to act as moral agents, it blocks action through indecision.

10 Researchers had previously distinguished fear and anxiety according to whether the threat was clear or indefinite, known or unknown. People could name what they feared and what they could do to protect themselves: that grizzly bear in front of me, play dead; imminent stock market crash, buy gold; and so on. It was thought that anxious people could not so neatly name the sources of their anxiety. But fears can involve uncertainty (it's spring; don't bears eat winter-killed carrion?) and the distinguishing uncertainty in anxiety needn't be *what* the threat is; instead, it may concern when it will occur, how severe its effects, or what to do about it. When it accompanies fear, uncertainty is not fear's *object*; a danger (harm, death, loss) is its object. Uncertainty is the object of anxiety, and this distinguishes it from fear.

11 Drawing on several empirical studies, Kurth (2015, 186) shows that anxiety motivates detailed hypothetical reasoning to help us discover how best to respond to a threat, and it enhances the epistemic virtues of open-mindedness—the anxious person is more receptive to new ideas and less likely to dismiss views that conflict with their own—and of valuing diverse perspectives.

12 Many reactions to illness and impairment occur without our conscious awareness of them. We may, for example, automatically adjust our gait or enlist others muscles to compensate for those weakened long before we realize something is amiss. It was only when someone whom I had not seen in a long time asked me why I was "walking like a duck" that I had any inkling, and that inkling certainly was not of limb-girdle muscular dystrophy.

13 Most drugs that stop attacks work better when taken early in the migraine cycle.

14 For an exemplar, consider the continual anxiety experienced by parents, but particularly by new parents who have not yet adapted to (discovered enough about) their new reality of parenthood.

15 Our selves are, on my view, essentially relational, both in the obvious senses of being thoroughly socially embedded and arising only through interaction with others (Brison 2017), and in the sense that they only exist as part of the evolution of human hypersociality. Thus, the individualistic notion of self-interest connected to prudential well-being is at best a misnomer. Our identities merge with others' and we cannot pry our loved ones' interests from our own, but we have shared interests with countless unknown others whom we depend on to lead the lives we live and others who likewise depend on us (Lindemann 2003). So, I do not intend every "we" and "our" in this paragraph as a mere collection of individuals.

16 HCPs define the "expert patient" not as someone who comes to understand their bodies through their singular ability to experience it, but as those who follow the approved medical scripts for "self-management." Ad-libbing, deciding off-script according to your personal values, labels you a failure, and has been found to lead to further stigmatization (Rosqvist, Katsui, and McLaughlin 2017).

17 The persistent disparities in the quality of healthcare chronically ill and disabled people receive reflect these attitudes. From access to contraception to pap smears and mammograms, disabled people receive deficient services concerning sexuality, childrearing, and reproductive care, indeed preventative care generally (NCD 2012; Andresen et al. 2013). These deficits bear serious consequences. For example, even when controlling for age, cancer stage, education, secondary conditions, and medical insurance status, disabled women with breast cancer are less likely to receive breast-conserving treatment. When they do receive it, their treatments more often go against medical guidelines and exclude radiotherapy; and far more often, the cancer kills them (McCarthy et al. 2006; Groß et al. 2020).

18 Following Link and Phelan (2001), I take stigma to be a social process of devaluation that includes these five components:

> In the first component, people distinguish and label human differences. In the second, dominant cultural beliefs link labeled persons to undesirable characteristics—to negative stereotypes. In the third, labeled persons are placed in distinct categories so as to accomplish some degree of separation of "us" from "them." In the fourth, labeled persons experience status loss and discrimination that lead to unequal outcomes. Finally, stigmatization is entirely contingent on access to social, economic, and political power that allows the identification of differentness, the construction of stereotypes, the separation of labeled persons into distinct categories, and the full execution of disapproval, rejection, exclusion, and discrimination (376).

Note that on this conception, stigma is not an *attribute* of persons. "Spoiled identities" are not discovered, they are created.

19 By global stereotype, I mean a social representation of "being impaired" that holds generally, regardless of the nature of any individual's particular physical or psychological condition.

20 Taking these ideas together, and in contrast to the individualistic *self*, a *person* is an "embodied subject whose identity is constituted in and through one's lived bodily engagement with the world and with others" (Mackenzie 2009, 119).

21 For a fuller and substantially more nuanced account of the complex interplay between first-person and third-person influences on personal identity, see Baylis (2012).

22 Micheline Mason poignantly describes internalized ableism this way: "Internalized oppression is not the cause of our mistreatment, it is the result of our mistreatment…. We harbor inside ourselves the pain and the memories, the fears and the confusions, the negative self-images and the low expectations, turning them into weapons with which to re-injure ourselves, every day of our lives."

23 The "hostile" in Thomas's hostile misfortunes intends to capture both "that the misfortune is owing to agency, and that the agency…is owing to morally objectionable attitudes" (1998, 363) toward members of marginalized social groups.

References

Andresen, Elena M., Jana J. Peterson-Besse, Gloria L. Krahn, Emily S. Walsh, Willi Horner-Johnson, and Lisa I. Iezzoni. 2013. "Pap, Mammography, and Clinical Breast Examination Screening among Women with Disabilities: A Systematic Review." *Women's Health Issues* 23(4): e205–e214. https://doi.org/10.1016/j.whi.2013.04.002.

Angner, Erik, Midge N. Ray, Kenneth G. Saag, and Jeroan J. Allison. 2009. "Health and Happiness among Older Adults." *Journal of Health Psychology* 14(4): 503–512. https://doi.org/10.1177/1359105309103570.

Barata, Anna, Rodrigo Martino, Ignasi Gich, Irene García-Cadenas, Eugenia Abella, Pere Barba, Javier Briones, et al. 2017. "Do Patients and Physicians Agree When They Assess Quality of Life?" *Biology of Blood and Marrow Transplantation* 23(6): 1005–1010. https://doi.org/10.1016/j.bbmt.2017.03.015.

Barlow, David. 2002. *Anxiety and Its Disorders: The Nature and Treatment of Anxiety and Panic.* New York, NY: The Guilford Press.

Baylis, Françoise. 2012. "The Self *in Situ*: A Relational Account of Personal Identity." In *Being Relational: Reflections on Relational Theory and Health Law*, edited by Jennifer Llewellyn and Jocelyn Downie, 109–131. Vancouver: UBC Press.

Bogart, Kathleen R. 2014. "The Role of Disability Self-Concept in Adaptation to Congenital or Acquired Disability." *Rehabilitation Psychology* 59(1): 107–115.

Bogart, Kathleen R. 2015. "Disability Identity Predicts Lower Anxiety and Depression in Multiple Sclerosis." *Rehabilitation Psychology* 60(1): 105–109.

Brison, Susan J. 2017. "Personal Identity and Relational Selves." In *The Routledge Companion to Feminist Philosophy*, edited by Ann Garry, Serene J. Khader, and Alison Stone, 218–230. New York, NY: Routledge, Taylor & Francis Group.

Bryson, Brooke A., and Kathleen R. Bogart. 2020. "Social Support, Stress, and Life Satisfaction among Adults with Rare Diseases." *Health Psychology* 39(10): 912–920. https://doi.org/10.1037/hea0000905.

Burch, Leah. 2018. "'You Are a Parasite on the Productive Classes': Online Disablist Hate Speech in Austere Times." *Disability & Society* 33(3): 392–415. http://doi.org/10.1080/09687599.2017.1411250.

Caillault, Pierre, Marianne Bourdon, Jean-Benoit Hardouin, and Leïla Moret. 2020. "How Do Doctors Perceive and Use Patient Quality of Life? Findings from Focus Group Interviews with Hospital Doctors and General Practitioners." *Quality of Life Research* 29(7): 1895–1901. http://doi.org/10.1007/s11136-020-02451-3.

Campbell, Fiona Kumari. "Inciting Legal Fictions: 'Disability's' Date with Ontology and the Ableist Body of Law," *Griffith Law Review* 42 (2001): 42–62.

Carbonell, Vanessa. 2018. "Sacrifice and Relational Well-Being." *International Journal of Philosophical Studies* 26(3): 335–353. https://doi.org/10.1080/09672559.2018.1489642.

Carel, Havi. 2016. *Phenomenology of Illness*. Oxford, UK: Oxford University Press.

Cassell, Eric J. 2004. *The Nature of Suffering and the Goals of Medicine*, 2nd ed. New York, NY: Oxford University Press.

Dolan, Paul. 1997. "Modelling Valuations for EuroQoL Health States." *Medical Care* 35(11): 1095–1108. https://doi.org/10.1097/00005650-199711000-00002.

Groß, Sophie E., Holger Pfaff, Michael Swora, Lena Ansmann, Ute-Susann Albert, and Anke Groß-Kunkel. 2020. "Health Disparities AMONG Breast Cancer Patients with/without Disabilities in Germany." *Disability and Health Journal* 13(2): 100873. https://doi.org/10.1016/j.dhjo.2019.100873.

Iezzoni, Lisa I., Sowmya R. Rao, Julie Ressalam, Dragana Bolcic-Jankovic, Nicole D. Agaronnik, Karen Donelan, Tara Lagu, and Eric G. Campbell. 2021. "Physicians' Perceptions of People with Disability and Their Health Care." *Health Affairs* 40(2): 297–306. https://doi.org/10.1377/hlthaff.2020.01452.

Krieger, Nancy. 2011. *Epidemiology and the People's Health: Theory and Context*. New York, NY: Oxford University Press.

Kurth, Charlie. 2015. "Moral Anxiety and Moral Agency." Essay. In *Oxford Studies in Normative Ethics* 5, edited by Mark Timmons, 171–195. Oxford, UK: Oxford University Press.

Lee, Keekok. 2012. *The Philosophical Foundations of Modern Medicine: Philosophy, Methodology, Science*. New York, NY: Palgrave Macmillan.

Lindemann, Kate. 2003. "The Ethics of Receiving." *Theoretical Medicine* 24: 501–509.

Lindemann Nelson, Hilde. 2001. *Damaged Identities, Narrative Repair*. Ithaca, NY: Cornell University Press.

Link, Bruce G., and Jo C. Phelan. 2001. "Conceptualizing Stigma." *Annual Review of Sociology* 27(1): 363–385. https://doi.org/10.1146/annurev.soc.27.1.363.

Mackenzie, Catriona. 2009. "Personal Identity, Narrative Integration, and Embodiment." In *Embodiment and Agency*, edited by Sue Campbell, Letitia Meynell, and Susan Sherwin, 100–125. University Park: Pennsylvania State University Press.

McCarthy, Ellen P., Long H. Ngo, Richard G. Roetzheim, Thomas N. Chirikos, Donglin Li, Reed E. Drews, and Lisa I. Iezzoni. 2006. "Disparities in Breast Cancer Treatment and Survival for Women with Disabilities." *Annals of Internal Medicine* 145(9): 637–646. https://doi.org/10.7326/0003-4819-145-9-200611070-00005.

Mello, Michelle M., Govind Persad, and Douglas B. White. 2020. "Respecting Disability Rights - Toward Improved Crisis Standards of Care." *The New England Journal of Medicine* 383(5): e26. http://doi.org/10.1056/NEJMp2011997.

Meyer, Geralyn A. 2002. "The Art of Watching Out: Vigilance in Women Who Have Migraine Headaches." *Qualitative Health Research* 12(9): 1220–1234. https://doi.org/10.1177/1049732302238246.

Nario-Redmond, Michelle R., Jeffery G. Noel, and Emily Fern. 2013. "Redefining Disability, Reimagining the Self: Disability Identification Predicts Self-Esteem and Strategic Responses to Stigma." *Self and Identity* 12(5):468–488. http://doi.org/10.1080/15298868.2012.681118.

Nario-Redmond, Michelle R., and Kathryn C. Oleson. 2016. "Disability Group Identification and Disability-Right Advocacy: Contingencies among Emerging and Other Adults." *Emerging Adulthood* 4(3): 207–218.

National Council on Disability. 2012. *Rocking the Cradle: Ensuring the Rights of Parents with Disabilities and their Children*. Edited by Robyn Powell. National Council on Disability, Washington, DC. Retrieved from http://www.ncd.gov/publications/2012/Sep272012/

Paul, L. A. 2014. *Transformative Experience*. Oxford, UK: Oxford University Press.

Rosqvist, Hanna, Hisayo Katsui, and Janice McLaughlin. 2017. "(Dis)Abling Practices and Theories?: Exploring Chronic Illness in Disability Studies." *Scandinavian Journal of Disability Research* 19(1): 1–6. https://doi.org/10.1080/15017419.2016.1269535.

Rubin, Emily B., Anna E. Buehler, and Scott D. Halpern. 2016. "States Worse than Death among Hospitalized Patients with Serious Illnesses." *JAMA Internal Medicine* 176(10): 1557–1559. https://doi.org/10.1001/jamainternmed.2016.4362.

Scambler, Sasha. 2020. "Long-Term Disabling Conditions and Disability Theory." In *Routledge Handbook of Disability Studies*, edited by Nick Watson and Simo Vehmas, 2nd ed., 172–188. London, UK: Routledge Taylor & Francis Group.

Silverman, Arielle M., Jason D. Gwinn, and Leaf van Boven. 2015. "Stumbling in Their Shoes: Disability Simulations Reduce Judged Capabilities of Disabled People." *Social Psychological and Personality Science* 6(4): 464–471. http://doi.org/10.1177/1948550614559650.

Slevin, M.L., H. Plant, D. Lynch, J. Drinkwater, and W.M. Gregory. 1988. "Who Should Measure Quality of Life, the Doctor or the Patient?" *British Journal of Cancer* 57(1): 109–112. https://doi.org/10.1038/bjc.1988.20.

Srikrishna, Sushma, Dudley Robinson, Linda Cardozo, and Juan Gonzalez. 2009. "Is There a Discrepancy between Patient and Physician Quality of Life Assessment?" *Neurourology and Urodynamics* 28(3): 179–182. https://doi.org/10.1002/nau.20634.

Thomas, Laurence. 1998. "Moral Deference." In *Theorizing Multiculturalism: A Guide to the Current Debate*, edited by Cynthia Willett, 359–381. Massachusetts; Oxford: Blackwell Publishing Ltd.

Toombs, S. Kay. 1992. *The Meaning of Illness: The Phenomenological Account of the Different Perspectives of Physician and Patient*. Dordrecht: Kluwer Academic.

Van den Berg, J. H. 1966. *The Psychology of the Sickbed*. Pittsburgh, PA: Duquesne University Press.

Further Reading

Barnes, Elizabeth. 2018. *The Minority Body: A Theory of Disability*. Oxford, UK: Oxford University Press.

Lambert, Enoch, and John Schwenkler, eds. 2020. *Becoming Someone New: Essays on Transformative Experience, Choice, and Change*. Oxford, UK: Oxford University Press.

Kukla, Elliot. "In My Chronic Illness, I Found a Deeper Meaning," *New York Times*, January 10, 2018, Disability, https://www.nytimes.com/2018/01/10/opinion/in-my-chronic-illness-i-found-a-deeper-meaning.html.

Nario-Redmond, Michelle R. 2020. *Ableism: The Causes and Consequences of Disability Prejudice*. Hoboken, NJ: Wiley Blackwell.

16

DISABILITY AND AGE STUDIES

Obstacles and Opportunities

Erin Gentry Lamb

Stigmatization of disabled people has influenced many older people to resist identifying as disabled despite facing some of the same social injustices and physical limitations. These groups would benefit from recognizing this overlap and working in solidarity. This chapter creatively combines insights from disability and age studies to examine the concept of "successful aging," assumptions about quality of life, and advance directives for people diagnosed with dementia. It offers a framework to better understand and reflect upon late life and the ability transitions it invariably occasions.

Disability and old age are categories that often overlap. According to the 2019 Disability Statistics Annual Report, in the USA more than a third (35.2%) of people 65 and older have a disability. Of the total population of people with disabilities, 41.4% are 65 and older (Kraus et al. 2018, 2). It makes sense that what affects one category of people may well be relevant to the other—that the prejudices of ableism and ageism may often intersect in people's experiences—and that scholars would think about age and disability alongside one another.

Age studies is an interdisciplinary field similar to disability studies in that they both "explore the meanings and consequences of embodied differences—of age and/or disability—within society, with the goal of understanding the mechanisms of and reducing experiences of ageism and/or ableism" (Lamb and Garden 2019, 148). Similar to the social model of disability, age studies considers how the challenges experienced by old people are not simply rooted in inevitable bodily decline (a medical model of aging) but are significantly influenced by social, cultural, political, historical, and economic factors. Thus, age studies tends to position itself as distinct from, and sometimes critical of, the fields that explore the biological mechanisms of aging, seek policy solutions for a growing aged population, and treat age-related medical concerns—namely, gerontology and geriatrics. This relationship may be akin to the way disability studies has at times been critical of rehabilitation medicine as a field looking to ameliorate disability (understood on a medical model of disability) instead of focusing on improving the environment in which disabled people live.

Despite a similar impulse to critically investigate the social, cultural, and other contexts that shape the experiences of disabled and old people, there has been surprisingly little dialogue between the fields of disability studies and age studies. Similarly, even though bioethics regularly

DOI: 10.4324/9781003289487-22

addresses disability and later life, the field has rarely done so through a critical disability studies lens, and the critical insights of age studies have had even less influence. This chapter illustrates the value of bringing insights from age studies into conversation with both disability studies and bioethics.

I Disability and Aging as Embodied Conditions and Identity Categories

If you live a long, full life, then the chances are excellent that you will experience both disability (whether temporary or permanent) and old age. Indeed, aging into old age is almost invariably accompanied by functional limitations that are physical and sometimes mental, and for many people, old age is conceptually synonymous with disability. As Susan Wendell (1996) has described it, "aging is a disabling condition" not only because of the increased likelihood of experiencing disability but also because of the reduction of opportunities faced by those who are perceived as old (18).

Just as some people try to hide their disability or refuse to identify as disabled, so, too, many aging people resist identifying as old. Both of these identities carry stigma and the likelihood that others will treat one differently, whether making assumptions about one's capacities, offering unwanted help, or discounting one's value. The multi-billion-dollar consumer market in anti-aging products—from hair dye to Botox to a vast supplement industry—offers clear evidence of widespread resistance in our culture to being perceived as old. As more evidence, consider how many euphemisms we use in order to avoid calling someone old: senior, senior citizen, elder, elderly, aged, older, well-seasoned, experienced, time-tested, mature.

People who have multiple marginalized identities due to age, ability, race, sex, etc., may experience unique modes of prejudice and discrimination above and beyond those that come with any one marginalized identity. It makes sense that someone who identifies as disabled may resist the additional stigma that accompanies identifying as old, or vice versa. For example, with people who are aging, it is most often disability, not appearance, that moves them into the ghettoized category of old. We can see this conflation of disability and old age in the common division within gerontology and age studies between a "third age" of "fit, healthy, and productive later life" and a "fourth age" that is "dogged by ill-health, incapacity, and neediness" (Gilleard and Higgs 2014). For people with disabilities, being additionally perceived as old may make others more likely to presume disability is a function of aging and thus less likely to see any need to accommodate for disability. For example, if a 70-year-old has a stroke that limits right-leg mobility, her family may take this change as a sign it is time for her to stop driving. However, if a 40-year-old member of that same family lost his right foot to diabetes, he might be encouraged to search out adaptive technologies to enable him to continue driving. "The impairments acquired in old age are seldom considered a disability in the cultural sense," and thus the disabled old may not benefit from the positive aspects of disability identity (such as belonging or activism) that younger disabled people might experience (Ljuslinder, Ellis, and Vikström 2020, 36).

The conflation of disability and old age and the fact that disabled and old are both stigmatized identities have been significant contributing factors to the lack of exchange between disability studies and age studies. The frequent inability to distinguish the consequences of old age from the consequences of disability, and thus to make claims about aging or disability in isolation, has kept age studies and disability studies scholars from seeing how fertile the intersection of these areas of study could be (Marshall 2014, 23). Similarly, efforts within each of these fields to provide more positivity about these terms—disabled and old—have led proponents to distance their chosen term from the added stigma of the other, effectively reinforcing the distinctions between them. It is easier to illustrate the social model of disability with a young wheelchair user than an old one who cannot wheel herself. Similarly, convincing someone not to fear growing old becomes simpler when the vision of old age features gray hair and wrinkles but no walkers or wheelchairs.

In addition, focusing on disability in old age may double the emphasis on the medical model—i.e., reinforce the impulse to see the "problem" of disability/old age as an individual's bodily difference and to see the "solution" as medically curing or managing that bodily difference (Lamb 2016, 316). This emphasis makes it more difficult to refocus attention on the sociocultural context of disability and age. One of the reigning concepts in gerontology—"successful aging"—can offer us a helpful illustration of the need for multiple critical perspectives at the intersection of disability, aging, and bioethics.

II "Successful Aging:" Disability Studies and Age Studies in Conversation

The model of "successful aging" was proposed by gerontologists John W. Rowe and Robert L. Kahn in the late 1980s, fully laid out in their 1998 book *Successful Aging*, and remains influential within and beyond gerontology today. Rowe and Kahn aimed to identify the factors that lead some people to an ideal version of aging marked by mental and physical vitality, as opposed to "usual aging" which involves some inevitable decline. Their model defines successful aging by three measures: avoiding disease and disability, maintaining high physical and cognitive functioning, and remaining actively engaged in life (Rowe and Kahn 1998). This model leaves no room for disabled people to grow old successfully, or old people who have disabilities to be seen as having aged successfully. The model's popularity suggests that popular and professional visions of unsuccessful aging are shaped by negative associations accompanying disability: a perceived lower quality of life, lack of independence, narrowing of social life, and vulnerability.

Despite its popularity, critiques of the theory of successful aging have come from within gerontology, age studies, and related fields. In the introduction to *Successful Aging as a Contemporary Obsession*, Lamb, Robbins-Ruszkowski, and Corwin (2017) summarize many of these critiques. The theory is ageist, they contend, "growing out of a deep cultural discomfort with what could (or should?) be regarded as the normal human conditions of frailty, (inter)dependence, vulnerability, and transience" (13). It overlooks social inequalities and their impact on health, assuming people are already healthy and must simply maintain that health as they age (15). It can also perpetuate gender stereotypes and "reinforce white, middle-class heterosexual norms of male performance and female beauty" (16). Additionally, it is ethnocentric, based on USA values, and does not hold up cross-culturally (16). Finally, it is uninformed by older people themselves, whose experiences might suggest different measures of success (16).

Some of the most interesting critiques of successful aging, however, have come out of disability studies and make the convincing case that the model of successful aging is both ageist and ableist. Hailee Gibbons (2016) focuses on how successful aging discourse falsely presents being old and being disabled as choices—the result of an individual's lifelong behaviors—and in so doing it creates a social mandate for people to remain both able-bodied and youthful into old age that she calls "compulsory youthfulness." She argues that rather than redefining old age in terms of health and functionality as was the goal of the theory, successful aging has paradoxically further entangled old age and disability as they form the joint antithesis of what is defined as successful aging. Joel Michael Reynolds (2018) similarly argues that one's capacity to age "successfully" is not an issue of individual effort or will alone, but rather depends on the individual's relationships to their environment:

If one lives in a just, caring society and in the presence of caring providers, family members, and friends, the meaning and lived experiences of these transitions [into different stages of age and ability] will more reliably tend toward the positive.

(S35)

Our focus, both Gibbons and Reynolds suggest, needs to be on social and environmental forces rather than on individual bodies, not just because these external factors shape one's capacity to age "successfully," but because they shape what we perceive as success in the first place. For example, we need to engage with the norms that make dependency so feared and keep us from recognizing the ways we are already interdependent; change the policies that limit our options when a loved one needs long-term care; and revise the popular rhetoric that positions old people as a "silver tsunami" ready to destroy the economy. Reynolds argues that "ableism is at the core of ageism" and that we "cannot conceptualize ageism without ableism" (S33). Gibbons instead cautions us against redefining ageism as ableism as it "prevents scholars from exploring how ageism and ableism intersect" (4). Regardless, both of these arguments reveal how we cannot think critically about aging into old age without also thinking critically about disability, and how age studies and disability studies have critical insights to offer one another.

Disability studies has already benefited from one of gerontology's key concepts, the life course perspective, "which considers how the culmination and timing of a person's experiences and connections with others…affect one's current situation, as well as the roles of age, period, and cohort effects in shaping expectations and opportunities" (de Medeiros 2018, S11). To date, however, life course perspectives on disability have primarily focused on the aging experiences of disabled people (Jeppsson-Grassman and Whitaker 2013; Priestley 2003) or on rethinking the normative life course itself (Ljuslinder, Ellis, and Vikström 2020). Scholars have not focused on how life course norms lead us to expect impairment in old age in ways not culturally labeled as disability and the resulting exclusion of old people from identifying with the disability community. Gilleard and Higgs (2013) raise a similar concern that "idealized models of disability stress both its lifelong nature and its unchanging presence," and call for a less essentialist narrative within disability studies that can encompass disability's changing and unstable nature—like that of aging—and embrace old people as a valued part of the disability community (81, 85). A life course perspective can also expand disability justice efforts by exploring how disadvantage may accumulate over the life course, affecting any given individual's likelihood and severity of, and capacity and resources to adjust to, disability experiences in old age.

Age studies can benefit from disability studies' strong social model approach; as Sally Chivers (2011) writes, "there is no need to chart a new way of thinking about aging as experienced in relation to social and cultural environments because disability studies already clearly, cogently, and consistently articulates that way of thinking" (22). Gerontology in particular, which insistently measures disabilities in old age via numerous indices—such as Activities of Daily Living (ADLs) and Instrumental Activities of Daily Living (IADLs)—can follow disability studies in exploring "what disability may mean to older people themselves, what struggles they conceal and what shame they hide" (Gilleard and Higgs 2013, 82). Age studies can also benefit from more recent critical models of disability that challenge the social model's limitations in accounting for pain and chronic illness, interdependence, care and caregiving—aspects of disability embodiment central to the lived experiences of many disabled old (Reynolds 2018, S32).

Both fields can helpfully converge on discussions of universal design and age-friendly communities, initiatives that create environments accessible by all people, including those who are disabled and old. Disability studies and age studies might also collaborate on rich theorizations of concepts like frailty and precarity. Frailty, which describes individual bodily vulnerability, can be used to interrogate how bodily and moral infirmity are often conflated (for example, how disability and old age are often used to mark characters as weak or evil). Precarity draws our attention to insecurity and risk in relation to economic, social, and political conditions; centering this term in analyses of disability and aging could help both fields attend to the cumulative effects of structural inequalities. The intersection of disability and aging is only one of many intersections providing important critiques of successful aging: critical race studies, women's and gender studies, trans studies, and more could all offer helpful critiques of the blind spots of a theory like successful aging. Exploring the interactions of ageism and ableism will facilitate more complex intersectional analyses.

III Disability Studies, Age Studies, and Bioethics

The field of bioethics has attended relatively little to critical disability studies perspectives, but it has attended even less to age studies. Berlinger and Solomon (2018) suggest that bioethics' attention to aging has not often looked beyond concerns with "the end of life and the care of patients with age-associated, progressive conditions" (S2). Even within these previously studied areas, however, age studies and disability studies insights can offer valuable new directions; two examples we will look more closely at include making judgments about quality of life and thinking critically about advance directives for dementia.

IV Assessing Quality of Life and Its Consequences

Quality of life (QOL) is a regular focal point of debate within bioethics and another area of similarity and overlapping interests between disability and old age. Writing in relation to disability, Ron Amundson (2005) critiques what he calls "the Standard View" of QOL—a view widely held by nondisabled people in which disabilities strongly and negatively impact the QOL of individuals who have them. Sometimes referred to as the "disability paradox," research has shown that many disabled people, including those with serious disabilities, self-report that their QOL is good or excellent, even though others might judge these individuals' daily lives as undesirable (Albrecht and Devlieger 1999). Amundson argues that the factors that cause low QOL among disabled people are the same as those for nondisabled people. Rather than stemming from impairment or abnormality, they are the result of social arrangements and environmental accessibility that can disadvantage old people, racial and sexual minorities, and other marginalized groups as easily as disabled people (Amundson 2005).

A similar "aging paradox" exists where emotional well-being improves from early adulthood to old age, even accounting for declines in physical health (Carstensen et al. 2000). Happiness and life satisfaction across the life span have been found—across 132 developed and developing countries—to map into a U-shaped curve, where people are happiest as children and older adults and experience a mid-life low (Blanchflower 2020; Carstensen et al. 2000).

Examining the intersection of these disability and aging paradoxes reveals important areas for further inquiry. Subjective well-being in old age remains high even in the presence of physical constraints and cognitive restrictions—i.e., physical and cognitive disabilities (Jopp and Rott 2006). However, these data generally do not hold true for individuals with dementia (Cooper, Bebbington, and Livingston 2011). Looking more closely at the relationship between cognitive decline and happiness, research affirms Amundson's point: supportive social networks and the ability to participate are the environmental factors that strongly affect life satisfaction, and such networks and access are often endangered by diagnoses of dementia (Cooper, Bebbington, and Livingston 2011; Shakespeare 2014). Many initiatives within age studies accordingly focus on addressing isolation and segregation among old people, particularly those with dementia (see, for example, Basting 2020).

Assessments of QOL have clear bioethical significance. Proxy decision-makers are "generally poor judges of a person's QoL [sic]," usually underestimating QOL across a range of contexts, although family members tend to offer ratings more similar to the person's self-reported QOL than do health-care professionals (Crocker, Smith, and Skevington 2015). However, family caregivers may also make decisions based on their own stakes in caregiving, as caring for dependent family members may be a financial, psychological, or other strain; countries like the USA which offer few if any federal resources to family caregivers may exacerbate such strains. In situations where health-care practitioners or family members must make life-altering or end-of-life decisions for an old and/or disabled patient, there may be a tendency to base such decisions on perceived QOL and thus to underestimate how that same individual would report their QOL.

The early months of the COVID-19 pandemic were full of examples of such flawed decision-making. In Italy and Spain, for example, when hospitals were overwhelmed and ventilators were in short supply, news stories reported on elderly patients being "left to die": while these reports were sensationalized, they speak to triage decisions made on the basis of greatest likelihood to live longest upon survival, an inherently ageist measure (Di Blasi 2020; Minder and Peltier 2020). Many hospital systems in the USA, too, have ventilator triage policies where age is a criterion (Antommaria et al. 2020), though such ageism is far subtler than the hashtag #BoomerRemover that trended in the early months of the pandemic (Kendall-Taylor, Neumann, and Schoen 2020). Disabled people also feared and experienced prejudice in resource allocation during the pandemic, as many allocation policies specifically disadvantage those with disabilities, and some people with disabilities had "Do Not Resuscitate" orders placed on their records without their consent (Alexiou 2020; Reynolds, Guidry-Grimes, and Savin 2020). The case of Michael Hickson, a physically and cognitively disabled Black Texan who was removed from ventilator support against his wife's wishes, drew attention directly to health-care providers' underestimation of QOL when a doctor told Michael's wife: "So as of right now, his QOL – he doesn't have much of one" (Cha 2020). Cases like these make clear that bioethicists need to develop more robust ways to account for ageist and ableist biases in estimating QOL and making allocation and care decisions.

V Dementia and Advance Directives

Another debate in bioethics in recent years has involved honoring advance directives that speak specifically to dementia, one of the most culturally feared intersections of old age and disability. For example, *The Washington Post* in 2020 featured a story on Susan Saran—a 64-year-old diagnosed seven years earlier with frontotemporal dementia who was told by her continuing care retirement community that they could not honor her wishes to withhold hand-feeding and fluids at the end of life (Aleccia 2020). An advance directive can allow an individual to set limits on the life-sustaining care received when they no longer have capacity to make their own medical decisions, including refusing treatments such as antibiotics and withholding nutrition and hydration at the end of life. In the case of dementia, determinations of both the end of life and capacity are complicated by the typically slow progression of dementia, which may span many years, and the accompanying uncertainties about the point at which an individual with dementia no longer has decision-making capacity. Saran, the article suggests, clearly demonstrates capacity seven years into her diagnosis (Aleccia 2020). Saran could choose to voluntarily stop eating and drinking now when she is still competent to make that choice but might lose out on several more good years of life.

Honoring an advance directive to withhold food and water by mouth in the case of advanced dementia becomes particularly tricky when the no-longer-competent individual still expresses a desire (or willingness) to eat and drink. The difficulty, often described as the "then-self/now-self" problem, is whether you honor the wishes of the person who wrote the advance directive, or interpret the wishes of the person who demonstrates continued interest in eating and drinking. The dominant view in such a situation is that the patient should always be offered water and food, residential care facilities' responsibility is to the now-self, and in fact many states prohibit the withdrawal of assisted feeding which they view as "comfort care" (Aleccia 2020; Wright, Jaggard, and Holahan 2019). Others, however, have pushed to expand the circumstances under which caregivers might follow an advance directive and stop offering food and water. For example, Menzel and Chandler-Cramer (2014) argue, following Dworkin, that it is possible to weigh the interests of the then-self (which continue to exist even after one stops experiencing them) against those of the now-self. Once the now-self shows little capacity to subjectively value their own survival (i.e., once "the self has withered"), the autonomy of the then-self should be honored (30). Across all of these arguments is the understanding that a diminishment in personhood takes place over time with dementia, and it is this perception that is particularly relevant to disability studies and age studies perspectives.

A disability studies critique of this issue raises two key related concerns: when making advance directive decisions, people are unable to imagine that they might live contentedly with dementia; however, a relatively happy life with dementia might be possible. Dresser (2018) argues that just as people are poorly able to judge the QOL of those with disability, the "tragedy discourse" that typifies public references to dementia means that people are likely unable to imagine that their future demented self might experience a good QOL (27). She refocuses our attention on the environment, arguing that "people with dementia *who live in supportive settings* rarely exhibit clear and persistent signs of suffering. Even people who cannot identify family and friends by name remain able to experience love and attachment to others" (my emphasis, 26). Menzel and Chandler-Cramer (2014) suggest that people may know they could end up being one of the "happy demented" yet still want to hasten their end because they do not want to be an emotional and financial burden (24). Combining Dresser's critique with an age studies perspective directs us to focus on the reasons that dementia is perceived as a burden on loved ones: long-term care in the USA is funded by Medicaid, not Medicare, requiring the individual to spend down their assets until they qualify for Medicaid, or leaving the financial burden of care on their families.

Similar to disability studies, age studies analyses are also invested in how representations fuel our fears of dementia, as well as in how we might think more expansively about the "self" in dementia. In considering portrayals of dementia, Basting (2009) makes the point that while the story of old age and memory loss is invariably told as a tragedy, popular culture is far more positive when portraying non-old characters with memory loss. For example, Dory in Finding Nemo and Drew Barrymore's character in 50 First Dates are still able to experience friendship, love, and an envisioned future. Through exploring the complexity of memory, she argues that "the formulas 'self = memory' and 'loss of memory = loss of self,' which inform so much of our thinking about people with dementia and determine how we care for them," are "terribly simplistic" (24). Viewing dementia as the withering of the self is, according to Basting, a fundamental misunderstanding because the "self" is relational; our response to memory loss should focus upon "creating a net of social memory around a person whose individual control of memory is compromised" (161). She urges us to insist on more complex stories of dementia that not only speak to grief and loss but also reveal that "one's 'self' persists until the end, that growth and learning are possible, that social memory remains when individual memory falters, and that relationships with a person with dementia are reciprocal" (Basting 2009, 156; see Gullette 2011 for an example of a more complex story of dementia). Age studies offers strategies for helping caregivers, family members, and people with dementia themselves think differently about the self in dementia. Basting's life's work, along with the work of many other artists and scholars, has focused on ways the creative arts can be used to connect with, to recognize the continuing personhood of, and to bring light and joy to those with dementia and the people who care for them.

These perspectives do not point us to a wholesale rejection of dementia directives. Rather, they suggest that the purview of bioethics in relation to advance directives for dementia needs to be far larger than just whether or not, or at what point, it is acceptable to withhold food and water. Bioethics needs to acknowledge and address the cultural conditions that have made these directives so appealing to so many people, and they need to examine critically the policies and environments which shape people's experiences of dementia. Disability studies and age studies both, and especially together, insist that bioethics expand its purview in order to adjudicate more equitably.

VI The Need for Intersectional Analysis

Given the commonality of experiencing disability in old age and given how central QOL and end of life have been to bioethics, it is evident that bioethics as a field needs to attend not only to disability studies' perspectives but also to age studies' perspectives and to the areas of fertile overlap between these approaches. A critical move that needs to be made across all three fields is greater attention to intersectional analysis, or the way multiple aspects of a person's social and political identities might combine, creating unique experiences of discrimination. What does "successful aging" mean to Black women, trans folx, or people with lifelong mobility impairments? How does one's race or socioeconomic status affect the QOL judgments others make when one is old and/or disabled? How does one's gender, race, or class affect the likelihood that their advance directive for dementia will be honored? Attending to the overlap of disability studies and age studies is an important step closer to such needed intersectional analysis.

References

Albrecht, Gary L. and Patrick J. Devlieger. 1999. "The Disability Paradox: High Quality of Life against All Odds." *Social Science and Medicine* 48, no. 8: 977–988.

Aleccia, JoNel. 2020. "Diagnosed with Dementia, She Documented Her Wishes for the End. Then Her Retirement Home Said No." *The Washington Post*, January 18, 2020. https://www.washington-post.com/health/diagnosed-with-dementia-she-documented-her-wishes-for-the-end-then-her-retirement-home-said-no/2020/01/17/cf63eeaa-3189-11ea-9313-6cba89b1b9fb_story.html.

Alexiou, Gus. 2020. "Doctors Issuing Unlawful 'Do Not Resuscitate' Orders for Disabled Covid Patients 'Outrageous.'" *Forbes*, June 23, 2020. https://www.forbes.com/sites/gusalexiou/2020/06/23/unlawful-do-not-resuscitate-orders-for-disabled-covid-patients-outrageous/.

Amundson, Ron. 2005. "Disability, Ideology, and Quality of Life: A Bias in Biomedical Ethics." In *Quality of Life and Human Difference*, edited by David Wasserman, Jerome Bickenbach, and Robert Wachbroit, 101–124. Cambridge: Cambridge University Press.

Antommaria, Armand H. Matheny., Tyler S. Gibb, Amy L. McGuire, Paul Root Wolpe, Matthew K. Wynia, Megan K. Applewhite, and Arthur Caplan et al., for a Task Force of the Association of Bioethics Program Directors. 2020. "Ventilator Triage Policies During the COVID-19 Pandemic at U.S. Hospitals Associated With Members of the Association of Bioethics Program Directors." *Annals of Internal Medicine*, M20–1738. Advance online publication. https://doi.org/10.7326/M20-1738.

Basting, Anne Davis. 2009. *Forget Memory: Creating Better Lives for People with Dementia*. Baltimore, MD: Johns Hopkins University Press.

Basting, Anne. 2020. *Creative Care: A Revolutionary Approach to Dementia and Elder Care*. New York: HarperOne.

Berlinger, Nancy and Mildred Z. Solomon. 2018. "Becoming Good Citizens of Aging Societies." *The Hastings Center Report* 48, no. 5 (S3): S2–S9.

Blanchflower, David G. 2020. "Is Happiness U-shaped Everywhere? Age and Subjective Well-being in 132 Countries." NBER Working Papers 26641, National Bureau of Economic Research, Inc. 1–63.

Carstensen, Laura L., Monisha Pasupathi, Ulrich Mayr, and John R. Nesselroade. 2000. "Emotional Experience in Everyday Life across the Adult Life Span." *Journal of Personality and Social Psychology* 79, no. 4: 644–655.

Cha, Ariana Eunjung. 2020. "Quadriplegic Man's Death from Covid-19 Spotlights Questions of Disability, Race and Family." *The Washington Post*, July 5, 2020. https://www.washingtonpost.com/health/2020/07/05/coronavirus-disability-death/.

Chivers, Sally. 2011. *The Silvering Screen: Old Age and Disability in Cinema*. Toronto: University of Toronto Press.

Cooper, Claudia, Paul Bebbington, and Gill Livingston. 2011. "Cognitive Impairment and Happiness in Old People in Low and Middle Income Countries: Results from the 10/66 Study." *Journal of Affective Disorders*, 130, no. 1–2: 198–204. https://doi.org/10.1016/j.jad.2010.09.017.

Crocker, Thomas F., Jaime K. Smith, and Suzanne M. Skevington. 2015. "Family and Professionals Underestimate Quality of Life across Diverse Cultures and Health Conditions: Systematic Review." *Journal of Clinical Epidemiology* 68, no. 5:584–595. https://doi.org/10.1016/j.jclinepi.2014.12.007.

de Medeiros, Kate. 2018. "What Can Thinking Like a Gerontologist Bring to Bioethics?" *What Makes a Good Life in Late Life? Citizenship and Justice in Aging Societies*, special report, *Hastings Center Report* 48, no. 5: S10–S14. https://doi.org/10.1002/hast.906.

Di Blasi, Erica. 2020. "Italians Over 80 'Will Be Left to Die' as Country Overwhelmed by Coronavirus." *The Telegraph (UK)*, March 14, 2020. https://www.telegraph.co.uk/news/2020/03/14/italians-80-will-left-die-country-overwhelmed-coronavirus/.

Dresser, Rebecca. 2018. "Advance Directives and Discrimination against People with Dementia." *The Hastings Center Report* 48, no. 4: 26–27. https://doi.org/10.1002/hast.867.

Gibbons, Hailee M. 2016. "Compulsory Youthfulness: Intersections of Ableism and Ageism in 'Successful Aging' Discourses." *Review of Disability Studies: An International Journal* 12, no. 2&3: 70–88.

Gilleard, Chris and Paul Higgs. 2013. *Ageing, Corporeality and Embodiment*. London: Anthem.

Gilleard, Chris and Paul Higgs. 2014. "Third and Fourth Ages." In *The Wiley Blackwell Encyclopedia of Health, Illness, Behavior, and Society*, First Edition, edited by William C. Cockerham, Robert Dingwall, and Stella R. Quah, 1–7. Hoboken, NJ: John Wiley & Sons, Ltd.

Jeppsson-Grassman, Eva and Anna Whitaker. 2013. *Ageing with Disability: A Lifecourse Perspective*. Chicago: University of Chicago Press.

Jopp, Daniela, and Christoph Rott. 2006. "Adaptation in Very Old Age: Exploring the Role of Resources, Beliefs, and Attitudes for Centenarians' Happiness." *Psychology and Aging* 21, no. 2: 266–280. https://doi.org/10.1037/0882-7974.21.2.266

Kendall-Taylor, Nat, Aly Neumann, and Julie Schoen. 2020. "Advocating for Age in an Age of Uncertainty." *Stanford Social Innovation Review*, May 28, 2020. https://ssir.org/articles/entry/advocating_for_age_in_an_age_of_uncertainty.

Kraus, L., Lauer, E., Coleman, R., and Houtenville, A. 2018. *2017 Disability Statistics Annual Report*. Durham, NH: University of New Hampshire.

Lamb, Erin Gentry. (August 11, 2016). "Age and/as Disability: A Call for Conversation (Forum Introduction)," *Age Culture Humanities*, no. 1. https://ageculturehumanities.org/WP/age-andas-disability-a-call-for-conversation-forum-introduction/;

Lamb, Erin Gentry and Rebecca Garden. 2019. "Age Studies and Disability Studies." In *Research Methods in Health Humanities*, edited by Craig Klugman and Erin Gentry Lamb, 148–164. Oxford University Press, 2019.

Lamb, Sarah, Jessica Robbins-Ruszkowski, and Anna I. Corwin. 2017. "Introduction: Successful Aging as a Twenty-first-Century Obsession." In *Successful Aging as a Contemporary Obsession: Global Perspectives*, edited by Sarah Lamb, 1–23. New Brunswick, NJ: Rutgers University Press.

Ljuslinder, Karin, Katie Ellis, and Lotta Vikström. 2020. "Cripping Time – Understanding the Life Course through the Lens of Ableism." *Scandinavian Journal of Disability Research*, 22, no. 1: 35–38. http://doi.org/10.16993/sjdr.710.

Marshall, Leni. 2014. "Ageility Studies: The Interplay of Critical Approaches in Age Studies and Disability Studies." In *Alive and Kicking at All Ages: Health, Life Expectancy, and Life Course Identity*, edited by Ulla Kriebernegg, Roberta Maierhofer, and Barbara Ratzenböck, 21–40. Bielefeld, Germany: Transcript Verlag.

Menzel, Paul T. and M. Colette Chandler-Cramer. 2014. "Advance Directives, Dementia, and Withholding Food and Water by Mouth." *The Hastings Center Report* 44, no. 3 (May-June 2014): 23–37.

Minder, Raphael and Elian Peltier. 2020. "A Deluged System Leaves Some Elderly to Die, Rocking Spain's Self-Image." *The New York Times*, March 25, 2020. https://www.nytimes.com/2020/03/25/world/europe/Spain-coronavirus-nursing-homes.html.

Priestley, Mark. 2003. *Disability: A Life Course Approach*. Cambridge: Polity Press.

Reynolds, Joel Michael. 2018. "The Extended Body: On Aging, Disability, and Well-Being." *The Hastings Center Report* 48, no. 5 (S3): S31–S36.

Reynolds, Joel Michael, Laura Guidry-Grimes, and Katie Savin. 2020. "Against Personal Ventilator Reallocation." *Cambridge Quarterly of Healthcare Ethics* Oct.: 1–13. https://doi.org/10.1017/S0963180120001103.

Rowe, John W., and Kahn, Robert L. 1998. *Successful Aging*. New York: Pantheon Books.

Shakespeare, Tom. 2014. "A Point of View: Happiness and Disability." *BBC*, June 1, 2014. https://www.bbc.com/news/magazine-27554754.

Wendell, Susan. 1996. *The Rejected Body: Feminist Philosophical Reflections on Disability*. New York: Routledge.

Wright, James L., Peter M. Jaggard, and Timothy Holahan. 2019. "Stopping Eating and Drinking by Advance Directives (SED by AD) in Assisted Living and Nursing Homes." *Journal of the American Medical Directors Association* 20, no. 11: 1362–1366. https://doi.org/10.1016/J.JAMDA.2019.07.026.

Further Reading

Basting, Anne. 2020. *Creative Care: A Revolutionary Approach to Dementia and Elder Care*. New York: Harper One.

Berlinger, Nancy, Kate de Medeiros, and Mildred Z. Solomon, eds. 2018. *What Makes a Good Life in Late Life? Citizenship and Justice in Aging Societies*, special report, *Hastings Center Report* 48 no. 5.

Gilleard, Chris and Paul Higgs. 2013. "Disability, Ageing and Identity." Chapter Five in *Ageing, Corporeality and Embodiment*. London: Anthem, 69–86.

Lamb, Erin Gentry and Rebecca Garden. 2019. "Age Studies and Disability Studies." In *Research Methods in Health Humanities*, edited by Craig Klugman and Erin Gentry Lamb, 148–164. Oxford University Press, 2019.

Marshall, Leni. 2014. "Ageility Studies: The Interplay of Critical Approaches in Age Studies and Disability Studies." In *Alive and Kicking at All Ages: Health, Life Expectancy, and Life Course Identity*, edited by Ulla Kriebernegg, Roberta Maierhofer, and Barbara Ratzenböck, 21–40. Bielefeld, Germany: Transcript Verlag.

Westwood, Sue, ed. 2019. "Part IV: Disabilities, Long-Term Conditions and Care." *Ageing, Diversity and Equality: Social Justice Perspectives (Open Access)*. Routledge: 223–290.

PART VI

Issues at the Edge and End of Life

17

DEATH, PANDEMIC, AND INTERSECTIONALITY

What the Failures in an End-of-Life Case Can Teach about Structural Justice and COVID-19[1]

Yolonda Wilson

In December 2013, Jahi McMath was declared dead by neurological criteria, commonly referred to as "brain death," after suffering complications from a tonsillectomy performed at Children's Hospital and Research Center in Oakland, CA.[2] McMath's family refused to accept the determination of death by neurological criteria and eventually transferred her to a care facility in New Jersey, the only US state that requires accommodation for rejection on religious grounds of determinations of brain death.[3] McMath experienced cessation of cardiopulmonary function, or cardiopulmonary death, in June of 2018. The circumstances of McMath's death—and her two separate death certificates issued in two different states—prompted much discussion about the nature of brain death: what is the hospital's obligation to families who do not accept determinations of death by neurological criteria for death (brain death)? How well is brain death understood among laypersons? How should medical provide proper space for religious views in decision-making about death and dying in healthcare environments? In this chapter I will offer an overview of Jahi McMath's case, situating the injustice she and her family experienced in the context of historic and continuing racial injustice. I will argue that bioethicists, in particular, should bring this understanding to clinical encounters. I will then highlight important lessons from the McMath tragedy and show how those lessons could apply to the current COVID-19 pandemic. Finally, I will argue that the bioethical principle of nonmaleficence, or doing no harm, is an important guiding principle for communities, especially those communities that have experienced injustice.

However, what is most relevant for this chapter is the role that race played, at least from the family's perspective, in the treatment of McMath and in subsequent discussions about her. I add "from the family's perspective" caveat not because I doubt the veracity of McMath's family's reports, but because the hospital denies that race played any role in McMath's care and consequently, they dispute the family's assessment of the events that unfolded at Children's. Indeed, I argue that race and racism did factor into the circumstances of McMath's case, and that bioethicists would ignore this reality at our peril. It is also relevant that McMath suffered from pediatric obstructive sleep apnea and that she was "obese," as the tonsillectomy was supposed to help alleviate some of the symptoms of her sleep apnea.[4]

DOI: 10.4324/9781003289487-24

Approximately an hour after her surgery, Jahi McMath began experiencing complications—specifically excessive bleeding. Her family alerted the nurses and were told that postoperative bleeding was to be expected. McMath's grandmother, herself a nurse, tried to convey to McMath's nurses that the amount of bleeding McMath was experiencing was excessive. Yet, despite filling multiple kidney basins with blood, four and a half hours elapsed from when the family first expressed concern about McMath's bleeding until a physician examined McMath. McMath was eventually taken into emergency surgery where she suffered cardiac arrest, and, eventually, brain death.

What followed was a battle between the hospital and the family over McMath's status as dead by neurological criteria, the sensitive issue of organ donation, and at least some speculation that the family was either too stupid to understand the reality of McMath's condition or was looking for a pay day—to get rich from their daughter's tragedy (as the USA is notoriously litigious). I continue to be haunted by the words of McMath's mother, Nailah Winkfield, "No one was listening to us, and I can't prove it, but I really feel in my heart: if Jahi was a little white girl, I feel we would have gotten a little more help and attention."[5] Note Winkfield did not say that a tragic outcome would not have occurred even with all of the help and attention in the world. But what is salient for Winkfield is that there was *no help*. And significantly, Winkfield believes that race mattered in the interaction, and, by extension, the outcome.

What brought Winkfield to her belief was not only the experience with her daughter, but also the weight of a history of racial bias and biased practices that continue into the present, including the recently published findings that Black patients in the USA are more frequently restrained than white patients in the emergency department[6]—a finding that echoes earlier findings for patients in mental health units in the UK.[7] It is important to remember that our human interactions, including our interactions with healthcare personnel, do not operate within a vacuum. Bioethicists would do well to think intentionally about the social and political contexts in which we do our work.

In her important book, *Medical Apartheid: The Dark History of Medical Experimentation on Black Americans from Colonial Times to the Present*, Harriet Washington dives into the extensive practice of medical experimentation on Black Americans, often without knowledge or consent, that has led to some of the most significant medical advances.[8] Far from heralding whatever medical knowledge has been gleaned from these practices, Washington is rightly critical of the damage done to Black people through both the injustice of these experiments, and the climate of distrust that has grown as a result of the historic and continuing unjust practices. For example, many southern US medical schools used enslaved Black people for medical experiments, even going so far as to rob cemeteries that held the remains of enslaved people for cadavers. The awareness that even in death Black people have been and can be the subject of unjust medical experimentation has remained part of the cultural milieu in many Black communities, and this awareness comes to the medical encounter.

Elsewhere I've written that, "Bioethicists can learn from this tragic [McMath] case, especially the ways that race may have shaped the interactions between the hospital and the family." As I wrote for the Hastings Center,

> It is possible to acknowledge the history of racist medical practices that underlie the mistrust that many African Americans feel toward the U.S. health care system and to acknowledge that McMath's family believed that race was a factor in the disrespect they felt without assigning racist intentions to the hospital personnel in McMath's case. In this

instance, bioethicists bring useful tools to the clinical environment. One of the roles that bioethicists occupy in hospital settings is mediating conflict within families and between patients and/or families and hospital personnel. To do this effectively in cases where at least one party believes race to be a factor in quality of care, bioethicists have to be aware of the history and basis for this belief and cannot shy away from uncomfortable conversations about race.

I am not suggesting that bioethicists should automatically presume that a charge of racism or racial bias is correct. I am suggesting that bioethicists can bring to the table a level of cultural sensitivity and awareness of how race can (and often does) shape perception. This understanding can be useful in gaining clarity about the issues at hand and helping to find a solution. This cultural awareness and willingness to address race, even when uncomfortable, wouldn't replace the role of emotional intelligence, but it could provide additional nuance in cases where race may be a factor.[9]

What I did not discuss in my previous work about McMath is the role that McMath's obesity played in popular conversations about what happened to her. There was some speculation that had McMath not been obese, she may not have experienced the complications that led to her brain death. Indeed, had McMath not been obese, she may not have needed surgery at all, so the thinking goes.[10] Stigma about obesity permeates discussions about bodies and health status in the West—whether in the form of paternalistic concern or disdain. Conversations about obesity and health status sometimes take the tone of "bringing negative health status on oneself" due to laziness and/or a lack of self-control.[11] Yet, while in many countries, including the USA, at least some forms of obesity are granted disability protections, the sense of whose "fault" obesity is doesn't go away.

The literature on stigma shows that to the extent that people are perceived to have brought illness on themselves, others exhibit less sympathy and fewer helping tendencies.[12] This perception tracks obesity, smoking, drug use, and sexual behavior. As far as I know, the family has not expressed that they were made to feel that McMath's care was shaped by her obesity. However, the discussion about McMath's weight in the popular press is instructive because it reflects broad cultural attitudes about what bodies are valuable and what bodies are deserving of medical resources. To that end, McMath's status as an obese young Black girl made her multiply vulnerable to potential bias.

What is the takeaway from Jahi McMath's tragic death? Or, better, what could have been important lessons for bioethicists in the aftermath of McMath's death, and why does it matter for our current moment of pandemic? For the remainder of the chapter, I want to explore a few avenues:

1   Patients and their families want to be heard and respected. This may sound simplistic and obvious. However, this need can get lost in the cultural dynamics at play at bedside, particularly when patients experience some form of vulnerability that complicates the illness that brought them to the clinical interaction.
2   When patients and their families do not trust their providers, the interaction suffers.
3   Although the nature of health care requires making difficult decisions, making these decisions without careful attention to the cultural and socio-political realities in which those decisions sit can reinforce structural inequality.

I think about these issues in light of McMath, but also how these same problems have complicated responses to the current global COVID-19 pandemic.

I Listening to Patients

Jahi McMath suffered brain death. This may have happened regardless of the care McMath received. However, her family believed that the quality of care she received was diminished due to bias (implicit or explicit) about McMath's race, her perceived class, and certainly after the determination of brain death, their religious sensibilities. It is partly that perception which contributed to the subsequent protracted legal battles and created further harm by hindering the family's ability to properly grieve their loved one.

A long-standing difficulty for bioethicists has been figuring out what it means to listen to patients and their families and, as a practical matter, what should follow from listening. Recall that McMath's family expressed the feeling of not being listened to as they watched her health decline. The feeling of being dismissed compounded the family's pain of their loss. It is important to remember that patients and their families often do bring unique insights to clinical encounters, and that these insights are not limited to symptom reports. When this insight is disregarded—either because the information that the patient brings is dismissed out of hand or because the patient themself is dismissed as a knower—epistemic injustice occurs. In the healthcare context, epistemic injustice is an additional harm.[13] The first harm is the harm of the illness itself (for which there may be no one responsible), and the second harm is the harm of epistemic injustice (for which someone is certainly responsible, even if they do not perceive themselves to be). Epistemic injustice may also lead to adverse effects as a result of inadequate medical care. In short, not being heard harms patients and families, sometimes with tragic outcomes. Due to the socio-cultural environment in which we operate, some patients, due to their social positioning, are less likely to be believed. Winkfield expressed that she felt unheard at least partly because of her race. Important lessons from the McMath case include the significance of being heard and regarded as a knower in one's own right. Conversely, this case demonstrates the harm that results when that does not happen, and how some are more vulnerable than others to not being heard.

What the data and the popular press are revealing during the various peaks of the current pandemic is that some populations of patients are being turned away from emergency departments in greater numbers than should be expected, all things being equal.[14] So-called racial minorities have been disproportionately impacted by COVID-19—in diagnosis and death. Some of this disproportionate impact occurs because these patient populations are turned away at first point of contact. Of course, this is more complicated in the USA, given its lack of universal health care; but that we still see this in the USA, UK, and Canada is (or should be) a matter of ethical concern as a health justice issue.

II Erosion of Trust

One consequence of epistemic injustice is the erosion of trust. While I am certainly not suggesting that epistemic injustice is the only (or even the primary) way that trust is damaged, having one's own knowing undermined or disregarded harms not only the sufferer of the injustice but also the relationship between parties. By trust, I mean the state of believing that one can rely on another. Epistemic injustice depends on mistrust. That is, x disregards y as a knower because x does not trust that y can or will accurately and reliably report y's experiences to x. So, mistrust lies at the heart of epistemic injustice. I submit that one who experiences the mistrust inherent in epistemic injustice returns the mistrust in turn. One who experiences epistemic injustice knows that they cannot rely on the perpetrator to take their knowing seriously.

Because at a certain point in the interaction, McMath's family no longer trusted the Children's Hospital staff, subsequent interactions were even more fraught than they otherwise might have been, and the provider/family relationship was irretrievably broken. This may have been, in the grand scheme of things, a one-off interaction. However, if enough of these interactions occur within a community over a sustained period, then an unfortunate one-off interaction is part of a series of interactions that become a community of potential patients who do not trust the area providers.

The relationship between McMath's family and the hospital personnel had deteriorated so significantly that, by the time the organ procurement team raised the possibility of organ donation, the family expressed suspicion that the hospital intentionally withheld care from McMath in order to procure her organs. Again, interactions between patients/families and providers do not occur in a vacuum. History was in that room, as much as the present circumstances were. To review, in conversations surrounding McMath's story, there was speculation whether the family was intelligent enough to understand brain death or whether they were using their daughter to "get over" on the hospital. These assumptions themselves carry the weight of historical racial bias and stereotypes against and about Black people as being inherently less intelligent. The (even unconscious) perception that the family was unintelligent also undermined their credibility as knowers who could make reasonable decisions about their loved one.

The failure to interrogate the role that mistrust played in the interaction with McMath was a missed opportunity for bioethics. Nevertheless, in the current pandemic, while vaccine hesitancy in Black, Brown, and Indigenous populations is, at times, rightly characterized as the result of mistrust, the undertone is that this mistrust is a failing of the populations who feel mistrustful, rather than a rational response to centuries of healthcare systems showing themselves to be untrustworthy. Furthermore, although trust is present in most successful interactions, and acknowledgment of the need to repair broken trust is laudable, equally important is thinking about how to operate through mistrust. In other words, where there is mistrust, the onus is on the healthcare institutions to rectify it. Acknowledging mistrust is a necessary first step in moral repair. However, in circumstances like the current pandemic, working through mistrust is even more pressing. Here is an opportunity for bioethics. In much of the world, the COVID-19 vaccine rollout could be improved. Globally, vulnerable communities find themselves disproportionately without equitable access to vaccines. Hiding behind often justified mistrust does not absolve one of the responsibility to think through how lack of equitable access reinforces structural injustice.

III Nonmaleficence and Cultural Sensitivity

There is a finite window within which organs are viable for transplant. As a result, specific institutional processes are triggered upon a determination of brain death, including immediately approaching a family about the possibility of organ donation. With the laudable goal in mind of maximizing the possibility of saving lives, it makes sense to initiate these conversations sooner rather than later. At the same time, this goal must be governed by a principle of nonmaleficence. Nonmaleficence is the bioethical principle to "abstain from causing harm to others." Yet, one must consider the context of institutional policies and even informal practices before creating and implementing them. While I understand that nonmaleficence is often thought of in the context of individual patients, entire communities can be harmed when broader considerations and their consequences are absent from the deliberation about institutional policies. This is the crux of structural injustice—the idea of disadvantage that accrues and persists, even in the absence of malice.

This pandemic has brought with it many difficult decisions and has exposed tremendous worry about allocation of medical resources, including ventilators, but also about the rationing of COVID-19 care more broadly. Members of marginalized communities have expressed worry that these decisions will reinforce (intentionally or not) hierarchies about whose lives are more valuable—leaving racial minorities, the disabled, and the aged among those who suffer unjustly. For example, algorithms and other mechanisms that assign quality-adjusted life years (QALYs) are sometimes heralded as being free of bias. Yet, it is clear that this is not the case. For example, pulse oximeters are devices that measure oxygen saturation. During the pandemic pulse oximeter readings have been taken as reliable indicators of a patient's decline, especially under emergency conditions with scarce resources, despite having been shown to read less accurately on darker skin.[15] That is, pulse oximeters can indicate greater oxygen saturation than is actually present, thereby giving the false impression that patients with darker skin are healthier than they actually are. This finding has been available for over ten years. As such, continued reliance on pulse oximeter readings as measures of the seriousness of a patient's condition, especially during a global pandemic, is an act of maleficence—and it is an act of maleficence, not solely against individual patients, but against entire patient populations. Therefore, it cannot be justified.

IV Conclusion

I have offered some ways how lessons from an individual tragic case that presented a variety of challenges for bioethics could be extrapolated to think about this current moment. I *do* think Jahi McMath and her family experienced injustice – even if no individual hospital staff person acted with malice, and even if she would have died anyway as a result of the complications she suffered. However, it is impossible to ignore the length of time it took for the family to receive any assistance as they helplessly watched Jahi McMath's condition deteriorate. The family, and most thoughtful observers who are sensitive to the weight of historic and continuous racial injustice are left wondering whether help would have arrived in time if McMath had been a little white girl. The disadvantages that accrued as a result of social categories like race, disability status, age, and class further challenged already difficult circumstances. Nonetheless, I think that bioethicists can learn from these individual moments. In that way, a commitment to justice remains at the forefront in the face of big moments.

Notes

1　An earlier version of this chapter was presented at the University of Oxford Medical Ethics, Law, and Discussion Group, Oxford, England (June 2021). I thank the audience for their helpful comments, questions, and suggestions.

2　Wilson (b), "Why the case of Jahi McMath is important for understanding the role of race for black patients," 2018. Death by neurological criteria is determined medically by "a clinical examination that demonstrates coma, brainstem areflexia, and apnea" (see Greer et al. "Determination of Brain Death/Death by Neurologic Criteria: The World Brain Death Project") and is defined legally in the Uniform Determination of Death Act as "death based upon irreversible loss of all brain functions" (see Uniform Determination of Death Act, 1997).

3　Katznelson, "Redefining Death in the Law," 2018.

4　Nelson, "Girl Predicts Devastating Outcome of Her Own Tonsillectomy," 2013.

5　Aviv, "What Does It Mean to Die?" 2018.

6　Wong et al., "Association of Race/Ethnicity and Other Demographic Characteristics with Use of Physical Restraints in the Emergency Department," 2021.

7　Denis Campbell, "Alarm Over Restraint of NHS Mental Health Patients," 2017.

8　See generally, Washington, *Medical Apartheid*, 2007.

9 Wilson (a), "Jahi McMath, Race, and Bioethics," 2018.
10 Nelson, "Girl Predicts Devastating Outcome of Her Own Tonsillectomy," 2013.
11 Please see Mollow, this volume.
12 R.M. Puhl and C.A. Heuer, "Obesity Stigma: Important Considerations for Public Health," *American Journal of Public Health*, 2010.
13 Please see Ho, this volume. See also, Havi Carel and Ian James Kidd. "Epistemic Injustice in Healthcare: A Philosophical Analysis." *Medicine, Health Care and Philosophy* 17, no. 4 (2014): 529–540, https://doi.org/10.1007/s11019-014-9560-2.
14 Eligon and Burch, "Questions of Bias in Covid-19 Treatment Add to the Mourning for Black Families," 2020.
15 Feiner et al., "Dark Skin Decreases the Accuracy of Pulse Oximeters at Low Oxygen Saturation: The Effects of Oximeter Probe Type and Gender," 2007.

References

Aviv, Rachel. 2018. "What Does It Mean to Die?" Annals of Medicine, *The New Yorker*, January 29. https://www.newyorker.com/magazine/2018/02/05/what-does-it-mean-to-die.

Beauchamp, Tom and James Childress. 2019. *Principles of Bioethics*. New York: Oxford University Press.

Campbell, Denis. 2017. "Alarm Over Restraint of NHS Mental Health Patients," The Observer Mental Health, *The Guardian*, December 9. https://www.theguardian.com/society/2017/dec/09/women-black-patients-physically-restrained-mental-health.

Carel, Havi and Ian James Kidd. 2014. "Epistemic Injustice in Healthcare: A Philosophical Analysis." *Medicine, Health Care and Philosophy* 17, no. 4: 529–540, https://doi.org/10.1007/s11019-014-9560-2.

Eligon, John and Audra D. S. Burch. 2020. "Questions of Bias in Covid-19 Treatment Add to the Mourning for Black Families," nytimes.com, *The New York Times*, May 10. https://www.nytimes.com/2020/05/10/us/coronavirus-african-americans-bias.html.

Feiner, John R., John W. Severinghaus, and Philip E. Bickler. 2007. "Dark Skin Decreases the Accuracy of Pulse Oximeters at Low Oxygen Saturation: The Effects of Oximeter Probe Type and Gender." *Anesthesia & Analgesia* 105, no. 6: S18–S23.

Greer, David M., Sam D. Shemie, Ariane Lewis, Sylvia Torrance, Panayiotis Varelas, Fernando D. Goldenberg, James L. Bernat, et al. 2020. "Determination of Brain Death/Death by Neurologic Criteria: The World Brain Death Project." *JAMA* 324, no. 11 (September 15): 1078. https://doi.org/10.1001/jama.2020.11586.

Ho, Anita. 2022. "Disability Bioethics and Epistemic Injustice." In *Disability Bioethics Reader*, edited by Joel Michael Reynolds and Christine Wieseler, pp. 324–332. Washington, DC: Routledge.

Katznelson, Gali. 2018. "Redefining Death in the Law," Bill of Health, Petrie-Flom Center at Harvard Law School, April 19. https://blog.petrieflom.law.harvard.edu/2018/04/19/redefining-death-in-the-law/.

Mollow, Anna. 2022. "Medicalization, Stigma & Health: Insights from Fat Studies." In *Disability Bioethics Reader*, edited by Joel Michael Reynolds and Christine Wieseler, p. XX. Washington, DC: Routledge.

Nelson, Shellie. 2013. "Girl Predicts Devastating Outcome of Her Own Tonsillectomy." In the Air, *WQAD-TV*, December 17. https://www.wqad.com/article/news/local/drone/8-in-the-air/girl-predicts-devastating-outcome-of-her-own-tonsillectomy/526-7a8afe03-0800-474a-894a-115f9d7b1f05.

Puhl, Rebecca M. and Chelsea A. Heuer. 2010. "Obesity Stigma: Important Considerations for Public Health." *American Journal of Public Health* 100: 1019–1028.

Uniform Determination of Death Act, 12 Uniform Laws Annotated (U.L.A.) 589 (West 1993 and West Supp. 1997).

Washington, Harriet. 2007. *Medical Apartheid: The Dark History of Medical Experimentation on Black Americans*. New York: Penguin Random House.

Wilson, Yolonda. 2018a. "Jahi McMath, Race, and Bioethics." The Hastings Center Bioethics Forum Essay. *The Hastings Center*, July 19, 2018. https://www.thehastingscenter.org/jahi-mcmath-race-bioethics/.

Wilson, Yolonda. 2018b. "Why the Case of Jahi McMath Is Important for Understanding the Role of Race for Black Patients." The Conversation Ethics + Religion, *The Conversation US*, July 12. https://theconversation.com/why-the-case-of-jahi-mcmath-is-important-for-understanding-the-role-of-race-for-black-patients-99353.

Wong A.H., Whitfill T., Ohuabunwa E.C., et al. 2021. "Association of Race/Ethnicity and Other Demographic Characteristics with Use of Physical Restraints in the Emergency Department." *JAMA Network Open* 4, no. 1: e2035241. doi:10.1001/jamanetworkopen.2020.35241

Further Reading

Pence, Gregory. 2021. *Pandemic Bioethics*. Ontario: Broadview Press.
Ray, Keisha. 2014. "The Case of Jahi McMath: Race, Culture, and Medical Decision-Making." *Voices in Bioethics*. July 8. https://doi.org/10.7916/vib.v1i.6509.
Toole, Briana. 2020. "From Standpoint Epistemology to Epistemic Oppression." *Hypatia* 34, no. 4: 598–618. https://doi.org/10.1111/hypa.12496.

18

DISORDERS OF CONSCIOUSNESS, DISABILITY RIGHTS, AND TRIAGE DURING THE COVID-19 PANDEMIC

Even the Best of Intentions Can Lead to Bias

Joseph J. Fins

This chapter is excerpted from: Joseph J. Fins, "Disorders of Consciousness, Disability Rights and Triage During the COVID-19 Pandemic: Even the Best of Intentions Can Lead to Bias," *The Journal of Philosophy of Disability* 1 (2021): 211–229. Reprinted with permission.

I Consultations and Commitments: Disorders of Consciousness Meets COVID-19

During the Spring (2020) surge of COVID-19 I directed an ethics consult service at an academic medical center in Manhattan (Fins and Prager 2020; Prager and Fins 2020). While I worked with my team doing consults on individual patients, an equally challenging part of my duties was trying to write institutional policies to allocate ventilators (Huberman et al. 2020). This was no easy task for two reasons. One was political and the other personal (Fins 2020e).

At the policy level, each hospital had to make its own determination about how to proceed because New York State never issued guidance for crisis standards of care, hoping that the expansion of hospital and ICU capacity could outrun the pandemic, making rationing unnecessary (Fins 2020f).

At a personal level, the need to respond to this crisis placed me in the uncomfortable position of having to make an ethical choice between the utilitarian needs posed by the pandemic and my long-standing scholarly commitment to the needs of people with disabilities (Guidry-Grimes et al. 2020), particularly those with disorders of consciousness and severe brain injury. These two goals were seemingly in opposition to each other and irreconcilable. And yet, both objectives were part of my academic workspace, having their origins in two publications I authored, or co-authored, five years earlier: *Rights Come to Mind: Brain Injury Ethics and the Struggle for Consciousness* (2015) and the *Ventilator Allocation Guidelines* (2015) drafted by the New York State Task Force on Life and the Law, of which I am a member.

In retrospect, it was ironic that these two documents were published in the same year. When they were written, these two domains seemed nicely separated in my life and work. Although we wrote to prepare for the pandemic, most of us were in a state of denial, much as Kubler-Ross had described when discussing stages of loss and grief (Fins 2009). Even as we planned for the future our hearts had not caught up with our minds. If we were prudential it was still

DOI: 10.4324/9781003289487-25

improbable and more an academic exercise than something we would have to operationalize in real time, with real consequences. My work on brain injury and disability rights stood safely apart from these efforts, protected as well by a hefty dose of discounting. Though public health officials warned of a pandemic, one would not come, and if it did the same sort of advances in medicine that were catalyzing a revolution in neuroscience and brain injury would protect us from the coming plague. But I was naïve. We all were.

I never truly imagined the confluence of intellectual and normative challenges that would arise as the pandemic hit New York City as I sought to reconcile my work on the ventilator guidelines, brain injury, and disability rights. This chapter tells that story, one that continues to evolve as the nation grapples with the pandemic and I reflect on my role in responding to the brutal surge of COVID-19 that struck New York City in the Spring of 2020.

II Understanding Disorders of Consciousness

To place my COVID-19 response into context, we need to step back and consider the origins and purpose of *Rights Come to Mind: Brain Injury Ethics and the Struggle for Consciousness* (Fins 2015). This volume sought to recount the progress that has been made in diagnosing, categorizing, and treating disorders of consciousness over the past 20 years, drawing upon family narratives and my own work collaborating with neuroscientists. I told this remarkable scientific story through the prism of disability rights, asserting that the needs of patients with disorders of consciousness have been unmet, in large part because of the nihilism that dates to the origins of the "right-to-die" in America (Fins 2003).

This is a complex story worthy of a longer exegesis (Fins 2020a) but simply put, the right to die was established in the context of severe brain injury, namely the vegetative state. The presumption of futility became the moral warrant to allow for the withdrawal of life-sustaining therapy in landmark legal cases (Fins 2006a), beginning with Quinlan[1] and continuing on through Cruzan (Fins 2020b) and Schiavo (Schiavo ex Rel. Schindler v. Schiavo 2005; Fins 2006b). But there was more complexity to the story: some patients might get better and some thought to be vegetative had covert consciousness and were sensate. To make sense of this, it is critical to first define our terms and categories (Fins 2019a).

Let us begin with coma, which is an eyes-closed state of unconsciousness. A coma after traumatic brain injury can last a week or two and can be a precursor to brain death or recovery. Comas can also be induced and prolonged with sedative medication, a therapeutic strategy sometimes used to promote recovery after brain trauma.

When a coma does not resolve to consciousness, patients progress to the vegetative state which represents the isolated recovery of the brain stem without higher cortical function. Patients in the vegetative state are clinically paradoxical to the untrained eye as theirs is an eyes opened state of unawareness. Because we often ascribe awareness to the opening of the eyes, this brain state can be very difficult for expectant families who expect that the opening of a loved one's eyes coming out of a coma heralds recovery and the person that they knew. But when a coma evolves into the vegetative state, the eyes are open but there is neither awareness nor responsiveness.

The vegetative state was first described – as the persistent vegetative state – in a 1972 Lancet publication by Bryan Jennett, the Scottish neurosurgeon (also responsible for the Glasgow Coma and Outcome Scale), and Fred Plum (the American neurologist who first described the Locked-in-State). Jennett and Plum described the vegetative state as one that "seems wakeful without awareness" (1972). Because they did not have functional neuroimaging to peer inside the injured brain, they were unable to definitively exclude the possibility of awareness (Fins 2019b).

In 1994, the Multi-Society Task Force published two articles in *The New England Journal of Medicine* which further classified the vegetative state into two distinct categories: persistent and permanent (Multi-Society Task Force on PVS 1994a, 1994b).[2, 3] According to this framework, a vegetative state became persistent if it persisted for a month. The vegetative state was designated as being permanent three months after anoxic brain injury, such as what would occur after a cardiac arrest, or 12 months after traumatic brain injury. This variable time course from persistence to permanence also reflects the more favorable prognosis of patients with traumatic versus anoxic brain injury.

A consensus definition of the minimally conscious state (MCS) entered the medical literature in 2002 under what was called the Aspen Criteria (Giacino et al. 2002). MCS patients demonstrate an awareness of self, others, and their environment. They may turn when they hear their name, look up when someone enters the room, or grasp an object presented to them, all signs indicative of consciousness according to the Aspen Criteria. MCS is further subdivided into MCS− and MCS+ reflective of whether patients can respond verbally or not (Thibaut et al. 2020).

The challenge with all behavioral manifestations in MCS is that these behaviors do not occur consistently or reliably, making diagnosis difficult. Because of this MCS can be easily confused with the vegetative state. When the behaviors are not manifest, these patients appear to be in the wakeful unresponsive state of the vegetative state. In one study over 40% of patients with traumatic brain injury in chronic care facilities thought to be in the vegetative state were actually in MCS (Schnakers et al. 2009). Patients who can reliably respond to commands are said to have emerged and are designated as MCS-E (Bodien et al. 2020).

Patients with a disorder of consciousness are best assessed by the validated Coma Recovery Scale-Revised (Giacino, Kamlar, and Whyte 2004), a neuropsychological bedside examination that assesses different domains reflective of consciousness, not simply motor function as evaluated by the older Glasgow scales (Fischer and Mathieson 2001). This is an important distinction for patients who lack motor output but who have normal cognitive function. This would be the case of a patient in the Locked-in-State who is paralyzed from the neck down. Such patients have low scores on the Glasgow scales because it assesses motor function not consciousness. Distinguishing the vegetative state from MCS is more than a diagnostic curiosity, it is scientifically, clinically, and normatively significant. At a scientific level, the neurocircuitry of the MCS patient is distinct from that of vegetative patients (Laureys et al. 2002). In neuroimaging studies, MCS patients have intact and distributed neural networks, which are the substrate for consciousness; this is unlike vegetative patients whose brains are functionally unable to achieve integrative function (Schiff et al. 2005). These networks, when they are activated, allow for the demonstration of behaviors indicative of awareness and consciousness. At a normative level, the presence of these distributed neural networks allows the patient to hear what is being said and to perceive pain (Chatelle et al. 2014). To quote a line from the musical Hamilton, the MCS patient "is in the room where it happens" (Miranda 2015; Fins 2016a). Unlike patients who are properly diagnosed as vegetative, MCS patients are aware and sensate, obliging us to be aware of their presence and to attend to their neuro-palliative care needs (Fins 2008; Fins and Pohl 2015).

Assessment is further complicated by what my colleague and I have previously described as non-behavioral MCS (Fins and Schiff 2006), a state in which patients respond to commands on neuroimaging scans without demonstrating associated behaviors. For example, a patient might be asked to imagine walking through their house, playing tennis, or disaggregating similar sounding words with different meanings and activate the areas in the brain responsible for spatial navigation, motor activities, or language processing (Owen et al. 2006; Bardin et al. 2011). These volitional tasks are manifest on the brain scans but not in overt behaviors, hence the notion of patients being in a non-behavioral MCS (Owen et al. 2006). They appear to be in the vegetative state at the bedside yet they have covert consciousness.

More recently, Schiff has described individuals with such discordances as having cognitive motor dissociation or CMD (Schiff 2015). CMD patients span a broader range of functional capacity than the non-behavioral MCS patients. This range includes patients in MCS as well as those who are in the Locked-in-State who have normal cognition but no motor output other than the cranial nerves.

In a milestone series of publications, the classification, diagnosis, and treatment changed in August 2018 when the American Academy of Neurology (AAN), the American Congress of Rehabilitation Medicine (ACRM), and the National Institute on Disability, Independent Living, and Rehabilitation Research (NIDILRR) came together to publish a systematic evidence-based review of this space (Giacino et al. 2018a, 2018b) and promulgate a practice guideline following upon this systematic review of the available data (Giacino et al. 2018c, 2018d). The 2018 AAN/ACRM/NIDILRR guideline revised the 1994 *New England Journal of Medicine* Multi-Society Task Force statements on the vegetative state[4, 5] and further endorsed the 2002 definition of the MCS (Giacino et al. 2002). Cognizant of the marginalization of this population, and how they are often neglected in chronic care, the practice guideline called for definitive standards of care for this vulnerable population (Fins and Bernat 2018).

They also offered a notable revision in diagnostic classification, replacing the old 1994 category of a "permanent" vegetative state with a "chronic" vegetative state. This was prompted by a review of data which suggested that some 20% of patients characterized as being in the permanent vegetative state might actually progress into a state of higher cognitive function such as the MCS. While 80% of the patients formerly described as being permanently vegetative would remain so, others would not based on the data. This empirical observation was testimony to the fact that some brain states – thought to be fixed and permanent – can and do evolve (Fins and Schiff 2017). Perhaps more critically that some patients thought to be permanently unconscious had unidentified or covert consciousness. The greatest proportion of these patients would be roughly 40% who had been misdiagnosed and had been minimally conscious all along (Schnakers et al. 2009).

III Disability Rights and Disorders of Consciousness

Patients with disorders of consciousness are among the most vulnerable and neglected of our citizenry. Saved by heroic measures from certain death only decades earlier, most are destined to receive what has been euphemistically called custodial care in chronic care facilities ill-equipped and ill-disposed to track their evolving brain states or attend to their pain and suffering. It is not an assertion of hyperbole that these patients suffer in isolation segregated from mainstream medical care undiagnosed (Fins 2015) and often with their pain both unrecognized and untreated (Berube et al. 2006; Chatelle et al. 2014).

I believe that this is a human rights violation under international law (Convention on the Rights of Persons with Disabilities 2006; Fins 2016b, 2016c; Wright et al. 2018). Domestically, as I have maintained in *Rights Come to Mind*, the neglect of this population is a pressing civil and disability rights issue that demands our attention (Fins 2017). The segregation question is particularly apt because this constitutes a stark violation of the Americans with Disabilities Act (ADA)[6] which mandates social integration of people with disabilities. As explained by a contemporaneous advisory memo from the George H. W. Bush era Department of Justice, the ADA was a "Mandate for the elimination of discrimination against individuals with disabilities" in large part because "historically, society has tended to isolate and segregate individuals with disabilities."[7] When Justice Ruth Bader Ginsburg later wrote the majority opinion upholding the ADA in *Olmstead v L.C.* – a case involving deinstitutionalization of two women in a Georgia psychiatric hospital – she explicitly pointed to Congressional intent in the drafting of the ADA. Justice Ginsburg noted, "Congress explicitly identified unjustified 'segregation' of persons with disability as a form of discrimination."[8] In lieu of segregation, the Olmstead court called for societal reintegration of individuals who have been placed out of the mainstream because of their disability.

In the context of brain injury, the issues are similar to what was contested in Olmstead. For some it is the possibility of deinstitutionalization. But for patients with disorders of consciousness who may not be able to live with family or in the community the question of reintegration is less about physical place and more about restoration of functional communication. Let me explain this key difference.

For people with a mobility disability, the ADA has allowed for societal integration by providing for a more accommodating physical environment. While certainly not perfect, reforms made because of the ADA have helped make it possible for many to go to work utilizing changes to the built environment like kneeling buses or a cut in the sidewalk that allows wheelchair access. But it is different for people with disorders of consciousness whose integration is often limited by their inability to communicate (Fins 2015). In this context, community is built (and indeed restored) by fostering functional communication, community's cognate (Fins 2017). This is the goal of all who work in disorders of consciousness and it is more than aspirational.

This was what my colleagues and I accomplished when we restored functional communication in an MCS patient with thalamic deep brain stimulation (Schiff et al. 2007). Before entering this phase I clinical trial, his highest level of function was inconsistent command following with eye movements. With bilateral stimulation of the intralaminar nuclei of the thalamus bilaterally, he was able to say six or seven word sentences, the first 16 words of the Pledge of Allegiance, and tell his mother he loved her. He could also express a clothing preference when his mother took him to Old Navy (Fins 2015). By returning voice to these patients through the restoration of functional communication we helped to overcome his isolation and restore the social and biological networks disrupted by his injury. While this work was at the proof of principle level, it is a promissory note for others in this condition and a step forward in fulfilling the reintegration mandate to which the ADA aspires.

IV New York State Task Force Report on Ventilator Allocation

It was against this backdrop that I entered the pandemic. I consider myself a disability rights advocate for patients made vulnerable by virtue of severe brain injury. They are deserving of the same respect and consideration as other individuals in the face of the pandemic, notwithstanding the utilitarian challenges posed by scarcity and crisis. Nonetheless, these commitments were tested when the pandemic hit New York City in the Spring of 2020. How does a clinical ethicist responsible for his hospital's ethics committee respond to a crisis while maintaining fidelity with his intellectual and normative commitments to people with disability?

To answer this question we need to again step back to 2015 and the *Ventilator Allocation Guidelines* developed by the New York State Task Force on Life and the Law,[9] appreciate how they came into being, the intent of Task Force and the metrics that we employed. The

Task Force came up with an allocation scheme that we hoped we would never use. That comforting sense of denial that has been shattered by the COVID-19 pandemic (Fins 2009). Despite this sense of improbability, we took our responsibilities seriously and after nearly of decade of research, hearings, and internal deliberations we came up with a plan that sought to be non-discriminatory using the Sequential Organ Failure Assessment (SOFA), a methodology score originally developed during the H1N1 influenza epidemic to assess a patient's need for ventilatory support and likelihood of survival should it be provided (Shahpori et al. 2011).

Triage decisions utilizing the SOFA framework were meant to be physiologically determined in a neutral fashion without regard to race, ethnicity, gender, age, disability, or other social determinants which might skew the triage process. Age did not explicitly play a role in the SOFA score. Instead, the omnibus score was used as a proxy for the patient's physiologic age which tracked the functionality of several organ systems.

The Task Force also sought to avoid discriminating against individuals with disabilities. Disability discrimination occurs when a disability that is irrelevant to the acute triage decision is used to bias analysis and deny or limit care. This can be the disability itself or a medical condition associated with a disability, even if neither has a bearing on whether or not the individual will survive the episode of respiratory failure which necessitates the provision of a ventilator. Previously, I have labeled the latter as crypto-discrimination—associated medical conditions that are comorbidities of a disability, such as the heart conditions associated with Down syndrome (Fins 2020c, 2020d).

After the authorities declared a public health emergency and invoked crisis standard of care which would replace "usual" care with "sufficient" care (Fischkoff et al. 2020), this methodology would be put into place and patients would be triaged based on their SOFA scores. These scores would be an aggregation of an assessment of the patient's blood pressure, lung, liver, and kidney functions, and the integrity of the blood's clotting system. In addition, the patient's neurological status would be assessed by using the Glasgow Coma Scale (GCS) (Jennett and Teasdale 1977). Each of these six metrics was divided into scores of 0 to 4, with progressive deterioration yielding a higher score for a maximal score of 24 points.

SOFA scores would be cohorted into four color-coded categories for triage: Blue, Green, Red, and Yellow. Patients in the Blue category (SOFA > 11) had the most dire prognosis and would not likely survive the acute infection despite maximal efforts. Those who were designated Green (No significant organ failure) were sick but not sick enough to require a ventilator. Red patients (SOFA < 7 or single organ failure) were very sick and in need of a ventilator and most likely to survive if they received one. Yellow (SOFA 8–11) was an intermediate category between Red and Blue. These patients were sicker than those in the Red category and would receive a ventilator after those designated as Red were allocated this scarce resource.

V SOFA and Brain Injury: A Critical Analysis

If we turn to the overall SOFA score we can see what an outsized impact the GCS can have on triage decisions. Patients with a GCS of 6–9 yield 3 points on the SOFA score while those with a GCS < 6 generate 4 points on the SOFA score. If the goal is to end up in the Red Zone, SOFA < 7, patients with lower GCS scores are at great risk of having scores outside of the triage range.

Consider this scenario: a patient who opens their eyes to sound and produces "inappropriate words" (the language used in the GCS) and did not have motor output would have a GCS of 6 and generate 3 points on the SOFA score. If they also had a slight decrease in their platelet count (1 point), some mild liver (1) and kidney (1) insufficiency, they would have a total of < 7 points. This would be their SOFA score before their need for supplemental oxygen was considered. That need, depending upon their level of oxygenation, could yield an additional 2 to 4 points bringing their SOFA score to 9–11, placing them in Yellow and on the cusp of ineligibility as they approached the Blue category.

This would be problematic enough and potentially discriminatory if the GCS were the correct metric to use for the assessment of patients with disorders of consciousness. While GCS is predictive of outcomes for patients who have had an acute traumatic brain injury,[10] it has no role in assessing the vast majority of patients who would enter a hospital during a pandemic with severe brain injury. Most would not be victims of acute trauma of the sort associated with a car accident or a cardiac arrest but rather residents of congregant living settings with chronic brain injuries. Those patients should not be evaluated by the GCS but rather by the Glasgow Outcome Scale-Extended (Jennett et al. 1981) if we sought to stay in the Glasgow family of assessment tools.

But even this is insufficient for patients with disorders of consciousness as all Glasgow assessment tools rely heavily on motor output as a marker of brain state. Thus, a patient who had inconsistent or scant motor output would yield low scores even when they might have higher levels of cognitive function. This would be the case of a patient in MCS or with CMD. These patients would have a low GOS-E score and would be potentially indistinguishable from a patient in the vegetative state devoid of consciousness. All of these patients should instead be evaluated by the Coma Recovery Scale-Revised which evaluates levels of consciousness (Giacino et al. 2018b). This instrument's inter-observer reliability and scientific utility was vetted in the aforementioned AAN/ACRM/NIDDLR evidence-based review and recommended as the neuropsychological assessment tool to be utilized in assessing patients with disorders of consciousness by the associated practice guideline, not the Glasgow scales (Giacino et al. 2018a, 2018b, 2018c, 2018d).

Beyond these methodological challenges there are outright errors in the SOFA "exclusion criteria," those conditions which would automatically place a patient into the Blue category making them ineligible for a ventilator. Relevant to our discussions, one of these criteria is "Traumatic brain injury with no motor response to painful stimuli (i.e. best motor response =1)."

This criterion is flawed for the reasons already enumerated regarding the fact that motor output, as assessed by the Glasgow Coma Scale, may not be an accurate reflection of a patient's cognitive state. There is a second more telling error as well: this exclusion criterion reflects poorly on the knowledge base of those who drafted it. The careful reader will note that the exclusion applies to patients with traumatic brain injury (TBI). What of those with anoxic brain injury – as would occur following severe oxygen deprivation during a cardiac arrest? It is well appreciated that patients with anoxic brain injury fare far worse than those with TBI (Posner et al. 2019). While cardiac arrest is one of the other exclusion criteria, that alone is not the same thing. There will be those who survive a cardiac arrest who are subsequently in an anoxic coma. Under the exclusion criteria for brain injury, patients in an anoxic coma would not be excluded while those in a coma secondary to TBI would be. This fails to acknowledge that 77% of patients with anoxic coma versus 50% of those with a trauma will remain in the vegetative state. Again, this is a limitation of the SOFA score which was apparently written without an adequate appreciation of the scholarly literature informing the care of patients with disorders of consciousness.

VI Good Facts Make for Good Ethics: Disability Advocacy in Context

Despite the intention of the New York State Task Force not to engage in discriminatory practices with respect to ventilator allocation, there were errors in our methodology that lingered
beneath the surface (Fins 2020c, 2020d).[11] While SOFA's reliance on the Glasgow Coma Scale
appears to be objective, it was misused when applied to patients with chronic disorders of
consciousness. As such it was ill-suited to the task of predicting survival from COVID-19-
associated respiratory failure. These limitations could lead to the denial of services to this population and compound vulnerability during times of crisis. While there might be an ethical
rationale to limit access to ventilator support to patients with disorders of consciousness, those
deliberations must first be predicated on solid data. Neither the SOFA score nor Glasgow Coma
Scale can provide that evidentiary base.

It has recently been asserted that society has been subject to a "tyranny of metrics" which can
simplify complex situations and promote reductionist thinking leading to erroneous conclusions
dressed up in a façade of objectivity (Muller 2019). Such is doubly the case when we consider
ventilator allocation and patients with disorders of consciousness. The seemingly objective conclusions drawn from SOFA scores applied to patients with disorders of consciousness can be especially insidious. They seem so logical and supported by the data. But in reality, they reinforce
a priori assumptions about the futility caring for these patients (Fins 2015). This makes patients
especially vulnerable during a pandemic.

It is highly unlikely that a busy ER or ICU doctor will drill down into the methodological
weeds of the SOFA score and appreciate the methodological peril that lurks beneath a veneer of
objectivity. That is probably too much to expect of busy clinicians trying to do their earnest best
during a pandemic. But is not too much to expect of disability advocates who must be vigilant
when reviewing allocation schema for methodological errors and/or implicit bias.

While the tale of this advocate is but one of many narratives which speak to the ways that
the disabled community has been further disadvantaged by the COVID-19 pandemic (Fins and
Bagenstos 2020), it is emblematic of a pervasive set of challenges that must be addressed by the
disability community and its allies. Advocacy resulting from this pandemic needs to be more
than sloganeering. It needs to be both granular and informed, ever cognizant that bias can find
its way into even the best of intentions.

Note: The views expressed are those of the author and do not represent the views of the
New York State Task Force on Life and the Law or any other organization with which he
is affiliated.

Notes

1 In the matter of Karen Quinlan, Supreme Court of New Jersey. 70 N.J. 10, 355 A.2d 677 (1976).
2 Multi-Society Task Force on PVS (1). *New England Journal of Medicine* 1994; 330(21): 1499–1508.
3 Multi-Society Task Force on PVS (2). *New England Journal of Medicine* 1994; 330(22): 1572–1579.
4 Multi-Society Task Force on PVS (1). *New England Journal of Medicine* 1994; 330(21): 1499–1508.
5 Multi-Society Task Force on PVS (2). *New England Journal of Medicine* 1994; 330(22): 1572–1579.
6 *Americans with Disabilities Act*, 42 U.S.C. 12101 (1990).
7 42 USC 1201 DOJ, Civil Rights Division.
8 *Olmstead* (527 U.S. 581 [1999]).
9 New York State Task Force on Life and the Law. *Ventilator Allocation Guidelines.* New York State
Department of Health, November 2015. Accessed on 9 April 2020 at: https://www.health.ny.gov/
regulations/task_force/reports_publications/docs/ventilator_guidelines.pdf.
10 Royal College of Physicians and Surgeons of Glasgow. The Glasgow Structured Approach to Assessment of the Glasgow Coma Scale. Accessed on 1 November 2020 at: https://www.glasgowcom
ascale.org.

11 New York State Task Force on Life and the Law. *Ventilator Allocation Guidelines*. New York State Department of Health, November 2015. Accessed on 9 April 2020 at: https://www.health.ny.gov/regulations/task_force/reports_publications/docs/ventilator_guidelines.pdf.

References

Americans with Disabilities Act of 1990, 42 U.S.C. § 12101 et seq. 1990. https://www.ada.gov/pubs/adastatute08.htm.

Bardin, Jonathan C., Joseph J. Fins, Douglas I. Katz, Jennifer Hersh, Linda A. Heier, Karsten Tabelow, and Jonathan P. Dyke et al. 2011. "Dissociations between Behavioural and Functional Magnetic Resonance Imaging-based Evaluations of Cognitive Function." *Brain* 134(3): 769–782.

Berube, J., J. J. Fins, J. Giacino, D. Katz, J. Langlois, J. Whyte, and G. A. Zitnay. 2006. *The Mohonk Report: A Report to Congress Improving Outcomes for Individuals with Disorders of Consciousness*. Charlottesville, VA: National Brain Injury Research, Treatment & Training Foundation.

Bodien, Yelena G., Geraldine Martens, Joseph Ostrow, Kristen Sheau, and Joseph T. Giacino. 2020. "Cognitive Impairment, Clinical Symptoms and Functional Disability in Patients Emerging from the Minimally Conscious State." *NeuroRehabilitation* 46(1): 65–74.

Chatelle, Camille, Aurore Thibaut, John Whyte, Marie Danièle De Val, Steven Laureys, and Caroline Schnakers. 2014. "Pain Issues in Disorders of Consciousness." *Brain Injury* 28(9): 1202–1208.

Convention on the Rights of Persons with Disabilities (CRPD). 2006. United Nations. Department of Economic and Social Affairs: Disability. https://www.un.org/development/desa/disabilities/convention-on-the-rights-of-persons-with-disabilities/convention-on-the-rights-of-persons-with-disabilities-2.html

Fins, Joseph J. 2003. "Constructing an Ethical Stereotaxy for Severe Brain Injury: Balancing Risks, Benefits and Access." *Nature Reviews Neuroscience* 4: 323–327.

———. 2006a. *A Palliative Ethic of Care: Clinical Wisdom at Life's End*. Sudbury MA: Jones and Bartlett.

———. 2006b. "Affirming the Right to Care, Preserving the Right to Die: Disorders of Consciousness and Neuroethics after Schiavo." *Supportive & Palliative Care* 4(2): 169–178.

———. 2008. "Neuroethics and Disorders of Consciousness: A Pragmatic Approach to Neuro-palliative Care." In *The Neurology of Consciousness, Cognitive Neuroscience and Neuropathology*, edited by S. Laureys and G. Tononi, pp. 234–244. New York: Academic Press-Elsevier.

———. 2009. When Endemic Disparities Catch the Pandemic Flu: Echoes of Kubler-Ross and Rawls. *Bioethics Forum*. The Hastings Center. April 30, 2009. https://www.thehastingscenter.org/when-endemic-disparities-catch-the-pandemic-flu/

———. 2015. *Rights Come to Mind: Brain Injury, Ethics and the Struggle for Consciousness*. New York: Cambridge University Press. https://doi.org/10.1017/cbo9781139051279.

———. 2016a. "We Can Rewrite the Script for Some Brain Injury Patients." *The Houston Chronicle*. January 15, 2016. page B9.

———. 2016b. "Bring Them Back." *AEON*. 10 May 2016. https://aeon.co/essays/thousands-of-patients-diagnosed-as-vegetative-are-actually-aware

———. 2016c. "Giving Voice to Consciousness: Neuroethics, Human Rights and the Indispensability of Neuroscience." *The Society for Neuroscience David Kopf Lecture on Neuroethics*. Cambridge Quarterly of Health Care Ethics 25(4): 583–599.

———. 2017. "The Civil Right We Don't Think About." Sunday Review. *The New York Times*. August 27, 2017. SR-10.

———. 2019a. "Disorders of Consciousness in Clinical Practice: Ethical, Legal and Policy Considerations." In *Plum and Posner's Diagnosis and Treatment of Stupor and Coma (5 edn)*, edited by Jerome Posner, Saper Cliford B, Claassen Jan, Schiff Nicholas. London; New York: Oxford University Press.

———. 2019b. "A Once and Future Clinical Neuroethics: A History of What Was and What Might Be." *Journal of Clinical Ethics* 30(1): 27–34.

———. 2020a. "The Jeremiah Metzger Lecture: Disorders of Consciousness and the Normative Uncertainty of an Emerging Nosology." *Transactions of the American Clinical and Climatological Association* 131: 235–269.

————. 2020b. "Cruzan and the Other Evidentiary Standard: A Reconsideration of a Landmark Case Given Advances in the Classification of Disorders of Consciousness and the Evolution of Disability Law." *Southern Methodist University Law Review* 73(1): 91–118.

————. 2020c. "Disabusing the Disability Critique of the New York State Task Force Report on Ventilator Allocation." *Hastings Center Bioethics Forum.* https://www.thehastingscenter.org/disabusing-the-disability-critique-of-the-new-york-state-task-force-report-on-ventilator-allocation/

————. 2020d. "The New York State Task Force on Life and the Law Ventilator Allocation Guidelines: How Our Views on Disability Evolved." *Hastings Center Bioethics Forum.* https://www.thehastingscenter.org/new-york-state-task-force-on-life-and-the-law-ventilator-allocation-guidelines-how-our-views-on-disability-evolved/

————. 2020e. Resuscitating Patient Rights during the Pandemic: COVID-19 and the Risk of Resurgent Paternalism. *Cambridge Quarterly of Healthcare Ethics.* Jun 24; 1–15. https://doi.org/10.1017/S0963180120000535.Online ahead of print.

————. 2020f. Sunshine Is the Best Disinfectant, Especially during a Pandemic. *Health Law Journal of the New York State Bar Association* 25(2): 141–146. Published on- line ahead of publication on 1 June 2020: https://nysba.org/sunshine-is-the-best-disinfectant-especially-during-a-pandemic/

Fins, Joseph J., and Samuel Bagenstos. 2020. The Americans with Disabilities Act at 30: A Cause for Celebration during COVID-19? *The Conversation.* July 26. https://theconversation.com/the-americans-with-disabilities-act-at-30-a-cause-for-celebration-during-covid-19-143399

Fins, Joseph J., and James L. Bernat. 2018. "Ethical, Palliative, and Policy Considerations in Disorders of Consciousness." *Neurology* 91(10): 471–475.

Fins, Joseph J., and Barbara Pohl. 2015. "Neuro-Palliative Care and Disorders of Consciousness." In *Oxford Textbook of Palliative Medicine,* edited by Nathan Cherny, Marie Fallon, Stein Kaasa, Russell K. Portenoy, and David C. Currow, pp. 234–244. Oxford: Oxford University Press.

Fins, Joseph J, and Kenneth M Prager. 2020. "The COVID-19 Crisis and Clinical Ethics in New York City." *The Journal of Clinical Ethics* 31(3): 228–232. http://www.ncbi.nlm. nih.gov/pubmed/32773405.

Fins, Joseph J., and Nicholas D. Schiff. 2006. "Shades of Gray: New Insights from the Vegetative State." *The Hastings Center Report* 36(6): 8.

————. 2017. "Differences That Make a Difference in Disorders of Consciousness." *American Journal of Bioethics-Neuroethics* 8(3): 131–134.

Fischer, Juliet, and Claranne Mathieson. 2001. "The History of the Glasgow Coma Scale: Implications for Practice." *Crit Care Nurse Q* 23(4): 52–58.

Fischkoff, Katherine, Mary Faith Marshall, Regina Okhuysen-Cawley, Preeti John, Sabrina Derrington, and Denise Dudzinski et al. on behalf of the Society of Critical Care Medicine. Society of Critical Care Medicine Crisis Standard of Care Recommendations for Triaging Critical Resources during the COVID-19 Pandemic. Accessed on 7 May 2020 at: https://www.sccm.org/COVID19RapidResources/Resources/Triaging-Critical-Resources.

Giacino, Joseph T., S. Ashwal, N. Childs, R. Cranford, B. Jennett, D. I. Katz, J. P. Kelly, et al. 2002. "The Minimally Conscious State: Definition and Diagnostic Criteria." *Neurology* 58(3): 349–353. https://doi.org/10.1212/wnl.58.3.349

Giacino, Joseph T., Kathleen Kalmar, and John Whyte. 2004. "The JFK Coma Recovery Scale-Revised: Measurement Characteristics and Diagnostic Utility." *Archives of Physical Medicine and Rehabilitation* Dec 85(12): 2020–2029.

Giacino, Joseph T., Douglas I. Katz, Nicholas D. Schiff, John Whyte, Eric J. Ashman, Stephen Ashwal, Richard Barbano, et al. 2018a. "Comprehensive Systematic Review Update Summary: Disorders of Consciousness: Report of the Guideline Development, Dissemination, and Implementation Subcommittee of the American Academy of Neurology; the American Congress of Rehabilitation Medicine; and the National Institute on Disability, Independent Living, and Rehabilitation Research." *Neurology* 91(10): 461–470.

————. 2018b. "Comprehensive Systematic Review Update Summary: Disorders of Consciousness: Report of the Guideline Development, Dissemination, and Implementation Subcommittee of the American Academy of Neurology; the American Congress of Rehabilitation Medicine; and the National Institute on Disability, Independent Living, and Rehabilitation Research." *Archives of Physical Medicine and Rehabilitation* 99(9): 1710–1719.

Giacino, Joseph T., Douglas I. Katz, Nicholas D. Schiff, John Whyte, Eric J. Ashman, Stephen Ashwal, Richard Barbano, et al. 2018c. "Practice Guideline: Disorders of Consciousness." *Neurology* 91(10): 450–460.

Giacino, Joseph T., Douglas I. Katz, Nicholas D. Schiff, et al. 2018d. "Practice Guideline: Disorders of Consciousness." *Archives of Physical Medicine and Rehabilitation*. 99(9): 1710–1719.

Guidry-Grimes, Laura, Katie Savin, Joseph A. Stramondo, Joel Michael Reynolds, Marina Tsaplina, Teresa Blankmeyer Burke, Angela Ballantyne, et al. 2020. "Disability Rights as a Necessary Framework for Crisis Standards of Care." *Hastings Center Report* 50(3): 28–32. https://doi.org/10.1002/hast.1128.

Huberman, Barrie J., Debjani Mukherjee, Ezra Gabbay, Samantha F. Knowlton, Douglas S. T. Green, Nekee Pandya, Nicole Meredith, et al. 2020. "Phases of a Pandemic Surge: The Experience of an Ethics Service in New York City During COVID-19." *Journal of Clinical Ethics* 31(3): 219–227. http://www.ncbi.nlm.nih.gov/pubmed/32773404.

In Re Quinlan. 1976. 70 N. J. 10, 355 A.2d 677.

Institute of Medicine. 2012. *Crisis Standards of Care: A Systems Framework for Catastrophic Disaster Response: Volume 1: Introduction and CSC Framework.* Edited by Dan Hanfling, Bruce M. Altevogt, Kristin Viswanathan, and Lawrence O. Gostin. Washington, DC: The National Academies Press. https://doi.org/10.17226/13351

Jennett, Bryan, and Fred Plum. 1972. Persistent Vegetative State after Brain Damage. A Syndrome in Search of a Name. *Lancet* 1(7753): 734–737.

Jennett, Bryan, and Graham Teasdale. 1977. Aspects of Coma after Severe Head Injury. *Lancet* 1(8017): 878–881.

Jennett Bryan, J. Snoek, M.R. Bond, and N. Brooks. 1981. Disability after Head Injury: Observations on the Use of the Glasgow Outcome Scale. *Journal of Neurology, Neurosurgery, and Psychiatry* 44(4): 285–293.

Laureys, S., M. E. Faymonville, P. Peigneux, P. Damas, B. Lambermont, G. Del Fiore, C. Degueldre, et al. 2002. "Cortical Processing of Noxious Somatosensory Stimuli in the Persistent Vegetative State." *Neuroimage* 17(2): 732–741.

Miranda, Lin-Manuel. 2015. Room Where It Happens. *Hamilton.* Act II. Internet Broadway Database. https://www.ibdb.com/broadway-production/hamilton-499521. Accessed on 1 November 2020.

Muller, Jerry. *Tyranny of Metrics.* 2019. Princeton, NJ: Princeton University Press.

Multi-Society Task Force on PVS. 1994a. "Medical Aspects of the Persistent Vegetative State." *New England Journal of Medicine* 330(21): 1499–1508. https://doi.org/10.1056/nejm199405263302107

Multi-Society Task Force on PVS. 1994b. "Medical Aspects of the Persistent Vegetative State." *New England Journal of Medicine* 330(22): 1572–1579. https://doi.org/10.1056/nejm199406023302206

New York State Task Force on Life and the Law. *Ventilator Allocation Guidelines.* New York State Department of Health, November 2015. Accessed on 9 April 2020 at: https://www.health.ny.gov/regulations/task_force/reports_publications/docs/ventilator_guidelines.pdf

Olmstead (527 U.S. 581 [1999]).

Owen, A. M., M. R. Coleman, M. Boly, M. H. Davis, S. Laureys, and J. D. Pickard. 2006. Willful Modulation of Brain Activity in Disorders of Consciousness. Detecting Awareness in the Vegetative State. *Science* 313(5792): 1402.

Posner, Jerome B., Clifford B. Saper, Nicholas D. Schiff, and Jan Claassen. 2019. *Plum and Posner's Diagnosis of Stupor and Coma,* Fifth Edition. New York: Oxford University Press.

Prager, Kenneth M, and Joseph J Fins. 2020. "Meeting the Challenge of COVID-19: The Response of Two Ethics Consultation Services in New York City." *The Journal of Clinical Ethics* 31(3): 209–211. http://www.ncbi.nlm.nih.gov/pubmed/32773402.

Schiavo ex Rel. 2005. Schindler v. Schiavo. F.3d 1289 (11th Cir. 2005).

Schiff, Nicholas D. 2015. Cognitive Motor Dissociation following Severe Brain Injuries. *JAMA Neurology* 72(12): 1413–1415.

Schiff, N. D., J. T. Giacino, K. Kalmar, J. D. Victor, K. Baker, M. Gerber, B. Fritz, et al. 2007. "Behavioral Improvements with Thalamic Stimulation after Severe Traumatic Brain Injury" *Nature* 448(7153): 600–603.

Schiff, N. D., D. Rodriguez-Moreno, A. Kamal, K. H. S. Kim, J. T. Giacino, F. Plum, and J. Hirsch. 2005. fMRI Reveals Large Scale Network Activation in Minimally Conscious Patients. *Neurology* 64: 514–523.

Schnakers, Caroline, Audrey Vanhaudenhuyse, Joseph Giacino, Manfredi Ventura, Melanie Boly, Steve Majerus, Gustave Moonen, and Steven Laureys. 2009. Diagnostic Accuracy of the Vegetative and Minimally Conscious State: Clinical Consensus versus Standardized Neurobehavioral Assessment. *BMC Neurology* 9: 35.

Shahpori, Reza, H. Tom Stelfox, Christopher J. Doig, Paul J. E. Boiteau, and David A. Zygun. 2011. Sequential Organ Failure Assessment in H1N1 Pandemic Planning. *Critical Care Medicine* 39(4): 827–832.

Thibaut, Aurore, Yelena G. Bodien, Steven Laureys, and Joseph T. Giacino. 2020. Minimally Conscious State "Plus": Diagnostic Criteria and Relation to Functional Recovery. *Journal of Neurology* 267(5): 1245–1254.

Wright, Megan S., Nina Varsava, Joel Ramirez, Kyle Edwards, Nathan Guevremont, Tamar Ezer, and Joseph Fins. 2018. "Justice and Severe Brain Injury: Legal Remedies for a Marginalized Population." *Florida State University Law Review* 45: 313–382.

Further Readings

Fins, Joseph J. 2015. *Rights Come to Mind: Brain Injury, Ethics and the Struggle for Consciousness.* New York: Cambridge University Press. https://doi.org/10.1017/cbo9781139051279.

———. 2017. "Brain Injury and the Civil Right We Don't Think About." *The New York Times.* https://www.nytimes.com/2017/08/24/opinion/minimally-conscious-brain-civil-rights.html.

Fins, Joseph J., and James L. Bernat. 2018. "Ethical, Palliative, and Policy Considerations in Disorders of Consciousness." *Neurology* 91(10): 471–475. https://doi.org/10.1212/wnl.0000000000005927.

Fins, Joseph J., Megan S. Wright, and Samuel R. Bagenstos. 2020. "Disorders of Consciousness and Disability Law." *Mayo Clinic Proceedings* 95(8): 1732–1739. https://doi.org/10.1016/j.mayocp.2020.02.008.

Wright, Megan S., Nina Varsava, Joel Ramirez, Kyle Edwards, Nathan Guevremont, Tamar Ezer, and Joseph Fins. 2019. "Severe Brain Injury, Disability, and the Law: Achieving Justice for a Marginalized Population." *Florida State University Law Review* 45(2): 313–382. https://ir.law.fsu.edu/lr/vol45/iss2/1.

19

BIOETHICAL ISSUES IN DEMENTIA AND ALZHEIMER'S DISEASE

Tia Powell

I Introduction

The central ethical challenge of dementia is how to support each person living with dementia and their care partners, a challenge that requires finding the right balance between liberty and safety. The right approach will be different for different people and different for the same person at different times. Our society intervenes and exerts control over decisions and actions made by many sorts of individuals. In the context of dementia, which interventions are ethically justified and which are unjustified? When might support be appropriate and helpful, and on what grounds?

This chapter will examine three domains that involve potential risk to the decision-maker and/or to others: guns, finances, and sexuality. I will consider whether and in what ways, some, all, or none of these activities should be subject to the supervision when a person is living with dementia. I will examine options that range from no intervention, to collaborative support, to limited access, to interdiction. If constraints are ethically supportable in some instances, can they be exercised in a way that promotes respect and dignity for a person with dementia? I will use the perspective of the capabilities approach, which conceives of dementia as an impairment resulting in disability in the sense of the social model.[1] This approach will serve as a lens to see which current practices meet criteria for respectful and helpful intervention and which fail to meet that standard.

Dementia is not universally viewed as a disability. Indeed, it was only in 2015 that dementia activists insisted on being included in the WHO conference on disability (Shakespeare, Zeilig, and Mittler 2019). Dementia is an umbrella category, like cancer, embracing various types of illness, all of which attack the brain and all of which are fatal. Genetic factors play a role in dementia, though a more prominent role in some types than in others. Dementia, like many illnesses, has a biological basis that is the focus of vigorous scientific research. Dementia entails far more than memory loss. By definition, people living with dementia undergo a decline from previous function in memory, but also learning, attention, problem-solving ability, language, perceptual-motor function, and social cognition (American Psychological Association 2013). There are no disease-modifying treatments for most dementia, only symptomatic approaches.

Yet neither dementia nor any other illness is a matter of biology only. Among other things, social determinants of health play a key role in the development of dementia and many illnesses (Powell et al. 2020). Dementia, its associated impairments, and the way it disables people are

DOI: 10.4324/9781003289487-26

most usefully viewed as complex entities with both biological and socially constructed aspects. The challenge of living with dementia arises in part because of biologically based changes in function, and in part because of how society sees, controls, and denigrates those with dementia. Forgetfulness may have a biological basis, but to see someone who is forgetful as stupid is a socially learned response with harmful ramifications. Difficulty understanding money is common as financial incapacity develops, but the anger of those waiting in line at a store, who see the struggling person as a nuisance, as someone in the way, is an additional and weighty burden for the person with dementia. The bias against those living with dementia is profound, for dementia frightens us – we fear the loss of our identity and of our dignity. Fear begets ugly behavior, and leads to biased actions toward those with dementia.

II The Capabilities Approach

The capabilities approach is a set of political and philosophical theories developed by a number of scholars, most notably Amartya Sen and Martha Nussbaum, that offers a holistic account of human well-being and development. Briefly stated, the capabilities approach argues that real freedom of choice is not merely a question of the individual but also of constraints imposed by society, including systemic factors like racism and poverty. Further, these scholars argue that successful human development requires basic material and interpersonal thresholds to be met for anyone to be happy and experience well-being, thresholds like shelter, food, access to healthcare, respect, etc.

Nussbaum specifically takes up the question of cognitive disability within a capabilities framework. She notes that people with cognitive disabilities are "fellow participants in human dignity, [...whose] needs, real and important, have not been adequately addressed by previous theories of justice" (Nussbaum 2009, 331). She asks what is required to demonstrate equal respect and equal dignity for all citizens, including those with cognitive disability. Nussbaum does not define cognitive disability in this essay, but her examples focus on young people with developmental disabilities like Down syndrome, rather than on older people with dementia. For instance, Nussbaum rightly points out the importance of appropriately adapted education for young people with cognitive disabilities, despite the expense of this education. While education is important for the younger group, it is of dubious benefit to older people with dementia, some of whom have already had extensive education.

Nussbaum's capabilities approach can offer us useful insights as we consider the domains of various risky decisions for older adults with dementia. Two of the ten capabilities she discusses are most relevant for the discussion at hand (Nussbaum 2011):

1   Bodily integrity: being able to move about freely, be free against assault, and having opportunities for sexual satisfaction
2   Affiliation: being able to live with others—to engage in social interaction having the social bases for self-respect and non-humiliation

The importance of these two capabilities must be accounted for when making decisions about the ethics of supervising or restricting the behavior of people with dementia.

III Decision Framework

I turn now to an examination of choices in domains fraught with risks: guns, finances, and sexuality. Making choices that include risk is part of what defines adulthood. Each poses different challenges in balancing liberty and safety, and thus maintaining human dignity, for those living with dementia. Society can support those living with dementia in related choices by listening to their perspectives, helping extend their capabilities, and offering support that permits liberty within the limits of safety, but without annihilating the freedom of choice and movement that are essential components of human dignity.

Decisional capacity is a cornerstone of autonomy for medical decisions. Those who can weigh risks and benefits make their own decisions. Those who cannot must have decisions made on their behalf by others. Two key points regarding decisional capacity are worth noting. First, a diagnosis of dementia does not automatically mean a person has lost the right or ability to make decisions. Dementia is a progressive illness; most people with dementia are adults in the early stages of the illness who can clearly express and act upon their life-long values and preferences. Second, capacity is decision-specific, meaning a person may be capable of making some decisions but not others. Fiscal capacity fades earlier in some people than other cognitive skills; a person may readily be able to choose whom they trust to make medical decisions for them even when they can no longer handle complex financial transactions.

This framework categorizes decisions according to the level of decisional capacity and risk, and then offers ethical justification for various degrees of intervention, ranging from none at all to fully taking over the decisional authority of the person with dementia. The framework takes J.F. Drane's concept of a sliding scale to assess decisional capacity for medical treatments, and applies it to non-medical choices as an ethical guide for when and how to offer decisional support (Drane 1984) (Figure 19.1).

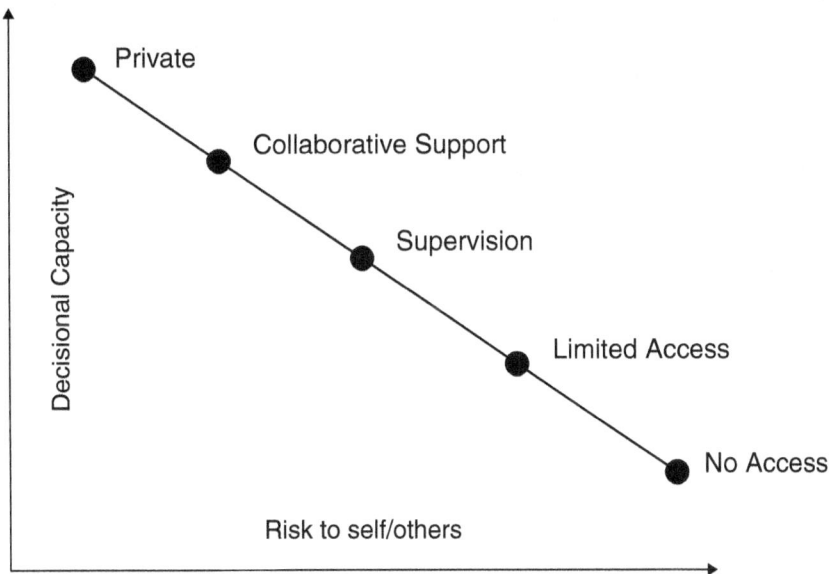

FIGURE 19.1   A Schema of the Relationship between Risk and Decisional Capacity

I propose a schema, represented graphically above, to help us think about capacity, risk, and intervention for decisions in three domains that are inherently risky: guns, money, and sexuality. Risks differ in these domains, and include potential threats to life, security, and dignity, both for one's self and for others. Those in early phases of dementia generally make their own decisions. Some worry about their current or future capacity and use their retained skills to plan for when they no longer have full capacity. As capacity declines, or as risks increase, there are options for supporting the person living with dementia. These supports should reflect the values and preferences of the person living with dementia – which makes advance care planning especially important. When capacity declines and/or risks to self and others increase, intervention may be justified and even required. Analogously to the Americans with Disability Act, we seek the least restrictive alternative; an ethically sound mechanism of support will consider not only safety but also the dignity of the person living with dementia.

For choices with a lower risk, decisions remain essentially private. Within this category, only minimal harm to either the person with dementia or others is raised, or such harm is relatively easy to prevent. Supportive friends may be moved to comment about choices in this category, as is common among adults who do not have cognitive impairments, but the decision remains within the scope of the person with dementia; intervention by others is not ethically warranted. As capacity declines and/or risk increases, the person with dementia expresses values and preferences, but faces greater challenges in making decisions that uphold those values. In this category, collaboration and assistance may help extend the period during which the person with dementia may act and still enjoy both safety and freedom. By analogy, for people with progressive difficulty walking, the best option is usually not to go from wholly unassisted walking to a wheelchair, but to try intermediate measures, such as a walker, that promote cardiovascular health and independence. So too in dementia we draw upon the capabilities approach to support both bodily integrity and non-humiliation. We seek ways to support decisions without canceling altogether the freedom to act.

As the ratio of capacity to risk grows less favorable, the person with dementia wishes to act in a domain, but risk to self and others and/or cognitive impairment have increased to an extent that limits are ethically justifiable. Yet even here, creative approaches to limits should be sought so as to maximize dignity and respect. Can supervision or limited access still enable the person with dementia to participate in this domain? Can innovative technology or other tools help extend autonomy while maintaining adequate safety? Finally, when capacity has severely declined and/or risk is high, adequate safety cannot be attained. Intervention to stop participation in a domain is warranted. However, in severe dementia, a person may no longer be able to act or express a desire to do so; cognitive disability itself imposes restrictions on action. Now is the time to strive to understand the person's wishes and provide comfort with dignity.

As dementia and its related cognitive impairments progress, the same type of decision may move from one category to another. In the earliest phase, cognitive impairment is modest and fewer decisions will need support of any kind. As impairment increases, this may change. Capacity to make decisions in one domain may not be the same as in another. As in all ethics challenges related to capacity, the easy questions fall at the two ends of the spectrum, for people who clearly can and can't act safely within a particular domain. The harder questions are in the middle range, in which a person may be physically able to participate and strongly desire to do so, yet others feel their choices are unwise and unsafe.

3.1 Guns

Americans have a lot of guns, and for some, those guns register heavily in their sense of identity and freedom. Roughly 50% of US households containing a person with dementia include a firearm (Betz et al. 2018). Older white men have especially high rates of gun ownership, dementia, and suicide – a deadly combination of factors (Betz et al. 2018). Guns pose risks not only to people with dementia but also to their family members. Larry Dillon of Princeton, West Virginia had undiagnosed Lewy Body dementia that caused him to hallucinate and believe his wife Sandy was an intruder. He shot her several times, killing her, but missed their visiting nine-year-old granddaughter (Aleccia and Bailey 2018).

Does the disability lens help us think about the challenging, highly politicized issue of gun safety in dementia? An early and crucial step must be to engage those living with dementia and their care partners to learn about their experience and views, and to draft potential solutions. If gun ownership is valued by some as a component of identity and dignity, how can safety measures take that into account? Are there ways to maintain gun access, at least in some form, while also protecting the safety of the person with dementia and those near them? And when gun access is no longer possible, can limits be set in a way that promotes dignity and respects autonomy?

The risk to others from gun violence shifts the balance in favor of a societal right to set limits on gun access for those with dementia. Yet, keeping in mind Nussbaum's capability for affiliation, which includes having a social basis for non-humiliation, those limits must also protect self-respect and dignity. Any limit to gun access should follow from an assessment of risk and a consideration of what steps best match the level of risk. Alas, there are no validated measures of capacity for gun safety; the development of such a tool is a crucial step in balancing both the right to possess guns and the responsibility to do so safely.

In applying the decision framework to firearms, we note there are people with dementia who handle guns safely. This group would not be struggling unduly with depression, irritability, or delusions, and would be likely to have long experience handling firearms safely. Just having a diagnosis of dementia does not justify broad intervention in an adult's choices and life. However, that diagnosis is an opportunity to think about preferences going forward and to discuss them with trusted partners. The Veterans' Administration, knowing that Veterans have a high rate of gun ownership, includes gun safety as a routine part of assessment following a diagnosis of dementia (Betz et al. 2018).

The person with dementia and their care partners may consider what steps might be taken now to improve safety, such as storing guns and ammunition separately, and locking guns in secure locations. Some have proposed addressing guns in advance directives, outlining when family members should limit firearm access (Betz et al. 2018). As dementia progresses, it may be possible to support limited access in a safe setting, such as in a shooting range, a simulator, or without live ammunition. Emergency physician Emmy Betz, MD and colleagues in Colorado have built a website called Safety in Dementia[2] that walks the reader through discussion points and questions related to guns. The tone is practical and non-judgmental, asking questions about who owns the guns and using that as a part of a decision tree about possible steps to take. The website stages advice based on the current perceived risk, and escalates as risk increases, including bringing in law enforcement for assistance. As dementia reaches the end stage, people will be less physically able to use firearms. Limiting access will become physically easier, less contested, and less relevant. In sum, a diagnosis of dementia does not automatically mean a person cannot responsibly use firearms, but it is a reason to think about how the diagnosis and appropriate gun use may evolve over time.

3.2 Money

Jonathan Swift, prolific author and scholar, died in 1745 of what we would recognize as dementia. Along the way, he lost the ability to care for himself and his affairs and the courts appointed a committee to make financial and other decisions for him (Damrosch 2013). Families still use the courts to wrest financial control from those living with dementia. Such an intervention is warranted for a person with the severe cognitive disability of late-stage dementia, who cannot pay bills for needed care. Yet there are people living with dementia who can make financial decisions, wish to do so, and object mightily to interference from relatives. The liberty to make financial decisions is a crucial part of adult identity for many; attempts to intervene may sunder trust between the person with dementia and others. The problem is not a trivial one. Financial abuse of older adults is a booming business, amounting to billions of dollars yearly in the USA (Burnes et al. 2017). Unfortunately, family members are not always protectors, and indeed are frequently involved in financial abuse of older relatives. Have we learned anything since Swift's era to help tailor supervision appropriately, balancing the rights of those with dementia to govern their finances versus the authority of others to intervene, whether for beneficence or self- interest?

As with other cases of decisional capacity, the easier questions fall at either end of the spectrum. The need for assistance is indisputable for those who live to dementia's finale. For those newly diagnosed, there may be no need for current intervention, though this is an important time for the person with dementia to record her preferences and make arrangements that will guide decisions later. There is real risk in loss of financial security. A scam can delete in a heartbeat a lifetime of savings needed for costly dementia care.

Unfortunately, financial capacity, defined as the "ability to independently manage one's financial affairs in a manner consistent with personal self-interest," is often among the earliest cognitive impairments caused by dementia (Widera et al. 2011). The person with dementia, as well as friends and family, may be unaware of these cognitive changes. Situations that lead to bitter disputes are those in which a person with dementia can act, believes they are making sound choices, and others strongly disagree with these decisions. A classic example is the older person who has a new romantic partner, one who arouses suspicion in adult children; in this narrative there are many variations. Adults of any age and cognitive status may choose partners that others dislike – there is nothing here that proves a cognitive disability. Then, too, adult children vary in their motivation and character; some are genuinely concerned for the dignity and welfare of an aging parent, while others are more motivated by their own financial interests.

There are unfortunately scammers who choose to exploit financial incapacity when it occurs in older people. Some of these actions are illegal; stealing someone's bank card and using it in a time and place where the person could not have been can be proven as theft. Earning someone's trust and pleading with them to give you money for alleged emergencies is much harder to prove as illegal. If a person has not been shown to lack capacity and chooses to make a gift, even one that strikes others as foolish, it may be impossible to prove a crime.

Jalayne Arias is a legal scholar and bioethicist who recommends an approach that can be adjusted as disability evolves. Rather than retaining all or none of one's financial authority, her model incorporates the preferences of a person with dementia and cognitive impairment by creating a limited guardianship (J Arias 2013). The goal is to support the choices of the person with dementia while limiting the risk of financial errors. Banks have also come up with helpful options. A joint checking account allows an appointed person to pay bills, and may make this person the sole inheritor of left-over funds. This option, however, does not prevent misuse of funds. There are also read-only accounts that permit a trusted person to observe a banking account without controlling it, leaving the person with dementia in charge while allowing early reporting of unusual activity.

As is the case with guns and other domains fraught with risks, what is needed is an approach that tailors intervention according to different levels of risk. People make mistakes, and that is a part of being an adult. Too much protection too soon undermines the ability of a person with dementia to live their life as they see fit – while safety is important, so is dignity. Interventions that are ethically sound must honor the values and preferences of the person with dementia. As Nussbaum reminds us, an essential part of the capability of affiliation is maintaining the social basis for self-respect and non-humiliation.

3.3 Sex

Sexual intimacy in the context of dementia is overshadowed by many unhelpful cultural biases, including those involving older people, disabled people, and gender. While we view old trees as magnificent because of their age, with their gnarled and twisted boughs, this view is seldom applied to older people. Older bodies, especially women's bodies, are viewed as vile, and unfit for anyone to see. This view of older bodies informs the biased commonplace view of older people and their sexual activity. Presumptions about the sexual activity of older people are strongly controlled by biased notions of gender norms and heterosexism. In the commonplace view, older men may have sexual ideas, but these are deeply wrong; older women have no such thoughts at all; and no older person could have a same-sex orientation. Moreover, any expressed interest in physical intimacy might lead toward sexual intimacy, which is assumed to be bad. In the setting of nursing homes, where three fourths of residents have dementia, these biased threads can weave together into restrictive and harmful attitudes and practices. Attempts at physical intimacy between residents are assumed to be problematic – "a lawsuit waiting to happen." Sexualized approaches to the staff by residents with dementia can result in distress for staff members. However, little training is offered for addressing sexual comments and acts by residents in a respectful and safe fashion.

In an excellent paper, Ward et al. note that most accounts of sexuality in nursing homes are given by proxy, i.e. told by staff, who are predominantly female and heterosexual. As a result, these accounts do not reflect the views and experience of residents with dementia. Sexualized approaches to staff by men, either verbal or physical, are seen as problematic, and are more likely to result in punitive actions, including rejection from a facility or transfer to a more restrictive section. Women with dementia are more likely to be seen as vulnerable and needing protection (Ward et al. 2005). Intake information rarely records sexual orientation. Institutions and family members may object and try to prevent sexual activity for an adult with dementia, especially one who lives in a supervised setting.

Nussbaum's capacities of bodily integrity and affiliation are useful in reflecting on sexuality within dementia. Until fairly recently, an expressed need for intimacy by a person with dementia, especially if living in a nursing home, but even sometimes in the community, was viewed as a "bad" behavior to be controlled and eradicated. Intimacy was not viewed as a normal human yearning for closeness, nor as a private domain of adult life that can promote happiness and health. By labeling the capacity to maintain sexual relations and human companionship as essential to human dignity, Nussbaum helps us see the role of sexuality within dementia in a different way than what the commonplace view permits. Sexual behavior and desire do not disappear magically as the diagnosis of dementia arises, any more than desires for food or kindness disappear. Certainly, sexual behaviors can include risks to both physical safety and dignity for any person, but it is possible to assess the risks in individual cases.

The presumption should be that sexual behavior among adults, irrespective of a diagnosis of dementia, is normal and healthful, and that willing and capable participation can be assessed. These decisions are of an essentially private nature; the intervention of others raises a significant threat to dignity and the freedom to act as other adults do. In sum, the starting point for decisions about sexual behavior is that they remain the province of the person with dementia. If the behavior is unwanted by a recipient, for instance, a staff member at a nursing home, then respectful, non-punitive reactions are appropriate, such as distraction and redirection. If the person with dementia is viewed as being at risk, by having a predatory partner for example, then intervention by family or even legal authorities may be appropriate, and must still be undertaken in a way that preserves dignity.

Intervention by others regarding sexual choices of persons with dementia falls on a spectrum. For a person with full capacity, no intervention is ethically justifiable within this intensely private domain. A next step along the spectrum might be the same sort of inquiry a concerned friend would make of a person who doesn't have cognitive impairment, but who appears to be making a risky choice. For someone with clear cognitive impairments associated with dementia, an assessment of their decisional capacity is warranted and is indeed standard in excellent nursing homes, where this situation arises often. As the cognitive deficits of dementia advance, a person may no longer be capable of expressing or acting upon a choice. A protective role is then ethically justified.

IV Conclusion

A defining feature of adulthood is the right to make risky choices and mistakes. Though dementia can undermine a person's ability to make choices that reflect life-long values, this does not happen in an instant, but gradually over time, and affects domains unevenly. The capabilities approach helps us see more clearly how we can ethically structure support for those living with dementia as they face risky decisions. Choices in the domains of guns, money and sex are all ones that stir fierce feelings of independence; these are private choices, and ones for which no healthy adult welcomes the interference of others.

Evaluating risk in these domains is complex and can be distorted by bias. A gun-related physical injury is easy enough to document, but it is harder to assess the injury to dignity for a lifelong gun owner whose access to guns – and his identity as a gun owner – is ended. Guns place not only the person with dementia but also those nearby at risk, and this must be taken into account in considering limits to gun access. Financial incapacity is common in dementia, and both the person with dementia and their family can be harmed by poor choices. However, an ethically sound intervention will not erase the interests and values of the person with dementia. Though many families provide devoted and unselfish dementia care, it is also true that family members are frequent perpetrators of financial abuse. Graduated steps to include financial supervision, with keen attention to the interests of the person with dementia, should guide attempts to protect. Collaborative efforts to help the person with dementia articulate and attain their goals are appropriate. In the domain of sexuality, biased beliefs about age and sexual behavior may lead well-meaning others to assume that sexual contact is necessarily dangerous and should be prohibited, but this may not reflect the experience of a person with dementia. Too much protection regarding sexual intimacy within dementia can be a mark of infantilization, and thus a harm to dignity. Supporting access to a dignified life with dementia requires us to devise ways to collaborate, to support, and to extend the time and ways of living with dementia according to the values of the person with dementia. Protection plays a role, but so does dignity, and each of these must be honored.

Notes

1 "Disability" as used in this chapter refers to the medical model of disability.
2 https://safetyindemntia.org.

References

Aleccia, JoNel and Melissa Bailey. 2018. "Unlocked and Loaded: Families Confront Dementia and Guns." *Kaiser Health News*. July 1, 2018. https://www.usatoday.com/story/news/2018/07/01/guns-mental-health-elderly-dementia/738845002/.

American Psychiatric Association. 2013. *Diagnostic and Statistical Manual of Mental Disorders: DSM-5*. Arlington, VA: American Psychiatric Association.

Arias, Jalayne J. 2013. "A Time to Step In: Legal Mechanisms for Protecting Those with Declining Capacity." *American Journal of Law and Medicine* 39(1): 134–159. doi: 10.1177/009885881303900103

Betz, Marian E., Alexander D. McCourt, Jon S. Vernick, Megan L. Ranney, Donovan T. Maust, and Garen J. Wintemute. 2018. "Firearms and Dementia: Clinical Considerations." *Annals of Internal Medicine* 169(10): 740.

Burnes, David, Charles R. Henderson Jr, Christine Sheppard, Rebecca Zhao, Pillemer Karl, and Mark S. Lachs. 2017. "Prevalence of Financial Fraud and Scams among Older Adults in the United States: A Systematic Review and Meta-Analysis." *American Journal of Public Health* Aug 107(8): e13–e21. doi: 10.2105/AJPH.2017.303821.

Damrosch, Leo. 2013. *Jonathan Swift: His Life & His World*. New Haven, CT: Yale University Press.

Drane, James F. 1984. "Competency to Give an Informed Consent: A Model for Making Clinical Assessments." *JAMA* 252(7): 925–927.

Nussbaum, Martha. 2009. "The Capabilities of People with Cognitive Disabilities." *Metaphilosophy* 40(3–4): 331–351.

Nussbaum, Martha. 2011. *Creating Capabilities: The Human Development Approach*. Cambridge, MA: Belknap Press.

Powell, W. Ryan, William R. Buckingham, Jamie L. Larson, Leigha Vilen, Menggang Yu, M. Shahriar Salamat, and Barbara B. Bendlin et al. 2020. "Association of Neighborhood-Level Disadvantage with Alzheimer Disease Neuropathology." *JAMA* Network Open 3(6): e207559. doi:10.1001/jamanetworkopen.2020.7559

Shakespeare, Tom, Hannah Zeilig, and Peter Mittler. 2019. "Rights in Mind: Thinking Differently About Dementia and Disability." *Dementia* 18(3): 1075–1088.

Ward, Richard, Anthony Vass, Neeru Aggarwal, Cydonie Garfield, and Beau Cybyk. 2005. "A Kiss Is Still a Kiss? The Construction of Sexuality in Dementia Care." *Dementia* 4(1): 49–72.

Widera, Erik, Veronika Steenpass, Daniel Marson, and Rebecca Sudore. 2011. "Finances in the Older Patient with Cognitive Impairment." *JAMA* 305(7): 698–706.

Further Reading

Powell, Tia. 2019. *Dementia Reimagined: Building a Life of Joy and Dignity from Beginning to End*. New York: Penguin Random House.

Shakespeare, Tom, Hannah Zeilig, and Peter Mittler. 2019. "Rights in Mind: Thinking Differently About Dementia and Disability." *Dementia* 18(3): 1075–1088.

20

BETWEEN "AID IN DYING" AND "ASSISTED SUICIDE"

Disability Bioethics and the Right to Die

Harold Braswell

Bioethical debates about end-of-life decision making have largely focused on one topic: the so-called "right to die." Most broadly, the "right to die" refers to the ability of incurably sick and/or disabled people to end their lives with the assistance of a medical provider. Defined as such, practically every bioethicist accepts some articulation of this right. Bioethical debates focus, rather, on its appropriate scope. These debates fall along—and at the intersection of—two axes: procedure and population.

Procedurally, bioethicists distinguish between voluntary active euthanasia (VAE), physician-assisted suicide (PAS), and the removal of life-sustaining treatment (LST). In VAE, the medical provider directly provokes the patient's death by injecting them with a lethal substance. In PAS, the provider prescribes this substance to the patient, who then takes it on their own. And in the removal of LST, the provider removes the patient from medical technology—such as tube feeding or mechanical ventilation—that was necessary for them to live.

What distinguishes these procedures is the agent that causes death. In VAE, death is caused by the medical provider; in PAS, by the patient; and in the removal of LST, by the underlying disease. Though bioethicists debate such distinctions, they make up a general spectrum along which bioethicists situate their own arguments. On this spectrum, VAE would be the most radical iteration of the right to die, the removal of LST the most conservative, and PAS the midpoint.

There is also a spectrum with regard to the distinct populations that might be eligible to seek the right to die. Most conservatively, the right to die could be limited to only people on LST. It could be extended to include individuals who are "terminally ill," meaning that they have an incurable medical condition that carries a prognosis of six months or less to live. Most radically, it can include anyone who has a chronic illness or disability.

Bioethicists generally debate which combination of procedure and population is ethical. To a degree, disability bioethicists—the term I will use for bioethicists who adopt a perspective grounded in disability studies—have done so as well: contesting, for example, the legalization of VAE or certain instances of LST removal. But, in practice, they have focused on one extremely specific iteration of the right to die: PAS, as made available to terminally ill people in the USA.

In this chapter, I will explain why this engagement has focused on this particular iteration of the "right to die." And I will summarize the debate that has ensued, among disability

DOI: 10.4324/9781003289487-27

bioethicists, about its appropriateness, and provide my own perspective on how it should be resolved. I will conclude with an intervention in the emerging question of whether it is more appropriate to refer to the form of the "right to die" that is legal in the USA as "physician-assisted suicide" or "medical aid in dying."

But first I will outline a disability bioethics approach to the right to die in its most general sense. Though disability bioethicists all support some iteration of this right, their debates about its scope are extremely different from those of bioethicists not primarily grounded in disability studies. To clarify these differences, I will begin by examining a tension that underpins both the larger bioethical debate and the disability bioethics response to it.

I Between Autonomy and Suicide Prevention

Bioethicists have generally been supportive of "patient autonomy": the ability of patients to make decisions about their medical treatment. But this support has not extended to decisions by patients to die. This is because bioethicists generally consider suicidal ideation as an indicator of an underlying mental illness. Patients are thus, by definition, decisionally incompetent to "choose" suicide. Physicians, in turn, are professionally obligated to treat the precipitating mental health condition. They must prevent suicide.

The "right to die" reverses this obligation: making it ethical for providers to assist patients in ending their lives. But it does not do so in general. Its focus is rather a specific population: patients with incurable medical conditions. Such conditions are presumed to provoke suffering; since this suffering originates in an incurable condition, it too is presumed to be relatively incurable. This makes it qualitatively different from suffering provoked by other causes, such as financial hardship or romantic breakup (Dworkin et al., 1997).

The putatively incurable nature of the individual's suffering changes the nature of the patient's decision to die. It makes this decision theoretically rational, since it is rational to choose death over incurable suffering (Werth 1996). This rationality makes the individual patient decisionally competent, and converts suicide assistance into a form of medical treatment that can be autonomously chosen. "Suicide prevention" thus becomes an impermissible violation of patient autonomy, while suicide assistance becomes obligatory.

The "right to die" debate within bioethics has thus focused on the degree to which incurable medical suffering reverses the professional obligation of medical providers to prevent suicide. Some bioethicists argue that it does not do so, often on the grounds of a Hippocratic obligation to "do no harm," or a theologically grounded affirmation of the sanctity of life. Others argue that it can do so, while debating the questions of population and procedure that I raised above. But, underpinning all such disagreements, there has generally been a presumption that the suffering that leads incurably ill individuals to want to die is provoked by their medical condition. It is this underlying presumption that has been criticized by disability bioethicists.

II The Disability Bioethics Critique of the Right to Die

Disability bioethicists agree with right to die advocates that people with incurable medical conditions suffer. But they disagree that this suffering is incurable. This is because its primary cause is not the incurable medical condition itself. Its origin, rather, is society. Individuals with incurable medical conditions suffer, and at times desire to die, because of ableism: social discrimination against disabled people.

This analysis is based on the distinction between the medical and social models of disability. In the medical model, disability is considered to be an inherently negative biological condition.

In the social model, the negative functions associated with disability are considered to be products of social discrimination. Replacing the medical model with the social model is foundational for the disability rights movement (Shapiro 1994). Disability studies scholars have sought to do so in the context of the right to die debate.

This application of the social model of disability to the right to die debate comes with a risk. The incurable medical conditions that are considered to be legitimate motivations for "rational suicide" can include the long-term disabilities that were central to the disability rights movement. But they also include many other conditions: progressive diseases, like amyotrophic later sclerosis, as well as terminal diseases like chronic obstructive pulmonary disease. There is thus the potential for overreach.

Nevertheless, there is also the opposite risk. The conditions that have typically provoked bioethical debates about the right to die are medically complicated. But they also exist in the context of an ableist society. And right to die proponents—and even opponents—have often ignored this context. This can lead to a fundamental misunderstanding of why people seek the right to die, and how bioethicists should respond.

Paul Longmore has illustrated this potential via an analysis of two cases of ventilator-dependent men who asked to be removed from LST (Longmore 2005). Bioethicists treated these requests as rational responses to disability. But Longmore found that, in each case, the individual's desire to die was largely a product not of their disability, but of their involuntary institutionalization in a nursing home. Their suffering was not biological. It was social.

Social suffering is curable. Such a cure does not require a medical cure of disability. It requires, rather, changing the social factors underpinning the suffering: in these cases, moving the individuals in question from a nursing home to a less restrictive environment, such as their homes. Such a change would largely ameliorate the suffering compelling them to die. That is, in fact, precisely what happened with one of the men, Larry McAfee. After overcoming significant political opposition, he received accommodations that allowed him to leave the nursing home. This change led him to recant his desire to die and advocate instead for improved funding for independent living.

In cases like McAfee's, the curable nature of this suffering renders the individual's decision to die not autonomous. It is not "rational"—even in the terms of rational suicide proponents—to want to die as a response to a suffering that could be alleviated through non-lethal means. It is suicidal ideation. It should be treated as any other form of suicidal ideation: with suicide prevention. The application of the social model of disability to the suffering experienced by individuals with incurable medical conditions thus affirms the normal ethical obligation to prevent suicide.

But it changes the nature of what "suicide prevention" normally is. Suicide prevention, as envisioned by Longmore, includes mental counseling. But the attending clinician must be trained in disability psychology. And such counseling is only a starting point. Though McAfee and Rivlin likely had untreated mental health conditions—including depression and post-traumatic stress disorder—the roots of these conditions were social. Changing these roots requires political transformation. And such political transformation is foundational to a disability bioethics model of suicide prevention. Such a model requires the radical transformation of an ableist society.

But such transformation is impossible when bioethicists misclassify suicidal ideation as "patient autonomy." By doing so, they draw attention away from the underlying social conditions driving people like Rivlin and McAfee to want to die. In the process, they ensure that these conditions will continue, thus indirectly provoking suicide. But their provocation is not solely indirect.

For Longmore and other disability bioethicists, the right to die creates a discriminatory double standard in suicide prevention: giving a lesser standard of prevention to incurably sick and disabled people. By doing so, it contains within it the suggestion that such people lead lives that are less valuable. When backed by the authority of a medical provider, this suggestion is highly manipulative. The very gesture of offering the "right to die" to a patient contains within it the suggestion that they should probably take it.

The right to die is thus not a neutral option that can be autonomously chosen. It is inherently coercive, and it reinforces the broader coercion of a society that denies incurably sick and disabled people real choices. It is, in the words of Carol Gill, "the false autonomy of forced choice" (Gill 1998). In their reclassification of suicidal ideation as "autonomy," bioethicists collude with a discriminatory society to drive incurably sick and disabled people to their deaths.

This is what happened in the case of the other man at the center of Longmore's study: David Rivlin. After his petition to leave the nursing home was denied, Rivlin discontinued LST, thus ending his life. After his death, however, it was discovered that he had, in fact, had another option available: he could have received the necessary support to return to his home. But his attending social worker had never pointed this option out to him. Death by social work error hardly seems a model of "autonomy" (Longmore 2005, 40). By celebrating it as such, bioethicists ensure that such unnecessary deaths will continue.

This result is a view of the "right to die" as, at best, indirectly abetting existing disability discrimination. More boldly, one might argue that it is ableism's most radical form: one that reclassifies violence against disabled people as the epitome of "autonomy" and "rationality."

This critique can be made of any iteration of the right to die. Longmore, for example, focuses on the most conservative articulation: the removal of LST. Other scholars have extended it to include VAE and PAS. Disability studies scholars show how underlying the seemingly reasonable desire to die among individuals with incurable medical conditions are a range of social factors that could be changed—but are ignored because of the "right to die" framework.

This critique is often extremely productive in the context of larger bioethical debates. Both supporters and opponents of the right to die, in such debates, tend to overlook how the social context provokes the desire to die in people with incurable medical conditions. The disability studies critique is thus an essential intervention.

But it is not definitive, even among disability bioethicists themselves. On the contrary, there are, and consistently have been, disability bioethicists who accept the core component of this critique—the danger of disability discrimination—while still supporting the right to die. This support is, they claim, consistent with the fundamental principles of the disability rights movement.

III The Value of Autonomy in a "Binocular" World

Defenses of the right to die within disability bioethics have begun with a recognition of the value of patient autonomy within the larger disability rights movement (Batavia 1997; Davis 2013, 95–107; Satz 2002). This value—encapsulated arguably in the slogan "nothing about us without us"—establishes that disabled people should have absolute control over their own medical decisions. The opposition of disability bioethicists to the right to die would, in this view, seem to go against the disability rights movement itself.

But such arguments run into a problem: if self-killing can be a legitimate expression of autonomy then why limit its exercise to individuals with incurable medical conditions? Why not depathogize suicide itself—as libertarian psychiatrist Thomas Szasz has suggested—thus enhancing autonomy for everyone? To legitimize it just for people with incurable medical conditions would seem—as Felicia Ackerman has argued—a double standard that implicitly de-values the lives of that group (Ackerman 1998). Thus, if autonomy is central to disability rights, disability bioethicists should oppose suicide prevention in general, and resist attempts to limit such opposition to individuals with incurable medical conditions.

But this objection can be overcome via a recuperation of the medical model of disability: an argument that incurable medical conditions can make their recipients' lives worse in a manner that is relatively autonomous of the social environment. Indeed, a wide array of dis-ability studies scholars have emphasized the negative aspects of disability that are relatively autonomous of ableism (Kafer 2013; Shakespeare 2013). Such aspects would seem even more salient with regard to the "incurable medical conditions" that might legitimate the right to die, since many such conditions are not normally analyzed with the social model. In fact, failure to consider the medical model in right to die cases can be disastrous, as is illustrated by the case of Sheila Pouliot.

Pouliot was an intellectually disabled woman who remained on tube feeding even after her body could no longer process the protein necessary to survive (Ouellette 2011, 285). While the tube feeding gave her the necessary calories to live, her lack of protein led her body to break down its own tissues for energy. The result inflicted severe pain, damaging Pouliot's organs and inducing swelling that stretched her skin until it began to fall off. But the disability rights group Not Dead Yet opposed removing Pouliot from tube feeding on the ground that it would be dis-criminatory. Such claims neglected to consider the severity of Pouliot's medical suffering—and prolonged it unnecessarily.

There is thus a need, in analyzing the right to die, to adopt what Erik Parens has called a "binocular view": one cognizant of both the social and the medical model of disability (Parens 2017). Such an approach creates a space for an analytically consistent support of the "right to die" within disability bioethics. One in which the biologically harmful aspects of certain in-curable medical conditions transform suicide into an expression of autonomy consistent with disability rights.

IV Maintaining the Tension

There is thus a conflict within disability bioethics between viewing the right to die as a form of disability discrimination and as a legitimate exercise of autonomy: one way to resolve this conflict would be to favor one position: to argue that the "right to die" is, from a disability bio-ethics position, fundamentally an expression of autonomy or of discrimination. But this would be a mistake.

Instead, disability bioethicists should maintain the tension between autonomy and disability discrimination in their analyses. In part the reason for doing so is theoretical: both resisting discrimination and protecting autonomy are equally central values to the disability rights move-ment. They must have a similar centrality in disability bioethics. One cannot trump the other in the abstract.

Rather, what is needed is an open-minded exploration of how particular instances of end-of-life decision making can be shaped by both ableism and relatively autonomous medical suffering. As this question requires addressing a broad range of factors—biological, psychological, social, and philosophical—it requires approach that is both interdisciplinary and applied. Contextual questions about particular applications of the right to die are thus, from a disability bioethics perspective, primary to abstract analysis.

Such questions should be asked of individual clinical situations in places where some form of right to die is legal. Not every exercise of the right to die would be legitimate in such instances, but some could be. For example, that the removal of LST, in the cases of McAfee and Rivlin, became an instrument of disability discrimination does not mean that it was in the case of Pouliot. Indeed, Longmore himself does not argue for banning the removal of LST. A sensitivity toward ableism in individual end-of-life decisions need not require—and in some cases cannot justify—banning the right to die.

But it can do so in some instances. For example, there is a relatively broad consensus, within disability bioethics, that the practice of VAE is problematic both in general and, more particularly, with regard to individuals with incurable disabilities. This consensus is not rooted in an abstract condemnation of VAE, much less the "right to die," but rather in an analysis of its particular dynamics in countries where it is legal. Perhaps disability bioethicists might disagree with this position in time, but if they do so, this disagreement will have to be based in an analysis that is contextually rooted.

There is thus relatively broad consensus, in disability bioethics, about the permissibility of the removal of LST and the impermissibility of VAE. But this consensus does not hold with regard to the midpoint on the "right to die" spectrum: PAS. The resulting debate has largely focused on one particular context: the USA.

V The American Debate

The debate among disability bioethicists about the "right to die" has in fact largely been a debate about PAS in America. This is not coincidental. The American debate is ideally suited to maximizing the tension between the "right to die" being a valid exercise of autonomy or a form of disability discrimination. This is because this debate is relatively moderate. In the USA, there is largely no debate about the legalization of VAE, a practice that even American right to die groups decry. There is similarly no debate about the potential banning of the refusal of LST, not even by critics who argue that it is routinely abused. Instead the debate has been about PAS.

This debate is itself highly constrained by the fragmented nature of US political structures. The US Supreme Court has neither banned nor legalized PAS (Chemerinsky 2006, 847–854). It has rather left the decision up to individual states, which can legalize it within their boundaries through either state court rulings or popular referendum. In this context PAS has been legalized in states with libertarian and, to a degree, liberal political constituencies, and will continue to be with increasing frequency in the years to come. It has not been legalized in states whose populations are more religiously conservative and may never be so. The result is, and will continue to be, a hybrid status.

The framework of this debate limits the potential for disability discrimination. PAS is itself a moderate iteration of the right to die. And this moderate iteration is, in the USA, only available to individuals who are terminally ill. This makes the American interpretation of PAS relatively conservative even relative to other countries where it is legal. This conservatism makes it possible for some disability bioethicists to justify supporting it, though such support is still relatively marginal within the field.

A key question is the degree to which the social model of disability can be applied to terminally ill people. Terminally ill people have medical complications that are extremely severe and frequently traumatic in their speed of onset. At the same time, they have significant disabilities and, in the US context, experience ableism in numerous ways (Braswell 2017, 2019). The degree to which such ableism impacts requests by dying Americans for PAS is a topic of significant debate among disability bioethicists—and it will and, indeed, should remain so in the future. But the very presence of this debate is only possible because of the relatively moderate nature of US right to die legislation.

VI What Is to Be Done

The case of PAS in America is thus reflective of the larger tensions in disability debates about the right to die. Yet it constrains these tensions in a unique context. This raises the question of how disability bioethicists should maneuver within these constraints. One strategy would be to resolve the debate one way or another: to argue that disability bioethicists should support or oppose PAS in America. But the wisdom of this strategy is limited.

In America, the disability bioethics perspective is politically marginal. Legislative changes on PAS are made by constituencies that consider disability discrimination in PAS to be secondary, if they consider it at all. PAS is not legal in Washington because disability bioethicists supported it. Nor is it illegal in North Carolina because of their opposition. Though disability bioethicists should strive to have influence on the general permissibility of PAS, we should also accept that this influence will likely be limited.

Our approach, in light of such limitations, should be to shape whatever view of PAS is dominant in a particular state in a manner that is more amenable to the goal of reducing ableism. If this goal is achieved, then any PAS policy becomes more permissible from a disability bioethics perspective. The resulting approach is fragmentary, but if we accept such fragmentation we can achieve significant results.

For example, North Carolina and Washington both recently passed legislation that significantly improves the home-based long-term care services accessible to dying people. These states have opposing policies on PAS—and will for the foreseeable future. While disability bioethicists can debate the merits of these policies, what is not debatable is that they are each better—from a disability bioethics perspective—because they come accompanied by benefits that will minimize disability discrimination.

Such benefits can be productively supported by disability bioethicists in individual states, and even nationally, through changes to the federal systems of long-term and end-of-life care. Their passage makes any PAS policy more amenable from a disability bioethics perspective. It thus provides a political path to advancing a disability bioethics agenda regarding PAS, and does so in a manner that can be broadly supported within the field.

Such an approach to the right to die is ideally suited to the American context because of the fragmented nature of American governance, the conservative nature of its right to die legislation, and the respective alignments of the country's two dominant political parties.

This context can however change dramatically, and such change—either in the favor of extending the right to die or curtailing it—should be viewed with caution. This caution should inform the way that disability bioethicists address the latest debate about PAS.

VII An Ambivalent Argument against "Medical Aid in Dying"

Recently advocates for PAS for terminally ill people have argued that the term "physician-assisted suicide" is inappropriate to describe the procedure that they support. Suicide, they claim, is an act by people who want to die. But terminally ill people who seek PAS do not want to die. They would, in fact, do anything to not be dying. But they have no choice on that score. They thus do not want to die. They want to hasten a death that is coming whether they want it or not. To categorize them as "suicidal," and the procedure they are pursuing as "physician-assisted suicide" is incorrect (Neumann 2016).

The procedure they are seeking is more accurately described as "medical aid in dying" (MAID). This term captures the fact that the people in question are already dying. They are thus not initiating their own deaths; they are rather expediting a process that is out of their control. As such, deaths that result from MAID are not human-caused. They are products of a natural disease process. Not suicide, they are—and should be categorized as—"natural deaths."

The shift from PAS to MAID is thus not just—or even primarily—terminological. It erases the human action inherent in an individual's self-administration of a fatal substance. "Physician-assisted suicide," in contrast, because of its emphasis on suicide, highlights the human element in the provocation of death. PAS and MAID do not, in this sense, describe the same procedure. They are rather opposite: one caused by human action, one by nature.

In my view, PAS is a far preferable term. It clarifies that the procedure in question is an exceptional form of suicide. Considered as such, it is still possible to support it as a valid exercise of individual autonomy; indeed, most bioethical arguments in favor of the practice have used this term (Weir 1997). But maintaining the term "suicide" draws attention to the possibility that this autonomy could in some cases be an example of suicidal ideation. It thus emphasizes the need to consider disability discrimination. PAS, as a term, thus maintains the tension between disability discrimination and patient autonomy.

MAID erases it. There is no space to consider disability discrimination when one is describing a biological process. MAID thus runs the risk of hiding disability violence under the guise of "nature." As Longmore has shown with regard to the removal of LST, the temptation to mischaracterize suicides as "natural deaths" is already present in bioethics. MAID expands the potential for such mischaracterizations to occur.

VIII Conclusion

I am thus supportive of disability bioethicists utilizing the term "physician-assisted suicide," rather than "medical aid in dying." But this support is ambivalent. This is because PAS does run the risk of stigmatizing a practice that can be a legitimate expression of autonomy in the face of a relatively autonomous biological process. It protects against disability discrimination. But it also encroaches on the individual autonomy of dying people. Like MAID, it too risks collapsing the tension that, as I have argued, should be constitutive of a disability bioethics approach.

In fact, a certain amount of ambivalence about the right to die is, in my view, a good thing. This ambivalence maintains the tension that should be constitutive of any disability bioethics approach. For this reason, my ultimate position is best described as in favor not of PAS, but of the necessity of maintaining the ambivalent space between PAS and MAID. This ambivalence, far from leading to an inability to intervene, can be the source of dynamic action in a bioethical debate too often characterized by certainty. When it comes to end-of-life decision making, such certainty erases the diverse, conflicting experiences of disabled people and the complexities of decision making at the end of life. Puncturing it is perhaps the most productive intervention that disability bioethics can make.

References

Ackerman, Felecia. 1998. "Assisted Suicide, Terminal Illness, Severe Disability, and the Double Standard." In *Physician Assisted Suicide: Expanding the Debate*, edited by Margaret P. Battin, Rosamond Rhodes, and Anita Silvers, pp. 149–162. New York: Routledge.

Batavia, A. I. 1997. "Disability and Physician-assisted Suicide." *New England Journal of Medicine*, *336*(23): 1671–1673.

Braswell, Howard. 2017. "Putting the 'Right to Die' in Its Place: Disability Rights and Physician-Assisted Suicide in the Context of US Hospice Care." *Studies in Law, Politics, and Society*, 76: 75–99.

———. 2019. *The Crisis of US Hospice Care*. Baltimore, MD: Johns Hopkins University Press.

Chemerinsky, Erik. 2006. *Constitutional Law, Principles and Policies*. New York: Aspen Publishers.

Davis, Lennard. 2013. *The End of Normal.* Ann Arbor: University of Michigan Press.

Dworkin, Ronald, Thomas Nagel, Robert Nozick, John Rawls, Judith Jarvis Thomson, and T.M. Scanlon. 1997. "Assisted Suicide: The Philosophers' Brief." *The New York Review.*

Gill, Carol. 1998. The False Autonomy of Forced Choice: Rationalizing Suicide for Persons with Disabilities. In *Contemporary Perspectives on Rational Suicide,* edited by James L. Werth, pp. 171–180. New York: Routledge.

Kafer, Alison. 2013. *Feminist, Queer, Crip.* Bloomington: Indiana University Press.

Longmore, Paul K. 2005. "Policy, Prejudice, and Reality Two Case Studies of Physician-Assisted Suicide." *Journal of Disability Policy Studies,* *16*(1): 38–45.

Neumann, Ann. 2016. *The Good Death: An Exploration of Dying in America.* Boston, MA: Beacon Press.

Ouellette, Alicia. 2011. *Bioethics and Disability: Toward a Disability-Conscious Bioethics.* Cambridge: Cambridge University Press.

Parens, Erik. 2017. "Choosing Flourishing: Toward a More 'Binocular' Way of Thinking about Disability." *Kennedy Institute of Ethics Journal* 27(2): 135–150.

Satz, Ani B. 2002. "The Case Against Assisted Suicide Reexamined." *Michigan Law Review,* *100*(6): 1380-1407.

Shakespeare, Tom. 2013. "The Social Model of Disability." In *The Disability Studies Reader,* edited by Lennard. Davis, pp. 214–221. New York: Routledge.

Shapiro, Joseph P. 1994. *No Pity: People with Disabilities Forging a New Civil Rights Movement.* New York: Three Rivers Press.

Weir, Robert F. ed. 1997. *Physician-Assisted Suicide.* Bloomngton, IN: Indiana University Press.

Werth, James L. 1996. *Rational Suicide? Implications for Mental Health Professionals.* New York: Taylor & Francis.

Further Reading

Braswell, Harold. 2018. "Putting the Right to Die in Its Place: Disability Rights and Physician- Assisted Suicide in the Context of US End-of-Life Care." *Studies in Law, Politics, and Society,* special issue "Accommodation, Recognition, Justice: Legal Treatment of Persons with Disabilities," 72: 75–99.

Braswell, Harold. 2011. "Can There Be a Disability Studies Theory of End-of-Life Autonomy?" *Disability Studies Quarterly,* *31*(4): n.p.

Gill, Carol J. 2010. "No, We Don't Think Our Doctors Are Out to Get Us: Responding to the Straw Man Distortions of Disability Rights Arguments against Assisted Suicide." *Disability and Health Journal* 3(1): 31–38.

Longmore, Paul K. 2005. "Policy, Prejudice, and Reality: Two Case Studies of Physician-assisted Suicide." *Journal of Disability Policy Studies* 16(1): 38–45.

Peace, William J. 2012. "Comfort Care as Denial of Personhood." *Hastings Center Report* 42(4): 14–17.

Riddle, Christopher A. 2017. "Assisted Dying & Disability." *Bioethics* 31(6): 484–489.

21

THEORIZING THE INTERSECTIONS OF ABLEISM, SANISM, AGEISM AND SUICIDISM IN SUICIDE AND PHYSICIAN-ASSISTED DEATH DEBATES

Alexandre Baril

> With the exception of the libertarian position that each person has a right against others that they not interfere with her suicidal intentions […] each of the moral positions on suicide we have addressed so far would appear to justify others intervening in suicidal plans […]. Little justification is necessary for actions that aim to prevent another's suicide but are non-coercive. […] The more challenging moral question is whether more coercive measures such as physical restraint, medication, deception, or institutionalization are ever justified to prevent suicide and when.
>
> *(Cholbi 2017)*

In his historical, philosophical and moral analysis of suicide, Cholbi (2017) illustrates a consensus among philosophers, bioethicists, as well as health professionals and researchers interested in suicide.[1] They agree that suicide should not only be discouraged, but should also be prevented, even if this sometimes involves coercion and violation of core values of bioethics, such as autonomy or the right to refuse treatment. As discussed by Cholbi, except for a few philosophers such as the libertarians who defend a right to suicide, very few throughout history have questioned the "logic of life," to reuse Améry's (1999) words, or what I call the "injunction to live and to futurity" (Baril 2017, 2018, 2020) that constitutes a dominant system of intelligibility. When this "logic of life" is questioned and when a right to suicide is put forward (Améry 1999; Szasz 1999; Stefan 2016), it is always formulated as a negative right, i.e. a right of non-interference with the liberty to complete a suicide. Although a positive right to suicide, involving duties and responsibilities from others or the state, has not yet been formulated (except, to my knowledge, in my previous work), an abundant literature has emerged in the past decades to propose a positive right to physician-assisted death (PAD) for those who are disabled, sick, ill and at the end of their life. In other words, philosophical and ethical discussions on the necessity of supporting some suicides, namely those considered "normal" and "rational," through PAD have only emerged in relation to disability, sickness, illness and old age, leaving unexamined the injunction to live and what I call "compulsory liveness" imposed on suicidal subjects (Baril 2020).

DOI: 10.4324/9781003289487-28

In this chapter, which is divided into five sections, I first describe the ontology of PAD (i.e. what PAD is and is not, what characterizes it and what its defining features are) and discuss the ableist, sanist and ageist presumptions in bioethical reflections on suicide and PAD. In the second section, I examine anti-ableist perspectives on suicide and PAD. In the third section, I show that the bioethical debates opposing "mainstream" bioethicists and disability activists/scholars/bioethicists[2] erase the voices of suicidal people and the oppression they experience, which I call suicidism (Baril 2018, 2020). In the fourth section, I argue that developing an intersectional disability bioethics on suicide and PAD requires taking into consideration not only ableism but also sanism, ageism and, importantly, suicidism. The fifth and last section demonstrates that an intersectional lens on suicide and PAD that includes an anti-suicidism perspective can bring a conversational switch, from arguing for a right-to-die for disabled/ill/sick people based on an individualist conceptualization of autonomy to advocating for positive rights and better support, including PAD access, for suicidal subjects from a structural and an anti-oppressive perspective that promotes relational autonomy.

I The Ontology of PAD: Ableist, Sanist and Ageist Presumptions in Bioethical Reflections on Suicide and PAD

> [...] Rational suicide is a subset of suicidality by people who possess medical decision-making capacity. These cases are sometimes argued in courts of law, where it is often determined that the autonomous actions by autonomous individuals are competent and rational. Those presenting in this fashion, are not the mentally distraught and oftentimes intoxicated persons who might be reacting to a recent or series of stressors in their lives, and clearly lack decision-making capacity or competence. This is not the impetuous patient who may be "acting emotional" and/or "seeking attention" after a recent break up from their significant other. Nor is it a person who suffers from a truly volitional or cognitive disorder. My arguments focus on those people who have acted in a competent and rational manner up until the point where they decided to end their lives. They may be suffering from an incurable ailment or have emotional and/or physical pain that no medication or therapy can control, or they may just feel that they have lived long enough and feel it is time to end their lives.
>
> *(Giwa 2019: 118–119)*

Al Giwa, in his "treatise on rational suicide," aptly represents the position of a majority of philosophers and bioethicists regarding suicide: it is fundamentally an irrational act (anchored in mental illness), except in a few marginal circumstances. Giwa states that among those exceptional circumstances that transform the irrationality of suicide into a rational act is the fact of being severely disabled, ill (physically or mentally) or old. While Giwa doesn't argue for PAD access for those who would like to complete what is considered a rational suicide, this distinction between irrational and rational suicide is often mobilized in bioethical discussions to justify PAD for the latter kind (Hewitt 2010; Creighton et al. 2017; Braswell 2018; Kious and Battin 2019).

Harold Braswell (this volume), in his review of the various positions of bioethicists on PAD, in particular regarding eligible populations and procedures for the right-to-die, establishes a spectrum, which ranges from a conservative position that targets only people on life-sustaining treatments to a more radical position that would include people living with non-terminal chronic disabilities and illnesses. In all these positions, what is clear is the focus of bioethicists on disability/sickness/illness, with rifts and fraught discussions emerging only when it comes to the type or severity of disability/sickness/illness.

For example, terminal illnesses tend to obtain more approval than non-terminal ones, since some think that PAD only changes the time of death instead of inducing it and that the closer one is to dying, the less PAD robs that person of a future (Kious and Battin 2019, 32). I argue that the criterion of terminality for PAD eligibility is both ableist and ageist. First, it is an ableist criterion since it targets only sick and ill people at the end of their life. Second, it is an ageist criterion because it creates a justification of PAD based on longevity. Indeed, the longer one has to live, the more their life is valued and protected. This message and insistence on longevity is detrimental to disabled/sick/ill people who have a shorter life expectancy and who are encouraged to pursue PAD regardless of their chronological age (they might even be minors). However, older adults are affected disproportionately, being in the later part of their lives and possessing, on average, a shorter life expectancy than younger people.

There are also disputes pertaining to the type of disability/sickness/illness that makes someone eligible for PAD, such as debates surrounding requests made solely on the basis of mental illness. On the one side is the idea that suicidality is caused by mental illness that impairs decision-making capacity and annihilates autonomy, and thus the desire to die in the absence of physical disability/sickness/illness is cast as abnormal or irrational (Creighton et al. 2017). On the other side, other bioethicists mobilize the parity argument to show that mental suffering can be as real and irremediable as physical suffering, therefore justifying the support for people with mental illness to access PAD without concomitant physical conditions (Appel 2007; Rich 2013; Kious and Battin 2019). These authors show that while terminality is a central criterion for allowing PAD in some national contexts, such as in the US, it is not the case in other national contexts such as the Netherlands or Belgium, which insist on unbearable suffering. However, when bioethicists support the extension of the right-to-die to mentally ill people, I contend that they do so on the basis of what they consider to be a severe and irremediable psychiatric diagnosis, thereby establishing a medical condition as the paramount criterion to justify PAD, leaving unexamined the ontology of PAD based on disability/sickness/illness, be it physical or psychological. In other words, the ontology of PAD rests upon a presumption that a medical diagnosis is necessary to justify assisted death.

I argue that whether they are for or against the inclusion of mental illness as a legitimate condition to qualify for PAD, bioethicists in each camp (and those in between) endorse sanist presumptions. On the one hand, those who refuse to extend PAD to mentally ill subjects adhere to some stereotypes about "craziness" and "madness," such as impulsivity, uncontrollability and dangerousness, and endorse unproven assumptions that equate mental illness with mental incompetence (Cholbi 2017; Creighton et al. 2017). A vast literature has emerged in the past decades in both Mad studies and disability bioethics to show how irrationality due to mental illness is not constitutive and that the decision-making capacity of people living with mental illness is not necessarily globally impacted (Appel 2007; Hewitt 2010; Burstow et al. 2014; Stefan 2016; Borecky et al. 2019). In other words, dismissing mentally ill people's requests for PAD sheerly on the basis that they are irrational and incapable of making decisions regarding their death is a form of sanism. On the other hand, those who argue for extending access to PAD for mentally ill people do not question the ontology of PAD and its targeted population: those who are at the end of their lives, disabled, sick or ill, including those who are mentally ill (Appel 2007; Rich 2013; Kious and Battin 2019). This PAD ontology targeting marginalized groups constructed on implicit forms of ableism, sanism and ageism remains unquestioned in mainstream bioethics discussions.

II Examining Anti-Ableist Perspectives on Suicide and PAD

> [Assisted suicide proponents'] ableism is so extreme that they want to carve a vaguely
> defined segment of old, ill and disabled people out of suicide prevention, enlist our health-
> care system in streamlining our path to death, and immunizing everyone involved from
> any legal consequences, thereby denying us the equal protection of the law.
>
> *(Coleman 2020)*

The exceptionalism, or the "right-to-die exception," to use Margaret Wardlaw's words (2010) regarding the suicidality of disabled/sick/ill people, in comparison with those regarded as able-bodied, healthy and sane, has framed the binary opposition between suicide and PAD (Braswell, this volume). This exceptionalism has long been critiqued by disability activists/ scholars/bioethicists (Coleman 2010, 2020; Ho 2014; Reynolds 2016; Wieseler 2020). Gill (1992, 1999, 2004) dedicated several papers to what she calls "selective" suicide intervention (1992, 37) that marginalizes disabled people based on the devaluation of their lives. Indeed, Gill (2004, 178–179) argues that PAD proponents adhere to three postulates that contribute to this devaluation: (1) disability causes despair and depression; (2) this despair is irreversible and irremediable; (3) suicide prevention should not be put forward when the despair is founded in disability/sickness/illness. As Gill contends:

> Those assumptions have triggered the charge from disability activists that assisted sui-
> cide is blatantly discriminatory. The practice, they point out, is not universally offered
> to all adult citizens but is offered only to persons who have incurable biological defects
> [I add, in some cases, what are considered psychiatric "defects"]. Moreover, the prac-
> tice calls for a two-tiered response from health professionals: if the individual has an
> incurable disabling condition, the wish to die can be judged rational and the individual
> can be helped to die, whereas "healthy" individuals who wish to die are given suicide
> intervention to save their lives.
>
> *(Gill 2004, 179)*

In other words, the PAD ontology founded in ableism/sanism (among other oppressive systems) and on the systemic dismissal of the quality of life of disabled/sick/ill people through epistemic oppression (Reynolds 2016; Wieseler 2020) creates "[…] two classes of suicidal subjects by considering physically disabled or ill people as legitimate subjects who should receive assistance in dying and suicidal people as illegitimate subjects who must be kept alive […]" (Baril 2017, 201). From an anti-ableist perspective, impairments are not the main nor the only source of suffering, but despair and the wish to die stem from the ableist oppression that pathologizes, marginalizes and discriminates against disabled/sick/ill people. In sum, oppression shapes suicidality and the remedy should not be to offer an individual solution through PAD but to address the social, political, medical, legal, and economic conditions at the root of the problem (Gill 1992, 1999, 2004; Coleman 2010, 2020; Wardlaw 2010; Ho 2014). While there is no consensus in disabled/ sick/ill communities, PAD is generally critiqued by or, at least, looked at from a critical stance by disability activists/scholars/bioethicists (Braswell 2018). Moreover, they insist on the fact that the notion of autonomy put forward by PAD proponents is individualist and does not take into account contexts, structures and forms of oppression that impact autonomy and choices (Ho 2014; Ho and Norman 2019; Braswell, this volume). As many argue, the autonomous decision to die for disabled/sick/ill people relies on a false notion of autonomy (Gill 1992, 1999, 2004; Ho 2014).

III Erasing Suicidal People's Oppression in Bioethical Debates on Suicide and PAD

Bioethical debates on PAD tend to conceal issues and difficulties experienced by suicidal people. On one side, bioethicists categorize suicidality as irrational because it stems from mental illness or, alternatively, mobilizes the mental illness diagnosis as the ultimate criterion to justify PAD. In both cases, however, they are leaving aside the realities and needs of suicidal individuals that don't fit this ontological PAD narrative. The result is either denying positive rights and support for suicidal people and casting their wishes to die as irrational or advancing a right-to-die for only a small subset of suicidal people based on ableist, sanist and ageist criteria. When suicidality is not filtered through that binary lens, philosophers and bioethicists usually discuss the (im)morality of suicide. While some argue, often based on a Kantian rationale, that suicide is immoral, other adopt a more existentialist or libertarian view on the topic (Cholbi 2017). But even among those who suggest that suicide is not immoral and should not be condemned or pathologized, the right to suicide is conceptualized as a negative right or a liberty to complete suicide, not as a positive right that requires external support to be fulfilled. For example, the psychiatrist Thomas Szasz (1999) denounces the social and medical mistreatments reserved for suicidal people. However, his ethics of suicide, anchored in a liberal conceptualization of liberty, choice and autonomy, only endorses suicide when it represents a "private" affair. In a similar fashion, Susan Stefan (2016), despite her demonstration that suicidal people are stigmatized, discriminated against and cast as irrational in our Western legal systems, does not argue for a positive right to suicide for suicidal people and claims that it is not the society's responsibility to help them to die. In sum, suicidal people don't appear to be a priority or are erased in bioethical discussions on PAD.

On the other side, some disability activists/scholars/bioethicists question the emphasis put on the right-to-die question for disabled/sick/ill people in bioethics discussions at the expense of other more pressing issues, while others condemn PAD and its ableist roots. From an anti-ableist perspective, since the causes of the wish to die among disabled/sick/ill people result from society's oppression, PAD doesn't constitute an appropriate response, and social and political solutions should be put forward (Gill 1992, 1999, 2004; Ho 2014; Ho and Norman 2019; Braswell, this volume). By doing so, disability activists/scholars/bioethicists either don't question or address suicidal people's realities and needs or reaffirm the importance, for all citizens including disabled/sick/ill people, of suicide intervention. For example, Gill (2004, 178–179) emphasizes the necessity of suicide prevention strategies, including "psychotherapy, dissuasion, hospitalization, or forms of protective vigilance." In their analysis of the double standard about the suicidality of disabled/ill/sick people, the well-known disability rights group Not Dead Yet is another example of this acritical endorsement of suicide prevention strategies. Not Dead Yet calls for "enforc[ing] laws requiring health professionals to protect individuals who pose a danger to themselves" (Coleman 2010, 44). Not only does Not Dead Yet not interrogate some of the harmful practices such as involuntary hospitalization (Szasz 1999; Stefan 2016; Baril 2017, 2018, 2020; Borecky et al. 2019) put forward in the suicide prevention script adopted by all official organizations, the State, and the Law, but adds that suicidal people, be they disabled or not, should be left to fend for themselves in their search for death: "[...] the law should leave them to their own devices. Any competent person, however disabled, can commit suicide by refusing food and water" (Coleman 2010, 49). This solution seems, to me, insensitive to suicidal people, by forcing them to die through solitary and violent means, such as starvation, poisoning, gunshots or hanging. Furthermore, this *laissez-faire* attitude toward a highly marginalized group such as suicidal people who are often criminalized, institutionalized, or stigmatized based on their perceived or actual mental illness (Szasz 1999; Stefan 2016;

Baril 2017, 2018, 2020) seems at odds with the structural analyses usually put forth by disability, crip and Mad movements.

Suicidal people are often forgotten or erased in publications that should, in my opinion, include discussions of them in the fields of disability, crip and Mad studies. While those fields of study have been quite vocal about PAD debates, the topic of suicide itself, outside the PAD context, has remained under-theorized. For example, in the edited volume *Psychiatry Disrupted* (Burstow et al. 2014, 10–13), the authors note seven groups (e.g. trans people, Indigenous people, older adults) who are not extensively discussed in their book. Despite the fact that suicidal people experience high rates of psychiatrization, pathologization and forced institutionalization and treatments, and despite the fact that many Mad people are subjugated to psychiatric treatments on the basis of the "danger" they pose to themselves, suicidal people remain invisible and are not mentioned among the under-analyzed groups. This is only one example among many that illustrate how suicidal people's intersecting realities and oppressions are erased when it comes time to discuss topics relevant to them, such as sanism, suicide and PAD.

Even in the field of critical suicidology, in which the individualist and medical approach to suicide is highly critiqued in favor of a social, political and structural anti-oppressive analysis of suicidality affecting marginalized groups, the oppression faced by suicidal people remains unexamined. As discussed in my previous work, the oppression of suicidal people is so pervasive, invisible and commonly endorsed – including by both sides of the heated bioethical debates on suicide and PAD – that until recently, this oppression didn't have a name (Baril 2017, 2018, 2020). In an era marked by diversity and in which intersectional analyses have provided nouns for almost every oppressive system (based on sex, class, race, gender identity, sexual orientation, age, language, ethnicity, religion, size and so on), the absence of a noun to name the specific oppression suicidal people experience is quite revealing. I therefore coined the neologism suicidism that I define as "[...] an oppressive system (stemming from non-suicidal perspectives) functioning at the normative, discursive, medical, legal, social, political, economic, and epistemic levels in which suicidal people experience multiple forms of injustice and violence [...]" (my translation, Baril 2018, 193). Erasing suicidal people's experiences of stigmatization and discrimination is a form of suicidism that can be found in bioethical discussions on suicide and PAD. I contend that an intersectional disability bioethics on suicide and PAD that wishes to take into account suicidal subjects' concerns needs to go beyond the opposing perspectives of bioethical debates on PAD that insist either on individual choices/liberties or on the overdetermination of choices informed by ableism.

IV Developing an Intersectional Disability Bioethics on Suicide and PAD

What distinguishes a bioethics on suicide and PAD from a *disability* bioethics on suicide and PAD is the anti-ableist perspective and agenda at the heart of the latter. While it seems crucial to build a strong disability bioethics on suicide and PAD, since most bioethical discussions on the topic either don't address or, worse, dismiss the ableist context in which those discussions are framed as well as the voices, concerns and claims of disabled/sick/ill people (Wardlaw 2010; Ho 2014; Reynolds 2016; Wieseler 2020), it also seems to me important to develop an *intersectional disability bioethics* on suicide and PAD. Not only would this allow us to take into consideration dis/ability and ableism in relation to suicide and PAD (for example, to examine how unemployment, poverty, institutionalization, etc., shape suicidality or PAD requests), but it would also enable us to better understand how dis/ability and ableism maintain complex relationships to other identities and forms of oppression, such as sexism, heterosexism, cisgenderism, racism, colonialism, classism, and ageism. Experiencing a terminal illness or degenerative disability when you are white or racialized, religious or not, poor or rich, transforms your perceptions, values and political views on suicide and PAD. I contend that the axis of suicidality/suicidism that has so far been absent from intersectional analyses (Baril 2017, 2018, 2020) must be added in order to develop a relevant intersectional disability bioethics on suicide and PAD. Indeed, as we can see currently in bioethics and even in disability bioethics, not including suicidism in the conversation leaves unexamined compulsory liveness, the injunction to live and to futurity, as well as current problematic suicide prevention strategies. For example, Braswell (this volume) states that suicidal ideations should be prevented due to a "normal ethical obligation to prevent suicide." Integrating suicidism into intersectional analyses allows us to question the "normality" of preventing suicides, which is assumed by both proponents and opponents to PAD, and the harmful practices that often come with that unexamined normative script.

As demonstrated above, suicidal people's voices have been systematically erased from the bioethical conversations on suicide and PAD (regardless of the position), an act that could be qualified as epistemic violence by denying their legitimacy and credibility as knowing subjects on the topic (Baril 2020).[3] Indeed, suicidal people are simply brushed aside in the bioethical debates on suicide and PAD, which are focused entirely on disabled/ill/sick people. Paradoxically, while disability activists/scholars/bioethicists are advocating in those debates for a better epistemic justice for disabled/sick/ill people, they often, as do "mainstream" bioethicists, forget the primary group of people who should be at the heart of those discussions on suicide and PAD: suicidal people.

This represents a missed opportunity for those disability activists/scholars/bioethicists, since being a suicidal and disabled/sick/ill person are not mutually exclusive categories and because ableism, sanism, ageism and suicidism are profoundly imbricated; analyzing one form of oppression without the others (as well as other oppressive systems) can provide only a partial insight into these issues. Indeed, to give but one example among many, several medical and psychiatric treatments, denounced as inhumane by critical disability and Mad activists/scholars and some bioethicists (Szasz 1999; Burstow et al. 2014; Stefan 2016; Borecky et al. 2019), such as forced hospitalization, physical and chemical restraint, are imposed not only on those considered "crazy" and "mad" but particularly on "crazy/mad" people who (might) represent a danger to others (threats of violence) or to themselves. In other words, in many national legal contexts, a person is often forcibly institutionalized, physically restrained or involuntary drugged because they are simultaneously suicidal or perceived as such (or have threatened somebody else).[4] As we can see, it becomes hard or impossible to distinguish between the systems of oppression: being suicidal itself is enough to be labeled as "crazy/mad"; being mentally ill requires a medical/psychiatric "cure" and coercive treatments are seen as the norm when it comes to people who represent a "danger to themselves." While the desire to cure disabled/sick/ill people is critiqued in disability/crip/Mad studies/movements, surprisingly, the need to cure suicidal individuals, a desire that emerges from the same ableist/sanist medical and psychiatric systems and that mobilizes the same kinds of narratives and coercive tools, is not questioned in those fields/movements. Integrating suicidality/suicidism into our intersectional analyses seems to be a unique opportunity to provide a richer, more complex and nuanced understanding of how ableism and sanism function and are embedded in our medical-industrial complex, policies, cultural values and representations, as well as in economic/social/political/legal systems that attempt to cure all of those considered as "broken," including suicidal subjects.

V Switching Conversations: From PAD for Disabled/Sick/Ill People to PAD for Suicidal People

Some bioethicists, such as Braswell (2018), suggest that, in order to overcome the unproductive debates on PAD, we need to refuse to take a clear position for or against PAD and switch the focus of our reflections and scholarship to more pressing issues that afflict the disability communities. While I cannot agree more with Braswell that we need to put more emphasis on what really counts in the daily experiences of the vast majority of disabled/sick/ill people, such as independent living, affordable housing, proper access to health care and employment, social, legal and economic support, I disagree that disability bioethicists should avoid taking a position in that debate. On the contrary, I contend that not taking a side is affirming a position. While this debate is complex and requires nuanced thinking in order to avoid dogmatic positions, it is clear to me, and to a vast majority of disability activists/scholars, that the PAD ontology is rooted in deep forms of ableism, ageism, and, as I showed, two forms of sanism. When it comes to PAD, it seems that allyship with disability/Mad communities would be for bioethicists to denounce the ableist/sanist/ageist violence that has structured and continues to shape PAD discussions, regulations, laws and social policies. I would add that avoiding taking a position or, worse, denying that this violence exists at the core of the PAD ontology as currently framed reproduces (micro-) aggressions toward disabled/sick/ill/Mad/old people. This is why in my work, in the spirit of disability activists/scholars, I firmly denounce PAD and its current ontological, social, political, and legal aspects.

Simultaneously, I firmly adhere to a positive right-to-die, not for disabled/sick/ill/Mad people but for suicidal people (be they disabled/sick/ill/Mad or not). It is a right-to-die, so far unthought, unintelligible and unacceptable, for both mainstream bioethicists and disability activists/scholars/bioethicists, who are all working under a suicidist regime, with its injunction to live and to futurity. As I wrote regarding the Canadian Law on Medical Assistance in Dying:

> Current laws, public policies, prevention strategies and models/discourses on suicide do not represent accountable, pragmatic or compassionate responses toward suicidal people. From a harm reduction approach, focused on the voices and well-being of suicidal people, my goal is not to reform the medical assistance in dying law to include suicidal people, but to propose an entirely different socio-politico-legal project. I suggest that this law should be repealed because it is doubly ableist and propose instead that, regardless of physical condition or imminent death, all people who wish to die, including suicidal people, should have access to medically assisted suicide.
>
> *(Baril 2017, 212)*

My position, often mistaken for the position of those proponents for the right-to-die who want to extend PAD access to people who experience psychological suffering (and/or other forms of suffering) but who continue to adhere to the ontological PAD ableist/sanist/ageist script, is radically different than what has been proposed so far. I do not believe in reforms of the current laws in any national context in which various forms of PAD are legalized. I propose instead an abolitionist social, political and legal agenda regarding those laws that are fundamentally violent and discriminatory for disabled/sick/ill/Mad/old and suicidal people.

Adopting a more robust intersectional perspective in disability bioethics discussions on suicide and PAD that considers suicidal people's realities and oppression, as well as the interlocking effects of ableism/sanism/ageism *with suicidism* will necessarily push us to change not only conversations but also laws, policies, practices and cultural perceptions regarding suicidal people. In my work, I propose a queercrip, non-coercive approach to suicidality. While it is beyond the scope of this chapter to elaborate on this approach, it is worth mentioning that it is anchored in anti-oppressive perspectives (anti-ableist, anti-sanist, etc.) and aims to highlight the suicidist violence that suicidal people experience in our societies in order to find better intervention strategies that would allow them to speak freely about their suicidality, and also to offer them, among many other well-considered options, "suicide-affirmative healthcare" based on a harm-reduction approach (Baril 2017, 2020).

This queercrip approach to suicide that includes PAD access for those who seriously consider ending their life is not founded on a liberal or individual notion of choice, liberty and autonomy but rather on a robust notion of relational autonomy and a structural anti-oppressive perspective recognizing that suicidal people, as an oppressed group, need to be listened to, believed, legitimized and supported through various mechanisms, including positive rights in relation to PAD. My approach also promotes a simultaneous robust work on the oppressive structures that shape suicidality among marginalized communities. In sum, instead of proposing an ethics of dying, I rather suggest an "ethics of 'living with' suicidal people" (Baril 2020). Recognizing suicidism for the first time in bioethical discussions on suicide and PAD therefore creates more than one conversational switch; it leads us to abandon the focus on disabled/sick/ill/Mad/old people when we discuss suicide and PAD and puts the emphasis on those who really want to die, i.e. suicidal people. Simultaneously, it invites us to devote our energy in disability bioethics to key issues that disabled/sick/ill/Mad people face daily. These crucial issues would include, but are not limited to, suicide and PAD.

Notes

1 Despite the stigma and disapproval surrounding suicide, I use the notions of suicide and physician-assisted suicide (or physician-assisted death) in this chapter rather than sanitized expressions such as "medical aid in dying" or "euthanasia," in order to highlight the importance of theorizing how suicidal subjects are excluded from discourses and practices surrounding medical aid in dying and euthanasia. As Stefan (2016: 133–134) demonstrates by mobilizing the results of quantitative surveys, the right-to-die activists reframed their discourses in the last decades by repudiating the stigmatized notion of "suicide" and favoring expressions that elicit more approval than suicide, such as medical aid in dying. Going against the grain of this approach favored by the right-to-die movement, I privilege the notion of suicide because I believe that it could be destigmatized and resignified to denounce the violence experienced by suicidal people in our societies. For definitions of various practices surrounding assisted death, such as voluntary active euthanasia, physician-assisted suicide or removal of life-sustaining treatment, see Braswell (this volume) and Quill and Sussman (2020). I use the expression "assisted death" as an umbrella term to refer to practices made on a *voluntary* basis.
2 I use the expression "disability activists/scholars/bioethicists" for those adopting an anti-ableist perspective.
3 On epistemic violence and injustice, see Dotson (2011), Reynolds (2016), Baril (2020), Wieseler (2020).
4 It is worth mentioning, from an intersectional perspective, that other marginalized identity components (e.g. being racialized or poor) play a central role in the arrests, "suicides by cop" (Stefan 2016), involuntary commitments, and other inhumane treatments. As demonstrated by Trans Lifeline (2020: 137), "the risk of harm or use of deadly force predictably increases when the person in [suicidal] crisis is a person of color or disabled." In other words, people who live at the intersections of multiple oppressions, such as racism, colonialism, classism, ableism, sanism, sizeism, and cisgenderism, experience suicidism in a more acute way.

References

Améry, Jean. 1999. *On Suicide. A Discourse on Voluntary Death*. Bloomington: Indiana University Press.
Appel, Jacob M. 2007. "A Suicide Right for the Mentally Ill?: A Swiss Case Opens a New Debate." *The Hastings Center Report*, *37*(3): 21–23.

Baril, Alexandre. 2017. "The Somatechnologies of Canada's Medical Assistance in Dying Law: LGBTQ Discourses on Suicide and the Injunction to Live." *Somatechnics*, 7(2): 201–217.

———. 2018. "Les personnes suicidaires peuvent-elles parler? Théoriser l'oppression suicidiste à partir d'un modèle sociosubjectif du handicap." *Criminologie*, 51(2): 189–212.

———. 2020. "Suicidism: A New Theoretical Framework to Conceptualize Suicide from an Anti-Oppressive Perspective." *Disability Studies Quarterly*, 40(3). https://dsq-sds.org/article/view/7053.v40i3.7053

Borecky, Adam, Calvin Thomsen, and Alex Dubov. 2019. "Reweighing the Ethical Tradeoffs in the Involuntary Hospitalization of Suicidal Patients." *The American Journal of Bioethics*, 19(10): 71–83.

Braswell, Harold. 2018. "Putting the 'Right to Die' in Its Place: Disability Rights and Physician-Assisted Suicide in the Context of US End-of-Life Care." *Studies in Law, Politics, and Society*, 76: 75–99.

Burstow, Bonnie, Brenda A. LeFrançois, Shaindl Diamond (Eds.). 2014. *Psychiatry Disrupted: Theorizing Resistance and Crafting the (R)evolution*. Montreal: McGill-Queen's University Press.

Cholbi, Micheal. 2017. "Suicide." In *Stanford Encyclopedia of Philosophy*, edited by Edward N. Zalta. https://plato.stanford.edu/archives/fall2017/entries/suicide/.

Coleman, Diane. 2010. "Assisted Suicide Laws Create Discriminatory Double Standard for Who Gets Suicide Prevention and Who Gets Suicide Assistance: Not Dead Yet Responds to Autonomy." *Disability and Health Journal*, 3: 39–50.

———. 2020. "The Extreme Ableism of Assisted Suicide." *Not Dead Yet*. http://notdeadyet.org/2020/01/the-extreme-ableism-of-assisted-suicide.html

Creighton, Colleen, Julie Cerel, and Margaret P. Battin. 2017. "Statement of the American Association of Suicidology: "Suicide" Is Not the Same as 'Physician Aid in Dying'." *American Association of Suicidology*. suicidology.org/wp-content/uploads/2019/07/AAS-PAD-Statement-Approved-10.30.17-ed-10-30-17.pdf

Dotson, Kristie. 2011. Tracking Epistemic Violence, Tracking Practices of Silencing. *Hypatia*, 26(2): 236–257.

Gill, Carol J. 1992. Suicide Intervention for People with Disabilities: A Lesson in Inequality. *Issues in Law & Medicine*, 8(1): 37–53.

———. 1999. "The False Autonomy of Forced Choice: Rationalizing Suicide for Persons with Disabilities (Con)." In *Contemporary Perspectives on Rational Suicide*, edited by James L. Werth, pp. 171–180. Philadelphia: Brunner/Mazel.

———. 2004. "Depression in the Context of Disability and the 'Right to Die.'" *Theoretical Medicine*, 25: 71–98.

Giwa, Al. 2019. "A Complete Treatise on Rational Suicide." *Bioethics UPdate*, 5: 107–120.

Hewitt, Jeanette. 2010. "Schizophrenia, Mental Capacity, and Rational Suicide." *Theoretical Medicine and Bioethics*, 31: 63–77.

Ho, Anita. 2014. "Choosing Death. Autonomy and Ableism." In *Autonomy, Oppression, and Gender*, edited by Andrea Veltman and Mark Piper, pp. 1–24. New York: Oxford Scholarship Online.

Ho, Anita, and Joshua S. Norman. 2019. "Social Determinants of Mental Health and Physician Aid-in-Dying: The Real Moral Crisis." *The American Journal of Bioethics*, 19(10): 52–54.

Kious, Brent M., and Margaret P. Battin. 2019. "Physician Aid-in-Dying and Suicide Prevention in Psychiatry: A Moral Crisis?" *The American Journal of Bioethics*, 19(10): 29–39.

Quill, Timothy E., and Bernard Sussman. 2020. "Physician-Assisted Death." *The Hasting Center*. https://www.thehastingscenter.org/briefingbook/physician-assisted-death/

Reynolds, Joel M. 2016. "The Ableism of Quality of Life Judgments in Disorders of Consciousness: Who Bears Epistemic Responsibility?" *AJOB Neuroscience*, 7(1): 59–61.

Rich, Ben A. 2013. "Suicidality, Refractory Suffering, and the Right to Choose Death." *The American Journal of Bioethics*, 13(3): 18–20.

Stefan, Susan. 2016. *Rational Suicide, Irrational Laws. Examining Current Approaches to Suicide in Policy and Law*. Oxford: Oxford University Press.

Szasz, Thomas. 1999. *Fatal Freedom: The Ethics and Politics of Suicide*. Westport: Praeger Publishers.

Trans Lifeline. 2020. "Why No Non-Consensual Active Rescue?" In *Beyond Survival: Strategies and Stories from the Transformative Justice Movement*, edited by Ejeris Dixon and Leah Lakshmi Piepzna-Samarasinha, pp. 135–139. Chico: AK Press.

Wardlaw, Margaret P. 2010. "The Right-to-Die Exception: How the Discourse of Individual Rights Impoverishes Bioethical Discussions of Disability and What We Can Do about It." *International Journal of Feminist Approaches to Bioethics*, *3*(2): 43–62.

Wieseler, Christine. 2020. "Epistemic Oppression and Ableism in Bioethics." *Hypatia* 35(4): 714–732. https://doi.org/10.1017/hyp.2020.38.

Further Readings

Baril, Alexandre. 2017. "The Somatechnologies of Canada's Medical Assistance in Dying Law: LGBTQ Discourses on Suicide and the Injunction to Live." *Somatechnics*, 7(2): 201–217. https://www.academia.edu/34640280.

_____. 2020. "Suicidism: A New Theoretical Framework to Conceptualize Suicide from an Anti-Oppressive Perspective." *DSQ: Disability Studies Quarterly*, *40*(3). https://dsq-sds.org/article/view/7053.

Stefan, Susan. 2016. *Rational Suicide, Irrational Laws. Examining Current Approaches to Suicide in Policy and Law.* Oxford: Oxford University Press.

Wedlake, Grace. 2020. "Complicating Theory through Practice: Affirming the Right to Die for Suicidal People." *Canadian Journal of Disability Studies*, *9*(4): 89–110.

PART VII

Disability, Difference, and Health Care

22

DISABILITY BIOETHICS AND RACE

Andrea J. Pitts

Disability studies and biomedical ethics have each been criticized for their respective omissions of the experiences, concerns, and contributions of people of color. Throughout the 1990s and 2000s, scholars began demonstrating that the first-person accounts, normative guidelines, and framings of embodiment within these interdisciplinary fields tended to prioritize the experiences of disability, pain, illness, health, and clinical treatment from a white perspective (e.g., Myser 2003; Bell 2006). In response, both fields have attempted to transform the focus of their research, activism, and clinical guidelines to include historical and present patterns of neglect, oversight, and harm impacting communities of color. Both fields have also begun to highlight insights, methods, and political strategies employed by people of color within disability activism and clinical settings.

The cross-pollination of these fields of study occurred during the mid-1990s through the development of disability bioethics, even prior to coinage of the term "disability bioethics." This new area of research urged practitioners to understand the importance of insights and methodologies from disability studies within the formation and regulation of ethical guidelines and clinical practices in biomedicine. Such insights included, specifically, that the experiences and contributions of disabled people are necessary in the field of bioethics.

Regarding race, founding theorists of disability bioethics engaged race and racism to varying degrees in their respective writings. For example, Rosemarie Garland-Thomson has been described by Sami Schalk as "one of the first disability studies scholars to provide sustained race and disability analysis in the final chapter of her book *Extraordinary Bodies*, published in 1997" (Schalk 2018, 10). Accordingly, disability bioethics, as this volume suggests, from its inception has attempted to examine the intricacies and structural conditions of race and racism within the biomedical sciences.

As a brief caveat, it is important to note that, although the focus of this chapter is on race, disability, and bioethics, the study of the intersection of race and disability exceeds the scope of the biomedical sciences. That is, as Shelley Tremain (2017) has cautioned her readers, to reduce disability to biomedical and health discourses is to potentially reify the broader societal and historical conditions that shape the pathologizing and/or normalizing relations of power that negatively impact disabled people. For this reason, although the focus of this volume is on disability bioethics, uncovering the broader relevance of the intersection of disability and race beyond medical and health discourses remains an important ongoing project. Accordingly, the

DOI: 10.4324/9781003289487-30

analysis offered here will be limited, but will serve as a piece of a broader scholarly and political movement focusing on the relationships between race and disability more generally.

To foreground a disability bioethics that grapples with the structural and intimate presence of oppression in its many valences and that emphasizes the lives and work of disabled people of color, this chapter focuses on two overlapping areas that thread together disability studies, bioethics, and critical approaches to race. First, the chapter examines work addressing the existential and conceptual relationships between race and disability. This section underscores writings by authors who analyze embodied experiences of community belonging, illness, pain, motility, madness, desire, futurity, and barriers to affirmation and care, all of which are in conversation with processes of racialization. Building on this discussion, the second section shifts to a focus on race and institutions. This section includes an examination of the functions of disability and race in contexts of clinical diagnosis and treatment, medical education and research, as well as disability and health activist movements that have challenged systemic oppressions impacting communities of color.

I Existential/Conceptual Analyses of Race and Disability within Biomedical Ethics

As noted above, recognizing the necessity and value of the perspectives of disabled people has been a critical turn that distinguishes disability studies more generally, and has surfaced within bioethics more recently. The common English-language refrain within disability rights activism since the 1990s "Nothing about us without us" presents a demand for policymakers, theorists, and organizers to prioritize the experiences, worldviews, and resources of people who identify as disabled or who may be socially marked as disabled or debilitated.

Likewise, critical race studies and Indigenous studies have long recognized the need to prioritize the perspectives of people of color and Indigenous peoples to avoid the exoticizing, commodifying, or otherwise misinformed views developed within the anthropological, biological, and medical sciences. Social epistemologist, Charles Mills (1998), for example, considers the novel *Invisible Man* by Ralph Ellison a classic illustration of the existential conditions of Black Americans who have experienced patterned epistemic erasure and invalidation by white Americans. Or, as Anishinaabe theorist and literary author, Gerald Vizenor (1994) proposes, the fact that vastly distinct Indigenous societies of the Americas have been placed under one homogeneous label "Indian" has required Native peoples to devise "postindian" strategies to subvert and survive the damage caused by settler colonial imaginaries.

Given this shared methodological stance across these fields, we find the study of race, disability, and the biomedical sciences similarly taking on such a priority. As such, we can turn to accounts of the relationship between race and disability that explore the existential and conceptual conditions of encounters with clinical, research, and other medical settings.

First, regarding existential conditions that frame the interconnected relationships between race and disability within biomedical contexts, a number of theorists have explored the differing forms of medical treatment, attitudes and expectations of medical providers, and the everyday forms disregard for people of color within clinical settings. Notably, women of color authors of the 1970s–1990s utilized descriptions of embodied pain, illness, and disability to critique heterosexual norms and post-racial framings of "color blindness" within medical settings. Works of fiction, nonfiction, and poetry by authors such as Gloria Anzaldúa, Audre Lorde, Evelyne Accad, and Leslie Marmon Silko (Laguna Pueblo) each respond, in differing ways, to conditions of pathologization and debilitation impacting communities of color.

For example, Lorde's *The Cancer Journals* (1980) analyzes the author's experiences of breast cancer, including biopsy, diagnosis, and mastectomy, and the text documents her refusal to be objectified or shamed within the clinical encounters that she endures. She describes her experience of breast cancer as stemming explicitly from a Black lesbian feminist perspective, and the text is framed throughout by excerpts from the author's journal from the years 1978–1980.

Seizing control of her narrative, Lorde opens *The Cancer Journals* by describing the "transformation of silence into language and action" (18). Specifically, she outlines the double binds often experienced by Black women, including their hypervisibility and simultaneous invisibility within allied political movements with white women. Within such settings, Black women, she writes,

> have had to fight and still do, for that very visibility which also renders us most vulnerable, our blackness. For to survive in the mouth of this dragon we call america [sic], we have had to learn this first and most vital lesson—that we were never meant to survive.
>
> *(21, 17)*

Against this deathly tendency, she states, survival depends on "teach[ing] by living and speaking those truths which we believe and know beyond understanding" (22).

In the context of *The Cancer Journals*, she intimates that writing about her experiences with breast cancer, including the patterned harms that she underwent within the clinical setting, is vital to her own survival and to the survival of other Black women. Additionally, "know[ing] without understanding" appears to refer to the inchoate existential awareness of one's own sensations, emotions, and other embodied experiences, experiences that are not fully transparent to oneself or others. According to Lorde, the act of bringing such pre- or non-discursive forms of awareness, including experiences of pain and illness, into expression becomes a method to reshape personal and collective possibilities.

On this point, Therí A. Pickens (2014) notes that Lorde "wrestles with the appropriate ethical response to her pain" and configures her writings as a means "to move toward awareness, discussion, and activism" (130–131). The importance of highlighting such a relationship between pain and expression is that Lorde is able to demonstrate the many ways in which Black lesbian women struggle through cancer and amputation. Among the themes that emerge in Lorde's work is the dominance of the medical establishment over patients' own narrative descriptions. Thus, describing her pain through evocations of joy and sensuality, including phantom pains, becomes a resistant act against the power of clinical medicine's force relations. For example, after her mastectomy, Lorde writes that "the pain returned home bringing all of its kinfolk" (Lorde 2014, 31–32). Through drawing on familiar vernacular, Lorde's description serves as a response to the insufficiency of medical discourses to describe the experiences of Black American women in pain (Pickens 2014, 132).

The Cancer Journals also underscores Lorde's decision not to wear a prosthesis following her mastectomy. Specifically, nurses and volunteer staff encourage her to wear a pink prosthetic breast following her surgery, and the reasons offered are that such a prosthesis will make Lorde and others around her feel more comfortable (1980, 42). The artificiality of the pink prosthetic contrasts with Lorde's descriptions of her own "soft brown skin" after surgery, and her remarks critique the white aesthetic norms that exist within biomedical technologies of the flesh (44).

Pickens also notes the heteronormative encounters Lorde describes while being encouraged to wear the prosthesis, including the suggestions by medical staff that wearing the prosthesis would allow her to remain attractive to men (2014, 135). In these ways, Lorde's writings refuse these tendencies to foreclose her own appreciation of her body and desires, and her descriptions reopen a discursive space for a distinctively Black lesbian response to the misunderstandings and erasures that stem from the medical world around her. As such, Lorde's writings offer a counternarrative to the whiteness of clinical medicine, and provide a descriptive opening for the existential conditions of illness, pain, and pathologization experienced by queer women of color.

Alongside the pivotal work by Black, Indigenous, and other women of color in the 1970s and 1980s, more recent scholarly and activist work operating through framings of disability justice has focused on the specific existential and conceptual tools necessary for disabled, queer and trans people of color. For example, although authors like Lorde and Gloria Anzaldúa wrote extensively of experiences of illness, medical intervention, pain, and blindness, these authors were often reluctant to interpret themselves in terms of "disability" or as "disabled."

Along these lines, Moya Bailey and Izetta Autumn Mobley (2019) and Aurora Levins Morales (2013) have respectively explored the relationships between race and disability, examining the reasons why people of color may choose not to politically organize around disability rights or through identification as "disabled." Namely, Bailey and Mobley note that tropes of the "crazy Black woman" are weaponized against Black women leading to high rates of violence committed against them, and that such violence occurs while simultaneously failing to address the needs of Black people with psychiatric or physical disabilities (2019, 31). Additionally, they point out that:

> Black women and other women of color do most of the labor in the service of disability despite the impact on their ability to care for themselves or their families. Caregivers are often engaged in debilitating work for disabled people and become disabled themselves.
>
> *(32)*

These concerns, along with others, may thus complicate political and personal mobilization through a framing of disability for people of color.

Bailey and Mobley also note, however, that growing work is currently being done among Black authors and activists to explore the terms of a disability within Black communities, including as their work attests, developing a Black feminist disability framework from which to theorize and mobilize. Along similar lines, Morales notes that for Gloria Anzaldúa, as a "dark-skinned working class Tejana lesbian," to have affirmed disability as an identity would likely have required "a strong, vocal, politically sophisticated, disability justice movement led by queer working class women and trans people of color who understood [her] life" (2013, 5).

From this insight, Bailey and Mobley, as well as disability justice activists have resisted individualist rights- and autonomy-based models of justice to focus on collective resistance to ableism in its many instantiations. Instead, disability justice activists and scholars demand a recognition of the interdependence of shared struggles against systemic oppressions and how disability functions across all forms of oppression—all of this while foregrounding the desires, joys, and networks of care among disabled people who are fighting against those systemic oppressions.

For example, disability justice organizers Talila A. Lewis and Dustin P. Gibson, through their social media, workshops, and collective public actions, offer powerful critiques of the forms of systemic ableism that undergird US prison systems and policing, while also dedicating time and energy to curate Black disability solidarity music, art, film, and history (Lewis 2020; Gibson 2020). Additionally, Sins Invalid, founded in 2005 by disability justice activists Patty Berne and LeRoy F. Moore, is a performance-based artist collective composed of artists of color and LGBTQ/gender-variant artists that explores themes of sexuality, the non-normative body, and social and economic justice through their performances and workshops (Sins Invalid 2020). These examples of disability justice activism thus illustrate the important point made by Bailey and Mobley that there are creative aesthetic, embodied, and political contributions founded in the experiences of disabled people of color, and that organizing work done within a disability justice framework celebrates both the lives and experiences of disabled people of color while also critiquing the very systematic patterns of harm that would seek to erase, reduce, or eliminate their very existence.

Lastly, regarding the relationship between disability and race, Nirmala Erevelles (2011) explores the "racialization of disability" and the "dis-abilization of race," a process that Erevelles offers to bring both categories into material relation. Specifically, she argues that whiteness, defended as a form of property right, is "the ideological discourse that has been used to justify the racial superiority of white people over people of color by using the logic of dis-ability (e.g., inferior genes, low IQ) to decide who has the rights to citizenship" (166). By this, Erevelles suggests that the racialization of disability is the process by which white, able-bodied heterosexual cisgender men become centered as the most productive (and thereby valuable) citizens within capitalist societies. In this, the productive possibilities of such white citizens are protected through institutions such as the law, education, and health care. Regarding education, for example, the criminalization and pathologization of Black children in k-12 settings, as well as the segregation of educational institutions by "cognitive ability" seeks to shore up protected educational resources for those deemed more deserving or more capable of making "positive" contributions to society.

Regarding the dis-abilization of race, Erevelles notes that people considered outside the productive metrics of advanced capitalism are often relegated to the welfare state, and thereby stigmatized as burdensome on the civic body. The responses to such so-called "dependency" on social welfare programs, however, is also a lucrative business, she argues. Testing companies, metrics and measurements for success, juvenile detention facilities, and other programs designed to manage or "uplift" children of color effectively foreground a life marked by pathologization and deviance for many people. Moreover, Erevelles proposes that the debilitating conditions of labor for many people of color in the USA also lead to increased marginalization, vulnerability, and medical precarity for such populations within advanced capitalism.

II Race, Disability, and Biomedical Institutions

Accordingly, such examinations of the existential and conceptual relationships between race and disability also reflect the shortcomings of institutional design and social organization, drawing readers' attention to the mutual imbrication of structural ableism and racism. As such, the next area of analysis draws lessons from these existential and conceptual analyses to frame how race and disability are mutually constituted within medical institutions.

One significant institutional dimension that disability scholars and activists (including Erevelles) have explored is the how incarceration functions across medical settings. That is, confinement and containment have long existed as strategies to control people deemed "deviant," "mad," "ill," or disabled. Although almshouses and poorhouses existed in the colonial Americas, stemming from practices in England and France, differentiated confinement and segregation is relatively recent. According to Ben-Moshe, Chapman, and Carey (2014), almshouses and poorhouses tended to confine many socially "undesirable" peoples together, including "poor, disabled, widowed, orphaned, and sick people" within the same institutional setting (3–4). Yet, by the nineteenth century, prisons and jails began to emerge across Canada and the USA to house persons deemed "criminal," and asylums built to house those deemed "mentally unfit" or "feebleminded."

Specifically, regarding race, Ben-Moshe, Chapman, and Carey note that Black and Indigenous populations were often not included within the undifferentiated forms of confinement prior to the nineteenth century. Institutional slavery, residential schools, indentured servitude, and forced displacement were all mechanisms for controlling and confining Black and Indigenous peoples outside of almshouses, poorhouses, prisons, and jails. For instance, the formal (legal) ending of slavery in Canada and the USA in the nineteenth century (1834 in Canada and 1865 in the USA) led to new societal functions for prisons and jails. Specifically, these institutions began to be considered for their "rehabilitative" effects, rather than serving simply as a means of punishment or deterrence. "Black Codes" in the USA, for example, penalized recently emancipated Black Americans for vagrancy and labor contract violations (e.g., refusing to work in accordance with the terms of a white employer) (Ben-Moshe, Chapman, and Carey 2014, 9). The rationale was that prisons and jails, alongside asylums, residential schools, and hospitals, could be used to socially, physically, and emotionally shape those who inhabited these institutions and make them ready to enter the labor force upon release.

Such a framing of "rehabilitation" also justified the convict leasing system throughout the US South, a series of programs that treated incarcerated peoples as slaves, and leased incarcerated workers to private companies for pay (Ibid., 9). By the 1970s and 1980s, welfare programming would shift to forms of "workfare" programming wherein the terms of inclusion for social support required proof of one's willingness to labor, including drug testing and work requirements for the receipt of medical, food, housing, and education benefits. Such institutional requirements remain with us today, and continue to shape access to medical and other social resources for many people.

Another institutional perspective from which we can consider the relationship between race and disability is through clinical research. For example, Harriet Washington (2006) offers a careful historical analysis of the exploitation and abuse of Black peoples across the medical sciences. Marking the cruel forms of experimentation and wanton suffering inflicted on enslaved peoples during the early development of the gynecological sciences, as well as the Tuskegee syphilis study in which hundreds of Black sharecroppers in Macon County, Alabama were lied to and actively prevented from accessing effective medical treatment by the US Public Health Service (among other horrific examples explored in the book), Washington's work provides an important lens through which to consider the institutional dependency between the exploitation and assumed expendability of Black people and advancements in the medical sciences. For example, the construction and diagnosis of disorders such as "drapetomania," "dysaesthesia aethiopica," and "dementia praecox in the colored race" utilized the epistemic authority of the biomedical sciences to defend the subjugation and enslavement of Black peoples in the USA.

Yet, interesting, as Bailey and Mobley note, Washington, as well as other researchers examining the history of institutionalized medical racism have not framed their work in terms of disability. As noted above, the conceptual and existential terms through which disability surfaces as a political form of mobilization among communities of color are often complex and contested. Thus, while Washington notes that "Enslavement could not have existed and certainly could not have persisted without medical science," and that physicians relied on slave labor and enslaved patients for their research and practices, Washington does not frame her analysis as a form of disability critique. To this point, Bailey and Mobley comment on the work of Washington and other studies of anti-Black racism and the medical sciences that "the specter of disability permeates the Black scholarship of enslavement and other studies of the Black experience" (2019, 25).

Thus, as with the writings of Lorde and Anzaldúa, understanding methods for transformation within the biomedical sciences that address the relationships between race and disability have not always stemmed from writing and activism employing the language of "disability." For example, the Black Panther Party's food, housing, and medical programs serve as substantive demands for universal health care and access to social services, all issues that intimately impact the lives of peoples at the intersections of ableism and racism. Additionally, the Young Lords' organizing efforts of the 1970s called attention to medical experimentation being conducted on Black and brown communities in the USA and environmental hazards like lead poisoning that were impacting racialized communities in the South Bronx at the time. The work of such activist organizations offered prescient critiques of housing and urban development projects that, to this day, continue to expose Black and brown communities to conditions of environmental toxicity, such as the lead poisoning crisis in Flint, Michigan, and the prevalent exposure to pesticides among Mexican and Central American migrant farmworkers. These issues, as Bailey and Mobley and other disability justice theorists attest, are now beginning to be addressed through the terms of disability critique.

Interestingly, the Black Panther Party and the Young Lords were also calling into question the financial motives of medical institutions, and the detrimental incentives that arise from generating profits from medical services. One column published in a 1970 newspaper of the Young Lords critiques the predatory tendencies of pharmaceutical companies and notes the lack of medical, dental, and optometry clinics in poor Black and brown neighborhoods. The same news column also announced the opening of free medical, dental, and eye health programs being supported by the Young Lords in their neighborhoods because "profit-making doctors refuse to take care of poor people and always open offices in neighborhoods where they can make money" (1970, 12). These concerns, voiced over 50 years ago, point to ongoing issues regarding the financial constraints that limit access to health care for many poor disabled and debilitated people across the USA.

Along similar lines, Annette Dula (2007), a founding figure in the development of African American bioethics, has critiqued the economic and political incentives that corporations and policymakers have in denying the existence of racial disparities in biomedicine (61–62). In particular, Dula argues that patterns of industry deregulation and the further perpetuation of individualizing narratives of responsibility effectively aid corporations, insurance companies, and their political beneficiaries who stand to gain from the widespread neglect of the health outcomes of communities of color in the USA. Dula's response, then, is to put discussions of the patient-provider relationship in clinical settings within a broader economic and sociopolitical context in which privatized healthcare systems, major corporations, and policymakers profit from a reduction of resources for people of color in the USA. Analyzing health care with these factors at play, then, she proposes, will aid in interpreting the motivations and measures necessary to reduce health disparities in the country. We could then add to Dula's concern that these same profit-motives, as well as the forms of exploitation of Black and brown people, and civic protections of whiteness as Erevelles notes, are ways in which poor and disabled people of color continue to be subject to a eugenicist constellation of social forces that seeks to erase or eliminate them.

This chapter has provided a brief survey of some of the connections between race, disability, and biomedical ethics. Accordingly, as disability and race continue to be explored by disabled people of color, and by nondisabled people as well as white people who are willing to examine their own relationships to racism and ableism, this area of study will continue to grow within and perhaps beyond bioethics.

References

Bailey, Moya and Izetta Autumn Mobley. 2019. "A Black Feminist Disability Framework." *Gender and Society* 33, no. 1: 19–40.

Bell, Christopher. 2006. "Introducing White Disability Studies: A Modest Proposal." In *The Disability Studies Reader 2*, edited by Lennard J. Davis, pp. 275–282. New York: Routledge.

Ben-Moshe, Liat, Chris Chapman, and Allison C. Carey. 2014. *Disability Incarcerated: Imprisonment and Disability in the United States and Canada*. New York: Palgrave.

Dula, Annette. 2007. "Whitewashing Black Health: Lies, Deceptions, Assumptions, and Assertions—and the Disparities Continue." In *African American Bioethics*, edited by Lawrence J. Prograis Jr. and Edmund D. Pellgrino, pp. 47–66. Washington, DC: Georgetown University Press.

Gibson, Dustin P. 2020. "Offerings." *DustinPGibson.com*. https://www.dustinpgibson.com/offerings.

Levins Morales, Aurora. 2013. *Kindling: Writings on the Body*. Cambridge: Palabrera Press.

Lewis, Talila A. 2020. "Blog." *TalilaLewis.Com*. https://www.talilalewis.com/blog.

Lorde, Audre. 1980. *The Cancer Journals*. San Francisco: Aunt Lute Books.

Mills, Charles. 1998. *Blackness Visible: Essays on Philosophy and Race*. Ithaca, NY: Cornell University Press.

Myser, Catherine. 2003. "Differences from Somewhere: The Normativity of Whiteness in Bioethics in the United States." *American Journal of Bioethics* 3, no. 2: 1–11.

Pickens, Therí A. 2014. *New Body Politics: Narrating Arab and Black Identity in the Contemporary United States*. New York: Routledge.

Schalk, Sami. 2018. *Bodyminds Reimagined: (Dis)ability, Race, and Gender in Black Women's Speculative Fiction*. Durham: Duke University Press.

Sins Invalid. 2020. "Mission & Vision." *SinsInvalid.org*. https://www.sinsinvalid.org/mission.

Tremain, Shelley. 2017. *Foucault and Feminist Philosophy of Disability*. Ann Arbor: University of Michigan Press.

Vizenor, Gerald. 1994. *Manifest Manners: Postindian Warriors of Survivance*. Middletown: Wesleyan University Press.

Washington, Harriet. 2006. *Medical Apartheid: The Dark History of Medical Experimentation on Black Americans from Colonial Times to the Present*. New York: Harlem Moon, Broadway Books.

Young Lords Organization. 1970. "Health Care is a Human Right." *Young Lords Organization* 1, no. 5 (January): 1–20.

Further Reading

Bailey, Moya and Izetta Autumn Mobley. 2019. "A Black Feminist Disability Framework." *Gender and Society* 33, no. 1: 19–40.

Bell, Christopher. 2011. *Blackness and Disability: Critical Examinations and Cultural Interventions*. East Lansing: Michigan State University Press.

Ben-Moshe, Liat, Chris Chapman, and Allison C. Carey. 2014. *Disability Incarcerated: Imprisonment and Disability in the United States and Canada*. New York: Palgrave.

Erevelles, Nirmala. 2011. *Disability and Difference in Global Contexts*. New York: Palgrave MacMillan.

Levins Morales, Aurora. 2013. *Kindling: Writings on the Body*. Cambridge: Palabrera Press.

Piepzna-Samarasinha, and Leah Lakshmi. 2019. *Care Work: Dreaming Disability Justice*. Vancouver: Arsenal Pulp Press.

23

BIOETHICS AND THE DEAF COMMUNITY

Teresa Blankmeyer Burke

This chapter is a reprint. The original chapter appears in *Signs and Voices: Deaf Culture, Identity, Language, and Arts*. Eds. Kristin A. Lindgren, Doreen DeLuca, and Donna Jo Napoli. Washington, DC: Georgetown University Press (2008), pp. 63–74. © Georgetown University Press. Reprinted with the permission of the author, editors, and Georgetown University Press.

When the editors of this volume [*Signs and Voices: Deaf Culture, Identity, Language, and Arts*] asked me to submit a chapter explaining the importance of bioethics to the deaf community, my first inclination was to refuse this daunting task.[1] Bioethics is a discipline with amorphous boundaries; defining the deaf community is similarly challenging. Given the difficulty of carving out a niche in which to situate my discussion, how could I possibly bring together these two disparate fields of bioethics and Deaf studies and, at the same time, offer a substantive discussion of the issues in just a few pages? After some reflection, I decided that this task was insurmountable—defining bioethics and the deaf community would take more ink than I was willing to spill, and there would be no space left to discuss the intersection of these two disciplines. The best that I could possibly do would be to designate working definitions, survey some important bioethical issues that affect the deaf community, and highlight some of the key concepts and arguments that accompany these issues. By reframing this chapter as an initial exploration into the questions of bioethics that might be raised by those working in Deaf studies scholarship, I hope to create a roadmap locating issues important to both of these domains and to suggest some starting points for cross-disciplinary discussion in Deaf studies and bioethics.

I Defining Terms

For the purposes of this chapter, I've designated bioethics as the discipline where ethical questions related to biological sciences emerge. The range of these questions affecting deaf and hard of hearing people is extensive—from concerns about the environmental impact of the manufacturing and disposal of technological devices used by deaf and hard of hearing people to questions about biomedical issues, such as the moral implications of genetic screening for deafness. Establishing a working definition for the larger deaf community is complicated; the

DOI: 10.4324/9781003289487-31

history of bioethics has impacted this community in discrete and different ways. Sometimes the issues affect a specific group within the group, such as members of the signing deaf community, the hard of hearing population, or people with cochlear implants. In other instances, a bioethics issue may affect the entire population, from the signing Deaf to those who consider themselves persons with hearing loss. Rather than establish one comprehensive definition, I will identify particular groups with each example I consider. In this way, I hope to avoid the pitfall of wrongly generalizing the population commonly referred to as "the deaf community." Having said this, I want to emphasize that the categories Deaf, oral deaf, hard of hearing, and so forth are fluid and not easily defined. By classifying issues as affecting certain groups, I merely intend to indicate a primary orientation from which to analyze an issue and not to exclude or dismiss other perspectives from within these groups.

In spring of 2006 I taught a course called Bioethics and Deafness at Gallaudet University. Although this was not the first course on bioethics offered at Gallaudet University, it was the first bioethics course specifically focused on ethical issues related to the signing Deaf community. These issues include the eugenics movement that began in the late nineteenth century and continued well into the twentieth century, the debate about the appropriateness of cochlear implant surgery for prelingually deaf children, genetic screening for hearing variation, and genetic selection. In addition to these issues, we also discussed the potential impact of future technology on the community, including nanotechnology, developments in cognitive and information science, and designer babies. Focused discussion on the ethical implications of emerging technologies has not been part of the everyday discourse at Gallaudet or in the greater signing Deaf community, but this is rapidly changing. One indication of increased interest in bioethics and biotechnology is the recently formed World Federation of the Deaf's ad hoc Task Force on Bioethics. Additionally, discussions about bioethics and biotechnology were featured for the first time at the 2007 Congress of the World Federation for the Deaf in Madrid.

As a philosopher and bioethicist, I chose to emphasize an Anglo-American philosophical approach to bioethics by using the method of argument analysis. Argument analysis is fairly straightforward: It involves identifying arguments, schematizing the premises and conclusion, assessing the truth of the premises, evaluating the strength of the link between the argument's premises and conclusion, and ultimately, deriving the argument's overall cogency. This approach comprises the core of analytic philosophy; the majority of American bioethicists trained as philosophers come from this background. This is the primary lens through which we evaluated bioethics in the Gallaudet course, although it was not the only one.

Determining the scope of bioethics for this course was much more difficult than determining the method of analysis. Typically, bioethics discussions related to specific communities within the broader population fall into two categories: (1) ethical issues that impact the specific community as a whole and (2) ethical issues that impact individual members of the community in particular ways. For example, the biomedical research agenda to eradicate deafness has the potential to affect the existence of the signing Deaf community as a whole, given the social construction of that community. The issue of whether to give a particular child a cochlear implant is an issue that impacts that particular child and her family. Of course, these two categories are not mutually exclusive. The decision of the parents to provide their child with a cochlear implant may result in the removal of that particular child from the signing Deaf community, but it is not necessarily the case that this will be the result.

Another way of looking at this is to consider bioethics from the macro-level of generally accepted institutional assumptions, values, and practices as well as to consider bioethics on a micro-level, where individual assumptions, values, and practices intersect. The fields of Deaf studies and bioethics typically posit two macro-level orientations at odds with each other; this is most commonly framed as the signing Deaf community that views itself as a cultural community versus the pathological definition of deafness espoused by the biomedical community. I believe that this framework omits an important consideration: Most, if not all, bioethics decisions are made at the level of the individuals directly involved and are not simply policy-driven. In order to consider bioethics as it plays out on the ground, we must pay just as much attention to the unique factors that influence each individual as we do to the viewpoints prescribed by the key institutions. Granted, this generates a messier picture of bioethics, but one that more accurately reflects the reality of bioethics in the greater deaf community.

II A Truncated History of Bioethics: Deaf Connections?

Before delving into the current relationship of bioethics to the Deaf community, a thumbnail sketch of the history of contemporary bioethics is in order. Some date the birth of present-day bioethics from the Nuremberg Doctors Trial, in which sixteen of the twenty-three German physicians tried were found guilty of euthanizing people and/or conducting human subject research without consent. The war crimes tribunal at Nuremberg is also notable for setting down a judgment of ten standards for morally permissible medical experiments, otherwise known as the Nuremberg Code, which established international conventions for the practice of medicine and protection of human subjects. Others date the beginning of bioethics from an American intellectual movement that began in the early 1960s with the advent of the so-called "God Squads," gatekeeper committees of community representatives who regulated access to kidney dialysis machines based on judgments of both the medical and moral worthiness of the candidates for dialysis. Still others consider that the discipline of bioethics coalesced when a series of egregious human subject experiments became public. This includes the infamous U.S. Public Health Services (USPHS) syphilis study at Tuskegee designed to track the natural outcomes of untreated syphilis in African American men. That study did not follow standard informed consent procedures established by the Nuremberg Code and did not provide information about standard antisyphilitic treatment available to them, such as penicillin, once it became available.[2]

Another government-approved experiment revealed to the public around the same time was the Willowbrook study on hepatitis, where children with mental retardation were injected with infected serum to induce hepatitis at the Willowbrook State Hospital in Staten Island, New York. Conditions at Willowbrook were such that all the children were expected to contract hepatitis as well as other infectious diseases at some point; part of the rationale behind the study was that children who were deliberately exposed to hepatitis would benefit by being carefully monitored in a special unit, where exposure to other infectious diseases would be less likely Veatch (1997, 275). Although this study purported to hold to a higher ethical standard than that of the USPHS syphilis study protocols in Tuskegee— because informed consent from the children's parents was obtained—several other ethical issues surfaced, including questions about the nature of informed consent when real or practical alternatives are beyond reach for the consenting parties.

What is striking about these noteworthy incidents in the history of bioethics is the documented and unexplored connections to deaf and hard of hearing people. While the Nazi euthanasia and human experimentation programs have been widely documented as affecting Deaf people, particularly the T-4 euthanasia program and the sterilization projects documented by Horst Biesold, scholars have overlooked the potential connections of these other incidents to the Deaf community Biesold (1999). In the case of Willowbrook State Hospital, there are at least two widely publicized instances of deaf adults mistakenly diagnosed with mental retardation who resided in Willowbrook during the time the hepatitis studies were conducted; it is not known if these people or other deaf people were participants in these studies. As for the Tuskegee syphilis experiment, one of the symptoms of advanced-stage syphilis is deafness; it would be worth investigating to determine whether any of the participants in this study were deaf or became deaf as a result of participation in this project. The suggested connection of the Tuskegee syphilis experiments to the deaf community may seem peripheral, given the standard summation of this case in which most emphasis is placed on how race and class issues played out to permit the continuation of the study long after it should have been discontinued. However, raising questions about participants' side effects, including deafness, offers the possibility of a new lens through which to analyze these cases. It goes without saying that the unethical practices that precipitated the development of contemporary bioethics are horrific on any reading— regardless of the presence or absence of deaf people. My point is that the hidden narrative of deaf and hard of hearing people as part of the history of bioethics is yet to be explored on many fronts. Stories of signing deaf people take center stage when people think of bioethics and deafness, but the stories about oral deaf and hard of hearing people are also part of this narrative and must be unearthed as well.

Although there is some debate over which events precipitated the birth of bioethics, by the 1970s, the National Commission for the Protection of Human Subjects of Biomedical and Behavioral Research was established to identify the guiding principles for ethically justifiable research on human subjects in the USA. The commission consisted of philosophers, theologians, community advocates, lawyers, researchers, and medical professionals. This commission authored the *Belmont Report*, one of the seminal documents of the American bioethics movement. The *Belmont Report* identified three ethical principles for evaluating human subject research: beneficence (do no harm), justice, and respect for persons. Shortly after this, philosopher Tom Beauchamp and religious studies scholar James Childress codified this approach into a school of thought now known as principlism, an amplification and explanation of the original principles identified in the *Belmont Report*. Beauchamp and Childress (2001) argue that all ethical issues in biomedicine can be addressed by consideration of the following four principles: autonomy, or respect for persons; beneficence, the duty to do good; justice, as in fair treatment; and nonmaleficence, the duty to do no harm. Of course, individual cases will differ. Some cases may appeal to one or two principles, whereas others may merit consideration of all four principles.

Principlism held sway over bioethics as the dominant approach for a couple of decades. In the 1990s, it was challenged on several fronts; most notable of the criticisms was that this mid-level approach did not have sufficient theoretical grounding. Philosophers argued that principlism, by listing several principles without providing a coherent and consistent framework that justifies each principle, was inconsistent and insufficiently rigorous. Principlism could not be justified from a single theory of morality, such as (to use a few historical examples) John Stuart Mill's utilitarianism or Immanuel Kant's deontological theory, where one foundational concept grounds everything. The attacks on principlism mirrored other substantive challenges playing out on the battlefield of twentieth century philosophy and applied ethics, including questions about the nature and justification of knowledge. The elegance of the top-down moral theories developed in the eighteenth and nineteenth centuries does not match up well to the real world of bioethics, where results tend to be messy and incomplete. Rather than using the top-down model where one foundational theory provides the supporting structure, Beauchamp and Childress's (2001) principlism bears more similarities to epistemic coherentism, where key mid-level concepts fit together supported in a web of beliefs. By weighing and balancing principles in the context of a particular case, principlism acknowledges the need for an approach that incorporates flexibility with some core concepts.

Due to the difficulties of establishing theoretical grounding and consistency for principlism, this approach fell out of favor with many philosophers and bioethicists, who offered a variety of other approaches to bioethics. Another criticism levied at principlism was the tendency for health care professionals to use it as a checklist; by working through the list of principles, ethics committees developed a false sense of security that all important moral issues had been covered. In worst case scenarios, this meant that robust discussion of complex moral dilemmas was sacrificed for brief conversations regarding each principle, sacrificing the nuances and particularities of the case at hand.

Although I am in agreement that there are deep concerns about principlism as a theoretical approach, the four principles offered by Beauchamp and Childress (2001) are useful for the layperson in that they offer a succinct summary of several major traditions in Western philosophy. The principles of nonmaleficence and beneficence, with their moral imperative to "do no harm" and to "do good" bring consequentialist or utilitarian theories to mind. Here, actions are determined ethical or unethical according to the net benefit or harm resulting from the action. The principle of autonomy, or respect for persons, comes straight out of the deontological tradition in moral philosophy associated with Immanuel Kant. This family of ethical theories places the highest emphasis on making decisions that honor the intrinsic worth of the person and do not treat the person as a means to an end. Contemporary ethical theory bears part of the responsibility for the concept of justice; the release of the Belmont Report coincided with John Rawls's groundbreaking work in political philosophy, which stressed the importance of justice as a fundamental principle for creating a society fair to all, an important consideration for ethical decision making.

In some ways, the question of whether principlism can be justified as a robust theory is beside the point and best left to moral philosophers who care about such arcane matters. Those who wish to think critically about bioethics and the deaf community would be well served to ask how these principles might help to clarify the relationship of bioethics to the deaf community. One approach is to think about the kinds of issues that arise under each principle, setting up a rough classification scheme. In the following sections, I attempt to do just this by identifying how a given principle raises different questions and issues for bioethics within the greater deaf community. Although this list is by no means exhaustive, it is my hope that it will form a skeleton on which to build further discussion of bioethical issues affecting members of the deaf community.

III First, Do No Harm

Primum non nocere, the command to do no harm, is one of the oldest in health care ethics. Yet what does harm mean in this context? Usually it is taken to mean that physicians and other health care providers have a duty not to worsen a person's state of health through their actions. A clear example of this would be performing a painful treatment that will worsen the patient's physical health. The harm in the paradigm case is twofold—the patient's health status is further compromised and the patient experiences pain and suffering. Health care providers have a duty not to increase a person's suffering when there is no benefit attached to the suffering, in contrast to short-term suffering in exchange for long-term health. Why should this be a pivotal issue for members of the signing Deaf community? This turns on how this community defines such concepts as health and harm; these are distinctly different than the definitions used by the dominant mainstream culture. The imperative of nonmaleficence occurs at the level of moral agents making decisions that affect individuals; the health care provider is charged with the responsibility of not harming her patients. At the most general level, nonmaleficence means avoiding actions that will likely bring about bad consequences overall. This differs from those actions involving pain or suffering that are expected to bring about a good result overall. For example, the child receiving a rubella vaccine briefly feels physical pain upon the act of immunization, but the morality of this action is generally not interpreted as harmful by mainstream culture—even if mild side effects occur or, more rarely, serious side effects occur.

When the focus shifts from the micro-level of particular individuals to the macro-level of a given population, the concept of harm takes on a different meaning. Asking a meta-level question related to harm raises the issue as to whether certain practices, such as the practice of medicine itself and the values subsumed under it, might be harmful to some and beneficial to others. Now, a connection to the signing Deaf community emerges. The medical approach that defines auditory status through the pathology of hearing loss brings with it a set of values that views individuals primarily through the lens of their disability; that is, people are seen as a broken set of ears in need of a fix. Defining a missing sense as harm drives all other determinations about harm and benefit related to hearing loss. According to this biomedical agenda, avoiding harm reduces to avoiding hearing loss. (This differs from taking actions to restore hearing, which would fall under the principle of beneficence for most health care workers.)

In general, health care providers subscribing to the definition of hearing loss as harm think of abiding by the principle of nonmaleficence for deaf people in a few ways, most notably through avoiding hearing loss. For example, this could be achieved by prescribing medications that do not have ototoxic side effects or by opting to prescribe ototoxic medications only when the risk of hearing loss is the lesser of two evils. Note that this position is not necessarily incompatible with a position of supporting a cultural conception of the signing Deaf community. One could simultaneously support the right of the signing Deaf community to continue to flourish and exist along with the right of hearing people to maintain their identity as "people of the ear." Also note that the issue of whether or not to prescribe ototoxic drugs to people who are already audiologically deaf is a different sort of issue than it is for hearing people. Additionally, the consequences of ototoxicity have different meanings for these communities and individuals residing within these communities. Hearing people, including most members of the medical community, equate the loss of a sense with being harmed; members of the signing Deaf community likely have a different orientation, especially if their experience of being in the world has never included auditory sensation.

Now, perhaps this is a bit too reductionistic. I'm not claiming that all medical professionals view members of the signing Deaf community as people whose ears must be fixed—certainly there are those who recognize that there is another aspect to this community. One important task is locating the putative harm. Hearing people would claim that the hearing loss itself is the harm; this however, is a privation, not a deprivation, to paraphrase Descartes. It is rare that one grieves for something that one has not lost. Consider gender as an analogy. A woman might wonder what it would be like to be a man, or vice versa; yet, this curiosity is not likely to be expressed in terms of loss. Granted, this is not a perfect analogy, since the range of hearing variation tends to be expressed as hearing or deaf (not hearing) by the dominant culture as a strictly audiological evaluation. In the case of the signing Deaf community, the range of hearing variation features, instead, cultural identification from Deaf to hearing-minded—a range that lessens the importance of audiological status. Many Deaf people would stake out a different claim about harm, arguing that the harms visited upon deaf people by the medical establishment are harms of bodily integrity—akin to those claimed by intersex people who are surgically forced into a particular social norm before they are old enough to decide for themselves. Another analogy would be the harm that is visited on the male infant who is circumcised—although medical reasons are offered for this procedure, it is also a procedure that has strong social (and sometimes religious or cultural membership) considerations.

At this point, I do not want to make the claim that a community is harmed. I think this is a difficult claim to establish philosophically, and I'll reserve this discussion for another essay. I do think it is clear that individuals within a community can be harmed; this issue must be considered in the context of how hearing loss or being Deaf is defined by professionals working in bioethics. Harms of bodily integrity are only part of this picture. Another issue important for bioethicists and health care professionals to consider when making decisions that affect people who are deaf or hard of hearing is the importance of broadening the definition of harm to include not just physical or bodily harm, but psychological harm. For example, is an individual harmed when he or she is not able to have full access to communication? Are decisions about treatment for hearing loss fully informed if this issue is not on the table? These questions are worthy of vigorous discussion and should be part of the informed consent process.

IV Beneficence

As any good utilitarian will tell you, sometimes a bit of harm is necessary to achieve the greater good. Health care providers do not set out to harm individuals with hearing loss, but rather to help them by doing good. Granted, some health care providers and scientists pursue their work from a paternalistic standpoint, but many others go into medicine because of a desire to ease suffering. The inability to hear is translated into potential suffering by most health care providers, who accordingly see their actions aimed at increasing hearing status as doing good.

Just as the notion of harm varies depending on who is defining the term, the notion of beneficence is also relative. A broad utilitarian definition of doing good rests on the concept of creating more benefit than harm. If we consider the dominant cultural view that the capacity to hear is a good, then it follows that the pursuit of actions toward this goal is beneficent. If we turn to the view that cultural Deafness is simply a human variation, a different kind of argument ensues. In biological terms, diversity is thought to be a good. The benefits of diversity are justified through both instrumental and intrinsic reasons. For example, environmentalists might offer both intrinsic and instrumental reasons for saving the Brazilian rain forests. An intrinsic reason would appeal to the value of the rain forest in itself. Here, an appeal is made by considering the nature and intrinsic worth of the rain forest as an object that deserves moral consideration. However, intrinsic arguments may not convince everyone. Sometimes instrumental arguments are also necessary for persuasion. In the case of preserving the rain forests, an instrumental reason commonly offered is the claim that plant species in danger of extinction might have the potential to cure disease; another instrumental reason relates to the importance of maintaining vast amounts of vegetation on the planet to counter the effects of greenhouse gases on global warming. Roughly stated, instrumental arguments can be persuasive to those who think in utilitarian terms; intrinsic arguments are more likely to convince those with a deontological bent.

The analogy of diversity can be extended to cultural diversity. For the person who accepts the idea that all cultural communities have intrinsic value, beneficent actions would include any actions that preserve a cultural good. For the signing Deaf community, this would include actions taken to preserve various aspects of the Deaf community, including its language and practices that encourage and maintain sufficient numbers of signing Deaf community members, thus benefiting individual members within that community. An instrumental argument supporting practices that encourage the flourishing of the signing Deaf community is to point to linguistic research on sign languages and the promise that sign languages offer the promise of gaining different kinds of knowledge about the development and structure of human language and cognition.

So far, I have posited a fairly strict duality between signing Deaf community members and the dominant mainstream culture, suggesting that research aimed at eradicating deafness is typically seen as a good by members of the dominant culture and that this same research agenda is seen as harmful by members of the signing Deaf community. In reality, it is not quite so simple. Those who occupy liminal space, such as hard of hearing people who sign, or culturally Deaf people who wear cochlear implants or hearing aids, must also be attended to. These people may find themselves in the position of supporting a research agenda aimed at improving technology that increases listening comprehension, but disavowing practices that threaten the existence of the signing Deaf community, such as genetic screening and selection. This position is not necessarily a contradiction, although it initially may appear to be so. Discussion of such complex positions is merited and likely to become more important as the numbers of signing Deaf adults with cochlear implants increase.

V Justice

Just as nonmaleficence and beneficence are opposite sides of the same coin, so do justice and autonomy fit together. The principle of justice refers to the treatment and constraints imposed on an individual from outside. The sordid history of eugenic practices under Nazi Germany is a classic example of injustice perpetuated against deaf people, among others. Much of the discussion in bioethics related to justice deals with unfair treatment regarding vulnerable populations; most commonly these are separated into the categories of children, prisoners, and people with cognitive disabilities. Although it is of course true that there are deaf people who fit into each of these categories, in the bioethics literature, little attention has been paid to the unique nature of hearing loss (I use this term to illustrate the dominant viewpoint) and the consequences of communication breakdown for people with decisional capacity. In this context, decisional capacity refers to one's ability to make health care decisions. As such, it requires the ability to comprehend the information provided and to appreciate this information in the context of the situation at hand. Ideally, when a person uses a language other than the dominant language of that medical institution, an interpreter will provide a translation to the person with decisional capacity so that the person can make an informed decision. In the case of a person with a hearing loss, the issue is not always one of providing language translation, but of providing communication access so that an informed decision can be reached.

This issue breaks down further. If we consider Deaf people who sign, one approach to making sure that justice is served is to provide equal treatment by ensuring equal access to communication through the use of sign language interpreters. If a patient or research subject is hard of hearing, amplified communication and the use of assistive listening devices allow that person to be treated on an equal footing with hearing people. The ideal of providing services to secure the basic conditions for just treatment depends on a number of factors, ranging from the awareness of the health care providers, researchers, and institutions of various laws governing access to communication, to the practical matter of availability of resources, both economic resources and scarcity or abundance of service providers and equipment. It is rare that the reality matches this ideal.

The population of late-deafened individuals, particularly elderly individuals, may be most at risk of injustice in medical and research settings. This is not to say that institutions deliberately set out to discriminate against such people. Rather, it is a statement about how existing stereo-types about elderly people and deaf people combine in a way that supports paternalistic social attitudes toward this population. Consider the process of informed consent, in which a person is given appropriate and relevant information about medical treatment options. If communication with a late-deafened individual is difficult due to a lack of residual hearing, a health care worker with time pressures may be more likely to presume choices or hurry through the process rather than take the time to ensure that effective communication has occurred. Unfair treatment has occurred through denying the late-deafened person the option of making his own choices. Typically, this occurs because the task of communicating with a late-deafened person is made more difficult; it can also occur because the late-deafened person has developed the habit of "bluffing" in an attempt to hide the extent of her hearing loss. Informed consent is a two-part process that requires both the act of notifying the patient about her treatment options and checking to be sure that the patient has understood this information. The process of checking for understanding with elderly late-deafened patients takes more time; this, coupled with the practice of bluffing, can lead to patient's granting consent without fully understanding or even misunderstanding what they have consented to.

VI Autonomy

The principle of autonomy, or respect for persons, goes hand in hand with justice. The central issue here is that one should have the freedom to make decisions about one's body and health. Unlike justice, the constraints here focus on the individual's right to make choices, even choices that may be out of step with mainstream medicine. A prototypical case is hearing parents faced with making a decision about providing their deaf child with a cochlear implant. In most cases, parents are presumed to be the best surrogate decision makers for their child because they have the child's best interests at heart. (An exception to this is surrogate decision making by parents who deny life-saving medical treatment to their child for religious reasons; typically, govern-ment intervention occurs in these cases because the state's interest in the child's right to con-tinued existence supersedes the parent's right to religious freedom.) Cochlear implant surgery is not a life or death matter—to the best of my knowledge, the state has not mandated medical intervention in these cases, but has left these matters to the parents. Yet, the parents are making a decision about their child's body that will have enormous consequences on that child's quality of life.

A more complex version of this problem exists with genetic screening and genetic selection against deafness. Now that it is possible to identify genes that have been associated with hearing variation, potential parents with a family history of deafness have the option to screen embryos for this genetic trait. Some Deaf people may want to have a Deaf child and will want to use the technology to increase their odds of having such a child; others will use the technology to screen out deafness. Imagine two sets of potential parents who have undergone in vitro fertil-ization and have learned that some of their fertilized eggs about to be implanted code for deaf-ness, and some of them do not. Do these sets of potential parents have equal freedom of choice regarding which fertilized eggs can be implanted? It is clearly the case that the potential parents who wish to screen out embryos carrying genes associated with deafness will be supported by the mainstream agenda of science and medicine. It is not so clear that this freedom to choose holds for Deaf parents who wish to have a Deaf child that will become a full-fledged member of their community. Yet, if we are to promote the principle of autonomy, with parents having the right to make decisions about the future of their offspring, it seems that both options must be permitted. This is without considering the question of autonomy for the potential person, which adds more complexity to this matter.

There are several other issues related to the principle of autonomy, or respect for persons. The twin issues of privacy and confidentiality take a new twist when a language is more public by its very nature, such as sign language in the case of culturally Deaf people. Privacy considerations are also an issue for late-deafened or hard of hearing people who may not realize that their health care provider is speaking loudly enough to be overheard or that they are projecting their own voice loudly in a very quiet environment. Confidentiality, or the duty to keep private information private, becomes more of a concern when other individuals are brought in to facilitate the communication process. Although most sign language interpreters are keenly aware of the ethical obligation of their profession to maintain confidentiality, there are differences in how this duty is perceived and practiced. Additionally, uniform standards of confidentiality for real-time captioning providers are unclear; some of these providers who defer to the standards of the court reporting profession may not realize how small the Deaf community is and may inadvertently breach confidentiality with the disclosure of one or two identifying characteristics. A right to autonomy includes the right to keep one's private information private; yet this may not always be the case.

VII Conclusion

In many ways, the history of bioethics parallels the recent history of the deaf community. Although the previous pages do not come close to providing a comprehensive list of the ethical concerns and issues related to bioethics and the deaf community, I hope that they will provide food for thought and a starting point for further discussion. The convergence of emerging technologies has put us at a crossroads; the future of the deaf community is in the hands of today's medical and scientific researchers. Positing the signing Deaf community as a cultural community that has resisted the biomedical establishment's attempts to eradicate it has opened people's eyes to a different viewpoint on hearing variation. The need for deaf people to engage in more discussion about bioethics with bioethicists and researchers is more critical today than ever. From the eugenics movement of the late nineteenth century to the current dialogue about the use of genetic technology in the deaf community, questions about the morality of curing, abating, or preventing hearing loss abound. Opening up dialogue between researchers and different members of the deaf community, whether hard of hearing, deaf-blind, oral deaf, late-deafened, or culturally Deaf, is imperative. Although these discussions may prove to be difficult and painful and may not result in universal agreement regarding a "deaf bioethic," encouraging this discussion to unfold in the realm of academic bioethics as well as Deaf studies scholarship offers the potential of better understanding, and one hopes, more thoughtful and ethical practices.

Notes

1 I have designated the lowercase version of *deaf* when it precedes *community* as an inclusive term that represents the variety of people with hearing variation (e.g., hard of hearing, oral deaf, late deafened, deaf-blind, cochlear implant users, and culturally Deaf). I maintain the standard practice of using uppercase *Deaf* to indicate those with a linguistic and cultural orientation.
2 Editors' note: Furthermore, researchers did not tell participants that they were part of a study; they were told that they were being treated for "bad blood." See Brandt 1978.

References

Beauchamp, Tom L., and James F. Childress. 2001. *Principles of biomedical ethics,* 5th ed. New York: Oxford University Press.

Brandt, Allan M. 1978. "Racism and research: The case of the Tuskegee Syphilis study." *The Hastings Center Report* 8(6): 21-29.

Biesold, Horst. 1999. *Crying Hands: Eugenics and Deaf people in Nazi Germany.* Washington, D.C.: Gallaudet University Press.

Veatch, Robert M. 1997. *Case Studies in Medical Ethics.* Cambridge, Mass: Harvard University Press.

24

HUNGER ALWAYS WINS

Contesting the Medicalization of Fat Bodies

Anna Mollow

Should fatness be defined as a disease? Should fat bodies be read as signs that we are in the middle of an epidemic of excessive eating? Can weight loss diets transform fat people into thin ones?[1] My answer to the preceding questions is "No." In this chapter, I argue that "hunger always wins." By this, I mean that the drive to eat can seldom be overcome, and that efforts to do so endanger, rather than help, one's health. In making this argument, I draw on insights from fat studies. Scholars in this field challenge the assumption that fatness is equivalent to illness and observe that there is no known way of making fat people thin.[2]

These arguments are highly relevant to disability studies, a field that, like fat studies, resists the medicalization of bodies deemed "abnormal." Anti-fat and ableist oppression take similar forms, including harmful and ineffective "cures," exclusionary constructions of physical attractiveness, workplace discrimination, and access barriers in the built environment (Herndon 2011; Mollow 2015). Yet fat studies and disability studies have for the most part developed separately from each other. I contend that close collaboration between the two fields is essential.

As I make connections between fat studies and disability studies, I critique one of disability studies' most influential assertions: the claim that disability should be understood as benign human variation rather than suffering. That argument has resulted in the field's marginalization of disabilities that do cause suffering. Yet such disabilities – for example, heart disease, diabetes, and cancer – are among the many conditions that fat people are told (incorrectly, as we shall see) that they will acquire if they do not become thin. Similarly, fat people who do have illnesses or pain are advised (again, incorrectly) that their disabilities will be cured if they lose weight. This means that if disability studies is to forge connections with fat people – including fat disabled people – we must be willing to talk about disabilities that cause suffering.

We must also be willing to talk about another form of suffering: hunger. Ninety-five percent of weight loss diets fail, in large part because dieters become too hungry to continue (Kolata 2007; Brown 2015, 36–37). Yet medical professionals routinely encourage fat people to restrict their food intake in ways that leave them feeling hungry. This should be surprising: healthcare providers often describe their jobs as alleviating suffering, and yet they pressure fat people to endure suffering in the form of hunger. This problem is overlooked in most medical discussions of body size. I maintain that hunger must become a focal point of any bioethics that aims to treat fat and disabled people justly. Such a shift will entail radical changes in the ways that healthcare providers treat

DOI: 10.4324/9781003289487-32

fat (and some thin) patients: instead of recommending weight loss, providers should take seriously patients' embodied experiences of hunger and should help ensure that patients are eating enough.

This chapter comprises three sections. First, I overview fat studies scholars' challenges to the medicalization of fat bodies. Second, I make connections between fat studies' and disability studies' respective approaches to bodily difference. Finally, I offer practical suggestions for healthcare providers who wish to avoid perpetuating anti-fat prejudice in their encounters with patients.

I Body Size and Health: Insights from Fat Studies

When fat studies scholars challenge fatphobia, we are often asked, "But what about health?" The implication is that because fatness is often assumed to cause disabilities, societies should try to discourage people from becoming or remaining fat. The medicalization of fat bodies is thus presented as a way of protecting the health of fat people. But as fat activists and scholars regularly point out, this formulation overlooks the damaging effects that weight loss dieting and anti-fat stigma have on mental health. It also relies upon several faulty assumptions: first, that it is every person's imperative to pursue physical health; second, that it is the job of other people to punish those whom they see as failing at this pursuit; third, that fatness causes illness; and fourth, that most fat people can choose to become thin.

The latter two assumptions may seem like common sense truths. After all, isn't there nearly unanimous agreement among scientists that fatness is a life-threatening disease, which can be cured by following the oft-repeated recommendation, "Eat less, move more"? Could these medical scientists all be wrong? In fact, researchers are not in unanimous agreement about these claims (Campos 2004, 4). The majority, however, do hold anti-fat beliefs. And yes, it is possible that they are mistaken. A quick glance at history makes it clear that medical authorities can be devastatingly wrong when they construct theories about marginalized people. Nineteenth-century physicians claimed that Black people had smaller skulls and less "intelligence" than white people; doctors of the same era warned that voting and intellectual activity would damage women's reproductive systems. Until 1973, medical experts labeled homosexuality mental illness; in the 1960s, psychiatrists described schizophrenia as a condition primarily affecting Black men, especially those engaged in political protest (Metzl 2009). Today, many healthcare providers believe, incorrectly, that Black people are less sensitive to pain than white people (Hoffman et al. 2016). Many doctors also believe, inaccurately, that women are prone to fabricating imaginary (i.e., "hysterical") symptoms.

Fatphobia is no different. Numerous studies have documented that medical professionals hold biases against fat people, using words like "lazy," "stupid," and "unattractive" to describe them (Puhl and Heuer 2009, 943–946; Brown 2015, 115–116). Healthcare providers, like the rest of us, live in a fatphobic world; so, it is no surprise they are influenced by anti-fat prejudice. In addition, medical researchers have financial incentives for constructing fatness as a disease. The weight loss industry funds most studies that purport to show fatness is a health risk (Campos 2004, 42–48; Brown 2015, 105; Harrison 2019, 43–49). Even "obesity" researchers who do not receive funding from the diet industry have financial conflicts of interest; their income and professional identities depend upon the belief that fatness is a disease in need of a cure. It is no wonder, then, that anti-fat researchers often manipulate data to create the impression that fatness endangers health (Campos 2004, 47–49; Kolata 2007, 202–203).

And when scientists publish research undermining anti-fat beliefs, "obesity" researchers often attack them. In 2005, Katherine Flegal, an epidemiologist and senior research scientist at the National Center for Health Statistics at the Centers for Disease Control, teamed up with three other highly regarded scientists (with expertise in statistics, epidemiology, medicine, and nutrition) to publish an article examining the relationship between weight and mortality. Flegal and her co-authors had no economic investment in the question of whether fatness imperils health. They also used more precise statistical methodologies and more reliable data than are typically used in "obesity" research.[3] Their results surprised them: instead of a clear correlation between larger body sizes and decreased lifespan, the data revealed a U-shaped curve, with people in the middle – those labeled "overweight" – living the longest (Flegal et al. 2005). As for most "obese" people, their lifespans were the same as those in the "normal"-weight group. At the extreme ends of the spectrum (very thin or very fat), there were higher mortality rates, but these differences were, in Flegal's words, "tiny" (Brown 2015, 15).

Anti-fat researchers reacted angrily, both to this study and to a follow-up study, published by Flegal and her co-authors in 2013, which had similar results. Of the latter study, Walter Willett, a professor of nutrition and epidemiology at the Harvard School of Public Health, said, "This study is really a pile of rubbish, and no one should waste their time reading it" (Aubrey 2013). Willett claimed that Flegal and her colleagues had erred by failing to exclude smokers and people with illnesses; since smoking and some chronic illnesses can lead to weight loss, these groups should not have been counted, he argued. In fact, Flegal and her colleagues had run multiple analyses of the data, examining the results with and without smokers, and with and without people with chronic illnesses; each time, the results were the same: except at the extreme ends of the weight spectrum, body size has little effect on longevity (Kolata 2007, 206).

But what about those extreme ends of the spectrum? As previously noted, studies such as Flegal's have found slight correlations between decreased lifespan and very high (as well as very low) weights. However, correlation is not causation. This crucial principle of statistical analysis is usually overlooked in studies about size and health, which seldom control for confounding variables such as race, socioeconomic status, access to health care, and anti-fat discrimination (Campos 2004, 26–27; Brown 2015, 30–31). Poor people, Latinx people, and Black people are disproportionately likely to be labeled "overweight" or "obese" by medical professionals; therefore, studies which appear to indicate that fatness adversely affects health may in fact be capturing the damaging health effects of poverty and systemic racism (Campos 2004, 81; Kolata 2007, 196). They may also be capturing the health risks of anti-fat bias among healthcare professionals who, rather than providing appropriate diagnostic tests and treatments, routinely send fat patients home with advice to "lose weight" (Kolata 2007, 68; First Do No Harm, n.d.).Weight loss dieting – a practice that fat people are disproportionately pressured to engage in – may also decrease longevity (Campos 2004, 29–33; Brown 2015, 59). Thus, it is not fatness itself, but rather social oppression, which is the more likely cause of the small differences between some fat and some thin people's life expectancies.

It is common to conflate fatness with unhealthful "lifestyle choices." But body size is not a matter of choice. To be clear, my point is not that fat people "do not eat a lot of food." People vary greatly in the amount that we need to eat to feel satisfied, and there is nothing wrong with having a large appetite. Yet two additional points should be made. First, how much one eats does not determine one's body size (Gard and Wright 2005, 44–45). Some fat people eat a lot; some eat a little. The same is true of thin people. Second, how much one eats is not a matter of choice. Hunger is a powerful force, and when one tries to eat less than one's body demands, the drive to eat often becomes uncontrollable.

Many researchers believe that everyone has a biologically determined "set point" (or "set range"), a weight or small span of weights that they can embody (Kolata 2007, 158–159). Although dieting may temporarily make fat people thinner, approximately 95% of diets fail. Within two to five years, almost every dieter regains all the weight they lost; many also gain additional weight (Campos 2004, 29; Brown 2015, 41–44). Set point is primarily hereditary; studies on twins and siblings show that the most important determinant of one's weight is the weight of one's biological relatives (Kolata 2007, 121–125).

Some readers may question whether it is advisable to point to hereditary causes of body size. This concern is understandable, since eugenicists have used specious claims about heredity to justify violence against disabled people, people of color, and other marginalized groups. Yet the problem with eugenics is not that its proponents talk about heredity – it is that they make false assertions about heredity and classify certain groups as deficient. By contrast, the observations that bodies come in a variety of shapes and sizes (all of which are equally valuable), and that these superficial human differences are influenced by heredity, serve as important counterarguments to eugenic thinking.

Unfortunately, eugenic thinking informs anti-fat discourse today. Much as early twentieth-century medical experts warned that the population was experiencing a decline in health because of ever-growing numbers of "defectives" (i.e., disabled people, homosexuals, poor people, and racial and ethnic minorities), contemporary medical authorities stoke fears about an alleged "obesity epidemic" that, they say, necessitates urgent societal intervention. Granted, people in wealthy countries have become heavier since the middle of the twentieth century; but this increase leveled off two decades ago (Brown 2015, 12). Nor is it clear that the shift is a bad thing. Americans have also become taller (Kolata 2007, 209; Brown 2015, 12); yet health authorities do not issue panicked warnings about an epidemic of excessive height. A wide variety of factors have been cited by researchers as possible reasons for population-wide increases in body weight: increased availability of nutritious food; lower rates of smoking; widespread use of SSRIs; environmental toxins; lower rates of childhood disease; an aging population; and epigenetic factors that are not yet understood (Kolata 2007, 209, 221–222; Brown 2015, 12–14). Since several of these factors reflect positive changes, some researchers have proposed that the so-called obesity epidemic may in fact be "a good thing" (Kolata 2007, 209).

II Connecting Fat Studies and Disability Studies

Regardless of the reasons for our collective increase in body size, one thing is clear: there is no known way of making fat people permanently thin. Over the past two centuries, hundreds of weight loss diets have been published, attracting legions of followers. Some are low-carbohydrate; some are low-fat; others extol the virtues of various "superfoods." These programs have one thing in common: they do not work. It does not matter whether one calls it a "miracle diet" or a "permanent lifestyle change," whether one advises emulating the eating patterns of ancient ancestors or employing cutting-edge scientific discoveries to calibrate one's diet. Again and again, the result is the same: large quantities of weight may initially be lost, but within a few years almost everyone regains the lost weight, and most dieters end up fatter (Campos 2004, 29; Kolata 2007, 158; Brown 2015, 41–42).[4]

One reason diets fail is that prolonged caloric deprivation can change metabolism, so that the body requires ever-smaller amounts of food to maintain the same weight (Kolata 2007, 114–119; Parker-Pope 2011). Another reason, familiar to almost anyone who has dieted, is that eventually one becomes too hungry to continue. It is one thing to use "willpower" (a concept fraught with ableist implications) to temporarily override the urge to eat; it is something else altogether to disregard hunger for years on end. Most people cannot do this. Nor should they; as discussed earlier, dieting appears to have negative health consequences.

Yet our culture is fascinated by stories about fat people who lose large amounts of weight. Dramatized on television programs like The Biggest Loser, these stories resemble what disability scholars call "overcoming narratives." Overcoming narratives show disabled people appearing to transcend the limits of supposedly defective bodies: a person with a mobility disability regaining the ability to walk, or a blind person making a cross-country bike trip. Overcoming narratives deflect attention away from the social oppression of disabled people, instead figuring disability as an individual problem that can be solved with perseverance and hard work. Similarly, narratives about fat people who (supposedly) solve all their problems by dieting obscure the social oppression that fat people face. Much as disabled people's participation in the social world is thwarted by buildings without ramps, lack of sign language interpretation, and the use of scented products in public spaces, fat people contend with airline, bus, and car seats that are too small for their bodies; chairs that are not sturdy enough to hold their weight; and a dearth of affordable clothing in their sizes. Workplace discrimination, stigmatizing medical language, cultural constructions of "beauty," and pressure to submit to ineffective and damaging cures are also hallmarks of both anti-fat and ableist oppression.

Despite these connections between ableism and fatphobia, disability studies and fat studies have tended to unfold separately. One barrier to conversation between the two fields is the argument, foundational in disability studies, that it is ableist to describe disability as suffering. To be clear, this argument has value: certainly, disabled people can have happy lives; and the dictum that "The problem is with society, not my body" is true for many disabled people. It is also true that many disabled people are oppressed by the widespread assumption that their lives constitute nothing more than suffering; for example, they may be told they are "courageous" for not attempting suicide. But for people whose disabilities involve pain and illness, oppression often takes a different form: others suspect that our suffering is not real. By spreading the message that disability is "not suffering," disability scholars make it more difficult for people with chronic pain and illness to be believed when we describe our embodied experiences.

Talking about suffering is also important in the context of fat justice. Imagine that you are a fat person with a disability that causes suffering – for example, back pain, fibromyalgia, or endometriosis. (Of course, some readers will not have to imagine this experience.) Your medical provider tells you your disability will diminish, will possibly even disappear, if you lose weight. This means you are being advised to trade one form of suffering (the hunger you will experience if you diet) for another form of suffering (the pain you are experiencing from your disability). As you consider this advice, disability studies' core arguments are unlikely to be helpful. The notions that it is ableist to want to be cured or that disability should not be described as suffering do not apply to your situation: you are suffering, and even if you do not expect to be cured, it is not ableist to look to healthcare providers for some relief. Fat studies, by contrast, offers crucial insights: scholarship in this field makes it clear that efforts to become permanently thin almost always fail and that diets likely harm, rather than help, one's health.

If foundational disability studies arguments are not useful to disabled fat people contemplating advice to lose weight, another strand of scholarship in disability studies, which does talk about suffering, may prove relevant. One example of this strand of scholarship is my own writing about "undocumented disabilities," a term that I coined to refer to impairments that mainstream Western medicine does not regard as legitimate (Mollow 2014). People with undocumented disabilities, many of whom are chronically ill or in pain, are frequently blamed for our disabilities, our suffering dismissed as "all in the head." When medical providers look at an undocumentedly disabled patient, they may think, "She looks fine, and all her tests are normal, so her symptoms must be psychogenic." Similarly, if a fat patient says, "I am unbearably hungry on the weight loss regimen that you prescribed," providers may think this cannot possibly be true. After all, the patient is fat, so how can she be hungry?

But make no mistake: doctors who counsel fat people to become thin, and to stay thin forever, ask them to engage in a struggle that they have virtually no hope of winning, a struggle whose result is hunger – that is, suffering – that will not end (Kolata 2007, 6). Many healthcare providers think being thin is simple: if you are fat, then just choose to eat less. This recommendation is cruel because it ignores the issue of hunger. When we attempt to disregard hunger, we suffer. Eventually, the drive to eat may become desperate. Those who have dieted know the experience intimately: almost always, hunger wins, and we eat.

III How to Treat Fat (and Thin) Patients

Health professionals typically describe their job as ending suffering, or at least alleviating it. Yet they cause unnecessary suffering when they instruct patients to lose weight. In what follows, I make the following eight suggestions about how medical providers can avoid perpetuating fatphobia in their practices.

1 *Read the literature.* Healthcare providers typically undergo years of training in which they are taught that fatness is a disease, which fat people can overcome by making healthful choices. This training is based on the work of researchers who, as discussed earlier, may be influenced by financial conflicts of interest and anti-fat bias. Medical professionals should therefore seek out alternative points of view. Numerous books and articles, many of which perform in-depth analyses of the scientific literature, contest the assumption that fatness is a health risk while highlighting the inefficacy of weight loss dieting. Healthcare providers should familiarize themselves with this literature. They should also learn about the "Health at Every Size" framework, which focuses on nutritious eating and other health-promoting practices, without regard to weight (Burgard 2009). In addition, providers should educate themselves about the experiences of fat people who reject weight loss dieting. A vibrant fat justice movement, centering fat people who accept their bodies, has been thriving for decades. An Internet search yields an abundance of fat-positive writing, videos, artwork, and activism.

2 *Create an Accessible Environment.* Many medical practices have access barriers that prevent fat people from receiving adequate health care. Offices should be stocked with gowns large enough to comfortably fit fat people, including those who are very fat. Waiting rooms should have wide, sturdy chairs without arms. Large blood pressure cuffs should be available. Examining tables need to be wide and strong enough to hold fat people.

3 *Skip the scale.* "Step on the scale" is one of the first instructions patients receive upon arriving at a medical office. For many fat people (and some thin people), being weighed is stressful at best, traumatic at worst. Moreover, obsessive attention to weight reinforces the misconception that thinness and health are equivalent. To be clear, it may occasionally be necessary to know a person's weight, for example, to calculate the correct dosage of medication. But weighing patients every time they come to the office – and recording that information at the top of their charts, as if it were the most important thing to know – creates an inhospitable environment for fat patients.

4 *Use Value-Neutral Language.* "I am a fat woman. Please don't call me 'obese,'" my wife, Jane, writes on patient intake forms when she sees a new provider. Far too often, she has requested copies of her medical records (for reasons unrelated to her size), only to learn that her provider's first impression of her is that she is a "pleasant obese woman." Such labels are dehumanizing. They are also irrelevant, since size is neither an accurate predictor of health nor a bodily characteristic one can realistically hope to change. When doctors use

terms such as "overweight," "obese," and – worst of all – "morbidly obese," they create the impression that variations in humans' body sizes constitute an essential element of who one is. These labels are stigmatizing, in much the same way that medicalized terms of earlier eras – "pervert," "idiot," "cripple" – dehumanize marginalized people. For this reason, fat studies scholars employ the term "fat," whose denotation is value-neutral. But because in fatphobic societies the word fat is seen by many as having negative connotations – and is sometimes used as a slur – I do not suggest applying the term to patients who do not themselves identify as fat. If it is necessary to refer to a patient's weight, non-stigmatizing words such as "large" or "heavy" can be used.[5]

5   *Do Not Assume Health Problems Are Caused by Weight.* When Jane once sought medical help for a splinter that was embedded in her toe, she was told, "If you'd lost weight, you would not have this splinter." Most fat people have had countless experiences like this. Sometimes, the consequences are deadly. For example, when Rebecca Hiles developed a lung tumor at age 17, her condition went undiagnosed for five years – her lung eventually became rotted and needed to be removed – because the numerous doctors with whom she consulted blamed her debilitating breathing problems on her weight (Dusenbery 2018). In 2010, a six-year-old girl, Claudialee Gomez-Nicanor, died from type 1 diabetes, her doctor having assumed that, because she was fat, she must have had type 2 diabetes (this condition is virtually unheard of in a six-year-old of any weight) (Brown 2015, 117–120). Fat studies scholars advise fat people to ask their medical providers, "How would you treat a thin patient with the same symptoms?" This is excellent advice; however, the burden of posing this question should not fall on fat patients. Instead, healthcare providers should ask themselves what they would do if the patient in front of them were thin – and then they should offer their fat patients the same quality of care.

6   *Do Not Assume Thin Patients Are Healthy.* Having accompanied Jane to many medical appointments, I have noticed that providers frequently ask her, "Do you have heart disease? Sleep apnea? Diabetes?" When she answers "No," they often look surprised – even though most fat people do not have these disabilities. By contrast, I have never been asked about the above-mentioned disabilities, presumably because providers do not (usually) perceive me as fat. Yet thin people also develop these conditions, along with many others that are erroneously assumed to be caused by fatness. When this happens, diagnoses are often missed.

Also, patients should not be complimented for being thin. When I was younger, medical providers often praised me for having a "healthy weight" – even when my thinness was the result of severe digestive disorders, which prevented me from eating enough. Now that my stomach problems are better (and I have entered middle age), I have become fatter. And doctors treat me differently: the compliments have stopped, and I receive occasional suggestions to "watch [my] portion sizes" or to follow low-fat or low-carbohydrate diets. (I reject these suggestions and refuse to engage in any form of weight loss dieting.) Because bodies often become fatter with age, many currently thin patients will one day be fat. To compliment a person for thinness is to implicitly criticize the fat person they may become in the future.

7   *Teach Patients That Diets Don't Work.* Many fat patients come to medical offices believing they need to lose weight and blaming themselves for not having been able to stay on weight loss programs. Providers should teach patients that diets don't work: not because fat people lack "willpower," and not because they have not yet found the right diet plan, but because, in the long run, hunger almost always wins. This is true regardless of what name one gives to a diet: even if an eating plan is dubbed a "permanent lifestyle change," the chances

that it will produce permanent, substantial weight loss are less than 5% (Brown 2015, 36). Similarly, diet drugs and bariatric surgery are largely ineffective for long-term weight loss; these treatments also have dangerous side effects. The good news is that none of these interventions is necessary. If fat patients are provided with decent medical care, their health outcomes will likely resemble those of their thin counterparts.

8   *Make Sure Patients Are Eating Enough.* As mentioned earlier, people vary in how much they eat. Some have small appetites and feel satisfied with a little food; others must eat much more to not feel hungry. These differences do not correlate with body size. For example, for most of my life I was thin and had a large appetite, but now that I am in middle age, my body is getting fatter while my appetite is decreasing. Meanwhile, Jane has been fat for most of her life and has always had a small appetite. Because Jane eats relatively little food, she and I have sometimes questioned whether she is getting adequate nutrition. But no healthcare provider has ever asked Jane if she is eating enough. Simply because she is fat, they assume she must be eating "too much." Such assumptions have damaging repercussions. Often, fat people engage in eating-disordered behavior, for example, obsessively counting calories, severely restricting food intake, and weighing themselves daily. If a thin person engages in these behaviors, she is likely to be diagnosed with anorexia. But fat people who starve themselves are usually encouraged to continue dieting (Burgard 2009, 47–48).

No person should be encouraged to eat in ways that leave them feeling hungry. Instead, patients should be asked, "Are you eating three meals a day?" Providers should also ask whether these are nourishing meals or unsatisfying snacks chosen out of fear of gaining weight. Perhaps most important, healthcare professionals should ask patients if they are hungry. Patients should be taught – and we should all teach ourselves – that hunger is not a weakness to be conquered but instead a potentially life-saving message from one's body – a message that, fortunately, almost always wins.

Notes

1   I use the value-neutral term "fat" in place of stigmatizing medical labels such as "obese" and "overweight" for much the same reasons that disability scholars eschew pathologizing labels (such as "idiot," "insane," or "cripple") which medical professionals have applied to disabled people.

2   In this chapter, I use the term "fat studies scholars" expansively, to encompass a wide range of authors, some of whom have connections to the fat justice movement and others of whom do not. Under this rubric, I refer to writing by epidemiologists, statisticians, medical researchers, journalists, psychologists, and others whose research has led them to question two pervasive assumptions: that fatness poses health risks, and that most fat people can make themselves thin by engaging in weight loss dieting.

3   Unlike previous studies, Flegal's used data that was representative of the entire US population, including a broad range of social classes (Kolata 2007, 202–203). And where other studies had used data based on participants' reported heights and weights (which are notoriously unreliable), Flegal and her colleagues gathered data that reported actual heights and weights.

4   In challenging the assumption that body size is a matter of individual choice, I am not voicing support for the "environmental" theory of fatness. Also known as the "foodscape argument," this theory claims that the supposed "problem" of fatness can be traced to food deserts and other racial and economic inequalities. In contrast to critics who make this argument, I do not believe that the existence of fat people constitutes a problem.

5   In most contexts, fat studies scholars avoid these terms as well, for the same reasons that disability scholars avoid labels such as "differently abled"; such euphemisms imply that there is something wrong with being fat or with being disabled.

References

Aubrey, Allison. 2013. "Research: A Little Extra Fat May Help You Live Longer." *NPR*. January 2. https://www.npr.org/sections/health-shots/2013/01/02/168437030/research-a-little-extra-fat-may-help-you-live-longer

Brown, Harriet. 2015. *Body of Truth: How Science, History, and Culture Drive Our Obsession with Weight – and What We Can Do about It*. Boston, MA: Da Capo Press.

Burgard, Deb. 2009. "What Is 'Health at Every Size'?" In *The Fat Studies Reader*, edited by Esther Rothblum and Sondra Solovay, pp. 42–53. New York: New York University Press.

Campos, Paul. 2004. *The Obesity Myth: Why America's Obsession with Weight Is Hazardous to Your Health*. New York: Penguin. [Reissued in 2005 as *The Diet Myth*.]

Dusenbery, Maya. 2018. "Doctors Told Her She Was Just Fat. She Actually Had Cancer." *Cosmopolitan*, April 17, 2018. Accessed February 15, 2021. https://www.cosmopolitan.com/health-fitness/a19608429/medical-fatshaming/.

First Do No Harm: Real Stories of Fat Prejudice in Health Care. N.d. Accessed August 14, 2020. https://fathealth.wordpress.com

Flegal, Katherine, and Barry Graubard, David F. Williamson, and Mitchell H. Gail. 2005. "Excess Deaths Associated with Underweight, Overweight, and Obesity." *Journal of the American Medical Association* 293: 1861–1867.

Flegal, Katherine, Brian K. Kit, and Heather Orpana et al. 2013. "Association of All-Cause Mortality with Overweight and Obesity Using Standard Body Mass Index Categories: A Systematic Review and Meta-Analysis." *Journal of the American Medical Association* 309(1): 71–82.

Gard, Michael, and Jan Wright. 2005. *The Obesity Epidemic: Science, Morality, and Ideology*. London and New York: Routledge.

Harrison, Christy. 2019. *Anti-Diet: Reclaim Your Time, Money, Well-Being, and Happiness through Intuitive Eating*. New York: Little, Brown Spark.

Herndon, April. 2011. "Disparate but Disabled: Fat Embodiment and Disability Studies." In *Feminist Disability Studies*, edited by Kim Q. Hall, pp. 245–262. Indiana University Press.

Hoffman, Kelly, Sophie Trawalter, Jordan R. Axt, and M. Norman Oliver. 2016. "Racial Bias in Pain Assessment and Treatment Recommendations, and False Beliefs about Biological Differences between Blacks and Whites." *PNAS* 113(16): 4296–4301.

Kolata, Gina. 2007. *Rethinking Thin: The New Science of Weight Loss – and the Myths and Realities of Dieting*. New York: Farrar, Straus and Giroux.

Metzl, Jonathan. 2009. *The Protest Psychosis: How Schizophrenia Became a Black Disease*. Boston, MA: Beacon Press.

Mollow, Anna. 2014. "Criphystemologies: What Disability Theory Needs to Know about Hysteria." *Journal of Literary and Cultural Disability Studies* 8(2): 185–201.

Mollow, Anna. 2015. "Disability Studies Gets Fat." *Hypatia* 30(1): 199–216.

Parker-Pope, Tara. 2011. "Behind the Fat Trap." *New York Times Magazine*, December 28. https://www.nytimes.com/2012/01/01/magazine/tara-parker-pope-fat-trap.html.

Puhl, Rebecca, and Chelsea A. Heuer. 2009. "The Stigma of Obesity: A Review and Update." *Obesity* 17(5): 941–964.

Further Reading

Clear, TaMeicka. 2015. "Fitness and Health Are Not the Same (No Matter What Fatphobia Attempts to Claim)." *Everyday Feminism*. March 23. https://everydayfeminism.com/2015/03/fitness-and-health-difference/.

Harding, Kate, and Marianne Kirby. 2009. *Lessons from the Fat-o-sphere: Quit Dieting and Declare a Truce with Your Body*. New York: Penguin.

Tovar, Virgie. 2018. *You Have the Right to Remain Fat*. New York: The Feminist Press at the City University of New York.

25

TRANS CARE WITHIN AND AGAINST THE MEDICAL-INDUSTRIAL COMPLEX

Hil Malatino

This chapter is an excerpt from Malatino, Hil. *Trans Care.* **Minneapolis: University of Minnesota Press (2020), pp. 61–72.**

I Denials of Care

The nexus of care most commonly associated with transness—that provided by the medical-industrial complex—has often offered not much more than cold comfort to those trans subjects seeking it. Economically inaccessible, geographically dispersed, and rigorously gatekept, access to those gender-confirming surgical procedures offered by physicians have been, for quite a long time, an indicator of relative privilege, most commonly open to White, well-educated subjects of considerable economic means. While the hard-fought and ongoing battles that have resulted in recent expansions in transition-related insurance coverage and (imperfectly) democratized access to hormones have shifted this terrain considerably, the legacy of gatekeeping persists into the present, as does the high out-of-pocket cost of surgery for those subjects not enrolled in an inclusive insurance program, who don't have insurance, or who utilize Medicaid and live in a state whose Medicaid policies contain specific trans exclusions.

The fact that the subjects that populate trans archives are implicated in our ongoing survival is made abundantly clear by the traces of medical contestation they've left. The battle to end medical gatekeeping began a long time ago, and the (still inadequate) representation that trans subjects have gained in the areas of trans healthcare and medical policy are both recent and hard won. The model of medical patriarchal benevolence that tracks from the era of Harry Benjamin forward has proven particularly recalcitrant—and trans folks have proven particularly ready to do battle with it.

In spring of 2018, I visited the Kinsey archives at Indiana University to look, specifically, at the materials associated with the Harry Benjamin International Gender Dysphoria Association (HBIGDA) and its later transformation into the World Professional Association for Transgender Health, the still-extant organization that has authored the official, widespread Standards of Care meant to guide medical providers in their providence of trans-related services. I wanted to

DOI: 10.4324/9781003289487-33

understand, in greater depth, how it was that HBIGDA transformed into WPATH and became the transnational standard-setting organization that it is today.

It turns out that, in the 1990s, HBIGDA was struggling considerably to cohere and operate, in large part because certain well-known members of trans activist communities began to publish critiques of the extant Standards of Care, honing in specifically on the problems that attended the so-called real life experience (or "real life test") that mandated that trans subjects live in their "preferred gender role" for an extended period of time (ranging from three months to a year) before they are able to access hormone replacement therapy and gender-confirming surgery. A paradigmatic example of this argument comes from a contributor named "Cheryl B.," who wrote in to *TV/TS Tapestry* (which would later become *Transgender Tapestry*) in 1994 in order to point to the danger of the real life test for trans women in public, single-sex spaces, pragmatically pointing out:

> During the real life test, a risk lies in the use of public restrooms. While some pre-ops may pass well as women, others may have some difficulty. The police may arrest the pre-op on the complaint of another patron. If the arrest occurs after 5pm on a Friday, she may be detained over the weekend in a men's jail. Like Dee Farmer, the offending TS may be exposed to unwanted rape and infection with HIV.
>
> The danger only exists during the real life test dictated by the tyranny of HBIGDA. Although I have repeatedly argued this point with my therapist and with other experts, the dictate remains in force.
>
> *(Cheryl B. 1994, 18)*

This argument should be, at this point, deeply familiar, because a variation of it has had to be routinely remade in response to mechanisms of medical gatekeeping as well as the sustained effort to prevent trans people from accessing single-sex spaces, witnessed most recently in the spate of so-called bathroom bills that seek to mandate that the sex marked on one's birth certificate is the determinant for which single-sex spaces they may access. The logic of it runs as follows: institutionalized transphobia and medical gatekeeping entwine to produce a necropolitical cascade of effects that threaten the lives of trans people, and trans women of color most intensely. The rhetorical force of Cheryl B.'s argument derives not only from her own experience but from her reference to Dee Farmer, a Black trans woman who was repeatedly raped and contracted HIV while imprisoned at the federal penitentiary in Terre Haute, Indiana. Farmer's case against prison officials—Farmer v. Brennan—hinged on her assertion that prison officials knew that she would be especially vulnerable to sexual violence; it went to the Supreme Court, who ruled—in their historic first direct address of rape in prisons—that prison officials who fail to make provisions for prisoner safety in such instances can be held responsible for the ensuing violence. They argued that it constitutes a violation of the Eighth Amendment (the one that prohibits cruel and unusual punishment). This ruling was not necessarily a victory for Farmer or for imprisoned trans people. There was no direct commentary on trans experience— Farmer was repeatedly and consistently misgendered by the court, treated and referred to as male throughout, and in much of the mainstream reporting on the case.

While the claim that this possibility for violence "only exists during the real life test dictated by the tyranny of HBIGDA" is overdetermined—certainly, trans women face violence both before and after accessing transition-related medical services—it does drive home the point that medical gatekeeping intensifies risk and compromises safety. This emergent critique of the Standards of Care would ultimately convulse the organization, resulting in the short term in the composition of a trans-led Advocacy and Liaison Committee that was consulted during each subsequent revision to the Standards of Care and in the long term produced a standing ethics committee in

WPATH (led, as of 2020, by trans activist and author Jamison Green). Importantly, the formation of these committees marks the first time in the history of trans medicine where trans folks were officially and actively consulted regarding the treatment they received. It's a landmark moment for trans patient advocacy and a crucial moment in the genesis of the push for depathologization.

Engaging in such struggle can come at a high cost. It's worth mentioning that when I encountered this letter to the editor in the archive, it was in the HBIGDA files because it had been photocopied and sent to Eli Coleman, the University of Minnesota–based sexologist who was the founding editor of the International Journal of Transgenderism and who, a few short years later in 1999, would become president of HBIGDA. The note scrawled in the top left-hand corner of the Xerox reads, "Eli—is she one of your pts [patients]" (HBIGDA, box 1, series 1, folder 1). The location tag beneath her name is "MN"—Minnesota. I presume that this Xerox was sent to Coleman by another medical specialist affiliated with HBIGDA, and it testifies to the smallness of trans worlds. It is a situation wherein medical professionals are actively reading the small handful of trans community publications and able to single out particular patients—those engaging in public critique and protest of medical gatekeeping—with ease. So much for anonymity and privacy. It also raises a red flag concerning the possibility of retaliation. What if a medical practitioner, displeased with such contestations, decided to actively withhold treatment? Discontinue their relationship with the patient? Given the paucity and geographic dispersal of providers, such a rejection might very well be tantamount to a full-on denial of transition.

This isn't far-fetched speculation. Such forms of retaliation indeed happen. Recurrently.

Denial can take many forms. One of the most heartbreaking and infuriating aspects of wading through trans medical archives has to do with the consistent appearance of letters from trans folks seeking treatment with limited funds, asking for long-distance diagnosis, or if sources of financial support for transition exist. One woman writes to Alice Webb, who ran a gender-affirming practice in Galveston, Texas, in the late 1980s, inquiring as to whether or not she can get "help at a low cost . . . via mail" (HBIGDA, box 2, series 4 B, folder 2) because she doesn't have money, time, or the ability to travel elsewhere for a consultation. This is but one example; such inquiries appear so often in Alice Webb's archives that she begins to use a form letter by way of response, one that states that no financial support exists and offers—if possible—information about local or regional trans support groups that the inquirer might attend. When a refusal of care is the best you can hope for, what do you do? Where do you turn?

Increasingly, we've turned to each other.

II Crowdsourcing Empathy, Building Solidarity

Each day, my social media feed is populated with crowdfunding requests for surgery. Often, it's for facial feminization surgery, which is nearly unilaterally denied coverage. Other times, it's a request for top surgery, from uninsured and underinsured trans masc folks.

Each day, my social media feed is populated with requests for rent money, for money to keep the power on, for funds to repair a car, or to fund some other necessary expense that ensures minimal forms of survival.

Sometimes, I can throw money at these requests. Sometimes, the most I can do is commiserate in frustrated empathy. Both of these responses are trans care praxis. We turn to social media for support that is simultaneously fiscal and affective, simultaneously practical (for advice about physicians, knowledge about underresearched side effects of exogenous hormones, about what clinics operate on an informed consent model, to seek legal advice) and ephemerally affirmative (to be told that we look hot, to bitch about quotidian transphobia). We hear so much about the purported echo chamber of social media, the way it has increasingly dissuaded political conversation across difference, the way it has contributed to intensified and polarized partisanship. This ostensible dilemma of the demos is structured by the assumption of a specifically dialectic ideal: that continual cultivation of political agonism leads to deliberation, compromise, and ultimately (at least provisional) consensus. Thus, the echo chamber effect makes nonpartisan consensus impossible, or, more hopefully, quite unlikely.

But what happens when your identity becomes a political wedge issue? From debates about "bathroom bills" to Republican outrage about trans-inclusive insurance coverage to continual fearmongering about the specter of trans women in sport, to continual conservative-led legal initiatives to reinterpret Titles VII and IX as trans-exclusionary, trans bodies and lives have been scapegoated again and again as a sign of the excess and irreality of the political Left. Then there are the debates about trans-exclusionary radical feminism currently convulsing the feminist Left and leading to forms of unlikely alliance between certain sectors of radical feminism and the religious right that we haven't seen since the height of the Sex Wars. In this cultural climate, the echo chamber afforded by social media might be better understood as a provisionary form of trans separatism that offers imperative reprieve. It's where we access forms of preservative love withheld in the popular domain, and too often scarce in our everyday interactions.

Sara Ruddick, in *Maternal Thinking* (1989)—a groundbreaking work in feminist care ethics—frames preservative love as one of the central acts of mothering, which is the relational position from which she derives an entire epistemology of care. It's important to note, as well, that Ruddick understands "mothering" to be a practice taken up by persons of any gender; rather, anyone who commits themselves "to responding to children's demands, and makes the work of response a considerable part of her or his life, is a mother" (xii). Preservative love is shorthand for all of those acts that keep a being alive and intact, and it is characterized by a specific response to the vulnerability of another. It means "to see vulnerability and to respond to it with care rather than abuse, indifference, or flight" (19). It doesn't require a particular affective orientation—we don't have to be cheerful or enthusiastic about it, and we may indeed feel deeply ambivalent about such forms of care. Ruddick: "what we are pleased to call 'mother-love' is intermixed with hate, sorrow, impatience, resentment, and despair" (64).

Of course, only some trans folks are children, and not all trans people engage in mothering. But if you're a person of trans experience and involved in trans communities, you know that intensified forms of vulnerability and exposure to violence and debility continue to inform trans lives across age groups. In addition to this, transition also scrambles normative temporalities of development. We have "second puberties" well into adulthood; we have "big brothers" or "big sisters" mentor us through transition because, though they may be younger in years, they've initiated transition long before us. We sometimes come from childhood homes that did not adequately provide the forms of preservative love and nurturance that form the crux of practices of mothering. Alternately, we may have these forms of motherhood reduced or withheld upon the revelation of our transness. This is all to say we remain in need of mothering (in the many-gendered, expansive sense of the word) well into adulthood.

Trans historian Morgan M Page has given us a golden rule as we navigate the spaces of social media, and it is deeply informed by the ethos of preservative love. The rule is simple. "I do not shit-talk other trans people in public. If I truly have a problem that must be addressed, I speak to them directly" (Page 2020). She goes on to unpack what motivates the rule: the high incidence of mental health struggle in trans communities means that call-outs and online harassment sometimes translate to self-harm and suicide. In addition to this, the rising tide of antitrans organizing has made a practice of solidarity across difference increasingly crucial. We can ill afford to be locked in self-aggrandizing battle with one another. This is doubly so when we consider that the online spaces wherein we congregate—from the Yahoo groups and chatrooms of yore to the networks we inhabit on Twitter, Instagram, and all of the closed groups on Facebook that effectively operate as both support groups and skillshares—are the only trans-majority spaces to which many of us have access. These spaces, despite their potential, often reify the forms of stratification and inequality that shape our experiences IRL.

Public health researchers Chris Barcelos and Stephanie Budge, in their recent work on inequities that manifest in the context of crowdfunding transition-related medical costs, point this out quite explicitly. While noting that crowdfunding medical care is a "response to health and social inequalities related to a disproportionate burden of ill health and lack of adequate insurance coverage for gender-affirming care" (2019, 84), and that very few trans crowdfunding projects meet—or come close to—their goal, it is nevertheless the case that "the majority of recipients were young, White, binary-identified transgender men" (84) and that relative success with crowdfunding is "related to having a large network of distant ties through which the fundraising page is shared" (86). The better networked you are, the more social media capital you have, the more successful your bid for funding will be. This means that crowdfunding favors folks with the time, the extroverted capacity for engagement, and an extant and well-received "brand." In other words, it makes health care access in the context of compounded inequalities tantamount to a popularity contest. As such, it is an intensifier of already-existing forms of biomedical stratification.

The work of Barcelos and Budge insinuates a broader point: that trans care can all too easily reproduce hierarchies of attention, aid, and deservingness and that such hierarchies exacerbate and amplify inequities. Any care praxis worth enacting must be attentive to such tendencies to reproduce injustice. This applies to forms of emotional support as much as it does to forms of financial support. Our energetic investments are subject to partage and apportioning, informed by economies of existential valuation that we must struggle to be conscious of—and to undo. We are all subject to forms of structurally produced and enhanced ignorance and elision, and these forms of unknowing and inattention are exacerbated, as Safiya Umoja Noble points out, by the algorithms that inform how and what we encounter in digital spaces. As she writes, "racism and sexism are part of the architecture and language of technology," from Google searches to crowdfunding initiatives (2018, 9). This is why Barcelos, in another article on the inequities of trans medical crowdfunding, calls for a "revolutionary ethic" that would transform the way this practice operates (2019, 7). Building on the concept of "revolutionary etiquette" developed by activist and performance artist Annie Danger (Danger and Nipon 2014), he suggests that such an ethic would call attention to the necessity of crowdfunding as a flawed work-around for the unjust neoliberal distribution of health and wellness services. He writes that "employing this etiquette would mean foregrounding a discussion not only of the healthcare inequalities facing individual trans people, but also an action plan that centers redistribution of financial and social benefits. This etiquette would prioritize a decentering of individual, normative transition narratives in favor of a collective vision of transgender liberation" (Barcelos 2019, 7).

There must be a dual movement wherein we highlight the imperfection and complicity that characterizes contemporary forms of trans care praxis as we push for collective redistribution. We need to address what constrains care, what marks certain bodies and subjectivities as (un)deserving of it, and call attention to the epistemologies, systems, and technologies that contribute to such unjust apportioning, even as we must navigate them in order to get (some of) our needs met. Care praxis is always within and beyond; forever prefigurative.

III Coda

To end, a story: when I was a kid, in the abandoned space of my childhood (which is a story for another time), I met another kid, similarly abandoned, and we lashed ourselves together in order to weather the sometimes devastatingly bad storms of our youth. We were both becoming genders we were never supposed to be, and we found home together. We built these homes, first, in each other. We added rooms through our zine trades and our late-night instant messaging and our lousy bands and our parking lot hangs outside of dyke bars and punk clubs. We found a broader network of folks who were similarly trans and queer and broke and traumatized and disassociated and trying desperately to find one another. We encountered the concept of prefigurative politics—building the new world in the shell of the old—pretty organically, through the anarcho-queer imaginary that animated the margins of punk scenes during those years. We came to realize that we were doing prefigurative political work whenever we made space for each other within our psyches, within our homes, within those spaces that felt like way less than homes, and within all those institutions and collectives through which we circulated—some durable, some ephemeral. We both became "voluntary gender workers," in addition to all the other hats we learned to wear, doing trans, intersex, nonbinary, and gender nonconforming advocacy work wherever we happened to find ourselves. We raised each other in the vacuum of care left by the overlapping economies of abandonment (Povinelli 2011) that shaped our days. We kept each other alive. We mothered each other through it, even though we'd both wind up way more daddy than mother. Our webs have spun out from the juncture of that decades-long intimacy, which is not a center so much as one significant t4t nexus in a constellation of so many, some thicker and tougher than others. The pattern of care and witness that we provided for one another is indelible, and I'm beyond lucky to have cocultivated it at such a young age; it has made it that much easier to identify and reject connections that fail to be characterized by this kind of commitment to making space for one another's becoming. When I think of care, I think first of him, and this, and the way it raised the bar for every significant encounter and intimacy to come. This is about a certain kind of faithfulness and a certain kind of obligation: about what we owe each other. Minimally, it is this: a commitment to showing up for all of those folks engaged in the necessary and integral care work that supports trans lives, however proximal or distant, in the ways that we can. This, along with an acknowledgment that it is precisely the recurrent, habitual, and mundane practice of showing up that makes us less and less willing to inhabit a world where we don't show up, and where whole systems fail to show up for us.

Note

From the Editors: As Malatino's chapter makes clear, there are many shared concerns across disability and trans experience, including those surrounding medicalization, prejudicial treatment, and the impact of politics and being made a target in political fights in particular. Having said this, readers will note that this piece does not directly discuss disability and the connections—as well as differences—between disability bioethics and trans healthcare more broadly. We had an author lined up and a piece nearly completed to take on that task, but, unfortunately, extenuating circumstances lead to that piece not being in the volume and did so at a stage too late to ask a different author to write such a piece. That unfortunate turn of events led to a fortunate one: we discovered Malatino's fantastic book, *Trans Care*, and jumped on the opportunity to highlight and incorporate part of Malatino's work here.

References

Barcelos, Chris A. 2019. "Go Fund Inequality: The Politics of Crowdfunding Transgender Medical Care." *Critical Public Health* (online first view): 1–10.

Barcelos, Chris A., and Stephanie L. Budge. 2019. "Inequalities in Crowdfunding for Transgender Health Care." *Transgender Health* 4 (1): 81–88.

Cheryl B. 1994. "Dispute over Real Life Test." TV-TS Tapestry 69: 18.

Danger, Annie, and Ezra Berkley Nipon. 2014. "Emily Post Capitalism and the Revolutionary Etiquette of Crowdfunding: A Conversation with Annie Danger." *Grassroots Funding Journal* (March–April): 11–14.

Noble, Safiya Umoja. 2018. *Algorithms of Oppression*. New York: New York University Press

Page, Morgan M (@morganmpage). 2020. "This seems like a good time to remind people of the rule I gave myself a few years ago that has really helped me immensely: I do not shit-talk other trans people in public if I truly have a problem that must be addressed, I speak to them directly." Twitter, January 2, 2020, 4:46 p.m. https://twitter.com/morganmpage/status/1212852732652077058.

Povinelli, Elizabeth A. 2011. *Economies of Abandonment: Social Belonging and Endurance in Late Liberalism*. Durham, N.C.: Duke University Press.

Ruddick, Sara. 1989. *Maternal Thinking*. Boston: Beacon Press.

PART VIII

Intellectual and Mental Disabilities

26

DEFINING MENTAL ILLNESS AND PSYCHIATRIC DISABILITY

Laura Guidry-Grimes

Defining illness and disability is a value-laden enterprise. Value judgments go into deciding what facets of a lived condition are pathological or within the range of healthy human diversity. Not all problems presented by living are pathological, so a line or threshold has to be drawn somewhere. This definitional work is complex, messy, and evolving in all of medicine, but the challenge in psychiatry can be especially immense. The history of psychiatric diagnostic systems is riddled with unfortunate social biases (including racist, sexist, heterosexist, cissexist, classist, and others). With any of these attempts, emotional states, desires, ways of thinking and processing, and patterns of behavior are identified that presumably would benefit from some sort of medical intervention. The stakes for psychiatric diagnosis are significant: patient complaints can be taken seriously by healthcare professionals and insurance providers only if they have a diagnostic label attached to their concerns, but the label of mental illness also frequently heaps on stigma, discrimination, and distrust.

This chapter begins by briefly describing the historical backdrop of nosology (the classification of disorders) in psychiatry. I summarize some controversial diagnostic categories as well as current criticisms of the *Diagnostic and Statistical Manual of Mental Disorders*, which is published by the American Psychiatric Association but has global influence. The purpose of this summary is neither to dismiss psychiatry as a medical field nor to demonize its practitioners. Rather, the aim is to highlight how problematic value judgments can creep into the construction of any nosological system.

This history and summary set the stage for discussing the consumer/survivor/ex-patient (c/s/x) movement (also called the user/survivor movement). Those who identify with this movement contest standard biomedical interpretations of mental illness. The c/s/x movement has numerous similarities with the disability rights movement (DRM). The DRM challenges standard conceptions of disability and shows how medically defined conditions are not necessarily conditions of suffering or any kind of lesser-than existence. In a similar vein, the c/s/x movement advocates for self-definition and celebrates "mad pride." Many who identify with c/s/x argue that their condition is not one of dysfunction or deficiency and that their so-called symptoms are sources of creativity, joy, and identity. They champion the notion of "generative madness," according to which "psychic difference is something to value, to be proud of, to channel, and even to enhance" (Lewis, 2012, 145). The very notion of "generative madness"

DOI: 10.4324/9781003289487-35

is conceptually incoherent according to standard interpretations of mental illness. This movement highlights the normative assumptions about well-being and meaningful agency that are embedded in these definitions.

For generative madness to be conceptually possible, first-person narratives have to be a central driving force in defining and classifying psychiatric disabilities. The remainder of this chapter will analyze a critical challenge to this proposal: individuals with psychiatric diagnoses are often doubted as reliable knowers and reporters of their condition. Unlike physical disability, the very nature of psychiatric disability is such that the individual might not be capable of understanding or appreciating their condition. First-person reports can end up being fallible, depending on the specific nature of the psychiatric disability. At the same time, discounting these narratives altogether raises concerns for epistemic justice. If first-person reports are to be centralized in the project of defining psychiatric disability, this challenge has to be carefully addressed. After laying out the nature of this challenge, this chapter ends by emphasizing the importance of shared decision-making and protecting a safe space for personal recovery.

I Nosology in Psychiatry: A Value-Laden Enterprise with Continuing Challenges

To demonstrate the value-laden nature of nosology, consider a question that is at the forefront of all these attempts: what is the aim of the diagnostic system? There is no way to answer this question that avoids value judgments. If the basic idea is to differentiate healthy and unhealthy mental states, we have to first understand the conception of health that should be guiding this work and how that relates to emotions, behavior, and cognition. A variety of value assumptions are loaded into any conception of health. For example, species-typical functioning could be the guiding value, but that requires having an ideal for the species and separating the atypical and exceptional from the atypical and undesirable. Alternatively, we could value functional abilities in everyday living, though we then fall into complex judgments both about abilities and everyday living in different contexts, which can quickly slide into capitalist values and ableism. Using a conception of health to assess emotions, behavior, and cognition requires additional value-laden steps, especially since how we think, feel, and react are culturally mediated and subjected to a host of norms (social, religious, economic, and others). Moreover, as evidenced in the history of the *Diagnostic and Statistical Manual of Mental Disorders* (DSM), the motivations and intended purposes for constructing a nosological system are more complicated than just searching for healthy versus unhealthy states.

Compiling a document like the DSM does not exist in a vacuum where "pure science" and "value-free objective discovery" lead to the development of diagnostic categories. Each iteration of the DSM is a product of its sociohistorical context, and the details of this context are telling indicators for why the system was created and revised over the years. Who is responsible for creating diagnostic categories, how they decide to create or revise categories, and what ends up being diagnosable also need to be investigated to understand what values are at play. The resulting nosological system will reflect whose value perspective has precedence, how values were prioritized and competing values adjudicated, and what the authors of the system believe to be a valuable set of officially legitimated conditions. Even the most well-intentioned nosological system with widespread scientific consensus will be value-laden in the ways described.

A glimpse into the history of psychiatry reveals a number of troubling biases that made their way into diagnosis. In the Antebellum South, the mental illness diagnosis of drapetomania was used to describe the pathology of enslaved people who tried to flee, and dyaethesia aethiopica was applied to explain perceived laziness among freed former slaves. During the 1960s, the diagnosis of schizophrenia went from being associated with a relatively harmless neurosis among middle-class housewives to being associated with violent aggression in Black men, so-called "protest psychosis" (Metzl 2010). Into the twentieth century, psychiatrists and other therapists offered the label of hysteria, based on the ancient notion of a "wandering uterus," to women who were perceived as overly emotional, sexually frustrated, or incapable of having biological children. Neurasthenia ("nervous exhaustion") was commonly associated with "unfulfilled" women during industrialization at the turn of the nineteenth century, and concerns were raised that educating women could also give rise to this problem (Tripathi, Messias, Spollen, and Salomon 2019). Heteronormativity and cisnormativity have influenced diagnostic categories as well; homosexuality, "transvestism," and "gender identity disorder" all appeared in the DSM. These examples show how groups who are vulnerable in their social context can have their vulnerability exacerbated through medicalization and ensuing presumptions about their mental health.

The DSM has considerable influence on psychiatric practice throughout the world, but criticism has followed the manual since its inception. Over the decades, one repeated concern is that the DSM contributes to "disease mongering" – the creation of conditions to fit psychotherapeutics, whether with respect to their development or sales. The diagnostic categories within the DSM expand significantly and become more inclusive with each edition, so much so that current estimates are that half of Americans qualify for a DSM diagnosis at some point in their lives (Kawa and Giordano 2012, 7). As a related concern: "the 'pathologization of deviance' and the 'medicalization of social ills' are potential effects of psychiatric diagnoses and treatment trends" (ibid.). Each new edition of the DSM has moved increasingly toward biomedical modeling, including attempts to incorporate neuroimaging and genetics. Research and insurance reimbursement dollars have also increased accordingly (ibid., 6). Concerns about biased diagnoses, disease mongering, expansive pathologization, and biomedical modeling culminated in a social-political movement for a group of mental health professionals, current and would-be patients, and the general public.

II Consumer/Survivor/Ex-Patient Activism: "Talking Back" to Psychiatry

Social justice movements of the 1960s challenged the historical disempowerment and marginalization of persons on the basis of their perceived race, sex and gender, nationality, and disability status. These movements ultimately contributed to public awareness of injustices at the interpersonal and structural levels, including social, political, and economic spheres of life. Within this historical context, the consumer/survivor/ex-patient (c/s/x)[1] movement disputed the foundational assumptions and methods of psychiatry. The groups are comprised of people who have experienced the mental health system in some form, their families, self-described radical mental health professionals, and other allies. In the beginning stages, R. D. Laing and Thomas Szasz provided the philosophical backbone for these activist projects. According to these antipsychiatry theorists, mental health interventions are generally more harmful than beneficial; mental illnesses are myths that are used for social control; and some interventions, such as electroshock therapy and involuntary commitment, violate human rights. Judi Chamberlin's memoir, *On Our Own* (1979), became a foundational text for the movement and convinced many people diagnosed with mental illnesses that peer-run alternatives had the potential to be far more effective and less stigmatizing than traditional therapies.

Although its philosophical roots are in anti-psychiatry, c/s/x activism has evolved over the years, becoming more nuanced as it branched throughout the world. Not all who identify with c/s/x activism are entirely opposed to receiving professional services in the mental health system, for example.[2] Prominent groups today include MindFreedom International, the Icarus Project, and INTERVOICE Hearing Voices Network.

MindFreedom International (MFI) is a global activist network of over 100 grassroots organizations with thousands of members. MFI argues against force and coercion in mental health care, abuses by the pharmaceutical industry, and electroconvulsive therapy. MFI protests outside the annual meetings of the American Psychiatric Association, calling attention to involuntary treatment and stereotypes attached to diagnostic labels. Their Choice in Mental Health Care campaign aims to "allow people's experiences to become more important than the diagnoses, validate the person and their experience, and encourage the person to come up with their own answers" (MindFreedom International, n.d.). Mad Pride is perhaps their most well-known event, which celebrates "mental diversity" and reclaims "the experience of madness and the language surrounding it" (Curtis, Dellar, Leslie, and Watson 2000, 7). MFI states: "All of us are different, no one has a grip on reality, and we're all on the edge. So celebrate!" (MindFreedom International, n.d.)

The Icarus Project, started in 2002, "envision[s] a new culture that allows the space and freedom for exploring different states of being, and recognizes that breakdown can be the entrance to breakthrough" (New York Icarus Project, n.d.). One of the founders describes "mental illness as a potential gift of great vision, creativity, and compassion that must be harnessed and respected, as well as an incredible hardship" (DuBrul 2014, 266). The Icarus Project emphasizes peer support, safe spaces for creative expression, anti-stigma, self-care, and self-determination. They also reject biomedical interpretations of their mental states and argue against standard support groups that try to "eradicate" mental illness. They argue that their ways of processing and thinking are often unique forms of brilliance, not pathology. The Fireweed Collective was created by former members of Icarus Project; they want to provide a healing space and mental health education that directly address the many forms of oppression that persons diagnosed with mental illness can experience (Fireweed Collective n.d.).

The INTERVOICE Hearing Voices Network began in 1997 with a meeting of "voice hearers" and mental health workers in Maastricht, the Netherlands. "INTERVOICE" stands for International Network for Training, Education, and Research into Hearing Voices. They argue: "hearing voices, seeing visions and related phenomena are meaningful experiences that can be understood in many ways," and "hearing voices is not, in itself, an indication of illness – but difficulties coping with voices can cause great distress" (INTERVOICE, 2020). Contrary to the standard medical interpretation of hallucinations as pathological and dangerous, this group emphasizes the diversity of experiences with these phenomena, including positive experiences. From their standpoint, medical interventions should not necessarily target voices; rather, voice hearers need support to choose the mode of being that best suits their values.

Some common aims of c/s/x groups are the following: (1) eliminate pervasive stigmas and social, economic, and political barriers facing those diagnosed with mental illness, (2) establish the right to self-definition, (3) significantly increase the inclusion of current and past patients in policy and treatment decisions, (4) abolish psychiatric interventions that violate human rights, and (5) overhaul the biomedical modeling of mental illness, allowing for accounts of flourishing in virtue of (not despite) psychiatric disability.[3] One of their reasons for coming together as a group is to "create an alternative space and narrative that is separate from psychiatry" and thereby "talk back" to psychiatry (Morrison 2005, 16). "Talking back" to psychiatry for this patient movement involves wrestling control and power away from the medical profession because, it is believed, outsider expertise has unfairly drowned out the insider expertise of patients.

When someone claims that they value their psychiatric disability and want it respected as a form of human difference, these claims are often met with skepticism and paternalism. C/s/x activists insist that psychiatric conditions do not necessarily diminish someone's agency or ability to flourish. They believe that their form of difference – mental diversity – deserves institutional and interpersonal recognition. They want healthcare professionals, policy makers, bioethicists, and others in society to respect their self-defined identities and interests. Given all these core aims of c/s/x activism, there is clear overlap with the disability rights movement. I argue that disability modeling can make sense of differing interpretations of mental illness and psychiatric disability.

III Disability Modeling and "Generative Madness"

An important contribution of the DRM and disability studies is to challenge traditional biomedical interpretations of disability. Anita Silvers (2011) explains:

> In contemporary Western culture, to be disabled is to be disadvantaged regardless of how much success one achieves individually. [...] 'disability' now is associated with physical or mental differences that compromise people's liberty to achieve typical levels of success in one or more areas of social participation, whether the relevant activities are learning, communicating, mobilizing, communication, being employed or some other important productive activity. The key phenomenon that informs this idea is the experience of disabled people's being more limited than other people in one or more seemingly important respects.
>
> *(23)*

According to the dominant cultural and medical narrative, disabilities are viewed as limitations that hinder worthwhile life pursuits or disrupt agency. On this pervasive interpretation, a disability is something wrong within the individual. Disability models differ in the extent to which they attribute suffering and life impediments to fixed conditions of the individual or to contingent social circumstances (discussed elsewhere in this volume in more detail).

Disability modeling has paradigmatically been used to unpack evaluative components of physical impairment and disability, and it provides useful insights for understanding psychiatric impairment and disability. Biomedical models interpret mental illness as disorders of the brain or genetics. According to these models, psychiatric disability necessarily threatens quality of life and diminishes flourishing potential. Based on this reasoning, mental health professionals can best treat these patients with medications and therapies that address the underlying biological problems and ameliorate symptoms. Psychotropic medications, electroconvulsive therapy, and other interventions would be intended to alter neural pathways, biochemistry, or other aspects of the brain's functioning (over time if not immediately).

Psychiatry's reliance on the biomedical model has received numerous critiques. C/s/x activists' objections to biomedical modeling are similar to the objections from the DRM. A reductionist causal explanation that centers on the brain or genetics means that psychiatry will primarily be centered on fixing perceived problems within the individual, such as through medications. Biomedical interpretations assume that these conditions can be described in a universal way that cuts across culture and context; this "framework does not ignore non-technical issues to do with relationships, meanings, or values, but these are understood as being of secondary importance" (Bracken and Thomas 2013, 124). C/s/x activists emphasize the role of stigma and discrimination in the challenges faced by people whose behavior and emotions do not fit the confines of certain norms. On the far end of the modeling spectrum, antipsychiatry theorists argue for a social constructionist interpretation; on this view, the alleged symptoms associated with psychiatric disabilities are solely caused by the sociocultural and political context, labeling the behaviors and mental states as disorders, and treating the behavior as abnormal. Others support a hybrid model that recognizes how environmental factors can have a substantial impact on biological factors underlying psychiatric disability; such a model, however, does not in itself resolve what should receive a diagnosis or where the "fix" should be aimed (for example, a biopsychosocial model does not legitimate homosexuality as a mental illness).

A common thread in c/s/x activism is the idea of "mental diversity" – that humans naturally have significant variations in how they reason and process their world, and what might be diagnosed as mental illness can be justifiably valued by the person experiencing it. Through the lens of "generative madness," psychiatric disabilities are interpreted in a way that runs "against the grain of most clinical models because they frame psychic difference as something positive rather than something negative" (Lewis 2012, 145). From a generative perspective, conditions that are clinically viewed as mental illnesses can be personally viewed as sources of personal identity, creativity, expressiveness, religious and spiritual experience, and social and political responsiveness (ibid., 155–158). Instead of learning to live with limitation, they want to cultivate these ways of being in the world and find support in doing so. The limitations purportedly inherent in their condition need not be viewed as such; what is limiting to others can be generative for them, and not all social and behavioral limitations are pathological or personally troubling. Awareness of one's psychiatric disability need not involve recognition of any form of dysfunction or suffering associated with it; it can be an outright rejection of seeing limitation as inherent to psychiatric disability. Eleanor Longden, a voice hearer and activist, describes her experience in the following manner:

> I am proud to be a voice hearer. It is an incredibly special and unique experience. I am so glad that I have been given the opportunity to see it that way because recovery is a fundamental human right and I shouldn't be the exception, I should be the rule. That is why I want to be a part of this movement to change the way we relate to human experience and diversity.
>
> *(qtd. in Woods 2013, 266)*

Traditional biomedical models leave little room for these first-person experiences of generative madness; this kind of narrative would have been to be reframed or distilled into recognizable data points for predetermined diagnostic categories (for example, effusively positive experiences with voice hearing could be documented by a psychiatrist as "auditory hallucinations with suggestions of mania or delusion").

IV A Challenge and Response

Individuals with psychiatric diagnoses are often doubted as reliable knowers and reporters of their condition. These doubts are often unjustified; psychiatric conditions have a vast range of abilities and do not always diminish capacities for knowledge or decision-making. If a person does have periodic poor reporting due to episodes of impairment, stigma and stereotype threats work together to suggest that that person is now in a lower class altogether as a knower – without taking into consideration that not all episodes lead to the same epistemic vantage point, and the person could be a reliable reporter with certain information and even on most occasions. Miranda Fricker describes how prejudice arising from negative stereotypes can distort someone's assessment of a person's credibility, which can follow that person in a wide range of social activities. The result is a form of epistemic injustice, the harm of which is to be "wronged as a knower […] to be wronged in a capacity essential to human value" (Fricker 2007, 44). There is increasing literature on epistemic injustice experienced by persons with psychiatric disabilities (cf. Sanati and Kyratsous 2015; Crichton, Carel, and Kidd 2017).

Consider when a voice hearer claims that she values her voices, insists that they are not a symptom of pathology, and tells her mental health professional that she does not want to take medications to eliminate her experience of them. A standard medical response is to deny her reports – to maintain that voice hearing is necessarily dysfunctional and any plan for recovery should include attempts to eliminate hallucinations. Biomedical interpretations of disability entail a clinical recovery paradigm, the "traditional understanding in the mental health system of recovery as socially valued outcomes, particularly symptom remission or alleviation and relatively independent functioning" (Slade 2012, 78). An implication is that any part of the patient's perspective that seems irrelevant or contrary to those traditionally valued outcomes would be excluded from therapeutic goal setting. Additionally, a patient who valued her symptoms or devalued symptom alleviation would appear not to want to be in recovery. When a person with psychiatric disability and a mental health professional disagree about what constitutes recovery (and even what a diagnosable problem is), a therapeutic impasse is inevitable. This disagreement halts shared decision-making, since the parties to the decision dispute what the therapeutic aims and value trade-offs should be. For instance, the effects of psychotropic medications may be intolerable for someone who values their voices or bursts of creativity, but the medical professional may view those side effects as preferable to ongoing hallucinations and episodes of mania. This impasse can further result in conflicts over non-adherence and concerns from the mental health professional about whether the patient should be trusted as a decision-maker.

A patient who rejects their diagnosis and care plan can be assessed to lack insight, meaning the mental health professional does not believe that the individual has self-knowledge or self-understanding. As long as poor insight is suspected, claims to patient expertise or self-definition are susceptible to refutation. This means that a patient who disagrees with standard biomedical interpretations can be at risk for losing their perceived epistemic standing in the therapeutic relationship. If the healthcare professional does not trust their patient as a knower, there is a barrier to genuine shared decision-making. This exacerbates the power asymmetry between patient and provider and makes it less likely that the person with unique epistemic access to their lived condition is authorized to make trade-off decisions with treatment.

The c/s/x movement contributed to the development of personal recovery as a therapeutic aim instead of clinical recovery: "Recovery is described as a deeply personal, unique process of changing one's attitudes, values, feelings, goals, skills, and/or roles. It is a way of living a satisfying, hopeful, and contributing life even with the limitations caused by illness" (Anthony 1994, 527). On this account, recovery will involve overcoming social obstacles such as the negative effects of stigma. Symptom alleviation would be neither necessary nor sufficient to be in recovery, since the symptoms might not be hindering the person from successful agency, happiness, or meaning-making.

To protect shared decision-making in the therapeutic relationship, healthcare professionals should recognize that reasonable interpretations of behaviors and mental phenomena do not necessarily fit the dominant biomedical framework. The better the provider can understand the patient's emotional and narrative knowledge, the more each party can come to understand and trust each other's perspectives. Through a diligent (and perhaps creative) shared decision-making process, the parties to the therapeutic relationship can come to feel less epistemically isolated from each other.[4]

For example, if a mental health professional has no evidence that a patient is harmed by hearing voices, then hearing voices should not be framed negatively in therapeutic goal setting.

Clinicians and therapists have to guard against overly narrow views of what it means to flourish and to have agency. Professional training about voice hearing should incorporate first-person narratives from people with these experiences. Fellow voice-hearers can also create safe, non-dismissive spaces for sharing their experiences of generative madness. Peer-run services may be more effective for helping voice hearers stay in recovery, since social supports are necessary to overcome social obstacles – as so much evidence in the medical and larger public health realm suggests.

Notes

1 Some activists believe that "patient" suggests a passive sick role, so they reject it; others prefer the term "consumer" to "survivor" because the latter implies traumatic experience which might not reflect all patients' experiences. Still other activists think "consumer" is a less-than-ideal label because consumers are thought to buy into the pharmaceutical industry or to be dependent on Supplemental Security Income and the like (Morrison 2005, p. 127).
2 It is also important to note that not all people with psychiatric disabilities would agree with all of the aims of c/s/x activism, including the endorsement of mad pride. Psychiatric disabilities are heterogeneous, even within any given diagnostic category, and this summary is not meant to overgeneralize.
3 I explore these aims of c/s/x activism more thoroughly in my dissertation, *Mental Diversity and Meaningful Psychiatric Disabilities*, Georgetown University, 2017.
4 An innovative approach is seen in the CommonGround program, founded by c/s/x activist and clinical psychologist Patricia Deegan, which incorporates a peer-run decision support center, decision support software, and specialized training of HCPs to support SDM in behavioral health (Deegan 2007).

References

Anthony, William A. 1994. "Recovery from Mental Illness: The Guiding Vision of Mental Health Service Systems in the 1990s." In *Readings in Psychiatric Rehabilitation*, edited by William Anthony and LeRoy Spaniol, pp. 521–538. Columbia: Boston University Center.

Bracken, Pat and Philip Thomas. 2013. "Challenges to the Modernist Identity of Psychiatry: User Empowerment and Recovery." In *Oxford Handbook of Philosophy and Psychiatry*, edited by K.W.M. Fulford, Martin Davies, Richard G.T. Gipps, George Graham, John Z. Sadler, Giovanni Stanghellini, and Tim Thornton, pp. 123–138. Oxford: Oxford University Press.

Chamberlin, Judi. 1979. *On Our Own: Patient-Controlled Alternatives to the Mental Health System*. New York: McGraw-Hill.

Crichton, Paul, Havi Carel, and Ian James Kidd. 2017. "Epistemic Injustice in Psychiatry." *BJPsych Bulletin 41.2*: 65–70.

Curtis, Ted, Robert Dellar, Esther Leslie, and Ben Watson. 2000. "Introduction." In *Mad Pride: A Celebration of Mad Culture*, edited by Ted Curtis, Robert Dellar, Esther Leslie, and Ben Watson. London: Chipmunka Publishing.

Deegan, Patricia E. 2007. "The Lived Experience of Using Psychiatric Medication in the Recovery Process and a Shared Decision-making Program to Support it." *Psychiatric Rehabilitation Journal 31.1*: 62–69.

DuBrul, Sascha Altman. 2014. "The Icarus Project: A Counter Narrative for Psychic Diversity." *Journal of Medical Humanities 35*: 257–271.

Fireweed Collective. n.d. Last accessed 14 July 2020. https://fireweedcollective.org/.

Fricker, Miranda. 2007. *Epistemic Injustice: Power & the Ethics of Knowing*. New York: Oxford University Press.

INTERVOICE: The International Community for Hearing Voices. 2020. Last accessed 14 July 2020. <http://www.intervoiceonline.org/>.

Kawa, Shadia and James Giordano. 2012. "A Brief Historicity of the *Diagnostic and Statistical Manual of Mental Disorders*: Issues and Implications for the Future of Psychiatric Canon and Practice." *Philosophy, Ethics, and Humanities in Medicine 7.2*: 1–9.

Lewis, Bradley. 2012. "Recovery, Narrative Theory, and Generative Madness." In *Recovery of People with Mental Illness: Philosophical and Related Perspectives*, edited by Abraham Rudnick, pp. 145–165. Oxford: Oxford University Press.

Metzl, Jonathan M. 2010. *Protest Psychosis: How Schizophrenia Became a Black Disease*. Boston, MA: Beacon Press.

Morrison, Linda J. 2005. *Talking Back to Psychiatry: The Psychiatric Consumer/Survivor/Exp-Patient Movement*. New York: Routledge.

MindFreedom International (MFI). n.d. Last accessed 14 July 2020. <http://mindfreedom.org/>.

New York Icarus Project. n.d. Last accessed 14 July 2020. <http://nycicarus.org/about/>.

Sanati, Abdi and Michalis Kyratsous. 2015. "Epistemic Injustice in Assessment of Delusions." *Journal of Evaluation in Clinical Practice 21.3*: 479–485.

Silvers, Anita. 2011. "An Essay on Modeling: The Social Model of Disability." In *Philosophical Reflections on Disability*, edited by D. Christopher Ralston and Justin Ho, pp. 19–36. New York: Springer.

Slade, Mike. 2012. "The Epistemological Basis of Personal Recovery." In *Recovery of People with Mental Illness: Philosophical and Related Perspectives*, edited by Abraham Rudnick, pp. 78–94. Oxford: Oxford University Press.

Tripathi, Samidha, Erick Messias, John Spollen, and Ronald M. Salomon. 2019. "Modern-Day Relics of Psychiatry." *Journal of Nervous and Mental Disease 207.9*: 701–704.

Woods, Angela. 2013. "The Voice-Hearer." *Journal of Mental Health 22.3*: 263–270.

Further Reading

Deegan, Patricia E. 2000. "Recovering Our Sense of Value after Being Labeled Mentally Ill." In *Readings for Diversity and Social Justice: An Anthology on Racism, Antisemitism, Sexism, Heterosexism, Ableism, and Classism*, edited by Maurianne Adams, Warren J. Blumenfeld, Rosie Castañeda, Heather W. Hackman, Madeline L. Peters, and Ximena Zúñiga, pp. 359–363. New York: Routledge.

Guidry-Grimes, Laura. 2019. "Ethical Complexities in Assessing Patients' Insight." *Journal of Medical Ethics 45,3*: 178–182.

———. 2020. "Overcoming Obstacles to Shared Mental Health Decision Making." *AMA Journal of Ethics 22.5* (May): 446–451.

Lewis, Bradley. 2010."A Mad Fight: Psychiatry and Disability Activism." In *The Disability Studies Reader*, 3rd ed., edited by Lennard J. Davis, pp. 160–175. New York: Routledge.

Romme, Marius, Sandra Escher, Jacqui Dillon, Dirk Corstens, and Mervyn Morris, eds. 2013. *Living with Voices: 50 Stories of Recovery*. Herefordshire: PCCS Books Ltd.

Shakespeare, Tom. 2010. "The Social Model of Disability." In *The Disability Studies Reader*, 3rd ed., edited by Lennard J. Davis, pp. 266–273. New York: Routledge.

27

RESEARCH ETHICS AND INTELLECTUAL DISABILITY

Finding the Middle Ground between Protection and Exclusion

Kevin Mintz and David Wasserman

1 Research as a Site for the Participation of People with Intellectual Disabilities

Contemporary research ethics in the USA evolved in parallel with the disability rights movement. In the 1970s, disability activists prioritized the deinstitutionalization of people with intellectual disabilities (PWIDs). We here define PWIDs as people with congenital or early onset intellectual disabilities expected to be lifelong (Carlson 2013, 303–304). Segregated residential facilities for PWIDs often served as the sites for gross medical research misconduct. For example, at the Willowbrook State School, children with intellectual disabilities were purposely injected with hepatitis in an effort to investigate the treatment and progression of the disease (Iacono and Carling-Jenkins 2012, 1124). Public outrage over this misconduct played a major role in the enactment of the first human subject protection laws and regulations in the USA. The Federal Policy for the Protection of Human Subjects, known as the "Common Rule," establishes standards, procedures, and institutions for protecting research participants.[1]

Public reaction to the treatment of institutionalized PWIDs also contributed to the passage of the first disability rights legislation in the USA, Section 504 of the Rehabilitation Act of 1973. Section 504 prohibits disability discrimination on the part of the federal government or any entity receiving federal funds. It requires entities to provide "reasonable accommodations" or individualized adjustments that enable people with disabilities to participate actively in federally funded programs, including research studies.

We focus on the Common Rule because it directly addresses the protection of research participants. It requires researchers and institutional review boards (IRBs) to evaluate the balance of individual risks and benefits to participants and society when designing and approving a study. The Rule also requires researchers to obtain informed consent from all participants or their legally authorized representatives. However, it provides no guidance about when consent may or must be obtained from a representative. Although each state has different rules for assigning representatives, they all require them to be assigned when the prospective participant lacks "competence" or "decision-making capacity" (National Institutes of Health 2009). In this chapter, we adopt the widely accepted view in bioethics that decision-making capacity involves four main

DOI: 10.4324/9781003289487-36

components: expression of a choice, understanding, appreciation, and reasoning (Grissso and Appelbaum 1998, 31).

In practice, a presumption of competence is not superseded simply because someone is intellectually disabled. PWIDs who are presumed or assessed to be competent face no more regulatory barriers to research participation than other individuals. However, many adults with IDs already have representatives or guardians assigned to them for other purposes. These "incapacitated" individuals face significant barriers to enrollment in clinical studies, even if they seek to participate with the approval of their representatives. PWIDs are generally permitted to enroll in research only when their participation (1) is "necessary" – when, for one reason or another, an IRB-approved study cannot be successfully conducted without enrolling them – or (2) holds out the prospect of direct medical benefit to them. We refer to these as the necessity and benefit requirements. Generally, only studies that have a non-negligible probability of preventing, mitigating, or curing a participant's health condition qualify as offering the prospect of direct benefit.

Disability bioethicists agree that PWIDs should be permitted to enroll in studies of direct benefit to them if they are clinically appropriate subjects and the risks are not excessive. Participation in this kind of research serves to promote their welfare, satisfying "the best interests standard" (Howard and Wendler 2020). This standard requires representatives to make decisions for PWIDs that conform to widely accepted notions of what is good for them. While the requirements of the Common Rule protect PWIDs from being exploited for the benefit of others, some PWIDs may ultimately be excluded from research in which they may have good reason to participate. A number of scholars have observed a "pendulum swing" from the exploitation of PWIDs in research to their purposeful exclusion (Rumney et al. 2015, 63). They argue that some PWIDs should be able to participate in research about intellectual impairment and associated conditions, even in studies offering them no prospect of direct benefit. They also recognize that representatives should promote not only the interests of the PWIDs, but their values and concerns as well.

Licia Carlson suggests that in assessing the participation of PWIDs, it is helpful to distinguish among three kinds of research: (a) "focused specifically on some dimension of intellectual disability (e.g., investigating etiology or associated symptoms)," (b) "investigating a health-related condition that is particularly relevant to this population (e.g., Alzheimer's in people with Down Syndrome)," and (c) "a study in which the disability itself is not a factor" (Carlson 2013, 307). Carlson argues that PWIDs can be ethically enrolled in studies within categories (a) and (b), but not (c). She argues that categorically excluding PWIDs from participation in research related to their intellectual disability or associated conditions with no prospect of benefit to them may be discriminatory. For Carlson, such exclusion assumes that PWIDs cannot value research participation for other reasons, such as helping others in their community.

However, she only endorses this altruistic participation in studies within her first two categories (research on ID and on associated conditions). In discussing the third category, Carlson maintains that "a surrogate decision maker placing someone incapable of consenting in harm's way for the sake of others is morally troubling" (Carlson 2013, 313). Her skepticism is understandable. Given the history of harmful exploitation of PWIDs in clinical research and the inequalities they confront in contemporary society, their inclusion in general health research warrants heightened ethical scrutiny.

While we agree with their rejection of the benefit requirement, we argue that writers like Carlson do not go far enough. They either accept or fail to challenge the necessity requirement – the restriction of PWIDs to research that requires their enrollment. PWIDs should not be considered valuable to the research enterprise solely because of their intellectual impairments. Those with appropriate values and concerns can and should be eligible to participate, with a representative's approval, in the same range of research as individuals without an impairment. To enable greater research participation for PWIDs, researchers should make protocols more accessible to them by providing reasonable accommodations and incorporating universal design.

II Respecting PWIDs for What They Can Do Rather than Penalizing Them for Their Limitations

Medical professionals may have difficulty reconciling the pathologies of intellectual impairment with the fact that many PWIDs can lead fulfilling lives with appropriate supports. They may believe that someone with an intellectual impairment should never be exposed to risks that other people willingly accept. They may assume that PWIDs cannot appreciate these risks, or are incapable of possessing the altruistic motivation to assent to them. Behind such generalizations about PWIDs lie paternalistic double standards regarding the level of risk they should be exposed to in comparison to non-disabled individuals of the same developmental age. Allowing PWIDs to take on risks with appropriate support from family, friends, and/or caregivers can be valuable for their maturation as it would be for most non-disabled people. For example, if researchers allowed a non-disabled 13-year-old to assent to participation in a study, it would be discriminatory to deny the same opportunity to a PWID whose developmental age was 13.

Our point is not that all PWIDs should be able to participate in any available study regardless of their developmental age. We recognize that in an intellectually rigorous and potentially harmful enterprise like clinical research, the inherent challenges of living with an intellectual impairment should not be ignored. However, researchers, IRBs, and representatives should not discount the possibility that some PWIDs may appreciate the benefits of research and have a desire to help others through participation. Katherine E. McDonald et al. present survey data that suggest PWIDs appreciate the benefits of research and that many of them want to contribute to the research enterprise (McDonald et al. 2016). Moreover, PWIDs responding to their survey surpassed the expectations of surveyed IRB members in their reported willingness to participate in research that would directly benefit other members of the disability community. These data support the claim that some PWIDs find value in helping others through research participation. Protecting PWIDs from harm and exploitation must not come at the expense of their opportunity to participate in research.

Yet, PWIDs potentially confront at least two barriers to participating in general health research. First, the additional procedures and paperwork needed to certify that someone with an intellectual disability is an informed and willing participant might make research teams hesitant to recruit PWIDs unless it is critical to the research. This pragmatic concern is legitimate and should ground further discussion on how to simplify consent and recruitment procedures. However, a less defensible barrier may be imposed by ill-informed and paternalistic beliefs about the incapacity of PWIDs to give meaningful assent to risks. These beliefs can make research teams, IRBs, and representatives overly cautious about permitting willing individuals with IDs from participating in clinical research. This excessive caution would be unfair to PWIDs. It would give excessive normative weight to the capacities they lack rather than those they possess.

Dana Howard and David Wendler have made a similar point, arguing that the "best interest" standard often fails to respect the agential capacities of PWIDs (2020). They maintain that representatives should give substantial consideration to the preferences PWIDs have expressed when making clinical decisions on their behalf. They argue that respect for PWIDs gives a representative reason to act in accordance with their explicit preferences, even if they conflict with their presumed best interests. Howard and Wendler do not claim that this should be the only consideration representatives take into account when evaluating whether or not a person should enroll, only that it should be a serious one. We take this argument further.

Those involved in assessing the capacity of PWIDs to participate in research should consider the particular abilities a PWID possesses. For example, suppose a representative knows that the person whom they are responsible for has significant emotional intelligence. The representative might draw on that knowledge to assist the potential participant in facilitated decision-making. They might help the person understand both how study participants will help others and the potential risks involved. Someone on the research team should be specifically trained to assist representatives in this process and ensure that they are not unduly influencing the potential participant. Those involved in capacity assessments might ask the potential participant to explain the particulars of the study in a level of detail consistent with that expected of a non-disabled person who shares their developmental age, if measurable. If someone with an intellectual disability is able to do this and expresses a willingness to participate in general health research, the representative should act in accordance with their wishes.

We acknowledge that there will be cases where other considerations override the potential participant's immediate preferences. For example, if a study involves exposure to conditions that a representative knows will cause the potential participant distress, they can reasonably decide against enrollment. Here, the potential risks to the individual would outweigh the benefits of facilitating the expression of their altruistic preferences. However, before making a final decision, the representative might involve the PWID in the decision-making process, pointing out that participation involves exposure to triggers and how the potential participant might respond. Moreover, the potential participant and their representative should be reminded that assent and consent are ongoing and can be revoked if circumstances change. To presumptively exclude PWIDs from participating in general health research, or to enroll them only as a last resort, assumes that they are all incapable of knowingly taking on a certain level of risk for the sake of others. This assumption, from our perspective, is unwarranted and demeaning.

Once this assumption is rejected, it is not clear why institutions should limit the participation of PWIDs to research of particular benefit to the intellectual disability community. We assume that Carlson does not think that the desire of PWIDs to help others is confined to others of "their own kind." They may also have, like many non-disabled people, a strong desire to assist their family, friends, and community.

It is more likely that Carlson is concerned that if PWIDs are recruited for general health studies, they risk the kind of exploitation common before the development of robust protections for research participants. As we explain later, it is important to keep in mind that participants in general health research with full capacity frequently incur considerable risks for the benefit of others (Stunkel and Grady 2011). We argue in the next section that the heightened risk of exploitation for PWIDs should not justify their presumptive exclusion from a valuable form of public service.

III Responding to Objections

A critic might grant that many PWIDs could be motivated to act for the sake of others and to assume significant risks in doing so. This critic might also grant that PWIDs might have compelling reasons to participate in general health research in order to benefit others. Finally, the critic might concede that with support from their representatives, PWIDs who are motivated to participate could understand enough about the risks to meaningfully assent. They might agree that it is appropriate for representatives to adopt a "substituted judgment" standard, making the decision they think PWIDs would make if they had full capacity – whether or not that decision aligns with the best interest standard (Torke et al. 2008, 1514).

Still, the critic might defend the necessity requirement (that PWIDs enroll in research only when their participation is necessary for the study) on two grounds. First, even if substituted judgment makes it more acceptable for the PWID to participate in clinical research with more than minimal risks, it is still a second-best alternative to actual consent (Wendler 2005). Second, the limited opportunities PWIDs have for community participation and paid work may make them especially vulnerable to exploitation. For both reasons, researchers should try to meet their enrollment goals with participants who can consent for themselves before enrolling those who require the consent of a representative.

There are two distinct concerns about the moral significance of the decisions made by representatives on behalf of PWIDs. First, decisions that individuals make for themselves generally have greater moral authority than those made for them by others. Second, there is strong evidence that the substituted judgment of representatives is often inaccurate (Rid and Wendler 2010, 37–39). We recognize that representatives may not always know what someone with an intellectual impairment would want if they had full capacity. However, we contend that this danger can be mitigated with procedures that maximize the involvement of PWIDs in informed consent processes (see Section IV).

We also agree that the participant's consent is preferable to a representative's consent in many contexts. With respect to minors, for example, it is sometimes preferable to delay participation in high-risk research rather than to allow a guardian to consent on a minor's behalf. For an adult patient about to undergo surgery, it is often preferable to obtain their consent to post-operative treatment options than it is to obtain their representative's consent should the patient become incapacitated. When it is possible to obtain consent from a participant rather than their surrogate, it respects their autonomy to obtain it.

Unlike minors and temporarily incapacitated adults, however, PWIDs may never be capable of actual consent. To give them any opportunity to contribute to general health research, researchers must give their assent greater normative weight than they would give to assent given by others lacking full capacity. Moreover, most PWIDs have comparatively few opportunities to contribute to their communities. Preventing them from participating in risky general health research with no prospect of direct benefit because they can only assent denies them one way of making this kind of contribution. Admittedly, participating in general health research is not the only way PWIDs can exercise their altruism, and it may not be attractive to many of them. Nonetheless, the option of research participation should be available to those who want to contribute in this way and can provide meaningful assent.

The danger of exploitation still exists for PWIDs, as it does for all research participants. However, we believe that the danger has been vastly reduced by the Common Rule and the various institutional mechanisms that operationalize it. PWIDs can only participate with their representatives' consent and other safeguards, which impose additional administrative burdens. These burdens are fairly placed on those involved in the research enterprise. They also serve to eliminate or greatly reduce any temptation to recruit PWIDs as a means of cutting costs or avoiding heightened ethical scrutiny.

However, potential for exploitation arises from the necessity condition itself. If PWIDs are always viewed as participants of last resort in general health research, they may face less careful scrutiny from research teams that are under heightened pressure to meet enrollment goals. Moreover, these studies may have had difficulty recruiting non-disabled participants because of the significant risks or burdens involved. Limiting PWIDs to participation in general health studies therefore potentially exposes them to risks or burdens greater than most people might be willing to take. Equally important though, this restriction disrespects PWIDs by conveying the belief that they are inferior research volunteers worthy of inclusion only when non-disabled volunteers are unwilling to participate. There would be less danger of harm or exploitation if all volunteers who meet study criteria were enrolled on a "first come, first served" basis.

We accept that heightened concerns about exploitation may arise for PWIDs in general health research. Because many of them are poor and unemployed (Bureau of Labor Statistics 2016), they may be willing to participate for far less compensation than that offered to better-situated volunteers. It would be exploitative to pay them less for their participation. This danger is easily avoided. If an IRB decides it is fair for participants with full capacity to receive a level of compensation, PWIDs should be entitled to that same level of compensation. Moreover, PWIDs confront significant barriers in accessing adequate healthcare. This lack of access increases the possibility that participants with intellectual impairments, or their representatives, might confuse clinical research with clinical care. They may wrongly assume that participating in general health research will result in their receiving access to healthcare. For these reasons, the safeguards in place to protect PWIDs are appropriate. However, as we have argued, these safeguards should not be used in a way that categorically excludes willingly assenting PWIDs.

Finally, representatives should be careful not to project their own values onto the individual they represent. They must also acquire all pertinent information to assure that a study does not pose special risks to the individual for whom they are responsible. As Carlson suggests, a lack of understanding of the procedure, or negative associations with medical settings, could make even a relatively minor procedure, such as a blood draw, traumatic.

Both the larger society and PWIDs would benefit from their participation in general health research. To the extent that such research excludes any significantly marginalized communities, findings might be less robust and generalizable, and of less value to the excluded groups. Unlike other marginalized groups, however, PWIDs may require reasonable accommodations over the course of the research process. Next, we consider what these accommodations might look like.

IV   Reasonable Accommodations and Universal Design in Research with PWIDs

Conceptualizing what reasonable accommodations are necessary for their participation in research and other domains of life is often more difficult than it is for people with physical disabilities. For example, someone with limited manual dexterity might need an accommodation that allows them to sign a consent form electronically or with assistance from a caregiver. This would be a critical adjustment for that person to accommodate their limitations. However, accommodating the physical inability to handwrite does not necessitate significant alterations to the research process.

Accommodating PWIDs may require more complex adjustments. By nature of their impairments, PWIDs will confront difficulties in understanding the particulars of medical terminology and procedures. Assuring that these limitations do not interfere with a person's participation in the research process requires thoughtfulness on the part of researchers, IRBs, and representatives. To illustrate, Rebecca H. Shields et al. have developed a series of adaptations for PWIDs whose cognitive functions are assessed through the NIH Toolbox Cognitive Battery, a series of electronic standardized assessments meant to measure intellectual abilities across clinical trials. The aim of these adjustments is to ensure that the assessments accurately measure how PWIDs respond to interventions meant to improve their cognitive function (Shields et al. 2020). These measures include modifying assessments so that they match a participant's developmental age or explaining testing procedures with pictures. We suggest that researchers could provide similarly thoughtful accommodations in recruiting PWIDs for general health research.

Assent forms may need to be modified to include more pictures as well as fewer and simpler words. A representative might be allowed to explain an assent form to the person with an intellectual impairment. In a study focusing on individuals with schizophrenia, William Carpenter et al. find that a variety of educational interventions in the informed consent process can contribute to more participants being able to provide consent (2000, 535). While schizophrenia is a different kind of impairment than the intellectual disabilities we have discussed, similar support should be put in place for PWIDs.

Additionally, a research team may need to strategize about how to minimize the distress some PWIDs might confront during a study. When working with a participant with moderate autism spectrum disorder, for example, researchers should strategize with the individual's representative and support staff to ensure that various stimuli in the research environment will not overwhelm the participant. For example, the participant might be allowed to wear headphones to block out loud noises in a clinic or undergo a certain procedure in a quieter room than other participants.

Admittedly, these kinds of adjustments could potentially compromise the integrity or usefulness of data in some studies. If a research team were to determine that this kind of accommodation would fundamentally alter the integrity of the study, the participant's discomfort would likely warrant their exclusion. Similarly, there might be situations where certain accommodations are infeasible because of factors such as cost and lack of necessary facilities or personnel. These considerations might provide ethical grounds for the exclusion of a PWID, assuming that research teams have made good faith efforts to find alternative adjustments. Nonetheless, the examples above illustrate the kind of creative problem-solving that would be necessary to include PWIDs in a wide range of clinical research. When appropriate, PWIDs should be given alternative means of participation that minimize the negative impact of their impairments on their progression through the study.

Our advocacy of these alternative means should not be understood as justifying categorically separate research practices for PWIDs. To the greatest extent possible, these individuals should be allowed to participate in the same research activities as all other participants. Our purpose in suggesting reasonable adjustments is to maximize their abilities to do so. In this way, an analogy exists between PWIDs receiving accommodations in research participation and their receiving special education services (Hehir 2005). While it is discriminatory to categorically segregate PWIDs into special classrooms, it is sometimes educationally beneficial to provide them services and supports not afforded to other students. The same should be true of enabling their participation in research.

However, making these adjustments for PWIDs should encourage research teams to consider similar adjustments for all participants, regardless of intellectual ability. Consent and assent forms that utilize simpler language and more pictures could make it easier for all participants to develop increased understanding of a particular study. Having individuals available to help PWIDs understand and complete a survey might also benefit non-disabled individuals by ensuring that they understand what the survey is asking.

Readers well-versed in disability studies will recognize these measures as exemplifying Universal Design (UD). UD is a framework in architecture, education, and other fields that seeks to minimize the need for individual disability accommodations by anticipating the limitations of a wide range of diverse individuals from the beginning of the design process (Mace et al. 1996). Universal research design would encourage those involved in the development of a protocol to avoid foreseeable barriers to research participation. Applying UD to research would not eliminate the need for individual disability accommodations. However, encouraging investigators to incorporate UD into the research process could enable a wider range of individuals to fully participate in clinical research. This would not only prevent the unnecessary exclusion of PWIDs from the research enterprise but make it more accessible to all.

V Toward Reconciling Medical Research with the Social Model of Intellectual Disability

We have advocated for the greater inclusion of PWIDs in clinical research. Nonetheless, there remains an ongoing tension between the medical research community and the staunchest advocates of the social model of disability. These advocates contend that intellectual disabilities should be considered a form of human diversity that should be accommodated rather than alleviated or cured. Medical researchers, by contrast, are often focused on developing interventions that alleviate or cure various illnesses or impairments. How, if at all, can disability bioethicists resolve this tension?

We cannot offer a definitive answer to this question in this chapter, which has focused on the distinct issue of inclusion. However, Carlson suggests a possible resolution that warrants further study. For studies of direct relevance to PWIDs she distinguishes ameliorative and eliminative research. Ameliorative research seeks to improve the functionality of PWIDs in their everyday lives. Eliminative research aims at preventing or curing intellectual disability. For Carlson, institutional support for eliminative research should be subject to heightened ethical scrutiny. This scrutiny is warranted because eliminating intellectual impairments might reinforce disability discrimination and stigma. It is important, from her perspective, to engage the intellectual disability community and other stakeholders in deliberations about how institutional support for either kind of research should be prioritized against other concerns (Carlson 2013, 312–313).

Carlson's suggestion has considerable merit. However, it is not clear that ameliorative research can avoid the concerns she raises about eliminative research. It may be difficult to distinguish research that aims to improve the functionality of PWIDs and research that aims to eliminate their disability by improving their intellectual functioning. Is the distinction merely one of degree, between mitigating and curing cognitive limitations? Does it depend on the means used, e.g., educational vs. pharmacological, or does eliminative research seek to normalize not only the functional but non-functional aspects of a disability, like the characteristic facial appearance or gait of individuals with Down syndrome? Disability bioethicists should further interrogate this ambiguity.

Yet, Carlson's distinction hints at an important point. Medical research should, in part, enable PWIDs to live full lives as disabled people. Including them in the research enterprise as much as possible is one way of accomplishing this goal. With increased opportunities to participate in clinical research, willing and able PWIDs can challenge the discriminatory assumption that an intellectual impairment denies someone the capacity to help others, or to assume significant risk in doing so.

Note

1 The "Common Rule" is Subpart A of 45 CFR Part 46.

References

Bureau of Labor Statistics. 2016. "Persons with a Disability: Labor Force Characteristics – 2015," *Economic New Release USDL-16-1248*. July 21, 2016. http://www.bls.gov/news.release/disabl.nr0.htm.

Carlson, Licia. 2013. "Research Ethics and Intellectual Disability: Broadening the Debates." *Yale Journal of Biology and Medicine* 86, no. 3: 303–314. PMID: 24058305.

Carpenter, Jr, William, James Gold, Adrienne Lahti, C.A. Queern, Robert R. Conley, J.J. Bartko, and Jeffrey Kovnik et al. 2000. "Decisional Capacity for Informed Consent in Schizophrenia Research." *Archives of General Psychiatry* 57, no. 6: 533–538.

Grisso, Thomas and Paul S. Appelbaum. 1998. *Assessing Competence to Consent to Treatment: A Guide for Physicians and Other Health Professionals*. New York and Oxford: Oxford University Press.

Hehir, Thomas. 2005. *New Directions in Special Education: Eliminating Ableism in Policy and Practice*. Cambridge, MA: Harvard Education Press.

Howard, Dana and David Wendler. 2020. "Beyond Instrumental Value: Respecting the Will of Others and Deciding on Their Behalf." In *The Oxford Handbook of Philosophy of Disability*, edited by Adam Cureton and David Wasserman, 522–540. New York: Oxford University Press.

Iacono, Teresa, and Rachel Carling-Jenkins. 2012. "The Human Rights Context for Ethical Requirements for Involving People with Intellectual Disability in Medical Research." *Journal of Intellectual Disability Research* 56, no. 11 (November): 1122–1132. https://doi.org/10.1111/j.1365-2788.2012.01617.x.

Mace, Ronald L., Graeme J. Hardie, and Jaine E. Place. 1996. *Accessible Environments: Toward Universal Design*. The Center for Universal Design. https://projects.ncsu.edu/ncsu/design/cud/pubs_p/docs/ACC%20Environments.pdf.

McDonald, Katherine E., Nicole E. Conroy, and Robert S. Olick. 2016. "Is It Worth It? Benefits in Research with Adults with Intellectual Disability." *American Journal on Intellectual and Developmental Disabilities* 54, no. 6 (December): 440–453. https://doi.org/10.1352/1934-9556-54.6.440.

National Institutes of Health. "Research Involving Individuals with Questionable Capacity to Consent: Points to Consider." *Grants & Funding*. Last modified November 2009. https://grants.nih.gov/grants/policy/questionablecapacity.htm.

Rid, Annette and David Wendler. 2010. "Can We Improve Treatment Decision-Making for Incapacitated Patients?" *Hastings Center Report* 40, no. 5: 36–45. https://doi.org/10.1353/hcr.2010.0001.

Rumney, Peter, James A. Anderson, and Stephen E. Ryan. 2015. "Ethics in Pharmacologic Research in the Child with a Disability." *Pediatric Drugs* 17, no. 1: 61–68. https://doi.org/10.1007/s40272-014-0102-4.

Shields, Rebecca, Aaron J. Kaat, Forrest J. McKenzie, Andrea Drayton, Stephanie M. Sansone, and Jeanine Coleman et al. 2020. "Validation of the NIH Toolbox Cognitive Battery in Intellectual Disability." *Neurology* 94, no. 12 (March): e1229–e1240. https://doi.org/10.1212/WNL.0000000000009131.

Stunkel, Leanne and Christine Grady. 2011. "More than the Money: A Review of the Literature Examining Healthy Volunteer Motivations." *Contemporary Clinical Trials* 32, no. 3 (May): 342–352. https://doi.org/10.1016/j.cct.2010.12.003.

Torke, Alexia M., G. Caleb Alexander, and John Lantos. 2008. "Substituted Judgment: The Limitations of Autonomy in Surrogate decision making." *Journal of General Internal Medicine* 23, no. 9: 1514. http://doi.org/10.1007/s11606-008-0688-8.

Wendler, David. 2005. "Projecting Subjects Who Cannot Give Consent: Toward a Better Standard for 'Minimal Risk'." *Hastings Center Report* 35, no. 5: 37–43.

Further Readings

Francis, Leslie P. 2009. "Understanding Autonomy in Light of Intellectual Disability." In *Disability and Disadvantage*, edited by Kimberley Brownlee and Adam Cureton, pp. 200–215. New York: Oxford University Press.

Litton, Paul. 2008. "Non-Beneficial Pediatric Research and the Best Interests Standard: A Legal and Ethical Reconciliation." *Yale Journal of Health Policy, Law, and Ethics* 8, no. 2: 359–420. https://digital-commons.law.yale.edu/yjhple/vol8/iss2/3.

McDonald, Katherine E., Nicole E. Conroy, Carolyn I. Kim, Emily J. LoBraico, Ellis M. Prather, and Robert S. Olick. 2016. "Is Safety in the Eye of the Beholder? Safeguards in Research with Adults with

Intellectual Disability." *Journal of Empirical Research on Human Research Ethics: An International Journal* 11, no. 5: 424–438. Accessed June 27, 2020. https://doi.org/10.2307/90012178.

McDonald, Katherine E., Nicole E. Conroy, and Robert S. Olick. 2017. "What's the Harm? Harms in Research with Adults with Intellectual Disability." *American Journal on Intellectual and Developmental Disabilities* 122, no. 1 (January): 78–92. https://doi.org/10.1352/1944-7558-122.1.78.

Millum, Joseph, and Michael Garnett. 2019. "How Payment for Research Participation Can Be Coercive." *The American Journal of Bioethics* 19, no. 9: 21–31. https://doi.org/10.1080/15265161.2019.1630497.

Rothman, David J., and Sheila M. Rothman. 2005. *The Willowbrook Wars: Bringing the Mentally Disabled into the Community.* New York: Taylor & Francis Group.

28

INCONVENIENT COMPLICATIONS TO PATIENT CHOICE AND PSYCHIATRIC DETENTION

An Auto-ethnographic Account of Mad Carework

Erica Hua Fletcher

I Introduction

I woke up on a Sunday morning to a series of missed calls from unknown callers. The block of red numbers on my cell, time-stamped between 3 and 4 a.m., filled me with dread. As I began listening to the voice messages, I soon discovered they had been left by Houston police officers. From what I gathered, neighbors hosting an outdoor party found my older sister Jessica walking barefoot near where we lived, and her odd interactions with guests led them to call law enforcement. The officers asked her a series of questions, which she did not answer. She did not speak at all, but eventually gave them the piece of paper that I insisted she carry with her while I was away. It contained her diagnosis—bipolar disorder, type 1, the name of the psychotropic medication that caused her only moderate side effects, and my cell phone number, as well as the number for her then separated, now ex-husband. After failing to reach either of us, the police dropped her off nearby at a private hospital in downtown Houston. I called the hospital, and after fumbling through its extensive phone tree, I reached the emergency department. A tired-sounding nurse told me that Jessica was admitted for being "unresponsive" and mute and that she had been placed on an involuntary hold.

I was on the other side of the country for my partner's grandmother's memorial service, to be held later that day. I had left Texas two days before, knowing that I would likely find myself in this predicament. In the two years following the onset of her mental crisis, Jessica had repeatedly refused psychiatric treatment; and given her heightened, sleepless state before I left for the memorial, I knew the situation would escalate. She needed to sleep, but it was very unlikely she would do so without forced sedation. It was the first time out of her many psychiatric hospitalizations that I had not been a part of decisions made to "treat" her against her will. Perhaps I was being a neglectful sibling or a dutiful partner. Perhaps my absence was a way to respect my sister's wishes to live as she saw fit.

This chapter offers an auto-ethnographic (first person) account of everyday forms of care and everyday decisions about mental health and safety. It is about the inconvenient complications I felt while grappling with the multiple professional and familial commitments I held while caring for those in mental crisis, myself included. It is also about mutual aid, reciprocity, and the environment which shaped the care my sister and I performed for each other during an intense

DOI: 10.4324/9781003289487-37

period. In what follows, I will first define the practice of psychiatric detention, then outline disability bioethics' and Mad Studies' critiques of the practice in relation to USA psychiatry and bioethics. Then I will describe my complex relationship to psychiatric detention, noting the particular socio-economic context which structured the way my sister's altered mental states could be addressed. I conclude that further attention to the relational and material conditions through which care occurs could helpfully reframe current scholarly and activists' focus on individual patient choice (autonomy) toward more politicized attention to carework and the mechanisms needed to sustain it.

II Debates for and against Psychiatric Detention

Psychiatric detention is the practice of treating patients against their will in a hospital, prison, or community setting. While mental health laws on psychiatric detention and its length of time vary by state in the USA, the medical-legal determinations that enable the practice are generally based on a clinical assessment of a patient's danger to self or others and a judicial finding of incompetency (which denotes a legal inability to make healthcare decisions).

The practice of psychiatric detention has long been a source of ethical controversy. Prominent psychiatrists such as E. Fuller Torrey and D.J. Jaffe have advocated for the practice based on supposed "neurobiological differences" in those with mental illness diagnoses and for the sake of medical beneficence, acting in the best interest of patients (see Wang 2019). A 2007 survey of USA psychiatrists indicates strong support for psychiatric detention in cases in which patients meet criteria based on the aforementioned clinical assessment of harm to self or others (Brooks 2007). Within USA medical and legal systems, a prevailing argument for psychiatric detention is that a patient loses their right to refuse care when their judgment is "clouded" as a product of their untreated mental illness (Gong 2017). In turn, bioethicists who focus on debates in psychiatry, like psychiatrists, tend to recognize the practice as one that is not ideal, but often necessary for patients' well-being. While bioethicists' debates on psychiatric detention have moved beyond medical-legal assessments of competency, they often limit ethical considerations relative to an individual patient's "decision-making capacity." This determination may be based on several factors, including a patient's understanding of her mental illness, the consequences of refusing psychiatric interventions, and her belief in psychiatric treatment as a potentially life-saving intervention (Schneider and Bramstedt 2006).

In contrast, some disability bioethicists critique the idea that medical authority can supersede a person's desire to refuse psychiatric treatment, an idea which they claim is rooted in an overly individualized approach to mental health care. Rather, disability bioethicists draw from the perspectives of Mad activists and Mad Studies scholars, who are a part of patient and scholarly movements that have radically questioned the assumptions, power, and force of the psychiatric and psychological disciplines, including the practice of psychiatric detention. These groups argue that psychiatric detention is traumatic and a form of violence that perpetuates psychiatric paternalism and other forms of medical overreach under the auspices of doing good for patients. Additional arguments against psychiatric detention are as follows: (1) Healthcare providers cannot establish a truly therapeutic alliance with patients if their patients are fearful of the repercussions for not following medical advice (Hunter 2018). (2) The overwhelming dominance of psychiatric frameworks to understand and treat "mental illness" fails to accept culturally- specific or personal understandings of mental distress, thus perpetuating a form of epistemic violence (Liegghio 2013). (3) The assumption of psychiatry as a universal science fails to recognize cultural biases, economic factors, and larger historic shifts associated with psychiatric diagnoses and treatment, including the field's historic ties to Western colonialism and imperialism (Mills 2014). (4) Mad activists and scholars point to the work of scholars in the fields of Critical Psychiatry and Science and Technology Studies who question the validity and reliability of psychiatric studies. Such critiques involve concerns over the role of pharmaceutical companies and the efficacy of high-dose and long-term psychotropic medications. These studies suggest that the risks of iatrogenic (treatment-caused) harm and the well-documented side effects of such treatments, in many cases, outweigh the potential for positive outcomes of their use (see Hunter 2018).

Disability bioethicists also draw upon Mad activists and scholars' work to uphold individual patient choice in psychiatric decision-making and to question the practice of psychiatric detention and forced treatment. Disability bioethicist Sarah Blanchette (2019) describes a troubling case in which a woman with a psychiatric diagnosis drank antifreeze, then refused life-saving treatment. Because Emergency Department physicians deemed her competent to make her own health decisions, no treatment was provided, and the woman died. Blanchette affirms this was the appropriate decision to make, writing "even if mental health users/refusers reject treatment that is potentially life-saving, like the woman who drank antifreeze, this action [not to provide renal dialysis] is still more reflective of their 'authentic' wishes than non-consensual 'treatment'" (8) Though this was not an ideal outcome, Blanchette upholds the principle of patient choice, writing, "A Mad Studies approach validates Mad experience and knowledge, which fosters autonomy before competency assessments, while also suggesting that these assessments and psychiatric paternalism are inherently flawed and harmful" (15). In other words, open communication between physicians and potential patients, without the threat of psychiatric detention, could enable patients to steer their medical treatment or lack thereof, even when those decisions lead to their death.

Despite this strong stance in support of patient autonomy, Blanchette also emphasizes the need for relational, intersectional perspectives that recognize the "complex social circumstances and structural oppression [that] can cause mental distress and undermine autonomy" (13). More broadly, many other disability bioethicists, Mad scholars, and activists acknowledge that patient choice inherently is structured by legal definitions, psychiatrists' interpretations of state laws, the availability of beds in a hospital, and the calculated risks associated with psychiatric detention or discharging patients with minimal treatment. Thus, "patient choice" is a contextual, situational, and relationally- driven concept, one that is inherently reliant on several human actors—such as psychiatrists' determinations, patient preferences, caregivers' capacities, law enforcement's determinations, and judicial rulings—and structural factors—such as hospital capacity, public policy, and social safety nets.

Here, we can begin to see a generative tension arise within clinical decision-making processes and Mad support for community-based alternatives. Inherent in disability bioethics' support of "fostering individual patient choice" is also a recognition of the limits of this approach, given the much more diffuse and distributed medico-legal and other structural forces that constrain psychiatric decision-making.

In what follows, I draw from my experience engaging in carework to explore the following questions: How does patient choice operate when "choices" take shape in relation to a whole host of decision-makers and decision-making structures—including patients, family supporters, psychiatrists, and public policies? How does the threat of psychiatric detention filter into the everyday decisions about carework and the ways that we approach mental crisis? How might a care-based approach to psychiatric decision-making disrupt current debates surrounding patient choice and psychiatric detention?

III On Carework

Here I will define carework as a practice of solidarity and as a mutual exchange of goods and services to address physical, mental, and/or emotional needs. Anthropologists and disability activists have long highlighted mutuality and reciprocity as a part of carework, emphasizing the dynamic, relational, interdependent, and multidirectional flows among those who care for each other. They have also emphasized the complexity of care webs and public policies that sustain and value certain lives over others, for example the fact that in the USA, many people's health insurance is tied to employment and so joblessness can result in worse or no health care (Heinemann 2014; Piepzna-Samarasinha 2018; Block 2019). This understanding of care as a distributed, reciprocal, yet structurally inequitable practice works against the idea that individual patient choice is ever truly possible. In re-framing ethical debates from patient choice toward a more expansive and politicized understanding of carework, I seek to situate everyday "patient choices," such as the one to refuse psychiatric treatment, within networks of care and constrained choices available to stakeholders. To that end, I will first contextualize my family's unique approach to Mad carework in our attempts to divert or stall psychiatric hospitalizations. I will then describe some of the interpersonal and structural factors that contributed to my sister's psychiatric detentions, our efforts to resist these processes, and limits to the "choices" available to us during this period.

IV Care Context

My sister and I grew up in the Greater Houston area, and we were raised by our Euro-American father and Taiwanese-Brazilian mother, who was diagnosed with paranoid schizophrenia and bipolar disorder. I myself have experienced periods of sadness and worry, which my sister had witnessed prior to her own mental crisis. My sister's altered mental states came about soon after our mother's untimely death, my sister's abortion, and a challenging separation from her partner. From 2016 to 2018, I engaged in care work with my sister; for over a year during this time, she lived with my partner and me at our home, where I performed the majority of household chores, such as meal preparation and cleaning. These shared and divergent experiences with altered mental states and madness gave us a unique way of recognizing and relating to each other's respective distress.

By the time Jessica moved in, I was an early career scholar, who had completed an ethnography with a group of radical mental health activists as a part of my dissertation. I considered myself to be a contributor to the Mad Studies movement and identified with its calls for Mad representation and inclusion in mental health reform and research. I was aware of various alternatives to mental health support that did not rely on psychiatric treatment in the Global North. At the time, I was conducting fieldwork at a peer respite, a short-term residential center and voluntary alternative to psychiatric hospitalization that was staffed by people with lived experience of mental distress, who are called "peer support specialists." My commitments to disability activism and Mad Studies scholarship made me acutely aware of activists' rebuttals against psychiatric detention. It also gave me a window into the incredible amount of social and financial support it would take to live life on the fringes of psychiatric intervention.

V Auto-ethnographic Methodology

In what follows, I will describe and analyze everyday moments of carework, drawn from my memory, a review of email correspondence and medical documents, and recent conversations with Jessica. As with any first-person account, memories fade and shift over time and are inherently subject to personal bias, including positive bias (Richards 2008). What is more, given disability activists' historic attention to the challenging power dynamics among families involved in carework, I have tried my best to portray Jessica's perspective and voice as being quite distinctive from my own (Piepzna-Samarasinha 2018). To that end, I asked Jessica for her permission to write about our experiences, solicited her feedback on drafts of this chapter, and confirmed that she felt comfortable with the narrative shared here.

VI Care beyond Choice

In what follows, I describe forms of care that served to stall psychiatric detention and resist an atomistic understanding of patient choice. In analyzing and participating in daily life with my sister, I became particularly attuned to the intersections of place, gender, ethnicity, class, and dis/ability that structured our lives, and their impact on our navigation of the social services available to us.

Our Queen Anne-style rental home sat across from a public elementary school in a predominately Mexican neighborhood, surrounded by noisy railroads, industrial plants, mechanic shops, and a strong police presence. The low-income, yet gentrifying neighborhood still had stray dogs and the occasional chicken or goat roaming the streets. The area was largely unscathed by the floodwaters of Hurricane Harvey, yet our proximity to the Houston Ship Channel and petrochemical plants nearby meant that we were often exposed to chemical smells and burn-off. Jessica characterized the area as one in which she often felt "unsafe" and "uneasy." It was in this setting that we found small workarounds to care for each other during distressing times.

Jessica and I grew up in an insular, immigrant family; and we had relied on each other for companionship and support in our childhood and through college. Our mother had instilled in us a strong sense of filial duty and interdependence, values that were often at odds with the rugged independence and individualism lauded in Texas. When Jessica first experienced non-consensus reality—what her psychiatrist termed psychosis—my tacit role as a daughter and sister was to ensure her survival. Jessica's resistance to her psychiatric diagnosis and prognosis meant that she considered herself quite capable of working in order to pay her share of our rent. She held a few different jobs for a while and kept trying to work, even in altered mental states, until we finally agreed that these attempts could derail her application for Social Security Disability Insurance. At that point, our father, who had entered the upper-middle class a decade prior, agreed to support her financially while we waited for her application to be approved.

When she had more energy, Jessica often spent time playing her violin, gardening, meditating, doing yoga, or making art. She would have some friends over or would go out to visit them. We would go on long walks in our neighborhood in the late afternoon after I had returned from work. Jessica knew our walks together helped to clear my mind, and I felt like this was a good activity for her to get out of the house. When she was feeling down, she had a hard time putting words together, remembering to eat or drink water, or preparing meals. I would make sure she had food, clean clothes, and some plan for the day before I left for work.

To honor my wishes, Jessica reluctantly agreed to participate in many of the supportive services provided by the community mental health center. She was assigned a case manager, a peer support specialist, and a counselor, as a part of an early intervention program. My logic for pushing her to attend meetings with her team was that I could not be her lone source of social interaction and that talking with others was a way for her to get feedback and support from people who were understanding of her mental state, even if they did see it in pathological terms. I wanted to respect her choice to not take medication, but I saw regular social interactions, including our family counseling sessions, as being potentially beneficial. Jessica agreed to these meetings to make me feel less guilty about my absence during the workday, but years later she responded that she found some of the encounters to be positive, particularly the ones with her peer support specialist. We also concurred that family counseling, though it felt unhelpful at the time, was a way for us to connect, to have difficult conversations with each other, and to show each other that we were trying to improve our relationship.

When I could see she was feeling worse, I felt myself growing more concerned, which she often read as my becoming hypercritical and overbearing. Yet it was hard for me not to be when confronted with the extent of her non-consensus reality: Once, when I was away for a conference, Jessica became very concerned about a heavy rainstorm, which she believed would flood our house, and she had placed one of our cats in a carrier with a plan to wade through the floodwaters over to one of her friend's houses. My partner found her before she left, and she argued with him about the need for everyone to evacuate the house. After consulting me, he took her to the peer respite to see if she could live there for a few days. She caused some stir there, and peer staff members told him that she did not meet the criteria for independent living skills to be eligible for a stay there. Later, my partner texted me that she had left our house in the afternoon and had been gone for several hours. When I called, her phone went straight to voicemail several times, but eventually, she answered while she was driving. She was quite confused and lost. It was the middle of the night. Her car was dangerously low on gas; her phone was also low on battery. She was quite slow to respond to my questions: "What are the cross streets? What do you see around you now?" After talking with her for what felt like an hour, I was able to identify her location and helped her find her way back. She had been just a 10-minute drive away from our house.

These forms of carework took a mental toll on me, just as it did for Jessica and our relationship as sisters. I wondered if my care was "enabling," as our family counselor believed it to be. Or was it sustaining her until she could care for herself once again? Was my ambivalent stance toward her refusal of psychotropic medication a help or a hindrance? What would the neighbors think of a slim, young woman of color walking shoeless, in her pajamas, on the unkempt sidewalks? Would her East Asian features save her from criminalization? Would the police give her a citation for loitering? What would happen if she got into a car accident while driving in an altered mental state? Would my colleagues in Mad Studies consider me to be a fraud if they knew my actions had led to her psychiatric detention? How could I prove myself as a scholar if I spent most of my time engaging in or recovering from carework? Was I Mad for trying to accommodate her altered mental states? If I had not been prone to bouts of sadness and catastrophic thinking myself, would I have been more able to support her? How long could I keep going as I had been? How long could she? I was tired—not just from a lack of sleep—but existentially tired from worrying that Jessica would decline as my mother had done. After I returned from my trip, I returned to care work with her, and when I reached a point of exhaustion, I called law enforcement trained in responding to mental health crises. Within a few minutes, police arrived at our house and quickly determined that she met criteria for psychiatric detention. They handcuffed her and led her outside. Tearfully, we rode together in the caged backseat of the police car to a hospital, where she was once again admitted.

In the emergency department, Jessica was once again subject to restraints securing her to a bed, and nurses injected her with sedatives. The medications caused her involuntary movements, blurred vision, weight gain, hives, memory loss, cognitive impairment, slurred speech, and hyper-sedation. Once in the psychiatric ward, male patients made unwanted comments toward her. She is now averse to syringes of any kind. Even today, I still tense up when I see missed calls from unknown numbers on my phone. She was traumatized by her hospitalizations, and I am traumatized from bearing responsibility for it and for bearing witness to it. Recently, Jessica told me, "I don't know what the purpose of the hospital was, necessarily, other than to help you cope, which is understandable—totally. And to maybe, give [you] a break from caregiving, I imagine that would be helpful." Even though it is hard for me to admit, I agree with her sentiment. The hospital gave me a break and forced her into a sleep, which she desperately needed, but the psychiatric staff also left her feeling crushed and disempowered and in need of recovery from the high-dose medications they administered. I was left with a sense of defeat and shock from seeing the effects of her overmedication. I was also left with her medical bill for several thousand dollars.

VII Discussion

The forms of care through which our mental states were felt, attended to, and lived were deeply affected by familial dynamics, our immediate surroundings, and my professional commitments (Ginsburg and Rapp 2013). Likewise, our small acts of care and mutual aid took shape within our home, neighborhood, and available social services.

Notably, my family's stratified socioeconomic position enabled a form of intensive care inside our home, yet prevented us from pursuing private nursing options, which could have led to much more intensive forms of surveillance and forced treatment (see Gong 2019). My profession afforded me the flexibility to spend a great deal of time at home, and enabled me to apply for charity care and to advocate for reduced medical bills; my financial position also allowed me to pay for some of Jessica's rent, transportation expenses, and medical bills. Our middle-class upbringing also afforded Jessica creative ways of coping with her mental distress; playing her violin, doing yoga, and gardening were all markers of leisure and past economic privilege, even while she experienced poverty during this time. Meanwhile, our father's ability to cover Jessica's living expenses during this time was crucial while we awaited a decision from the Social Security Administration.

With regard to my academic privilege, my study of the local community mental health system and local professional connections enabled me to point my partner to a peer respite as a potential resource when I was out of town. However, it was also a professional risk for me to have a family member served by a field site in which my research participants may come to see me as an "enabling" caregiver or a "biased" social researcher. Regardless, the peer staff's determination of her "lack of independent living skills" served to perpetuate the notion of her "mental disability" rather than provide a meaningful alternative to psychiatric hospitalization.

Likewise, my involvement with Mad Studies scholarship and disability activism often worked toward constraining and liberatory ends. It helped me support Jessica's wishes to avoid psychiatric treatment up to a point: Without medication, she would stay awake for several consecutive nights and start experiencing altered mental spaces that could have resulted in criminal justice involvement. In those moments, I was faced with the pragmatic reality that it would be far worse for her to experience the violence of criminalization than the violence of psychiatric detention. For these reasons, I began to question the kind of feats being a family caregiver to a psychiatric refuser required in terms of time, mental fortitude, financial resources, educational background, and professional networks. I began to question the anarchist and libertarian bent of certain branches of Mad Studies and Mad activism. I wondered how Mad Studies scholars' consistent calls for alternatives to psychiatric detention squared with the lived reality of economic precarity, community failures to support people through crisis, and the need for public social welfare programs, such as housing, health care, and home health support (Diamond 2013; Jones and Kelly 2015).

In short, my family and I tried our best to create a space in which Jessica's altered mental states could be accepted to a higher degree than it would be outside of our home. This space enabled my family and I to acknowledge my mental distress from engaging in carework, and together we learned to accommodate each other's peculiarities and anxieties. Our forms of care were often subject to scrutiny by mental healthcare providers, while our care for each other helped us secure resources and stall out the process leading up to psychiatric detention. Even so, my actions in the name of "care" upheld an imbalanced power dynamic between us as sisters: Ultimately, she was the one who was forced into psychiatric detention. It was her body that reacted against psychiatric medication. It was her life that would be lived with the stigma associated with receiving federal assistance. I still wrestle with these tensions: Care sustains lives. Care diminishes vibrant ways of being. Care justifies violence.

VIII Conclusion

This auto-ethnographic portrayal of carework presents a much more diffuse understanding of patient choice, beyond medical and legal frameworks, and its complex entanglements with psychiatric detention. Here, everyday decisions about care are seen in their true complexity: distributed among a number of actors, networks, and systems and caught up within larger webs of familial ties, mutual aid, and public policies. The available social services, our citizenship status, and our standpoint as young, college-educated, multiethnic women living in a low-income neighborhood—all factored into our engagement with psychiatric detention and our limited ability to access meaningful alternatives to psychiatric services. Likewise, the "choices" available to us were also products of our respective positions to psychiatric power, class, educational privilege, ethnicity, and gender. The inconvenient complications presented here call disability bioethicists, Mad activists, and Mad Studies scholars to consider the limits of individualized understandings of patient choice in psychiatric detention toward a more diffuse, relational, and structural approach to carework.

Coda

More than two years have passed since Jessica's last hospitalization. A few months after she was discharged and received SSDI, I moved with my partner to Los Angeles, where I am now working as a lecturer. Jessica started living on her own, adopted a Chihuahua, and opened a small business serving tea. Separately, and together, we are mending our sisterhood. We show care for each other in new ways. She hosts me when I visit. I bring her out to California when I can. We videochat and message each other with updates on our animal companions and gardening projects. We phonebank and canvass on behalf of Bernie Sanders's 2020 presidential campaign. We fight for social welfare programs, universal health care, free college, and a greener economy. We vote. Our carework is political, as it has always been.

References

Blanchette, Sarah. 2019. A Feminist Bioethical and Mad Studies Approach to Resisting an Increase in Psychiatric Paternalism to Competent Mental Health Users/Refusers. *Journal of Ethics in Mental Health* 6(10): 1–18.

Block, Pamela. 2019. "Activism, Anthropology, and Disability Studies in Times of Austerity." *Current Anthropology* 61(S21): S68–S75.

Brooks, Robert. 2007. "Psychiatrists' Opinions about Involuntary Civil Commitment: Results of a National Survey." *The Journal of the American Academy of Psychiatry and the Law* 35(2): 219–228.

Diamond, Shaindl. 2013. What Makes Us a Community? Reflections on Building Solidarity in Anti-Sanist Praxis. In *Mad Matters: A Critical Reader in Canadian Mad Studies*, edited by Brenda A. LeFrancois, Robert Menzies, and Geoffrey Reaume, pp. 64–78. Toronto: Canadian Scholars Press.

Gong, Neil. 2017. "'That Proves You Mad, Because You Know It Not': Impaired Insight and the Dilemma of Governing Psychiatric Patients and Liberal Subjects." *Theory and Society* 46(3): 201–228.

Gong, Neil. 2019. "Between Tolerant Containment and Concerted Constraint: Managing Madness for the City and the Privileged Family." *American Sociological Review* 84(4): 664–689.

Ginsburg, Faye, and Rayna Rapp. 2013. "Disability Worlds." *Annual Review of Anthropology* 42(1): 53–68.

Heinemann, Laura. 2014. "For the Sake of Others: Reciprocal Webs of Obligation and the Pursuit of Transplantation as a Caring Act." *Medical Anthropology Quarterly* 28(1): 66–84.

Hunter, Noël. 2018. *Trauma and Madness in Mental Health Services*. New York: Palgrave Macmillan.

Jones, Nev and Timothy Kelly. 2015. "Inconvenient Complications: On the Heterogeneities of Madness and Their Relationship to Disability." In *Madness, Distress and the Politics of Disablement*, edited by Helen Spandler, Jill. Anderson, and Bob Sapey, pp. 43–56. London: Polity Press.

Liegghio, Maria. 2013. "A Denial of Being: Psychiatrization as Epistemic Violence" In *Mad Matters: A Critical Reader in Canadian Mad Studies*, edited by Brenda A. LeFrancois, Robert. Menzies, and Geoffrey Reaume, pp. 122–129. Toronto: Canadian Scholars Press.

Mills, China. 2014. *Decolonizing Global Mental Health: The Psychiatrization of the Majority World*. London: Routledge.

Piepzna-Samarasinha, Leah L. 2018. *Care Work: Dreaming Disability Justice*. Vancouver: Arsenal Pulp Press.

Richards, Rose. 2008. "Writing the Othered Self: Autoethnography and the Problem of Objectification in Writing about Illness and Disability." *Qualitative Health Research* 18(12): 1717–1728.

Schneider, Paul L., and Katrina Bramstedt. 2006. "When Psychiatry and Bioethics Disagree about Patient Decision Making Capacity." *Journal of Medical Ethics* 32: 90–93.

Wang, Esmé W. 2019. *The Collected Schizophrenias*. Minneapolis: Graywolf Press.

Further Reading

Brodwin, Paul. 2012. *Everyday Ethics: Voices from the Front Line of Community Psychiatry*. San Francisco: University of California Press.

Hansen, Helena, Philippe Bourgois, and Ernest Drucker. 2014. "Pathologizing Poverty: New Forms of Diagnosis, Disability, and Structural Stigma under Welfare Reform." *Social Science & Medicine* 103: 76–83.

Luhrmann, T. M. 2007. "Social Defeat and the Culture of Chronicity: Or, Why Schizophrenia Does So Well Over There and So Badly Here." *Culture, Medicine and Psychiatry* 31: 135–172.

Mattingly, Cheryl. 2014. *Moral Laboratories: Family Peril and the Struggle for a Good Life*. San Francisco: University of California Press.

Mol, Annemarie. 2008. *The Logic of Care: Health and the Problem of Patient Choice*. New York: Routledge.

29

DISABILITY BIOETHICS, ASHLEY X, AND DISABILITY JUSTICE FOR PEOPLE WITH COGNITIVE IMPAIRMENTS

Christine Wieseler

I Introduction

In this chapter, I apply disability scholar and bioethicist Rosemarie Garland-Thomson's notion of misfitting in order to examine ways of thinking about a woman referred to in the bioethics literature as Ashley X. This concept is important for directing our attention to the inherently relational and dynamic nature of disability as opposed to conceptions that would characterize disability as an individual, static phenomenon. My primary goal is to provide an overview of some of the central moral issues and disagreements related to the medical interventions performed on Ashley and children with similar diagnoses.[1] As is often the case, the language we use to describe an individual or situation indicates our moral stance and the position we believe others should adopt. This is clearly the case when discussing Ashley's story, and throughout my discussion I will attend to the terms various narrators use. I will draw on the work of Eva Kittay, who, in addition to having been involved in the Seattle Growth Attenuation Working Group[2] and having written about the case independently, also has personal experience relevant to the moral and epistemic issues raised by this treatment.[3] I will conclude by considering the marginalization of people with cognitive impairments within disability activism and by offering recommendations to advance disability justice.

Throughout the history of the USA, medical professionals have been responsible not only for promoting well-being but also for violence and abuse toward disabled children and adults. Before turning to discussion of Ashley, I want to attend to the story of another young woman in order to provide context regarding mistreatment and abuse, in the guise of medical care, enacted upon people diagnosed as having cognitive impairments. In May of 1924 Carrie Buck, a poor white woman, was involuntarily institutionalized at the State Colony for Epileptics and Feebleminded in Lynchburg, Virginia after her adoptive family claimed that she was "feebleminded." Earlier that year, Virginia had enacted the Eugenic Sterilization Act, which legalized state-mandated sterilization of anyone deemed "unfit" to reproduce. Those in power applied the category of "unfit to reproduce" as they deemed appropriate; typically, involuntary sterilization was performed on disabled people, people of color, poor people, those considered immoral (often accused of promiscuity or theft), and people who purportedly fell into more than one of these classifications. The USA Supreme Court affirmed the constitutionality of forced sterilization in its 1927 ruling in *Buck v. Bell*, which takes its name from Carrie Buck, her mother, and her

DOI: 10.4324/9781003289487-38

daughter. Buck was one of approximately 70,000 people who were forcibly sterilized in the USA, a practice that reportedly continues to this day in Irwin County Detention Center, a facility incarcerating immigrants in Ocilla, Georgia (O'Toole 2021). There are no federal laws against involuntary or nonvoluntary sterilization, though many states have enacted laws prohibiting or regulating sterilization of people who do not or cannot provide consent. Involuntary sterilization is one of numerous examples of medical professionals engaging in egregious treatment of members of marginalized groups.[4] Although medical professionals tend to endorse the principles of beneficence and nonmaleficence (physicians do so explicitly when taking the Hippocratic Oath), unfortunately they do not always live up to these ideals. One of the reasons for this is that they are not immune to the influence of ableism, classism, racism, and sexism, which may inform their judgments about what types of treatment are appropriate. All of these types of oppression involve narratives regarding what types of bodyminds[5] are valued and which are devalued or considered "defective" (Clare 2017, 23). Disability theorist and activist Eli Clare draws attention to historians such as Paul Lombardo who locate the injustice of Carrie Buck's forced institutionalization and sterilization in the discovery that she "wasn't *really* feebleminded," as if what was done to her—and tens of thousands of other people—was otherwise defensible (2017, 110).

Disability bioethicists are suspicious of such double standards, namely, when the same medical intervention is given a different moral evaluation depending on whether the patient is disabled or nondisabled. The medical model of disability locates disability at the level of the individual bodymind, while the social model of disability considers individuals with atypical bodyminds to be disabled—that is, disadvantaged—by social norms and practices. This background provides important context for consideration of the medical interventions performed on Ashley as well as broader concerns disability bioethicists have about medical treatment of disabled people.

II Ashley and the Positions of Mainstream and Disability Bioethicists on "the Ashley Treatment"

Ashley was born in 1997, and when she was a few months old, doctors diagnosed her with "static encephalopathy," a chronic nonprogressive disorder of the brain. The etiology of her condition is unknown, meaning that physicians do not know its underlying cause (Kittay 2011, 611). At the age of six, Ashley's cognitive development was assessed to be at the level of a typical infant; she was unable to walk, sit up, or use language (Kafer 2013, 47). In 2004, her parents opted to have physicians at Seattle Children's Hospital begin growth attenuation therapy as well as multiple types of surgery (also referred to as "the Ashley treatment," hereafter AT) on her.

Doctors performed three types of surgical interventions: a hysterectomy, bilateral mastectomy, and appendectomy, followed by two and a half years of "high doses of estrogen in an attempt to stunt her growth" (Ibid, 49). The primary aim cited by Ashley's parents as well as the main physician and pediatric bioethicist involved in her case was to facilitate the provision of care for her in the family's home. They maintain that if she remains approximately the height and weight of an average nine-year-old girl, then her parents—and possibly others in the future—will be able to move her around easily and include her in more activities than they would otherwise be able. They claim that growth attenuation will prevent the need to institutionalize her, to, in the words of her parents, "put [her care] in the hands of strangers" (Gunther and Diekema 2006, 1014).[6]

Ashley's parents, with the aid of a physician (Daniel Gunther), and a bioethicist who worked closely with the family (Douglas Diekema), made additional arguments for the hysterectomy. Ashley's physicians deemed surgery to remove her uterus to be advisable to prevent potential adverse effects of the high-dose estrogen treatment, such as uterine bleeding. They also suggested that what they termed "pretreatment hysterectomy" would "reduce the long-term complications of puberty in general" (Ibid., 1014). Disability theorist Alison Kafer notes that Gunther and Diekema focus on their *intention* in performing the hysterectomy—as discussed above—rather than the foreseeable *effect* of ensuring that Ashley will not be able to have children (2013, 51). They generally avoid using the term sterilization; when they do use this term, they place it in scare quotes. Nonetheless, the Washington Protection and Advocacy System (WPAS) reported in 2007 that sterilization of Ashley violated state law; Seattle Children's Hospital was legally obligated to obtain a court order since Ashley was unable to provide informed consent for the procedure (Kafer 2013, 49). Ashley's parents and their lawyer suggest that it was unproblematic to sterilize her, given the expectation that she will not gain capacity to make the decision to have children.

Gunther and Diekema did not mention mastectomy in their initial discussion of growth attenuation therapy (2006). Ashley's parents claim that bilateral mastectomy—in their words, removal of her "breast buds"—was performed to eliminate the risk that she would experience any of the following: pain or discomfort from developing large breasts, cancer, or fibrocystic growth—all of which are present in her family (Kafer 2013, 52). They suggest that it would also prevent "sexualization toward [her] caregiver" (Ibid., 65). Kafer observes the strangeness of this wording and the fact that "[a] lack of breasts does not render one safe from sexual assault or abuse" (65).[7] S. Matthew Liao, Julian Savulescu, and Mark Sheehan concur, stating "someone might sexually abuse Ashley whether she has breasts or not. The focus should be on the potential sex offenders" (2007, 18). They also note that prophylactic mastectomy in children is not recommended, even in families in which there is a greater than average likelihood of developing breast cancer due to the presence of BRCA1 and 2 genes. Kafer and Liao et al. point out that one of the arguments her parents advance in favor of removal of Ashley's breasts assumes that the only reason to have breasts is for breastfeeding—an assumption these authors reject (Kafer 2013, 56; Laio et al. 2007, 18). Thus, they find the arguments her parents make insufficient to justify the bilateral mastectomy that was performed on Ashley.

Her parents refer to Ashley and children like her as "pillow angels" (Kafer, 66). This is the final term I want to highlight. Clare remarks:

> I imagine Ashley as a cushion giving you comfort, a cherub resting passively among pillows, a spiritual abstraction. Does love mean denying your daughter's earthbound, every day, messy body-mind? Do you love her angry, ornery, profane self too?
>
> *(2017, 154)*

Clare's words reflect concern that the term "pillow angel" has the effect of idealizing Ashley in a way that ignores her embodiment and the range of her ways of being. He also notes that this concept evokes "associations between disability, whiteness, innocence, passivity, and blessing" (Ibid., 154). Similarly, Patricia Williams notes that race and class are relevant to how Ashley is portrayed, musing, "I wonder if a poor black child would have been so easily romanticized as a 'pillow angel'" (2017, N.P.). It is important to consider how this term functions not only to shape perception of Ashley but also to shape how we as a society think about and treat other disabled children.

Alicia Ouellette observes in *Bioethics and Disability* (2011) that some mainstream bioethicists initially expressed opposition to AT, but that as time went on, the number of bioethicists who came to support growth attenuation (if not the other elements of AT) grew. "The Ashley Treatment: Best Interests, Convenience, and Parental Decision-Making" (2007) by Liao et al. played a significant role in this shift.

Liao et al. note that the moral issues that arise from AT have public policy implications, as Ashley's parents and physicians advocate this treatment be available for children who meet certain criteria. They argue that growth attenuation is a morally permissible medical intervention insofar as they believe that it is in the best interest of both Ashley and her parents. They suggest that growth attenuation is consistent with respecting Ashley's human dignity, while rejecting other types of surgical intervention that might make it easier for her parents to care for her (e.g., amputating her legs). In contrast, bioethicists George Dvorsky and Peter Singer claim that it is not possible to violate Ashley's dignity because her cognitive impairment prevents her from having dignity (Dvorsky 2006; Singer 2007). Liao et al. conclude that "decisions of this kind should be made on a case by case basis, such as occurred in this case through a hospital's clinical ethics committee" (2007, 18). However, they reject the arguments for two other components of AT: bilateral mastectomy (as noted above) and hysterectomy. They maintain that less invasive treatments than a hysterectomy were available to address any discomfort Ashley may have felt due to menstruation, and they also assert that hysterectomy presents its own risks (Ibid).

Ouellette identifies two primary types of arguments mainstream bioethicists provide in support of AT or growth attenuation: the first focuses on "parental rights and value neutrality" and the second "rests on disability status and the effect that status has on caregivers" (2011, 179; 181). The first approach suggests that growth attenuation, if not the other medical interventions included in AT, is not relevantly different from procedures such as tonsillectomy, insertion of a feeding tube, or a surgical procedure to treat acid reflux (fundoplication) (Diekema and Fost 2010, 34–37). The main idea is that parents are the ones who ought to have the authority to make medical decisions on behalf of their children and that physicians should provide information and recommendations that are not value-laden. The second type of argument relies on the assumption that treating Ashley and children with similar diagnoses differently from non-disabled children and disabled children with different types of diagnoses is morally acceptable. The Seattle Growth Attenuation and Ethics Working Group (SWG) engaged in both types of arguments in their endorsement of growth attenuation.

In 2010, SWG—comprised of 20 bioethicists, pediatricians, professors, and community members—published their position on growth attenuation. Most group members concur that it was morally acceptable to perform growth attenuation on Ashley and express support for allowing parents to choose this set of procedures for children with similar diagnoses, claiming that "the potential impacts of growth attenuation are no more profound than the impacts of other decisions that parents routinely make" (2010, 29). They argue that parents are the ones who should decide what is in the best interest of their children. They also claim that these procedures were intended to benefit Ashley—to improve her quality of life by making it easier for her family members to care for her and include her in family activities.

One concern that SWG and some mainstream bioethicists raise is the possibility that AT may be applied more broadly in the future: although current recommendations suggest limiting eligibility to children who are have severe cognitive impairments and are non-ambulatory and non-communicative, some bioethicists worry that eligibility may be expanded to disabled children who do not fit this category.[8] In other words, they posit that allowing these medical interventions to be performed on a narrow category of disabled children is the beginning of a slippery slope (Wilfond et al. 2010, 33).

Disability bioethicists and theorists perceive the situation quite differently. Their perspectives are grounded in the diverse experiences of disabled people, including ways that medical practice and research has contributed to our oppression. Rather than raising concerns about a future slide down the slippery slope, they claim that we have already gone over the metaphorical cliff by allowing even one child to be subjected to medical interventions when the long-term effects are unknown and they were not medically indicated.[9] One of the reasons some people within the disability rights community have voiced opposition to growth attenuation therapy is that it drastically changes the body of a child who is unable to give consent when changes could be made to her environment to allow her to receive the care that she needs in her parents' home. In addition, feminist philosophers, among others, maintain that justice requires that society support caregivers rather than leaving them to their own devices.

Garland-Thomson's notions of fit and misfit are useful in making sense of the relational and dynamic aspects of disabled people's experiences. She distances her conceptions from those that would consider a bodymind itself to be fit or misfit, asserting, "A fit occurs when a harmonious, proper interaction occurs between a particularly shaped and functioning body and an environment that sustains that body" (2011, 594). When Garland-Thomson speaks of the environment, she refers to physical as well as social components of a person's situation. A misfit:

> describes an incongruent relationship between two things: a square peg in a round hole. The problem with a misfit, then, inheres, not in either of the two things but rather in their juxtaposition, the awkward attempt to fit them together. When the spatial and temporal context shifts, so does the fit, and with it meanings and consequences.
>
> *(596)*

A misfit cannot be attributed solely to a bodymind or the environment but results from a mismatch between them.

To create a better fit, either Ashley's surroundings need to be accessible and the family must obtain and use equipment to lift and transfer her for activities of daily living or physicians must employ surgical and hormonal intervention in order to facilitate moving her with little or no reliance on such equipment. Bioethicist Arthur Caplan asserts:

> a decent society should be able to provide appropriately sized wheelchairs and bathtubs and home-health assistance to families like this one. Keeping Ashley small is a pharmacological solution for a social failure—the fact that American society does not do what it should to help severely disabled children and their families.
>
> *(cited in Ouellette 2011, 174)*

Here Caplan attends to the ways in which social values and policies harm disabled children and their families as well as the potential to support them, rather than assuming that changing Ashley's body is the best or only way to create a better fit.

It is important to note that the notion of a misfit may be appropriated in the service of interventions that are rejected by members of disability communities. Advocates for AT appeal to growth attenuation and mastectomy as medical means to reduce the disjunct between social expectations for nondisabled women—such as being employed, involved in a romantic relationship, and having children—and Ashley's anticipated future bodymind (Kafer 2013, 54). For example, her parents state: "given Ashley's mental age, a nine and a half year old body is more appropriate and provides her with more dignity than a fully grown female body" (Ashley's Mom and Dad 2007). They frame the effects of Ashley's static encephalopathy, namely the current and predicted future low level of cognitive development, as entailing a type of misfit between her body and mind that is seen as in need of correction.

In an interview with Christopher Mims in *Scientific American* bioethicist Norman Fost remarks:

> [H]aving her size be more appropriate to her developmental level will make her less of a "freak." ...I have long thought that part of the discomfort we feel in looking at pro- foundly retarded [sic] adults is the aesthetic disconnect between their development and their bodies. There is nothing repulsive about a 2 month old infant, despite its limited cognitive, motor, and social skills. But when the 2 month old baby is put into a 20 year old body, the disconnect is jarring.
>
> *(2007)[10]*

Fost seems to assume that the discomfort some feel in the presence of adults with severe cogni- tive impairments is appropriate, and that, rather than working to change this, it would be best to alter the appearance of such adults so that their developmental level will be readily apparent to observers. In the same article, he goes on to state: "If children like Ashley could magically retain the appearance of an infant, they would not only be easier to care for in the physical sense, but the emotional reaction to them would probably be more favorable" (Mims 2007). Lest the reader should think Fost is the only one who holds this position, I turn to a related remark made by bioethicist George Dvorsky: "The estrogen treatment is not what is grotesque here. Rather, it is the prospect of having a full-grown and fertile woman endowed with the mind of a baby" (2006).[11] It is unclear what harm being "full-grown and fertile" could cause Ashley. However, Dvorsky reveals concerns for the dissonance *others* may experience upon encounter- ing her (perhaps as a result of finding her sexually attractive). Fost and Dvorsky fail to identify their own reactions as well as the possible reactions of the public as a component of the misfit, which means that they do not consider whether that is the problem that needs to be addressed rather than changing Ashley's body.

Garland-Thomson does not hold that the goal should be to eliminate all misfits or that the effects of misfitting are only negative. According to her, misfits may facilitate novel ways of negotiating unsupportive environments. Garland-Thomson holds, "[a]cquiring or being born with the traits we call disabilities fosters an adaptability and resourcefulness that often is underdeveloped in those whose bodies fit smoothly into the prevailing, sustaining environ- ment" (2011, 604). She supports reduction in the frequency of misfits through development of more sustaining environments characterized by "accessibly designed built public spaces, welcoming natural surroundings, communication devices, tools, and implements, as well as other people," some of which have been made more likely through civil rights legislation (594). However, she contends that the imperative to decrease the number of misfits is harmful when it is taken as a mandate to eliminate "the particularities of embodiment we think of as disability" through technological means such as preimplantation genetic diagnosis, prenatal testing and selective abortion, and medical normalization (602). Thus, on Garland-Thomson's account, not only are misfits not always something to be prevented, but they also have the po- tential to provide epistemic resources. She suggests that misfitting may lead to a burgeoning awareness of social injustice and the formation of activist communities (597). For example, Ashley's parents could have used their knowledge and experiences in order to further the activist work of disability communities rather than advocating medical intervention for chil- dren like Ashley.

Fits and misfits also involve social norms and responses to a person's bodymind, which play a role in how one is enabled or disabled in ordinary interactions. In the context of discussing Ashley, Clare asserts:

I know women who can only move one finger; women who operate their wheelchairs by sipping and puffing; women who get lifted to and from the toilet, shower, bed; women who never leave their beds; women who speak with computers or alphabet boards or not at all; women who lost their words at age seventy; women who never had words; women who as girls were thought to be without communication. Busty women, bleeding women, women—all disabled—who are safe, comfortable, and happy.

(2017, 153)

This portrayal of what is possible for disabled women is unintelligible within the ableist imaginary in which a body-mind considered atypical is in itself assumed to entail risk, discomfort, and unhappiness.

Joel Michael Reynolds refers to this way of thinking about disability as "the ableist conflation" (2017). In his words, "the ableist conflation flattens communication about disability to communication about pain, suffering, hardship, disadvantage, morbidity, and mortality" (2017, 152). As should be clear by now, this is a very limiting and problematic perspective that distorts the lived experiences of disabled people.

III Medical Uncertainty, Double Standards, and Epistemic Humility

Parents are often in the position of making decisions on behalf of their children that will have lasting effects without knowing what their children themselves would choose. For example, the parents of children who are born intersex, meaning they have a combination of physical characteristics typically used to classify people as being female or male that defy easy categorization (e.g., ambiguous genitalia, XXY chromosomes, and/or ovo-testes), are forced to decide about medical interventions as well as how to raise their children in relation to gender norms. In the recent past, physicians typically pressured, or at least advised, parents to consent to genital surgery to "normalize" children with ambiguous genitalia. Anne Fausto-Sterling asserts, "[d]ogma has it that without medical care, especially early surgical intervention, hermaphrodites [intersex people] are doomed to a life of misery. Yet there are few empirical investigations to back up that claim" (2000, 93). Indeed, much of the empirical research suggests the opposite conclusion: surgical and other medical interventions can actively cause harm (Fausto-Sterling 2000, 2008). Many parents believed that such interventions were in the best interest of their children and consented. In some cases, they were advised to prevent their children from knowing that they were intersex and to socialize them as girls or boys, even if this involved lying to them. Indeed, medical manuals and research typically recommended that physicians withhold information from parents rather than revealing that the sex of their child is ambiguous (Ibid., 64). The underlying assumption was that these actions would lead to a better quality of life for intersex children.

Many intersex adults have spoken out about the ways in which medical interventions and lies about their bodies traumatized them and decreased their quality of life, in spite of the good intentions of their parents and clinicians. Harms of early genital surgery include pain, scarring, psychological trauma due to follow-up treatment,[12] and lack of ability to experience sexual pleasure and/or orgasm (Fausto-Sterling 2008, 128). This example makes clear that good intentions are insufficient for determining the best way to support the well-being of one's child.

Care ethics centers the importance of relationships as well as the context in which decision-making occurs. Kittay applies care ethics as a framework for examining what is required in order to make decisions on behalf of a child unable to voice their wishes, given the highly asymmetrical relationship between the child and parents as well as health care providers. She recognizes the epistemic limitations of this situation—namely, that the latter are likely to have ableist biases, asserting: "The physician and the parent, in order to be fully capable of caring, need to understand themselves as requiring information from those better situated to provide a perspective from a life lived with disability" (2019, 227). For this reason, Kittay encourages epistemic humility and learning from those "whose lives are closer to the child's likely adulthood" in order to have a clearer picture of what will support a particular child's flourishing (Ibid., 228). This example points to the need for physicians to better understand the lived experiences of disabled people and their families. This is one among many examples demonstrating why I believe it is essential for critical disability theorists to be involved in the theory and praxis of bioethics in its many forms, including medical education.[13]

Kittay manages to be critical of AT without questioning the devotion of parents who explore or have had their children undergo aspects of this treatment. She states,

> I refrain from arguing that Ashley's parents did not accept her as she was. There is every reason to think that their love was unconditional. My argument instead is that its expression in attempting to keep her small was misguided.
>
> *(2019, 222)*

She notes that the judgments of parents—all parents—are fallible. She raises the question of whether AT should be an option for parents of children "whose prognosis is severe cognitive disability and no ambulation (SCDN)" (222). Kittay expresses concern that there is a double standard at play here: children who fit into this category are eligible for AT while other disabled children whose caregivers' work may be facilitated by growth attenuation are not. She also notes that advocates of this treatment make the following problematic assumptions: first, the body is solely of instrumental value, and second, a cognitively impaired individual can be classified at an unchanging developmental level (e.g., cognitive function of a typical two-year-old). Kittay rejects both of these assumptions.

Supporters of the medical interventions performed on Ashley show a high degree of confidence regarding the static nature of her condition. For example, Larry Jones, an attorney Ashley's parents hired, suggested that laws against sterilizing disabled people were designed to protect people with the potential to raise children in the future. Given his assumption that she would not have this ability, he considered her to be without need of such protections and thus an exception to any such laws (Kafer 2013, 49–50). Her parents' blog states "While we support laws protecting vulnerable people against involuntary sterilization, the law appears too broadly based to distinguish between people who are…capable of decision making and those who… [like] Ashley…will never become remotely capable of decision making" ("Updates on Ashley's Story" May 8, 2007). Similarly, Gunther and Diekema state that Ashley is "an individual who will never be capable of holding a job, establishing a romantic relationship, or interacting as an adult" (2006, 1016).

We might question whether this level of confidence is warranted. After all, the prognoses of physicians sometimes turn out to be wrong. As noted above, doctors gave Ashley the vague diagnosis of "static encephalopathy," which indicates that they do not know the cause for her symptoms and that they do not expect her condition to change. It's worth considering Kittay's observation of her daughter, Sesha, here: "Her development will not show up on an IQ test, but she has become increasingly mature emotionally. (…) the decades have altered her tastes, her understanding, and her responses to the world" (2011, 614).

Yet, if we judge AT to have been a morally objectionable course of action, I would argue that this should not be on the basis of the fallibility of prognoses. We veer dangerously close to the problematic conclusion that the forced sterilization and institutionalization of Carrie Buck was only a tragedy if she wasn't "really" "feebleminded." The view that medical interventions that would be morally unacceptable if performed on nondisabled children, but they are morally acceptable if performed on disabled children is a dangerous double standard when the treatment in question is not intended to ameliorate symptoms related to the child's diagnosis (Kittay 2011). At the very least, it should give us pause when a child who can communicate in typical ways is categorically prevented from assenting to a set of procedures that is readily available to the child who cannot indicate their wishes.

IV Disability Justice and People with Cognitive Impairments

Historically, the focus within disability communities centered around disability rights has primarily been on physical accessibility and has tended to assume disabled people who are consistently capable of making their own decisions and participating in the workforce if given necessary accommodations. This emphasis results in excluding the experiences of those whose impairments and illnesses—especially, but not only, those involving cognitive limitations.[14] In "When Caring Is Just and Justice Is Caring" Kittay notes that people with severe cognitive impairments have benefitted the least from the disability community's advocacy. In her words:

> They have argued that their impairments are only disabling in an environment that is hostile to their differences and that has been constructed to exclude them. Yet, the impairment of mental retardation [sic] is not easily addressed by physical changes in the environment.
>
> *(2002, 258)*

Although social understandings of cognitive impairments are malleable and could help to create better living conditions for individuals with cognitive impairments, physical changes in the environment are minimally effective. While physical accessibility is important, it is insufficient for meeting the needs of people with cognitive impairments such as appropriate education, transportation, and care. Of course, these needs will vary greatly, given the diversity of the individuals included in this group.

A transformation of social practices and attitudes has the potential to benefit all disabled people, but this will not be possible if the needs of people with cognitive impairments are not taken into account. Kittay suggests that moving toward a more inclusive and just society requires that disability advocates stop appealing to independence, cost-effectiveness, and productive futures—in other words, cease buying into a liberal paradigm that excludes people who are unable to approximate these ideals.

In addition to omitting consideration of people with cognitive impairments, disability activists and theorists sometimes contribute to their othering. Clare demonstrates agreement with Kittay's position on personhood and cognitive impairment, namely that personhood—a term associated with having a socially recognized moral status (e.g., having fundamental rights)—must be more inclusive. However, Clare concedes "Yet when faced with allegations or assumptions of stupidity or diminished mental capacity, many of us respond by asserting our intelligence and distancing ourselves from intellectual disability" (156). Given dominant narratives of disability as global incapacity and the seeming choice between being an object of pity or a disabled hero, it is understandable that we would highlight our abilities and prefer to be perceived as the latter. However, insisting on our rights in connection with our cognitive abilities comes at too high of a cost—contributing to the oppression of those who do not have this option. Rather than falling into this false dichotomy, we need to promulgate a wide range of narratives that better capture the everyday lives of disabled people and what enables us to flourish.

Kittay makes important contributions to this project in sharing reflections on her experiences of life with Sesha that are both philosophical and deeply personal. She notes

> [w]hile disability can offer an identity, no disabled person wants to be defined exclusively by their impairment. No one wants to be the object of the metonymic mistake of identifying their whole body by means of a singular bodily part, aspect, or function.
>
> *(248–249)*

Kittay is careful not to allow her descriptions of Sesha's limitations to overshadow her uniqueness: the joy she finds in Beethoven's and Mahler's symphonies, how intensely present she is, and her laughter. These rich details allow the reader to imagine the distinct pleasure of being in the presence of Sesha, rather than the impoverished picture many mainstream bioethicists tend to present when portraying Ashley and other people with cognitive impairments.

Notes

1 This is in no way a comprehensive account of the bioethics literature on Ashley X.
2 The Seattle Growth Attenuation Working Group wrote an influential early assessment of the medical interventions performed on Ashley.
3 Kittay states,

> [t]hose who have not raised a severely cognitively disabled and nonambulatory child into adulthood may feel diffident about expressing opposition to growth attenuation because they have not walked in the parent's [sic] shoes. I *have* walked in them, or at least in very similar ones. My daughter Sesha is now a woman of forty. She too, does not toilet herself, speak, turn herself in bed, or manage daily tasks of living, and has no measurable IQ.
>
> *(Wilfond et al. 2010, 32)*

4 Other examples include intentionally infecting children with intellectual impairments with hepatitis in order to conduct medical research at Willowbrook State School and the development of triage protocols that make people with intellectual impairments ineligible for scarce medical resources during the COVID-19 pandemic.
5 Margaret Price introduces the term "bodymind" in "The Bodymind Problem and the Possibilities of Pain" (2015). Sami Schalk clarifies "[t]he term *bodymind* insists on the inextricability of mind and body and highlights how processes within our being impact one another in such a way that the notion of a physical versus mental process is difficult, if not impossible to clearly discern in most cases" (2018, 5).
6 Alison Kafer calls this a "false choice," adding that AT does not guarantee that Ashley will never be institutionalized (2013, 62).
7 Arthur Caplan remarks "[t]rue, it may be better if Ashley does not become sexually developed in terms of protecting her from attack. But that can be said of any woman" (as quoted by Ouellette 2011, 174; see Kaplan, 2007).
8 Eva Kittay dissented from the position articulated by the group, stating:

> The compromise position rests on the assumption that the constraint [that growth attenuation should only be available to "severely cognitively disabled and nonambulatory children"] will avoid many of its possible abuses. The problem is that the limitation itself is already an abuse. If growth attenuation should not be done on children without these impairments, then it should not be done on any children. To do otherwise amounts to discrimination."
>
> *(Wilfond 2010, 32)*

Norman Fost suggests that the Seattle Growth Attenuation Working Group gave "too much deference" to third parties, including the Washington Protection and Advocacy System (30). He frames the objections to growth attenuation of third parties, including disability advocates, as stemming from "emotional distress on becoming aware that one's moral or political views are not shared by everyone" (Ibid). Disability advocates know that their views are not shared by everyone; indeed, this is why advocacy work is necessary. I would argue, along with Patricia Williams, that what is at issue is not whether disabled people are offended by growth attenuation, but whether our social, legal, and medical institutions should permit physicians to enact this intervention on children with cognitive impairments with parental consent (Williams 2007).

9 I would like to thank Joel Michael Reynolds for encouraging me to make explicit the point that these medical interventions were not medically indicated.

10 As quoted by Kafer (2013, 55).

11 Also quoted by Kafer (2013, 55).

12 Some examples of follow-up treatment include physicians masturbating intersex children in order to measure their erections or parents repeatedly inserting a dildo to preserve their child's surgically created vagina (Fausto-Sterling 2008, 128).

13 Not everyone agrees that bioethics can serve the interests of disabled people. For example, feminist philosopher of disability Shelley Tremain asserts:

in philosophy the field of bioethics significantly contributes to the problematization of impairment and disability and thus to their naturalization and materialization; that is, as I argue, the field of bioethics, rather than merely describing impairment, actively participates in its constitution and derogation as a disadvantageous natural human attribute.

(2017, 162)

Tremain holds that disability produces impairment and that accounts of bodily features that seem to be descriptive are, in fact, normative. Elsewhere, she describes "natural impairment" as akin to sex in Judith Butler's conception of gender performativity: constituted through the performative acts of disabled subjects (2002, 34).

14 This is not to say that the continued focus on physical accessibility is unwarranted; however, there need to be additional priorities.

References

Ashley's Mom and Dad. "The 'Ashley Treatment': Towards a Better Quality of Life for 'Pillow Angels.'" March 25, 2007. http://pillowangel.org/Ashley%20Treatment%20v7.pdf

Caplan, Arthur. 2007. "Is 'Peter Pan' Treatment a Moral Choice?" January 4, 2007. https://www.nbcnews.com/id/wbna16472931.

Clare, Eli. 2017. *Brilliant Imperfection: Grappling with Cure.* Durham, NC: Duke University Press.

Diekema, Douglas and Norman Fost. 2010. "Ashley Revisited: A Response to the Critics." *American Journal of Bioethics* 10(1): 30–44.

Dvorsky, George. 2006. "Helping Families Care for the Helpless." Institute for Ethics and Emerging Technologies. November 6, 2006. http://ieet.org/index.php/IEET/more/809/.

Fausto-Sterling, Anne. 2000. *Sexing the Body: Gender Politics and the Construction of Sexuality.* New York: Basic Books.

Fausto-Sterling, Anne. 2008. "Should There Be Only Two Sexes?" In *The Feminist Philosophy Reader.* Edited by Alison Bailey and Chris Cuomo. Pp. 124–144. New York: McGraw-Hill.

Garland-Thomson, Rosemarie. 2011. "Misfit: A Feminist Materialist Disability Concept." *Hypatia* 26: 591–609.

Gunther, Daniel and Douglas Diekema. 2006. "Attenuating Growth in Children with Profound Developmental Disability." *Archives of Pediatrics and Adolescent Medicine* 160(10): 1013–1017.

Kafer, Alison. 2013. *Feminist, Queer, Crip.* Bloomington: Indiana University Press.

Kittay, Eva. 2011. "Forever Small: The Strange Case of Ashley X." *Hypatia* 26(3): 610–631.

———. 2019. *Learning from My Daughter: The Value and Care of Disabled Minds.* New York: Oxford University Press.

Liao, Matthew, Julian Savulescu, and Mark Sheehan. 2007. "The Ashley Treatment: Best Interests, Convenience, and Parental Decision-Making." *Hastings Center Report* 37(2): 16–20.

O'Toole, Molly. 2021. "ICE to Close Georgia Detention Center Where Immigrant Women Alleged Medical Abuse." *Los Angeles Times,* May 20. https://www.latimes.com/politics/story/2021-05-20/ice-irwin-detention-center-georgia-immigrant-women-alleged-abuse.

Ouellette, Alicia. 2011. *Bioethics and Disability: Toward a Disability-Conscious Bioethics*. Cambridge: Cambridge University Press.

Price, Margaret. 2015. "The Bodymind Problem and the Possibilities of Pain." *Hypatia* 30(1): 268–284.

Reynolds, Joel Michael. 2017. "'I'd Rather Be Dead than Disabled'—the Ableist Conflation and the Meanings of Disability." *Review of Communication* 17(3): 149–163.

Schalk, Sami. 2018. *Bodyminds Reimagined: (Dis)ability, Race, and Gender in Black Women's Speculative Fiction*. Durham, NC: Duke University Press.

Singer, Peter. 2007. "A Convenient Truth." *New York Times*, January 26. https://www.nytimes.com/2007/01/26/opinion/26singer.html.

Tremain, Shelley. 2002. "The Subject of Impairment." In *Disability/Postmodernity: Embodying Disability Theory*. Edited by Mairian Corker and Tom Shakespeare. Pp. 32–47. New York: Continuum.

Tremain, Shelley. 2017. *Foucault and Feminist Philosophy of Disability*. Ann Arbor: University of Michigan Press.

Wilfond, Benjamin, Paul Steven Miller, Carolyn Korfiatis, Douglas Diekema, Denise Dudzinski, Sara Goering, and The Seattle Growth Attenuation and Ethics Working Group. 2010. *The Hastings Center Report* 40(6): 27–40.

Williams, Patricia. 2007. "Judge Not?" *The Nation*, March 26. https://www.thenation.com/article/archive/judge-not/.

PART IX

Disability Bioethics

Connections and New Directions

30

FEMINIST THEORIZING AND DISABILITY BIOETHICS

Lauren Guilmette

What is feminism, and in what ways might the conceptual tools of feminism be taken up and reworked in the emerging field of disability bioethics?

First, we ought to unpack the word "feminist," a task as daunting as unpacking the word "bioethics," so we need to choose a way in. Sara Ahmed (2017) frames feminism not as a discipline or a method of inquiry but as a way of life. Being a feminist is not about adopting a fixed set of ideals or norms, but about existing in a kind of ongoing disruption of the status quo. This includes "asking ethical questions about how to live better in an unjust and unequal world (in a not-feminist and antifeminist world)," finding "ways to support those who are not supported or are less supported by social systems," and pushing against stuck habits of privilege that would resist imagining these familiar relations otherwise (1). Ahmed's definition offers an ideal to which feminism can aspire, but it is also important to keep in mind that feminists have often failed to be inclusive. We need to figure out how to live a feminist life without replicating these historical errors and harms. At its best, feminism has been one among many liberation efforts (including anti-racist, decolonial, and other movements) that defamiliarize stuck habits and work toward the flourishing of those living under them. This chapter provides a brief overview of the history of feminist theory, including what are commonly called the "first," "second," and "third" waves. I conclude by developing four feminist conceptual resources productively taken up in critical disability studies and, increasingly, disability bioethics. These include: (1) questioning inherited binaries like mind/body, reason/emotion, culture/nature, and theory/practice, according to which the latter term tends to be devalued and feminized in relation to the former; (2) situating knowledge claims in a social and historical context and recognizing that only some voices surface as intelligible; (3) considering the personal as political, thereby interrogating the Western public/private distinction as well as the historical devaluation of care work; (4) analyzing cultural representations for their racialized and sexualized as well as ableist biases, with concern for how these representations shape our reality, both politically and aesthetically.

I First-Wave Feminism

The history of feminist theory to date is commonly described as happening in three "waves" over the last couple centuries. Work from the nineteenth and early twentieth centuries is

DOI: 10.4324/9781003289487-40

typically categorized as "first wave," in which feminists took up tools of liberalism and Marxism, respectively, to describe their oppression and to claim participation in the public sphere.

The liberal feminist movement was sparked by people like Mary Wollstonecraft, who argued that even in the "State of Nature," we have the guiding law of human rationality and that this capacity for reason means that we bear innate rights and also duties. In *A Vindication of the Rights of Woman* (1792), Wollstonecraft claims that women trained only to attract men cannot be good citizens, and that women must be cultivated to take on societal *duties* so they can also become worthy of *rights*. In the mid-nineteenth century, Elizabeth Cady Stanton's "Declaration of Rights and Sentiments" (1848) at the Seneca Falls Convention makes a parallel claim for the inclusion of women within existing rights of citizenship. These first-wave liberal claims culminated in legal and social reforms, such as the 19th Amendment to the USA Constitution, which was ratified in 1920. Liberal feminism thus offers a model of "immanent critique"—of claiming something is unjust about existing conditions, not by some outside ("transcendent") standard, but by the terms of the dominant culture creating those conditions. More recent liberal advances, such as the Americans with Disabilities Act (ADA) of 1990, follow this model of inclusion within existing socio-political and legal frameworks.

A different set of feminist strategies came into focus through the lens of Marxist thinking in the mid-nineteenth into the early twentieth centuries. While liberal feminists sought to reform the existing order on its own terms according to the liberal ideals of universal human rights and freedoms, Marxist theorists focus on changing the societal structures and institutions that produce and reproduce inequalities. They insist that formal rights alone do not address the economic basis of inequality in the exploitation of labor. Following on Marx and Engels' (1848) critical claim that respectable "bourgeois" (middle- and upper-middle-class) marriage ought to be compared to prostitution—namely, insofar as it is a means of ownership over women's sexual and reproductive capacities—the Russian feminist Alexandra Kollontai (1909) observes a class division between women. Kollontai distinguished the liberation of working "proletarian" women from the aims of bourgeois feminism, and she declared that women's true emancipation would be achieved only with the revolutionary reorganization of societal norms and economic conditions. Society should offer not only negative freedoms but also the positive freedoms of state support, like parental leave from work and access to childcare.

Emma Goldman's "A New Declaration of Independence" (1909) supplies a different vision than that of the Seneca Falls Convention: "When, in the course of human development, existing institutions prove inadequate to the needs of man… the people have the eternal right to rebel against, and overthrow, these institutions." With Goldman (1909), we are called to consider not only the "external tyrants" that obstruct women's full flourishing, for example, access to voting and other legal rights, but also the "internal tyrants" that one has internalized and lives out unreflectively. Among these are "the sacredness of property… the stupid arrogance of national, racial, religious, and sex superiority, the narrow puritanical conception of human life." These Marxist feminist arguments offer a predecessor for *disability justice* discourses in the twentieth and twenty-first centuries, which demand not just inclusion in the legal order but a reexamination of both legal and social norms and necessary changes to existing social structures.

In the previous paragraphs, the reader may have noticed a glaring absence. What about women of color, especially Black women in the USA during these years? Black feminism does not fit neatly into the traditional division between liberal and Marxist theories for a number of reasons, not least of which because Black women were not actively included in these movements. Black feminists appreciated both the negative freedoms of liberal rights (i.e. freedom from state interference) and the structural criticisms of Marxist feminists concerning the injustice of the existing order. As bell hooks (1981) observes, movements for gender and racial justice could well have been aligned in these decades, but were instead set in competition for the right to vote: should suffrage be extended to Black men *as men*, or to white women *as white*? The first and second waves of feminism have been criticized for failing to recognize the limits of their own social location as typically white and middle class, and for leaving out the intersecting oppressions of sexism and racism faced by Black women.

While these early Black feminists were not included in the mainstream first-wave feminist movement in its own time, they have been retroactively added to the feminist canon and their insights bear comparison with their historical contemporaries. From nineteenth- and early twentieth-century Black feminism, we learn strategies of claiming space for experiences not addressed in dominant or mainstream accounts. Brittney Cooper (2017) writes of late nineteenth-century Black feminist Anna Julia Cooper's descriptions of incongruent spaces and demands—for instance, choosing between the public restroom options of "for ladies" and "for colored people." Cooper thus critically inserts the fact of her own embodiment and how it matters for projects of knowledge production. Black women's early activism against the segregation of bathrooms can offer inspiration to present-day coalitional efforts between disability and trans activists, such as PISSAR (People in Search of Safe and Accessible Restrooms) founded at UC-Santa Barbara, which surveys and shares data about the accessibility of bathrooms in the area (Kafer 2012). As Cooper understood already in 1892, this incongruence reflected a problem not with her body but with structures of systemic oppression that insisted upon these divisions.

II Second-Wave Feminism

The second wave of feminism can in many ways be seen as a development of the claim of French feminist Simone de Beauvoir (1949) that one is not born but *becomes* a woman. Even if *sex* is a biological given, *gender* is constructed by cultural values and practices and could conceivably be constructed otherwise. This feminist distinction has been important for feminist disability studies in framing the difference between *impairment* and *disability* according to the social model of disability; thus, Susan Wendell (2010) frames disability as "socially constructed from biological reality" (338). Following the social model, one's sex and/or one's impairment may describe a fact about one's body, but disability and gender alike are constructs that gather meaning from cultural practices, such as styles of dress, gestures, the expectations built into our various buildings and institutions. We will return to reconsider the sex-gender distinction with third-wave feminists, but let us here focus on a key site of divergence between feminists in the second wave on values of equality and difference. Should women have the *same* training and access to existing institutions as men have, so that they can be *equal* participants in cultural production? Or, do women have *different* capacities than those valued by Western patriarchy, perhaps cultivated by the historical responsibility for care of dependents, and should these be upheld as new and alternative ideals?

The latter position of second-wave "difference" feminism can be seen in the work of feminist care ethics. This tradition begins with Carol Gilligan's (1982) psychological studies of differences in women's moral reasoning. Her observations began in the 1960s while assisting Harvard psychologist Lawrence Kohlberg with his research into moral maturity in boys and girls. According to Kohlberg, the highest expression of moral reasoning is the application of abstract and universal principles of morality, independently of context or personal relationships. As Gilligan critically observed, however, girls and women in his study appeared to be stuck at a lower moral maturity than boys and men, but she argued this was due to Kohlberg's framing of the stages.

Gilligan's alternative model began from the question, what if women's moral reasoning is not inferior to men's but, instead, appeals to *different* ideals? Nel Noddings (1984) goes one step further than Gilligan, arguing that the "ethics of care" historically associated with women's moral choices is in fact *superior* to the abstract and de-contextualized justice claims of "masculine" models. Eva Feder Kittay (1999) has challenged this emphasis upon the "feminine," instead reframing the conversation around "mothering persons" and "dependency workers" as well as around the observation that care relations typically occur between unequal, yet interdependent persons. Developing Carol Hanisch's (1969) feminist claim that "the personal is political," Kittay (2010) draws on her own experience as a caregiver to a disabled daughter and claims that this experience has *philosophical* implications. Here, she finds herself debating utilitarian bioethicists who make cost-benefit calculations against the worth of her daughter's life. Issues of disability will impact virtually all of our lives in some form or another; thus, with respect to long-term care and aging, Kittay (2013) argues in favor of federal insurance and social welfare programs to ensure this care as a human right. In summary, toward a disability bioethics, "difference" feminism offers a willingness to question hierarchies and to envision alternative, non-exclusionary models, as well as to revalue vulnerability, dependence, and care.

The ethos of second-wave "equality" feminism, by contrast, is well summarized in Betty Friedan's (1963) observation that the unhappy housewife had recently burst into public view, even though "pretty housewives still beamed over their foaming dishpans" in TV commercials. Second-wave equality feminism challenged the fantasy figure of Woman in the domestic sphere, in part by demanding that women be granted better access to education and employment. Advances made by equality feminists include access to birth control (1965, *Griswold v. Connecticut*) and legal and safe abortion (1973, *Roe v. Wade*), as well as the extension of the 1964 Civil Rights Act to address sexual harassment in the workplace, sexual abuse, and domestic violence (1994, *Violence Against Women Act*). However, much like the first wave of feminism, the second-wave feminism been called out for tending to center the voices of white, heterosexual, middle-class women; as Ahmed notes, the liberation of some women from the private sphere required that others, most often working-class women of color, had to step in "to take over 'the foaming dishpans'" (51).

Furthermore, second-wave feminist arguments for reproductive "rights" emphasized the right to prevent or end pregnancy but not the right to give birth, though the latter had been coercively and forcibly taken from women of color, especially Black and Indigenous women, Latinas, and those deemed "feeble-minded" (for more on eugenics, see Robert Wilson's chapter in this volume). The latter demands are typically referred to as *reproductive justice*, demands which clearly exceed the scope of "choice" as claimed by a predominately white feminist movement (Ross and Solinger 2017). In the twenty-first century, disability activists have made a similar move in shifting from the liberal language of rights to structural questions of justice.

III Third-Wave Feminism

The third wave of feminism in the late twentieth and early twenty-first centuries pushes back against the second wave's over-generalization of woman's experience as such, especially with the rise of Black feminist theory and theories of intersectionality, queer theory, transnational feminisms, and feminist disability studies. Third-wave feminist theories begin from the observation—by bell hooks (1981), Kimberlé Crenshaw (1991), and others—that women of color could not be subsumed in a predominately white feminist movement or by a male-dominated civil rights movement but instead demand an intersectional perspective. Others, such as Gloria Anzaldúa (1987), have explored how negotiating multiple identities can enable critical distance from the assumptions of a given culture. Feminist attention to patriarchy demands consideration of structural racism, transphobia, and other forms of oppression at work, so that we do not unthinkingly perpetuate these through a narrow focus on patriarchy. As part of this third-wave attention to intersectionality, feminist disability studies challenged the metaphorical use of disability language to describe women's oppression in feminist theory. Kim Q. Hall (2002) writes that this at once stigmatizes and erases the experiences of disabled women and further reinforces "norms of embodiment that have been used to justify the oppression of those marked as different" (x). Third-wave feminists have also explored the complications of "speaking for others," especially when privilege is the condition of doing so (Alcoff 1991). Third-wave concerns about speaking for others take on an additional dimension when we reflect on the disability rights mantra, "nothing about us without us." Such speaking demands careful engagement with those about whom one speaks and concern for the impact (not only the intent) of our words.

With the rise of queer theory, Judith Butler and others shifted to consider how compulsory heterosexuality has delimited not only the social construction of gender but even the sexed body itself. In 1990, Butler claims that to "be" a *gender* is a matter of continuous "doing," continuous *performances* through dress, gesture, speech, habit, etc. In more recent work, she turns to interrogate *sex* as the supposed material basis of gender difference; "sex" too, she argues, is consolidated through repeated performances of a dominant cultural understanding (1993). Placing Butler's insights in direct relationship to contemporary research in transgender studies, Susan Stryker (2017) explains that understanding gender or sex as doings, as performances, does not make them somehow illusory; rather, all genders (cis, trans, non-binary) and all sexes are constituted by the "innumerable acts" of performing them, "like a language we use to communicate ourselves to others and to understand ourselves" (163). Furthermore, this doesn't mean that gender and sex are merely questions of nurture, as opposed to nature. Instead, the point is that how we are is in a mutually reciprocal relationship with how we act as well as with how a given society structures the possibilities of being and doing in the first place.

In the previous section, I noted that some feminist disability theorists found a resource in the second-wave distinction of sex and gender to describe cultural constraints of disability apart from medical questions of impairment. Alternatively, recent feminist philosophers of disability (Tremain 2006) follow Butler to argue that impairment must also be interrogated as a social construction. On this view, Licia Carlson (2016) argues that the social model of disability is "guilty of reifying" and thereby naturalizing impairments that are themselves "nothing more than departures from a socially defined norm" (543). When departure from a given cultural norm figures one as necessarily lacking, we might call that norm "compulsory"; this claim that norms are "compulsory" requires a bit more explanation.

The concept of "compulsory heterosexuality" was already circulating among 1970s feminists, but tends to be associated with Adrienne Rich (1980), who uses this term to name the bias through which (her own) lesbian experience is "perceived on a scale ranging from deviant to abhorrent, or simply rendered invisible" (632). Drawing on Rich's concept, Butler challenges the naturalization of "sex" as a self-identical substance as distinct from "gender" as a form of being, to reinterpret gender as a *doing* that draws upon pre-existing norms. While we cannot wholly escape the gender norms through which we have been raised, we can gain critical distance from their compulsory status. Robert McRuer (2017/2002, 2020) draws on Butler's theory of *heterosexualized* gender to develop the theory of "compulsory able-bodiedness," which explores how norms of the able body and the heterosexual body are not only mutually illuminating but deeply intertwined. Alison Kafer (2012) expands McRuer's account to discuss "compulsory able-mindedness," and M. Remi Yergeau's (2018) recent work on "neuroqueerness" offers an affirmative self-identifying term and a site of resistance to ableist disciplinary practices. Specifically, Yergeau draws deep connections, historically and methodologically, between some popular forms of autism therapy and "reparative" therapies used on queer and gender-variant children; both are designed to make their child-patients "indistinguishable" from their peers, and were undertaken by some of the same UCLA researchers in the 1960s. As Yergeau writes of these methods, "in order to normalize one's brain, or to train one's cerebral self into productive mental health, one …[is] compelled to rewire our neural circuits for the social good of the body politic, *ad infinitum*" (131). Thus, in queer theory's naming of these "compulsory" norms, the point is to offer more leeway by recognizing their contingent histories, to make them just a bit *less* compelled.

Third-wave feminism also features insights from transnational and decolonial feminists who challenge the narrowly Western perspectives and colonial baggage of the normative construction of "Woman" (Mohanty 1991). Nirmala Erevelles (2011) importantly brings together global feminist concerns with those of feminist disability studies, underscoring the need for a "common platform of resistance" (130). She observes that (1) "disability is conspicuously missing in third world feminist analyses of difference," and (2) feminist disability studies cannot simply celebrate disability as "the most universal of human conditions" *if it is unjustly acquired*—namely, through oppressive and violent conditions, including police brutality, neocolonial violence, structural poverty, and/or inadequate access to health care.

Jasbir Puar's (2017) concept of "debility" builds on Erevelles's transnational concerns by triangulating the distinction of ability and disability, so as to include the debilitation produced by "the war machines of colonialism, occupation, and U.S. imperialism" (xvii). As Puar observes, "exploitative capital and imperial structures of the global north: generate much of the world's disability yet contribute unruly source material for rights discourses that propagate visibility, empowerment, identification, and pride" (65). A transnational disability studies lens reveals the coexistence of disability empowerment alongside this unjust distribution of debilitation and precarity. The point is not to counter or diminish disability rights or empowerment but to make these more widely meaningful by shifting attention to the four-fifths of disabled people who live in the "global south" (xvii). This perspective shift is important for disability bioethics, which has long been critiqued for its narrow Western scope of disabled experiences.

IV Feminist Conceptual Resources for Disability Bioethics

Having briefly reviewed the three waves of feminism, I conclude with four theoretical commitments that this history and its many traditions have to offer for disability bioethics. I intend for these to highlight some unique methodological contributions arising from the feminist tradition writ large. Hopefully, they support the reader to bring feminist insights into new areas of inquiry.

First, feminism is committed to questioning inherited hierarchies that shape our experiences, like mind and body, reason and emotion, culture and nature, according to which the latter tends to be devalued and feminized in relation to the former. Thus, Rosemarie Garland-Thomson (2011) writes that a feminist disability theory denaturalizes hierarchies "by unseating the dominant assumption that disability is something that is wrong with someone" (18). Bodily, cognitive, and other expressive differences could on such a view be *celebrated* rather than ranked; here, we might also recall Yergeau's (2018) concept of neuroqueerness, as an affirmative self-identifying term that resists compulsory able-mindedness.

Second, feminism is committed to situating knowledge claims in a social and historical context of structural oppression, recognizing that only some voices surface as intelligible under these circumstances. Feminist standpoint theorists had already claimed a distinctive and devalued knowledge apart from the male-identified Western canon, and feminist philosophers of disability such as Jennifer Scuro (2017) add that disability also bears such knowledge, one that has been all the more minimized and silenced in relation to expert discourses and the terms of diagnosis (21; for more on epistemic injustice, please see Anita Ho's chapter in this volume).

Third, feminism is committed to interrogating the Western public/private distinction that has led to the devaluation of care work and to the glorification on the self-sufficient individual. The legacies of feminist care ethics and concern for dependency workers could be readily placed under this subheading. On this point, the filmed dialogue between Judith Butler and Sunaura Taylor in *Examined Life* (dir. Astra Taylor, 2008) is a rich resource, as Taylor and Butler move through public spaces—city streets, cafes, a thrift shop—and ponder together what it means to "take a walk" from queer and disability perspectives.

Fourth, feminism is committed to analyzing the racialized, sexualized, ableist, and other biases at work in cultural representation, with special concern for how these representations shape our political, social, and aesthetic realities. For example, Aimi Hamraie (2013) develops the concept of the "normate template" to engage with the expectations of furniture, architecture, and other products. Hamraie draws on Garland-Thomson's concept of the "normate" (1996), which refers to the ideal embodiment presumed by ableism. These design choices are not politically neutral; they are matters of accessible world-building, of who is expected and welcome to enter.

To conclude, these feminist commitments push *disability bioethics* to attend to hierarchies of power and objectification at play in health care, biomedicine, and beyond. Importantly, some philosophical work on disability (Barnes 2016; Tremain 2017) turns to ontological and political questions and resists the narrowing of philosophical work on disability to biomedical questions that all too often position disability as a suboptimal way to life. Philosophy of disability is, importantly, a diverse field that extends beyond disability bioethics, much as feminism extends beyond feminist bioethics. Still, insofar as bioethics remains a vital theoretical field with concrete effects for policy, disability perspectives can offer much to counter ableist and eugenicist histories in medicine. Disability bioethics approaches this field with what Garland-Thomson (2017) calls *disability cultural competence*: to "enlarge our shared understanding of what it means to live with disabilities and be counted as disabled" (325). For Garland-Thomson, feminist and critical disability insights must enter into concrete medical practices, as a "knowledge-translation project" by which biomedicine can become more substantively ethical.

References

Ahmed, Sara. 2006. *Queer Phenomenology.* Durham: Duke University Press.
———. 2010. *The Promise of Happiness.* Durham: Duke University Press.
———. 2017. *Living a Feminist Life.* Durham: Duke University Press.
Alcoff, Linda Martín. 1991. "The Problem of Speaking for Others." *Cultural Critique* 20: 5–32.
Anzaldúa, Gloria. 1987. *Borderlands/La Frontera.* San Francisco: Aunt Lute Books.
Barnes, Elizabeth. 2016. *The Minority Body: A Theory of Disability.* Oxford: Oxford University Press.
Butler, Judith. 1990. *Gender Trouble: Feminism and the Subversion of Identity.* London and New York: Routledge.
———. 1993. *Bodies that Matter: On the Discursive Limits of Sex.* London and New York: Routledge.
Butler, Judith and Sunaura Taylor. 2008. Untitled Conversation. In *Examined Life*, dir. Astra Taylor.
Carlson, Licia. 2016. "Feminist Approaches to Cognitive Disability." *Philosophy Compass* 11(10): 541–553.
Cooper, Brittney. 2017. *Beyond Respectability: The Intellectual Thought of Race Women.* Urbana: University of Illinois Press.
Crenshaw, Kimberlé. 1991. "Mapping the Margins: Intersectionality, Identity Politics, and Violence against Women of Color." *Stanford Law Review* 43(6): 1241–1299.
Erevelles, Nirmala. 2011. *Disability and Difference in Global Contexts: Enabling a Transformative Body Politic.* London: Palgrave Macmillan.

Friedan, Betty. 2001/1963. *The Feminine Mystique*. New York: W.W. Norton.

Garland-Thomson, Rosemarie. 1996. *Extraordinary Bodies: Figuring Disability in American Culture and Literature*. New York: Columbia University Press.

———. 2001. "Integrating Disability, Transforming Feminist Theory." *Feminist Disability Studies*, ed. Hall. Bloomington: Indiana University Press, 13–47.

———. 2011. "Misfits: A Feminist Materialist Disability Concept." *Hypatia* 26(3): 591–609.

———. 2017. "Disability Bioethics: From Theory to Practice." *Kennedy Institute of Ethics Journal* 27(2): 323–339.

———. 2020. "Misfitting." *50 Concepts for a Critical Phenomenology*, ed. Weiss et al. Evanston: Northwestern University Press, 225–230.

Gilligan, Carol. 1982. *In a Different Voice: Psychological Theory and Women's Development*. Cambridge, MA: Harvard University Press.

Goldman, Emma. 1909. "A New Declaration of Independence." *Mother Earth* IV(5): https://www.marxists.org/reference/archive/goldman/works/1909/declaration.htm.

Hall, Kim Q. 2002. "Feminism, Disability, and Embodiment." *NWSA Journal* 14(3): vii–xiii.

———, ed. 2011. *Feminist Disability Studies*. Bloomington: Indiana University Press.

Hamraie, Aimi. 2013. "Designing Collective Access: A Feminist Disability Theory of Universal Design." *Disability Studies Quarterly* 33(4): 1–33.

———. 2017. *Building Access: Universal Design and the Politics of Disability*. Minneapolis: Minnesota University Press.

Hanisch, Carol. 1969. "The Personal is Political." In *Notes from the Second Year: Women's Liberation in 1970*, ed. Firestone and Koedt: http://www.carolhanisch.org/CHwritings/PIP.html.

hooks, bell. 2014/1981. *Ain't I A Woman: Black Women and Feminism*, 2nd ed. Boston: South End Press.

Kafer, Alison. 2012. *Feminist, Queer, Crip*. Bloomington: Indiana University Press.

Kittay, Eva Feder. 1999. *Love's Labor: Essays on Women, Equality, and Dependency*. New York and London: Routledge.

———. 2010. "The Personal Is Philosophical Is Political: A Philosopher and Mother of a Cognitively Disabled Person Sends Notes from the Battlefield." In *Cognitive Disability and Its Challenge to Moral Philosophy*, eds. Kittay and Carlson. Wiley-Blackwell, 393–413.

———. 2013. "Caring for the Long Haul: Long-Term Care Needs and the (Moral) Failure to Acknowledge Them." *International Journal of Feminist Approaches to Bioethics* 6(2): 66–88.

Kittay, Eva Feder, and Licia Carlson, eds. 2010. *Cognitive Disability and Its Challenge to Moral Philosophy*. Wiley-Blackwell.

Kollontai, Alexandra. 1909. *The Social Basis of the Women's Question*. Marxists.org: The Alexandra Kollontai Archive.

McRuer, Robert. 2017/2002. "Compulsory Able-Bodiedness and Queer/Disabled Existence." *The Disability Studies Reader*, ed. Davis, 5th Ed. New York and London: Routledge, 396-405.

———. 2020. "Compulsory Able-Bodiedness." *50 Concepts for a Critical Phenomenology*, ed. Weiss et al. Evanston: Northwestern University Press, 61–67.

Mohanty, Chandra Talpade. 1991. "Under Western Eyes: Feminist Scholarship and Colonial Discourse." *Third World Women and the Politics of Feminism*, ed. Mohanty et al. Bloomington: Indiana University Press, 51–80.

Noddings, Nel. 1984. *Caring: A Relational Approach to Ethics and Moral Education*. Berkeley: University of California Press.

Puar, Jasbir. 2017. *The Right to Maim: Debility, Capacity, Disability*. Durham: Duke University Press.

Rich, Adrienne. 1980. "Compulsory Heterosexuality and Lesbian Existence." *Signs* 5(4): 631–660.

Ross, Loretta and Rickie Solinger. 2017. *Reproductive Justice: An Introduction*. Berkeley: UC Press.

Scuro, Jennifer. 2017. *Addressing Ableism: Philosophical Questions via Disability Studies*. Lanham, MD: Rowman & Littlefield.

Stryker, Susan. 2017. *Transgender History: The Roots of Today's Revolution*, Revised Edition. New York: Seal Press.

Tremain, Shelley. 2006. "On the Government of Disability: Foucault, Power, and the Subject of Impairment." In *The Disability Studies Reader*, edited by Lennard Davis. New York: Routledge, 185–196.

———. 2013. "Introducing Feminist Philosophy of Disability." *Disability Studies Quarterly* 33(4). 1–37.

_____. 2017. *Foucault and Feminist Philosophy of Disability*. Ann Arbor: University of Michigan Press.

Wendell, Susan. 2010 (1996). "Toward a Feminist Theory of Disability." *Hypatia* 4(2): 104–124.

Wollstonecraft, Mary. 1792. *A Vindication of the Rights of Woman*. Project Gutenberg: http://www.gutenberg.org/ebooks/3420.

Further Readings

Ahmed, Sara. 2019. *What's the Use? The Uses of Use*. Durham, NC: Duke University Press.

Hall, Kim Q., ed. 2011. *Feminist Disability Studies*. Bloomington: Indiana University Press.

Kafer, Alison. 2012. *Feminist, Queer, Crip*. Bloomington: Indiana University Press.

Kim, Jina B., and Sami Schalk. 2020. "Integrating Race, Transforming Feminist Disability Studies." *Signs* 46(1): 31–55.

Scuro, Jennifer. 2017. *Addressing Ableism: Philosophical Questions via Disability Studies*. Lanham, MD: Rowman & Littlefield.

Tremain, Shelley. 2017. *Foucault and Feminist Philosophy of Disability*. Ann Arbor: University of Michigan Press.

Valentine, Desiree. 2020. "Shifting the Weight of Inaccessibility: Access Intimacy as a Critical Phenomenological Ethos." *Puncta* 3(2): 76–94.

Yergeau, M. Remi. 2018. *Authoring Autism: On Rhetoric and Neurological Queerness*. Durham: Duke University Press.

31
DISABILITY BIOETHICS AND EPISTEMIC INJUSTICE

Anita Ho

In Western bioethics, the intersecting ethical principles of respect for autonomy, beneficence, non-maleficence, and justice have generally been accepted in determining the right actions, practices, and policies in various health care scenarios. From therapeutic relationships to reproductive decisions to end-of-life care, these ethical considerations explicitly or implicitly inform clinical research and medical treatment, as well as broader funding and policy decisions. They highlight the conflicting duties borne by health care providers (HCPs), administrators, and policy makers as they navigate various care scenarios and practices.

Nonetheless, discussions regarding bioethical dilemmas that use these principles or adopt the assumptions underlying these principles often privilege non-disabled positions, actively dismissing disability perspectives and reinforcing ableist assumptions. They neglect the different epistemic powers at play in therapeutic encounters that can compromise the ability of people with disabilities to negotiate appropriate care that meets their goals and values. It is also noteworthy that race, gender, and other aspects of social identity that intersect with disability can exacerbate problematic epistemic hierarchies between patients and HCPs and further marginalize people with disabilities in health care encounters.

Utilizing genetic reproductive technologies as the main example, this chapter will illustrate how Western bioethical principles are often defined, interpreted, and applied through an explicit or implicit ableist lens. Perspectives based on disability experiences and concerns are often prematurely or even categorically dismissed as being incredulous in health care encounters (Ho 2014), ironically violating some of the fundamental tenets of these aforementioned principles. Despite the rhetoric of person-centered care, which presumably allows people to determine their own conceptions of the good and make health care decisions accordingly, this chapter will argue that much of the bioethics discourse around reproductive decisions commits epistemic injustice by privileging non-disabled perspectives over the lived experiences and perspectives of people with disabilities. The chapter will end by suggesting that in order to truly promote a just health care system that respects people's autonomy and well-being, bioethicists must do more than simply include diverse voices in the discourse. We need to also critically examine and correct various faulty assumptions that underlie our ethical approaches to various care dilemmas, thereby transforming the culture and practice of medicine to truly respect and serve the well-being of all.

DOI: 10.4324/9781003289487-41

I Objective Bioethical Principles?

The four canonical bioethical principles of respect for autonomy, beneficence, non-maleficence, and justice have been widely adopted in Western teaching and evaluation of ethical dilemmas arising in various health care scenarios.[1] Briefly, respect for autonomy requires that mentally capable patients and surrogates of people who lack such capacity be allowed to make voluntary treatment decisions without undue influence from others. Non-maleficence describes the obligation of health systems and care providers to not inflict unnecessary harm on patients, requiring providers and policy makers to balance the risks and benefits of various procedures, and to avoid preventable injuries in their practice and policies. Beneficence refers to a duty to promote patients' well-being, which is a basic tenet of clinical medicine. Justice deals with fairness, equality, and equitable treatment of all patients, emphasizing nondiscrimination of various populations and forbidding unequal treatment based on morally arbitrary factors. In the form of distributive justice, this principle demands equitable allocation of scarce medical and social resources to support patients' needs.

These mainstream bioethical principles have generally been considered objective approaches that neutrally apply to all persons without privileging or delegitimizing any particular perspective or population. For example, respect for autonomy often takes the form of promoting one's right to informed consent, which requires HCPs to give patients (or their surrogate decision makers) relevant information regarding available care options and then allow them to decide among these options according to their values, whatever they may be. As technological advances offer multiple clinical options with various levels and types of risks and benefits, person-centered care encourages HCPs to solicit and incorporate patients' heterogeneous goals and priorities into care decisions. This is particularly important for culturally and experientially diverse societies, where people may have different concerns or notions of the good that can affect how they would consider prospective care options. While HCPs may advise patients about the respective clinical risks and benefits of different options, manipulating or coercing patients, even for their own good, is to paternalistically treat them as if they lacked the capacity to shape their own lives – it is to deny them their moral status as persons.

The bioethical principle of justice also appears to uphold objectivity and impartiality, particularly since it explicitly advocates for fairness in health care decisions by promoting equal treatment for persons in similar contexts. This principle purports to guard against discrimination or unequal distribution of scarce health care resources or different treatments of individuals and groups based on morally irrelevant factors, both at the bedside and at the system level. Once the morally relevant criteria for service or resource qualification have been set (e.g., age requirement for Medicare), justice requires that everyone who is eligible be granted equal access.

However, a closer look at how these bioethical principles are frequently applied to justify various biomedical decisions or practices reveals the following. Western bioethics, which is influenced by the concepts of individual self and autonomy in moral philosophy, generally adopts the norms of independence and control. It idealizes the mind and the body, demanding that we control them at all times and painting dependency as a type of misfortune or harm that should best be avoided (Wendell 1989). Medical and bioethics discourses often portray people with disabilities as "the other" who failed to protect their own well-being or unfortunate victims, who are less capable than non-disabled people in determining or promoting their own good. Western bioethics uncritically privileges the non-disabled perspective as paradigmatic in addressing various bioethical dilemmas, creating and reinforcing epistemic injustice.

II Epistemic Hierarchies or Epistemic Injustice?

In specialized areas such as clinical medicine, there is a general presumption and acceptance of knowledge or epistemic hierarchy between professionals and lay patients. By virtue of their intensive education and practice in evidence-based medicine, HCPs are expected to have superior ability to use scientific methods to diagnose and determine a range of possible clinical solutions for various medical problems relative to laypersons. The presumed epistemological gap gives prima facie reasons for patients who cannot adequately assess medical evidence to accept care providers' clinical judgment (Ho 2011).

However, even granting that HCPs are more qualified in using scientific methods in clinical diagnosis and decision making, whether these methods are the superior or only relevant methods in determining suitable care approaches cannot be determined a priori, i.e., ahead of time and without evidence. What constitutes risk and how much risk is acceptable, what types of benefits are favored, and what costs are more worrisome from a patient's perspective depend on various socio-relational factors that go beyond scientific or clinical considerations. These can include what immediate and long-term goals patients are trying to achieve, what negative outcomes they want to avoid (e.g., financial, relational, physical, psychological), and the different trade-offs among various possible outcomes, given the patient's broader context.

When one accepts a particular framework as absolute and presumes that the perspectives of those who employ different methodologies or prioritize divergent goals are less worthy of careful consideration, one may overlook other potentially relevant or legitimate viewpoints. As both a moral and epistemological concept, epistemic injustice is a type of wrong and harm that is done to individuals or social groups when their capacity as knowers, reasoners, or collaborative learners is undeservedly undermined (Fricker 2007). It happens when some people are unfairly excluded from the epistemic community because of their social backgrounds, such as their ethnic or cultural background, socio-economic status, gender, or disability, leaving them disadvantaged in their effort to contribute to or reap benefits from knowledge creation (Fricker 2007; Ho 2014).

Epistemic injustice can take the form of testimonial injustice, which occurs when someone's assertions or testimonies are dismissed because the person is unfairly deemed by the hearer to lack what is required to be a credible informant. In these situations, the hearer pre-establishes their own criteria for credibility and rejects critical evaluation or challenge of these criteria by others. The hearer prematurely neglects or denies the need to gather or consider accessible counter evidence, and forms a prejudice against others' epistemic credibility due to their social identity.

In therapeutic relationships, testimonial injustice places culpability on those who have an obligation to assess the speaker's credibility fairly but fail to do so. In the age of patient-centered care, HCPs are expected to acknowledge and understand patients' perspectives in co-determining appropriate care plans. Clinicians' recommendations, decisions, or actions can significantly impact patients' well-being. Patients, thus, have a lot at stake in whether their testimony is given due consideration and can rightfully expect their care providers to respect their experience and claims.

Testimonial injustice may occur when patients' reports of their symptoms, experience, or concerns are not believed by their HCPs because of the patients' social status and other prejudicial reasons. A paradigmatic situation of epistemic injustice is one where an individual's testimony is dismissed as categorically less credible or relevant due to stereotypes around morally irrelevant factors such as the person's skin color, gender, sexual orientation, socio-economic background, or disability status.

III Epistemic Injustice and Disability

In the context of various disabilities, civil rights legislation such as the Americans with Disabilities Act is designed to ensure equal participation in society by requiring relevant organizations and agencies to provide necessary services and accommodations for people with disabilities. However, as the epistemic structure of Western allopathic medicine utilizes frameworks of clinical correlation and pathological anatomy (Buchman, Ho, Goldberg 2017), individuals' reports about their needs and the best means to meet those needs are sometimes met with suspicion, if their symptoms (e.g., pain, fatigue) do not correspond with what clinicians believe to be the accepted physical sign. Various services for people with disabilities (e.g., accessible parking, flexible or frequent break times) are made available based on physician approval, signaling that it is only when a person's experience has been verified by an "objective" expert and formally labeled according to their specific methodology that the validity of the person's claim is accepted.

Such prejudice is systematic and exacerbated in clinical medicine when multiple social identities intersect. For example, people of color with disabilities experience compounded disadvantages. In the USA, Black bodies and minds often either receive less medical attention, or are more frequently prescribed a sick role (Boswell et al. 2019). Many have inadequate health care access even when they are more likely to have a disability, possibly due to a lack of health insurance or culturally safe/appropriate services. At the same time, Black people are more likely than white people to be pathologized and subjected to involuntary treatments, particularly around cognitive or mental issues, as racial stereotypes may affect clinicians' assessments of a patient's state of being regardless of the person's own experience. Black patients consistently report having poorer therapeutic encounters with HCPs, including having their concerns dismissed, compared to their white counterparts. When these patients hesitate to follow professional recommendations that do not fit with their own experience, their actions are also more likely to be characterized as non-compliance rather than an exercise of autonomy or the result of distrust that needs to be addressed accordingly.

Epistemic injustice is particularly problematic in the health care settings because it affects a patient's ability to negotiate or access appropriate care (Ho 2014), thereby compromising their autonomy and well-being. In the context of contemporary scientific and medical discourses, the non-disabled perspective is the core ideology underlying the conceptualization and application of the aforementioned ethical principles, pre-determining the lens with which these principles would be utilized in addressing various bioethical dilemmas. Stereotypical or ableist assumptions about people with disabilities have been privileged but disguised as objective practices (Wieseler 2016). These assumptions shape the generation and application of scientific knowledge as well as the promotion of various medical practices.

People with disabilities have been systematically considered less credible in bioethics and medical discourses compared to medical experts, ironically violating the aforementioned ethical principles. Medical professionals generally attend to patients under highly specific circumstances for very short periods, such as when patients present at the clinics and hospitals. They usually do not have opportunities to directly observe their patients' daily activities outside of these controlled settings. Nonetheless, clinicians' "objective" assessments are often considered categorically superior to and more trustworthy than patients' narrative accounts about their own experiences and needs. In categorizing scientific objectivity as trustworthiness, judgments of clinical medicine are presented as perspectives that can be rationally trusted and favored over other approaches (Scheman 2001). While clinicians are generally benevolent toward their patients, people with disabilities have long been subjected to social exclusion and marginalization. They are often seen as globally less credible and less competent in assessing their own experiences and needs, subjecting them to epistemic injustice.

IV Epistemic Injustice and Reproductive Genetic Technologies

Epistemic injustice can be illustrated by how biomedicine is practiced in reproductive genetics. As many have argued, reproduction and raising children are some of the most significant processes of one's life, and can have a tremendous psychological, physical, and financial impact on potential parents, especially for women. Bioethics discussions often follow the common argument for reproductive autonomy or liberty, integrating genetic technologies into the realm of choices among which potential parents can decide (Harris 2005).

In medical and bioethics literature, reproductive genetic technologies such as prenatal and preimplantation genetic diagnoses are generally touted as valuable scientific procedures that can promote the autonomy and well-being of people who have an increased probability of having a child with various genetic traits that are deemed undesirable. In addition to the few common conditions related to potential parents' age, family history, or ethnic background, an increasing variety of genetic tests are now available to screen hundreds of conditions, including rare diseases. Potential parents can undergo a blood or saliva test to see if they may be a carrier of autosomal recessive diseases (e.g., cystic fibrosis). They can also utilize in vitro fertilization, produce multiple embryos, screen or test each of them for various genetic conditions, and only implant embryos that are deemed to be free of certain undesired genetic characteristics. Those who are already pregnant can also undergo prenatal tests to screen for chromosomal and monogenetic deviations.

Critics who are concerned about these reproductive genetic technologies have mostly focused on whether individuals who utilize these technologies have and understand the relevant information of the available options, have control over how their data may be used or shared with parties outside of the immediate clinical treatment context, and have the opportunity to voluntarily consent to or refuse these tests according to their value system. Discussions of respect for autonomy and harm prevention in the debate both take for granted that the ideology underlying the development and implementation of reproductive genetic technologies is itself unproblematic or even desirable, as long as qualified professionals (e.g., genetic counselors) are available to explain the results and ease potential parents' psychological distress, and that appropriate data governance structures can ensure data confidentiality and security. Rapidly expanding panels for different genetic conditions suggest that any abstract, impartial, and rational decision maker – whether it is a HCP, policy maker, or prospective patient – would opt to adopt appropriate means to prevent or mitigate genetic conditions that may accompany any corporeal or cognitive limitations.

Taking the benefits and widespread desirability of these technologies for granted, mainstream bioethics discussions generally adopt the medical model of disability, which conceptualizes disability as a genetic or natural disadvantage that is the inevitable outcome of biomedical facts (Amundson 2005). This model assumes that the harm of disability should and can only be reduced or eliminated through biomedical interventions. Even though one in four[2] adults in the USA has some type of disability, our society is built mostly to serve non-disabled people. The medical model neglects how it is often the social structure that is imposing disadvantages on people with various impairments. Justice-related discussions of reproductive genetic services thus often focus on resource allocation questions around equal access. As biomedical innovations often privilege the few who can afford these technologies, distributive justice considerations in reproductive genetics have generally been focusing on whether everyone, including people with lower socio-economic backgrounds or lacking comprehensive health insurance, may have equal access to such technology. In recent years, many who consider the intersecting ethical considerations of reproductive autonomy and distributive justice have proposed increasing public funding for genetic tests so as to expand the number of options for potential parents of different backgrounds who are concerned about passing on various genetic traits (Buchanan, Brock, Daniels, and Wikler 2000).

However, underlying the rhetoric of respect for autonomy, prevention of harm, and promotion of distributive justice in the mainstream bioethics discussion of reproductive genetics is a persistent ableist paradigm. This paradigm continues to consider non-disabled perspectives and presumptions as default positions over the experiences of people with disabilities or disability perspectives, perpetuating epistemic and related injustices. Reproductive genetic screening and diagnostic processes are not simply clinical or technical exercises – they are also social and political processes that can exacerbate inequalities. People with disabilities report a much higher quality of life than projected by non-disabled people and can fully integrate into the society when appropriate accommodations and arrangements are available (Amundson 2010). Nonetheless, prominent biomedical and bioethical discussions of the quality of life of people with disability continue to discredit these people's experience as subjective, mistaken, unreliable, or simply the result of lowered expectations due to disabilities (Amundson 2005).

The social domains continue to treat life without disability as the standard of "normal" or even ideal living, weighing against people with disabilities and potential parents who choose to continue with pregnancies regardless of whether they may or may not be affected by various genetic traits (Shakespeare 2005). Encounters between HCPs and people with disabilities usually occur only in the clinical setting which focuses on bio-physical symptoms in the individual, some of which may not be related to the person's disability. Even as medical professionals rarely have opportunities to directly observe patients' daily activities in their broader contexts, clinicians generally exhibit very little modesty about their knowledge (Wendell 1989). Despite their non-disabled ignorance (Wieseler 2016), professionals presume that their "objective" assessments are categorically superior to and more trustworthy than people's own narrative accounts regarding living with or caring for someone with disabilities.

Most well-meaning professionals have not experienced life with a significant disability and are devoting their career to maintaining or restoring various forms of functioning under the medical model. Unfortunately, many inadvertently adopt and reproduce skewed impressions of the disability experience as they counsel their patients (Ho 2008). Even though people with varying types and levels of impairment have vastly different experiences, available information regarding people's quality of life and medical descriptions of various conditions and experiences are generally one-sided, selectively representing these conditions in static, absolute, negative, and stereotypical terms. In reproductive genetics, probabilities of having different genetic traits are typically presented as inherent risks that ought to be avoided at all costs. As one of the co-discoverers of the molecular structure of DNA notes, genetic diseases are "random tragedies that we should do everything in our power to prevent" (Watson 2000, 225).

The dominance of the technological imperative and the ableist socio-cultural framework is pervasive and structural. Reproductive genetic technologies are increasingly institutionalized within standard protocols for "routine" maternal and prenatal care. For example, an increasing number of international professional organizations (e.g., the Society of Obstetricians and Gynecologists of Canada) now recommend that pregnant women of all age groups be offered screening for assessing their likelihood of having a child with Down syndrome, even in the absence of family history of this condition. Such recommendation sends the normative message that diagnostic technologies are legitimate, inherently good, desirable, or necessary in promoting prospective parents' well-being, and that acceptance is expected and/or recommended as part of prenatal care, even when fetal interventions for these conditions are unavailable. Language regarding various conditions – including widely variable manifestations of Down syndrome – is generally negative, and pregnant women often do not receive information on the socio-relational experiences of people with these conditions.

As screening is routinized, it shapes how professionals communicate about such technologies, thereby affecting how potential parents may interpret, accept, and act on the information.

A desire to know about the status of an embryo or fetus does not necessarily translate into an intention to refrain from getting pregnant or to terminate an existing pregnancy. Nonetheless, in an ableist environment that sees disabilities as medical tragedies to be avoided, an agreement to screening is often taken by professionals to imply a plan to prevent or stop an otherwise desired pregnancy. In recent years, some bioethicists have also argued that people who may pass on various genes that can contribute to certain disabling conditions have a duty to get tested and not bear children if they are in "high-risk" groups. In other words, people are not only encouraged or expected to undergo genetic tests. Further pressures ensue if a test result indicates the potential presence of presumably undesired traits (Shakespeare 1998, 676).

Despite the promise of reproductive autonomy, routine screening in the name of the fetus' or pregnant woman's good discourages women from questioning professionally recommended tests. This pressure may be compounded for women coming from a lower socio-economic background, as their intersecting social disadvantages may lead some HCPs to prejudicially question their understanding and ability to care for their offspring with various genetic conditions.

The informed consent provision gives the illusion of autonomy by giving potential parents the opportunity to accept clinical recommendations they are socially expected to choose. Even though women are not legally required to seek genetic tests and selective termination for various genetic conditions, these expectations are formalized when supported by socially and epistemically dominant medical professionals. They frame the meaning of becoming pregnant and make it difficult for women to opt for or continue a pregnancy without genetic testing, since their perspectives and concerns may be dismissed by professionals. When reproductive recommendations are delivered in the language of clinical science that presumes value neutrality and objectivity, they potentially shield HCPs from critically evaluating their own assumptions or the non-disabled paradigm, thereby allowing them to silence patients who employ more "subjective" approaches to critique or challenge their determinations (Ho 2011).

Uncritical acceptance of an entrenched epistemic hierarchy that precludes or at least discourages challenges from those with different perspectives or methodologies can have significant impacts on people's ability to receive the most appropriate reproductive care that corresponds to their goals and values. Without meaningful changes to dismantle ableist barriers in society, these practices set the stage for social control and for blaming those who do not follow professional advice regardless of their reasons for such decisions.

V Counteracting Ableism and Epistemic Injustice

Health care practices and policies often take the non-disabled paradigm as the default and foundational framework, while denying that these perspectives may unreflectively generalize these decision makers' own interests, values, and prejudices (Wieseler 2016). However, as the implicit purpose and role of various genetic screening programs is to reduce disability incidence, declarations of neutrality by various professionals and policy makers may not be self-substantiating.

Systematic exclusion or invalidation of disabled people's lived experiences as capable of producing knowledge can lead to a limited understanding of disability-related issues. It can also further disadvantage these people in the construction of knowledge. To promote epistemic justice for all, a commitment to epistemic humility may help to minimize the power asymmetry that can harm people with disabilities in health care settings. Epistemic humility signifies professionals' acknowledgment that scientific and clinical expertise is necessary but insufficient in determining appropriate care regarding people with disabilities. Epistemic humility in this context means a commitment to make a realistic assessment of what one knows and does not know, and to restrict one's confidence and claims to knowledge accordingly. It is a recognition that knowledge creation is an interdependent and collaborative activity, requiring the perspectives of both the HCPs and people with disabilities to obtain a full picture of the person's experience and determine the most appropriate care pathways.

People with disabilities, who have been disadvantaged by ableist social structures, are better suited to understand and report their own experiences of embodiment than non-disabled people who may lack that experience (Wendell 1989). They can be an invaluable source of information for prospective parents who are considering various reproductive genetic technologies. This insight, a core of feminist standpoint theory, reminds us that epistemic privilege can be drawn from the position of the marginalized. To counteract epistemic injustice, bioethicists, clinicians, and policy makers can actively participate in groups where the subjective experience of patients is explored, and learn to become more aware of their own unconscious bias regarding various impairments, which can turn into prejudices toward people with disabilities (Crichton, Carel, and Kidd 2017). This intentional testimonial exchange may help relevant decision makers to recondition their testimonial sensibility so that they can assess patients' (and their own) credibility more accurately. It may also enhance good will and facilitate trustworthiness of professionals in therapeutic encounters and decision-making processes.

In recognizing and correcting how power influences knowledge production in medical practices, epistemic humility has both epistemological and ethical significance. By critically examining the power structures that downplay the credibility or relevance of the disability perspectives, epistemic humility situates knowledge in local experiences. In addressing how the historical exclusion of the perspective of people with disabilities may have led to an impoverished view of their needs in designing health care and social services, epistemic humility may help well-meaning clinicians and policy makers to correct the impact of inadvertent prejudice in their credibility judgments.

Notes

1 Clinical research is beyond the scope for this chapter. Here, I will only focus on clinical and policy decisions that are directly related to patient care delivery.
2 "CDC: 1 in 4 US adults live with a disability," Center for Disease Control and Prevention (2018).

References

Amundson, Ron. 2005. "Disability, Ideology, and Quality of Life." In *Quality of Life and Human Difference: Genetic Testing, Health Care, and Disability,* edited by David Wasserman, Jerome Bickenbach, and Robert Wachbroit, pp. 101–124. New York: Cambridge University Press.

————. 2010. "Quality of Life, Disability, and Hedonic Psychology." *Journal for the Theory of Social Behavior,* 40(4): 374–392.

Buchanan, Allan, Dan Brock, Norman Daniels, and Dan Wikler. 2000. *From Chance to Choice: Genetics and Justice.* Cambridge: Cambridge University Press.

Buchman, Daniel, Anita Ho, and Daniel Goldberg. 2017. "Investigating Trust, Expertise, and Epistemic Injustice in Chronic Pain." *Journal of Bioethical Inquiry,* 14: 31–42.

Boswell, Barbara, Zimitri Erasmus, Shanel Johannes, Shaheed Mahomed, and Kopano Ratele. 2019. "Racist Science: The Burden of Black Bodies and Minds." *The Thinker: A Pan-African Quarterly for Thought Leaders,* 81: 4–8.

Center for Disease Control and Prevention. 2018. "CDC: 1 in 4 US Adults Live with a Disability." Accessed September 29, 2021.

Crichton, Paul, Havi Carel, and Ian J. Kidd. 2017. "Epistemic Injustice in Psychiatry." *BJPsych Bulletin,* 41(2): 65–70.

Fricker, Miranda. 2007. *Epistemic Injustice: The Power and Ethics of Knowing.* Oxford: Oxford University Press.

Harris, John. 2005. "Reproductive Liberty, Disease and Disability." *Reproductive Biomedicine Online*, 10(Suppl 1): 13–16.

Ho, Anita. 2008. "The Individualist Model of Autonomy and the Challenge of Disability." *Journal of Bioethical Inquiry*, 5: 193–207.

———. 2011. "Trusting Experts and Epistemic Humility in Disability." *International Journal of Feminist Approaches to Bioethics*, 4(2), 102–123.

———. 2014. "Epistemic Injustice." In *Encyclopedia of Bioethics*, 4th ed., edited by Bruce Jennings. Farmington Hills, Michigan: Macmillan Reference.

Scheman, Naomi. 2001. "Epistemology Resuscitated: Objectivity as Trustworthiness." In *(En)gendering Rationalities*, edited by Nancy Tuana and Sandra Morgen, pp. 23–52. Albany, NY: SUNY Press.

Shakespeare, Tom. 1998. "Choices and Rights: Eugenics, Genetics and Disability Equality, *Disability & Society*, 13(5): 665–681. doi:10.1080/09687599826452.

———. 2005. "The Social Context of Individual Choice." In *Quality of Life and Human Difference: Genetic Testing, Health Care, and Disability*, edited by David Wasserman, Jerome Bickenbach, and Robert Wachbroit, pp. 217–236. New York: Cambridge University Press.

Watson, James. 2000. *A Passion for DNA: Genes, Genomes, and Society*. Plainview, NY: Cold Spring Harbor Laboratory Press.

Wendell, Susan. 1989. "Toward a Feminist Theory of Disability." *Hypatia*, 4(2): 104–124.

Wieseler, Christine. 2016. "Objectivity as Neutrality, Nondisabled Ignorance, and Strong Objectivity." *Social Philosophy Today*, 32: 85–102.

Further Readings

Blease, Charlotte, Havi Carel, and Keith Geraghty. 2017. "Epistemic Injustice in Healthcare Encounters: Evidence from Chronic Fatigue Syndrome." *Journal of Medical Ethics*, 43: 549–557.

Buchman, Daniel, Anita Ho, and Daniel Goldberg. 2017. "Investigating Trust, Expertise, and Epistemic Injustice in Chronic Pain." *Journal of Bioethical Inquiry*, 14: 31–42.

Kidd, Ian James and Havi Carel. 2017. "Epistemic Injustice and Illness." *Journal of Applied Philosophy*, 34(2): 172–190. doi:10.1111/japp.12172.

Peña-Guzmán, David M., and Joel Reynolds. 2019. "The Harm of Ableism: Medical Error and Epistemic Injustice." *Kennedy Institute of Ethics journal*, 29(3): 205–242. doi:10.1353/ken.2019.0023.

Reynolds, Joel. 2020. "'What If There's Something Wrong with Her?'-How Biomedical Technologies Contribute to Epistemic Injustice in Healthcare." *Southern Journal of Philosophy*, 58: 161–185. doi:10.1111/sjp.12353.

32
DISABILITY STUDIES MEETS ANIMAL STUDIES

David M. Peña-Guzmán

I Introduction

The relationship between critical disability studies and critical animal studies is complicated and contentious.[1] In recent years, however, a handful of experts have sought to reimagine this relationship in the hope that these fields might become mutual allies in fights against injustice. Some have done this by thinking about ableism and speciesism as closely interconnected structures of oppression that reinforce one another. Others have contributed to this reconciliatory effort by using the resources of critical disability studies and critical animal studies to analyze social phenomena related to disability and animality.

This chapter offers a brief account of three areas of research that highlight the value of combining critical disability studies and critical animal studies: (1) theories of justice that exclude neither disabled humans, nor non-human animals, (2) accounts of disability in other species, and (3) controversies over "service animals" and "emotional support animals."

II Disability Justice, Animal Justice: The Myth of the "Normal Human Animal"

As structures of oppression, ableism and speciesism have much in common. For example, the oppression of people with disabilities and the oppression of animals have historically hinged on privileging specific "capacities" that neither (certain) disabled humans, nor (most) non-human animals are thought to possess, such as spoken language, rationality, and autonomy. At the heart of both ableism and speciesism, we find an image of the normal human animal as the unspoken archetype that sways our thinking about who matters and who does not. People with disabilities have historically been excluded from this image on account of being deemed "not normal," while animals have been excluded on account of being "not human." While we can, and should, understand that there are differences between the strategies and goals of the disability and animal liberation movements, both have a common enemy: the assumption that there is a hierarchy of moral value that rules the animal kingdom and that normal humans (read: abled-bodied Homo sapiens) are at the top of it, with everyone else occupying the "lower" ranks.

DOI: 10.4324/9781003289487-42

In a co-authored 2015 essay, "Rethinking Membership and Participation in an Inclusive Democracy: Cognitive Disability, Children, Animals," Canadian political philosophers Sue Donaldson and Will Kymlicka argue that this the myth of the normal human has guided theories of citizenship for a very long time and been used to exclude three vulnerable social groups who fall short of this ideal: (1) children, (2) the cognitively disabled, and (3) non-human animals. Their plea for "new models of inclusive citizenship" provides an example of how scholars who identify as both disability rights and animal rights advocates strive to forge a "fellowship" between critical disability studies and critical animal studies.

In political theory, the concept of citizenship is fundamental since citizenship entitles individuals to full protection under the law. Since the seventeenth century, the core idea behind social contract theory (as articulated by John Locke, Thomas Hobbes, and Jean-Jacques Rousseau) is that humans are rational enough to understand that the state of nature is too dangerous and unpredictable to live a decent life. Thus, humans agree to enter into a social contract with one another whereby they collectively agree to give up some of the freedoms they previously enjoyed in the state of nature in exchange for the security of living in a state of civil society. Because the state of nature is a state of ruthless anarchy in which "man is wolf to man," any rational human will see this tradeoff as a rational choice and thus agree to form a common government (monarchy, democracy, etc.). As soon as said government is created, the signatories to the social contract are transformed into "citizens." By definition, a citizen is a member of a civil society who has all the rights and enjoys all privileges that the system of law in that society can afford.

According to Donaldson and Kymlicka, when thinking about citizenship we must ask two questions. The first is: who counts as a citizen and who doesn't? The second is: how do we decide what makes someone a citizen in the first place? Unfortunately, the theories of citizenship we have inherited hold that only "normal" humans can be full-blown citizens because citizenship demands rationality and language.

> In traditional political theory, the citizen has been conceived as a person with capacities for public reason or *logos* or Kantian autonomy or rational reflection and deliberation—complex language-mediated capacities which we will call (following Gary Steiner) 'linguistic agency.' Linguistic agency has operated not just as an ideal, but as a threshold capacity. Those seen as lacking this capacity have been relegated to the margins of political community, situated as passive wards to whom society owes duties of care rather than as co-citizens with equal rights.
>
> *(2015, 234)*

Groups that are thought to lack these properties are relegated to the margins and denied a wide array of rights and privileges. At different historical moments, different social groups—e.g. Black people, women, indigenous peoples, children, disabled people—have been seen as lacking these capacities and consequently treated as "second-class citizens." Even if they are entitled to protection from some forms of abuse, they are not seen as entitled to full participation in the political life of the community—in its practices, rituals, and institutions.

But why should rights of citizenship, including what Donaldson and Kymlicka call "rights of participation," be tied either to one's "capacity for public reason" or to one's "linguistic agency"? They shouldn't because that is fundamentally unjust. Citing recent developments in international human rights law, such as the UN Committee on the Rights of Persons with Disabilities and the 1989 UN Convention on the Rights of the Child, Donaldson and Kymlicka engage new ways of thinking about citizenship and propose that we use these developments as a springboard for co-creating a model of citizenship that is inclusive rather than exclusive.[2]

"Citizenship," they say, "isn't a select club for linguistic agents; it's a commitment to include and empower all members of society, across the whole spectrum of diversity, on their own terms" (234). This raises an important question. If citizenship doesn't require reason or language, what is its basis? What makes someone a candidate for citizenship?

Donaldson and Kymlicka's answer is norm responsiveness, which is the ability to grasp norms in intersubjective relations. Norm responsiveness is "the ability to moderate behavior in accord with internalized norms when relating to other selves" (235). Let's unpack this complex quote one step at a time. Philosophers use the phrase "norm responsiveness" to describe how people behave when they understand that their interactions with others are mediated by certain norms (i.e. rules or expectations), even if said norms are never explicitly articulated but only apprehended in a tacit way. For example, from a young age most children understand that there are rules they must follow when playing with others, even if they cannot express these rules in language. For instance, when playing "tag" it is important to take turns, even if one kid is obviously faster than the rest. Similarly, when wrestling someone smaller, one must not wrestle too hard. Otherwise the game ceases being fun, and one runs the risk of becoming a bully. When these norms are broken, violators quickly learn there is a price to pay as other kids regularly refuse to play with those who "don't play nice." All of this is norm responsiveness. The children's actions—at the park, in the playground, at home, etc.—suggest that they moderate their behavior in accordance with internalized norms that affect their relations to others. Norm responsiveness is a social back and forth, a continual give and take. What is essential, for Donaldson and Kymlicka, is that children do this before having a sophisticated grasp of language and before reaching the so-called "age of reason," which is around seven or eight years old.

When thinking about whether someone should be recognized as a citizen under the law, we should look at whether they are capable of norm responsiveness. Children are entitled to citizenship not because they have an impressive mastery of language (they don't) nor because they are paragons of rationality (they aren't), but because they enter meaningful social relationships. Like adults, but in their own child-like way, children can act in accordance to norms and express their own agency in relationships. They move and thrive in social settings "involving cooperation, trust, and intersubjective recognition." But if we take norm responsiveness to be the active ingredient of citizenship, it is not only children who should be included. Cognitively disabled humans and some non-human animals (especially domestic companions) should too. People who have no first-hand experience living or interacting with cognitively disabled people are likely to have a distorted view of this community, assuming they must be incapable of norm responsiveness. Nothing could be farther from the truth. Yes, norm responsiveness looks different for the cognitively disabled than for neurotypicals—and, to be clear, can look quite different from one person to the next within that group of people as well—but this hardly matters. What matters is that cognitively disabled individuals recognize others and modulate their relations to them based on shared norms.

Following work in critical disability theory (especially research on non-linguistic modes of communication with and among the cognitively disabled), Donaldson and Kymlicka explain that the only difference is that disabled people tend to only display norm responsiveness when they find themselves in a safe space, in a relationship (say, with a parent, family member, a caregiver, a close friend) that is already mediated by trust. A stranger who interacts with a person with severe cognitive disabilities on a one-time basis or without taking the time to get to know them might walk away with an impoverished view of the kind of life this person leads. But that is only because the cognitively disabled person is not putting the full range of their norm-responsive capacities on display in the presence of the stranger. Perhaps, if the stranger took the time to interact with that person regularly, to engage in a meaningful back and forth for an extended period of time, to learn to interpret their behaviors and responses, the stranger would perceive them in a very different light—as someone who is norm-responsive and capable of expressing their agency, even if in an atypical way. Donaldson and Kymlicka admit that the participation of people with severe cognitive disability may not look the same as that of neurotypical human adults. But there is a massive difference between acknowledging this and saying, as nearly all political philosophers have historically, that the cognitively disabled cannot participate in social life at all.

Pushing the boundaries of the concept of citizen even further, Donaldson and Kymlicka extend their analysis to domestic animals. Yes, your dog "Fido" should be recognized as a "citizen" under the law, and he should have the rights associated with this label. This may sound bizarre to many, but when we keep in mind that the cardinal goal of a theory of citizenship is to include rather than exclude the most vulnerable around us and that the key to citizenship is norm responsiveness (rather than reason or language), it is clear that many animals are good candidates for citizenship, even if they don't have human-style reason or human-style language.

> If citizenship is indeed about recognizing membership, voice and agency within socially meaningful relationships involving cooperation, trust and intersubjective recognition – rather than threshold capacities for linguistic agency – then [domestic animals] qualify. Indeed, the process of domestication is precisely about the incorporation of animals into such relations. Domestication has presupposed, and further developed, capacities for trust, cooperation and communication, in ways that lay the behavioral foundations for relationships of co-citizenship. Having incorporated them into our society, and bred them to be dependent on us (or interdependent with us), we are morally obliged to recognize the membership of [domestic animals] in society, and to enable their participation in shaping the norms that govern that shared society.
>
> *(237)*

Donaldson and Kymlicka develop their theory of animal rights more fully in *Zoopolis: A Political Theory of Animal Rights* (2011), where they distinguish between three classes of animals and the legal rights each should be granted: "wild animals" should have rights of sovereignty; "liminal animals" that live at the intersection of the wild and human society, such as pigeons and rats, should have "immigrant" rights, such as the right of free movement; and "domestic animals" should have rights of citizenship, including the right to have their interests be treated as equal to those of human beings when making political decisions that affect them.

Donaldson and Kymlicka are not the only ones who have connected disability justice and animal justice,[3] but two points are worth emphasizing about their approach. First, they concede that their approach brings a host of conceptual challenges, but insist that we must never cease asking ourselves the two core questions mentioned above: Who is in or out? And how are we reaching this conclusion? The moment we stop asking ourselves these questions, we risk assuming that our present conception of citizenship is the only plausible one and we deny ourselves the opportunity to create a more fair, inclusive, and just political order. Second, they present their analysis of citizenship as an attempt to bring together critical disability studies and critical animal studies. In this case, they do so by identifying a shared enemy: the classical theory of citizenship that, in one breath, excludes children, many people with disabilities, and non-human animals. By targeting this theory, they expose the ways in which speciesism and ableism are interconnected and the extent to which the struggle for global justice requires an intersectional approach.

III Disability in the Animal Kingdom: Beyond Homo Sapiens

In the "Foreword" to *Disability and Animality: Crip Perspectives in Critical Animal Studies* (2020), animal ethicist Lori Gruen recounts her encounters with two disabled primates and how they changed her way of thinking about the nature of disability.

> Over 15 years ago, I met a chimpanzee named Knuckles, who lives at the Center for Great Apes, a sanctuary for chimpanzees and orangutans in Wauchula, Florida. Knuckles has cerebral palsy. Mari, an orangutan, who lost her arms in an accident as an infant while living in a cognition laboratory, also now lives in the sanctuary. Mari moves around easily (even though the sanctuary has taken care to make structures safe for her while also insisting that she doesn't "need concessions"). She uses her legs and her chin to get around and manipulate objects and she lives happily with other orangutans. Knuckles requires more elaborate care, particularly when he was younger, but after physical and occupational therapy, he is able to feed himself, climb up and down, and play. He too is now able to live with others of his kind who recognize that he is different.

Meeting Knuckles and Mari made Gruen question what we mean when we say that something is "natural" or "normal," and to appreciate that disability is always a product of the relationship between an organism and its environment. In less accommodating and accessible spaces, Knuckles and Mari would have lived more limited and "disabled" lives—lives purged of physical stimulation, social interaction, the possibility of friendship, and so on. But in spaces intentionally designed to facilitate their flourishing, they led exciting lives filled with projects, relationships, and emotional bonds. Echoing what disability studies scholars call "the social model of disability,"[4] Gruen insists that we need to change our perspective. Knuckles is not "abled" or "disabled" by his brain, but by his environment. Mari is not "abled" or "disabled" by her body, but by her social settings. The same is true of all other animals, independently of whether their impairments are sensory, physical, cognitive, or neurological.

Unfortunately, the stories we usually hear about disabled animals are very different. Often, they reek of what disability activists call "inspiration porn," which refers to an exploitative way of portraying disability that emphasizes triumphing over adversity, overcoming one's disability, and inspiring the non-disabled to cherish their non-disabled lives. As with all forms of representation, the stories we tell about disabled animals matter. These stories are political. They say a lot about how we conceptualize disability and animality. Many of them, for instance, feature only "cute" domestic animals (e.g. cats and dogs), whose disabilities confirm pre-existing notions that non-disabled people have about what disability looks like (e.g. blindness and wheelchair use). Moreover, these animals are portrayed as living life to the fullest despite their wretched existence, which is to say, in spite of their pitiable and tragic condition. These stories adopt what disability studies scholars call "the medical model of disability," which sees disability as personal tragedy rather than a product of social relations.

Disability activists have been fighting this model since at least the 1960s. The medical model ignores that even if animals are born with atypical embodiments or have various physical or cognitive impairments, what their bodies and capacities mean—that is to say, whether they count as "disabilities" or not, whether they are deemed "natural" or not, whether they are seen as "normal" or not, whether they are rejected or embraced—depends on the social environment. The social model tells us that if we pay more attention to this environment, the line separating ability and disability crumbles. Is a dog who has access to a large yard where he can freely move and play but who happens to use a wheelchair any more "disabled" than a dog who moves without a wheelchair but who is confined to a small space and is not allowed to exercise any control about her own life? Is Mari, the orangutan who uses her legs and chin to move around, more "disabled" than the many great apes who have lived their entire lives under horrifying conditions of captivity? Disability is not a property of individuals. It is a dynamic of social relations.

In her chapter for the volume, *Disability and Animality: Crip Perspectives in Critical Animal Studies*, Sunaura Taylor highlights the value of embracing a social rather than medical understanding of animal disability with a story of her own. She writes:

> A few years ago, I found a story about a fox with arthrogryposis, which is the disability I was born with. According to the Canadian Cooperative Wildlife Health Center, a wildlife conservation and management organization, the fox was shot by a resident of the area because "it had an abnormal gait and appeared sick." The animal, whose disabilities were quite significant, had normal muscle mass, and his stomach contained a large amount of digested food, which suggested to researchers that "the limb deformity did not preclude successful hunting or foraging." The resident seems to have shot the animal out of pity (a sort of mercy killing) and fear (perhaps assuming the fox was sick with a contagious disease).
>
> *(2020, 13)*[5]

Two facts are worth highlighting here. First, the fox appeared sick because he had an atypical gait. In the eyes of the resident, this impairment equaled sickness. And second, the fox was not killed for conservation, food, or sport. He was killed out of "mercy." Convinced that being disabled must be worse than death, the resident decided that the fox should die—as if it were a self-evident truth that living with arthrogryposis is worse than not living at all.[6]

Taylor's story is, in some ways, the inverse of Gruen's. In it we encounter an animal that thrives in nature on its own, but that becomes disabled under the gaze of a human observer, an observer who happens to have at his disposal the technical means to end the animal's life. In what sense is this animal "disabled" before this fateful encounter? To what extent does the human not only kill the fox, but render him killable by seeing him as tragically disabled? Might the resident himself be the ultimate source of the fox's dis-ablement? How different things are in Gruen's story! There, we run into animals that are en-abled rather than dis-abled by their surroundings, animals that—thanks to thoughtful accommodation and intentional inclusion—came to lead rich and happy lives in the company of members of the same species.

IV Service Animals and Emotional Support Animals

In the USA, the American with Disabilities Act (ADA) bans discrimination against disabled people who live with "service animals" (SAs), which are animals that help disabled people achieve all kinds of tasks. Although the passage of the ADA in 1990 was a milestone in the fight for disability liberation, its definition of SAs continues to be a source of controversy. Title III of the Act defines an SA as an animal that (a) benefits a person living with a physical, sensory, psychiatric, intellectual, or other mental disability and (b) has been trained to perform specific tasks related to the handler's disability that the animal would not otherwise perform. It also stipulates that only dogs and miniature horses can be SAs.

Rebecca Huss, a law professor at Valparaiso University, has argued that this definition is overly restrictive. On the one hand, this definition makes problematic assumptions about species membership. Already, this has become a problem for people with disabilities who, for any number of reasons, cannot be around dogs or miniature horses (e.g. because of allergies). On the other hand, since the act requires that an animal be formally trained to perform certain tasks, it ends up excluding a large category of animals that provide invaluable support to people with all kinds of disabilities but that have not received formal training. This includes "emotional support animals" (ESAs) that provide emotional support to people who struggle with conditions like anxiety and depression but that, again, do not meet the ADA's definition of an SA due to the lack of formal training. In the relevant sections, it seems, the ADA framers had a rather narrow view of SAs as animals that help someone physically, such as by fetching objects, opening doors, and guiding them while crossing the street. As a result, animals that help disabled people emotionally are not protected under the ADA, especially if they aren't formally trained in some way. This, Huss worries, only fuels societal skepticism about the "legitimacy" of some disabilities.

Now, there are other statutes that define SAs differently and even incorporate language that is more accommodating of ESAs, such as the Fair Housing Act (FHA) and the Air Carrier Access Act (ACAA). But even in these cases, there is a cloud of confusion hovering over a group of related categories. What is the difference between an SA and an ESA? What about the difference between an ESA and a regular pet? The lack of clarity has created significant social tension as different actors disagree about where the lines between these categories should be drawn and, as a result, about which kinds of disability-related accommodations should be honored and on what grounds.

Take the example of air travel and consider the following newspaper headlines about ESAs on planes, all drawn from major news outlets:

- "Horse Joins Owner on Flight from Chicago as Service Animal," The Guardian.
- "Peacocks, Ducks and Doomed Hamsters: The Wildest Emotional-Support-Animal Travel Stories," *The Washington Post*.
- "Goats? No. Miniature Ponies? Maybe: American Airlines Debuts New Emotional Support Animal Policy," *NBC News*.
- "Iowa Man Wants Coyote Back As Emotional Support Animal," *ABC News*.
- "Emotional Support Pig Kicked Off US Airways Flight," *CNN*.

What can one say about these headlines? To start, they point to a sense of unease about what the rules on airports are concerning ESAs. They are, however, also indicators of a larger social discomfort that goes beyond the need for clear rules and is about something much more fundamental: the relationship between disability and animality and their visibility in public spaces.

These headlines, moreover, are signs of problematic social attitudes about the very notion of an ESA. It is not an accident that all of them allude to the presumably self-evident ridiculousness of ESA accommodations. In a 2018 article entitled "When Pigs Fly: Emotional Support Animals, Service Dogs and the Politics of Legitimacy Across Species Boundaries," Justyna Wlodarczyk analyzed media accounts of ESAs and found that the stories most likely to get traction are those that focus on "limit cases"—the emotional support pig, the emotional support kangaroo, the emotional support peacock. By sensationalizing limit cases, the media perpetuates the idea that there are legitimate and illegitimate disability claims and, what's more, that non-disabled people are the ones who get to judge which are which. The legitimate ones involve "real" animals (e.g. dogs) offering "real" (i.e. physical) support to people with "real" disabilities (e.g. blindness), not "fake" animals (e.g. birds) giving "fake" (i.e. emotional) support to people with "fake" disabilities (e.g. anxiety). According to Wlodarczyk, this fosters a culture of suspicion about disability that has disastrous effects for people with disabilities.

As it has on airports, this controversy has played itself on college and university campuses, where applications for accommodations pertaining to ESAs have been steadily on the rise for more than a decade (Sidhu 2008). Increasingly, students are asking to bring their ESAs to campus, either to provide emotional support during class or to live with them in university housing. In making this demand, they cite a growing body of research that highlights the therapeutic benefits of interacting with animals (Kogan et al. 2016, 275). They also cite the language of the federal statute that governs housing (the FHA) since, as noted above, it includes less restrictive language about what counts as a reasonable accommodation for animals who offer emotional support. How colleges and universities respond to these requests tells a great deal about how they understand disability. Are some disabilities less worthy of recognition than others just because they are not sensory or physical? It also reveals some of the implicit assumptions they might have about the place of animals in public space and about the difference between various classes of animals. What, in the end, is the difference between an SA and an ESA, or between them and a typical animal companion?

Some folks—and this includes some disability rights advocates—have argued that all ESAs should be excluded from anti-discrimination statutes such as the ADA. They worry that this category is too fuzzy and that forcing public or private establishments to accommodate people with ESAs may lead to a more general backlash against the accommodation claims of disabled people who live with ADA-approved SAs. Others—and this also includes some disability rights advocates—have responded that ESAs should be protected under federal law across the board because they offer essential emotional and mental support to people with certain kinds of disabilities who shouldn't have to fall through the cracks of the ADA and its inadequate language concerning SAs and ESAs (Ligatti 2009). Still others have tried to find a middle ground between these positions, with a handful of people suggesting that the ADA should recognize ESAs but limit the kinds of animals that can receive this recognition to those animals already included as possible SAs (i.e. dogs and miniature horses).

V Conclusion

In this short chapter, I have only discussed three areas in which experts on critical disability and critical animal studies are currently working. There are others, but these underscore the point that these discourses can be mutual allies. Together, they can help us gain a better understanding of how ableism and speciesism move through the arteries of society, how they mold the lives of disabled humans but also of non-human animals, and how they create problems that can only be tackled collectively and intersectionally.

Notes

1 Historically, these two fields have had a tense relationship given that (a) the concept of "animality" has been used to dehumanize the disabled and (b) the concept of "disability" has also been used to delegitimize the claims of animal rights activists, who are often "medicalized" (Wrenn et al. 2015).

2 Above all, what Donaldson and Kymlicka find appealing about these developments is that they apply the concept of citizen to individuals traditionally excluded from it, thereby challenging the classical assumption that citizenship requires advanced rational or linguistic capacities.

3 See the chapters under Part II of Jenkins, Struthers, and Taylor (2020).

4 This is typically opposed to the "medical model of disability," which locates disability *inside* the body of the disabled rather than in *the relationship* between this body and its environment (Shakespeare 2006).

5 According to the original report from the Canadian Cooperative Wildlife Health Center from 1999, at the time of death the fox had a normal weight and body mass, and his stomach contained "the remains of two rodents and bones from a larger mammal mixed with partially digested apple."
6 See Reynolds (2018) for an analysis of mercy.

References

Donaldson, Sue, and Will Kymlicka. 2011. *Zoopolis: A Political Theory of Animal Rights.* Oxford: Oxford University Press.

———. 2017. "Rethinking Membership and Participation in an Inclusive Democracy: Cognitive Disability, Children, Animals." In *Disability and Political Theory*, edited by Barbara Arneil and Nancy Hirschmann, pp. 168–197. Cambridge: Cambridge UP.

Jenkins, Stephanie, Kelly Struthers Montford, and Chloë Taylor, eds. 2020. *Disability and Animality: Crip Perspectives in Critical Animal Studies.* New York: Routledge.

Kogan, Lori R., Karen Schaefer, Phyllis Erdman, and Regina Schoenfeld-Tacher. 2016. "University Counseling Centers' Perceptions and Experiences Pertaining to Emotional Support Animals." *Journal of College Student Psychotherapy* 30.4: 268–283.

Ligatti, Christopher C. 2009. "No Training Required: The Availability of Emotional Support Animals as a Component of Equal Access for the Psychiatrically Disabled under the Fair Housing Act." *Thurgood Marshall Law Review* 35: 139.

Reynolds, Joel Michael. 2018. "Killing in the Name of Care." *Levinas Studies* 12: 141–164.

Shakespeare, Tom. 2006. "The Social Model of Disability." In *The Disability Studies Reader*, 2nd edition, edited by Lennard J. Davis, pp. 197–204. New York: Routledge.

Sidhu, Dawinder S. 2008. "Cujo Goes to College: On the Use of Animals by Individuals with Disabilities in Postsecondary Institutions." *University of Baltimore Law Review* 38: 267.

Taylor, Sunaura. 2020. "Animal Crips." In *Disability and Animality: Crip Perspectives in Critical Animal Studies*, edited by Stephanie Jenkins, Kelly Struthers Montford, and Chloë Taylor, pp. 13–34. New York: Routledge.

Wlodarczyk, Justyna. 2019. "When Pigs Fly: Emotional Support Animals, Service Dogs and the Politics of Legitimacy Across Species Boundaries." *Medical Humanities* 45.1: 82–91.

Wrenn, Corey Lee, Joanne Clark, Maddie Judge, Katherine A. Gilchrist, Delanie Woodlock, Katherine Dotson, and Riva Spanos et al. 2015. "The Medicalization of Nonhuman Animal Rights: Frame Contestation and the Exploitation of Disability." *Disability & Society* 30.9: 1307–1327.

Further Reading

Bentley, Judy K.C., et al. 2017. *The Intersectionality of Critical Animal, Disability, and Environmental Studies: Toward Eco-ability, Justice, and Liberation.* Lexington, KY: Lexington Books.

Grandin, Temple, and Catherine Johnson. 2009. *Animals in Translation: Using the Mysteries of Autism to Decode Animal Behavior.* Albany, NY: SUNY Press.

Kim, Claire Jean. 2015. *Dangerous Crossings.* Cambridge: Cambridge University Press.

Nussbaum, Martha C. 2009. *Frontiers of Justice: Disability, Nationality, Species Membership.* Cambridge, MA: Harvard University Press.

Oliver, Kelly. 2016. "Service Dogs: Between Animal Studies and Disability Studies." *PhiloSOPHIA* 6.2: 241–258.

Taylor, Sunaura. 2017. *Beasts of Burden: Animal and Disability Liberation.* New York: The New Press.

PART X

The Ends of Medicine

Caring, Curing, and Justice

33

IMPROVING ACCESS WITHIN THE CLINIC

Nicole D. Agaronnik and Lisa I. Iezzoni

Now the other myth that gets around is the idea that legislation cannot really solve the problem … because you've got to change the heart and you can't change the heart through legislation. You can't legislate morals…. [W]hile it may be true that morality cannot be legislated, behavior can be regulated…. There is a need for civil rights legislation…

Martin Luther King Jr.
December 18, 1963
Western Michigan University (Dawes 2020)

Fifteen years after passage of the Americans with Disabilities Act (ADA), the USA Surgeon General issued a Call to Action to Improve the Health and Wellness of Persons with Disabilities, reinforcing the "need to promote accessible, comprehensive health care that enables persons with disabilities to have a full life in the community" (Office of the Surgeon General 2005). The Surgeon General underscored the imperative for people with disabilities to receive not only necessary healthcare services but also wellness care to promote overall health and well-being. The Call to Action recognized that, despite the ADA's mandates for equity, people with disability continued to receive inequitable care.

Health care is essential for everyone, including people with disabilities. However, many complex factors contribute to the healthcare disparities frequently experienced by people with disability, even inequities in the routine screening and preventive services recommended by the US Preventive Services Task Force. Nearly 61 million Americans have a disability (Okoro et al. 2018), and this population is expected to rise given aging of "baby boomers," the growing longevity of persons with disabling congenital or early-life impairments, and increasing prevalence of chronic conditions even among younger people. The population of people with disability is diverse, and the nature and extent of healthcare disparities vary across subgroups. Specific healthcare needs also differ depending upon one's underlying condition(s) and their relationship to functional impairment(s).

This chapter explores issues relating to improving healthcare access for people with disability and mitigating current disparities. As noted, multiple and diverse factors underlie these inequities. At various points throughout this chapter, we introduce findings from our nationwide survey of 714 practicing physicians across seven specialties, which was conducted from October

DOI: 10.4324/9781003289487-44

2019 through March 2020. This was the first-ever survey of USA physicians about their perceptions of and experiences with caring for patients with disability, and the results provide timely insights about current factors affecting access to care for this population.

I Healthcare Disparities and People with Disability

Healthy People 2010, the 2000 installment of the decennial initiative which sets USA public health priorities for the upcoming decade, was the first federal report to recognize disparities in health care for people with disability, attributing them to erroneous assumptions about health and wellness preferences among people with disability. Ten years later, Healthy People 2020 also recognized persons with disability as a population experiencing disparities, attributing these inequities to disadvantages that people with disability experience in what are called "social determinants of health." People with disability are more likely to experience disadvantages in these social determinants, including lower education and income levels, higher rates of unemployment, lower rates of health insurance, and higher rates of inability to afford health care (Table 33.1).

TABLE 33.1 Sociodemographic Characteristics and Social Determinants of Health by Disability Status

	Disability status						
	None	Any	Mobility	Vision	Hearing	Cognitive	Self-care
Sociodemographic characteristic	*Age-adjusted prevalence (%)*						
Education level							
Some high school or less	9.7	21.3	24.5	27.6	23.2	23.8	30.3
High school or less	57.3	64.1	62.5	60.5	61.4	63.7	59.3
College graduate	32.9	14.6	13.0	11.9	15.4	12.5	10.4
Employment status							
Employed	66.4	41.8	30.5	39.9	49.4	37.7	22.2
Out of work	4.2	8.8	8.9	9.1	6.8	9.7	8.7
Unable to work	1.7	21.1	34.8	24.3	17.0	25.4	47.9
Other	27.7	28.3	25.7	26.6	26.7	27.1	21.2
Income level							
<$15,000	6.6	20.7	26.6	27.1	18.2	24.9	31.3
$15,000 to <$25,000	12.9	25.5	28.2	27.8	23.3	27.2	30.1
$25,000 to <$35,000	9.5	11.8	11.4	12.5	11.3	11.6	11.5
$35,000 to <$50,000	13.0	12.4	11.1	10.3	13.3	11.3	9.4
$50,000+	58.0	29.6	22.7	22.2	33.8	25.1	17.8
Marital status							
Married/unmarried couple	58.3	43.7	43.9	41.5	50.7	39.1	40.1
Divorced/separated	10.8	18.0	21.6	19.5	18.1	19.8	24.4
Widowed	4.7	7.8	8.7	8.9	7.9	8.3	8.6
Never married	26.2	30.5	25.8	30.1	23.3	32.9	26.8
Access to care							
Have healthcare coverage	88.1	84.2	85.7	80.5	83.9	84.0	—
Could not see a doctor due to cost in past 12 months	9.7	25.6	28.0	29.3	24.8	29.4	—

Data source: 2018 Behavioral Risk Factor Surveillance System Survey.

In addition to socioeconomic disadvantages, people with disability have higher rates of health problems or risk factors. People with disability are more likely to report smoking, obesity, depression, and histories of chronic conditions (Table 33.2). Federal survey data from 2010 to 2017 found that people with pre-existing disability have higher rates of four common cancer types (ovarian, prostate, colorectal cancers, and non-Hodgkin's lymphoma) than non-disabled people (Iezzoni et al. 2020). Reasons for this remain unclear, but some research suggests that people with disability may experience earlier physiological aging than others.

Health disparities are further amplified for people with disability with intersectional identities (having a disability and belonging to one or more population[s] experiencing health disparities). According to the 2018 Behavioral Risk Factor Surveillance System data (BRFSS), disability prevalence is highest among non-Hispanic American Indian/Alaska Natives (40%) and lowest among non-Hispanic Asians (16%) (Centers for Disease Control and Prevention 2018). People with disability who identify as a racial/ethnic minority and/or sexual orientation/ gender identity minority may experience greater barriers to health care and social disadvantages that are unique to their multi-identity backgrounds.

The relative dearth of information on this population hampers efforts to quantify and understand healthcare disparities experienced by people with disability. Notably, standard diagnosis codes and enrollment information in large administrative data files – which informed early studies on disparities by race and ethnicity – do not indicate whether individuals have a disability, or if so, its level of severity. Administrative markers for disability, such as whether someone's original entitlement for Medicare resulted from disability (i.e., having Social Security Disability Insurance), do not provide insights into the current disability status of the individuals. Most information about healthcare disparities for people with disability comes from federal surveys, but these data sources focus almost exclusively on use of screening and preventive services or participants' experiences with care. Relatively little information is available about other problems in care, such as delays in diagnosis or treatment differences.

That said, notable examples of healthcare disparities for people with disability, over time – since shortly after passage of the ADA to recent years – include the following:

- **Papanicolaou (Pap) tests.** The U.S. Preventive Services Task Force (USPSTF) recommends Pap tests (Grade A level recommendation) for cervical cancer screening every three years for women aged 21–65. Analysis of 1994 National Health Interview Survey (NHIS) data found that, compared to non-disabled women aged 18 and over, women were less likely to receive Pap testing within the previous three years (76% versus 65% for women with one or two functional limitations and 61% for women with three or more limitations, respectively) (Centers for Disease Control and Prevention 1998). Analysis of 1998–2010 NHIS data demonstrated that Pap testing rates had not substantially changed over time, with screening rates for women without disability remaining around 84–87%; among women with more severe movement difficulty, 65% reported Pap testing in 2010 (Iezzoni, Kurtz, and Rao 2016a). Analysis of 2018 BRFSS data indicates that 78% of women with disability aged 21–65 years are up to date with cervical cancer screening recommendations, compared to 84% of women without disability (Table 33.3).
- **Screening mammography.** The USPSTF recommends biennial screening mammography (Grade B level recommendation) for women aged 50–74. Analysis of 1998–2010 NHIS data found that screening rates have remained approximately 50% lower for women with movement disability than for non-disabled women, failing to decrease over time (Iezzoni, Kurtz, and Rao 2015). A systematic review suggests that mammography screening rates

TABLE 33.2 Health Risks and Other Conditions by Disability Status

	Disability status					
	None	Any	Mobility	Vision	Hearing	Cognitive
Health risk or behavior	Age-adjusted percent (%)					
Smoking status						
Current smoker	12.7	26.2	29.3	27.1	26.8	29.3
Former smoker	22.1	24.9	26.2	22.4	28.4	24.1
Never smoker	65.2	49.0	44.5	50.5	44.8	46.6
Body mass index category						
Underweight	1.7	2.7	2.2	2.5	1.9	3.1
Normal weight	34.5	28.3	21.5	31.0	26.1	30.0
Overweight	36.3	29.8	27.2	28.6	32.7	29.9
Obese	27.6	39.1	49.1	37.9	39.4	37.0
Health condition						
Depression[a]	11.6	42.0	44.7	37.5	34.0	57.5
Heart disease[a]	4.0	11.0	14.5	13.8	12.3	11.9
High blood pressure[a,b]	25.9	41.9	51.2	43.7	43.6	41.3
High cholesterol[a,b]	26.8	38.1	43.6	40.2	38.8	40.4
Stroke[a]	1.5	6.6	9.3	9.6	7.0	8.6
Arthritis[a]	17.2	39.9	56.8	36.9	39.8	41.0
Diabetes[a]	7.4	16.7	22.5	19.7	15.7	17.0
Cancer[a,c]	5.2	8.7	11.1	9.5	8.6	9.1
Chronic obstructive pulmonary disease[a]	3.3	13.8	19.7	16.9	14.2	16.0

Data source: 2018 Behavioral Risk Factor Surveillance System (BRFSS) Survey.

[a] Participant reports ever having had health condition.

[b] 2017 BRFSS data.

[c] Excluding skin cancer.

Note: This table appears at the end of the chapter, after Table 33.1.

TABLE 33.3 Cancer Screening Test Rates by Disability Status

	Disability status					
	None	Any	Mobility	Hearing	Vision	Cognitive
Screening test	Percent (%)					
Mammogram in the past two years among females 50–74 years of age	81.1	74.4	74.6	75.9	70.7	71.9
Up-to-date on cervical cancer screening among females 21–65 years of age	83.9	78.1	78.0	77.3	75.3	77.9
Up-to-date colorectal cancer screening among adults 50–75 years of age	68.9	67.3	68.1	70.6	59.8	65.5

Data source: 2018 Behavioral Risk Factor Surveillance System Survey.

Note: This table appears at the very end of the chapter, after Table 33.2.

among women with disability decrease with increasing disability severity (Andresen et al. 2013). Analysis of 2018 BRFSS data shows that 74% of women with disability aged 50–74 years have had a mammogram in the past two years, compared with 81% of women without disability (Table 33.3).

- **Colorectal cancer screening.** USPSTF recommends colorectal cancer screening (Grade A level recommendation) for people aged 50–75 years. Research examining disability disparities for colorectal cancer screening has produced inconsistent findings. For instance, 1998–2010 NHIS data suggest a decrease in colorectal screening disparities over time (Iezzoni, Kurtz, and Rao 2016b). However, 2013 NHIS data identified an association between increasing severity of mobility disability and lower screening rates (Gofine et al. 2018). Analysis of 2018 BRFSS data found that 67% of people with disability aged 50–75 years are up to date with colorectal cancer screening, compared to 69% of people without disability – not a meaningful difference (Table 33.3).

- **Care during pregnancy.** Analysis of the 2002–2011 Rhode Island Pregnancy Risk Assessment Monitoring System survey suggests that women with disability are more likely than non-disabled women to report stressful events and medical complications during pregnancy, experience lower likelihood of receiving first trimester prenatal care, and have higher likelihood of preterm births (Mitra et al. 2015). In-depth interviews with women with disability identified unmet needs for pain relief during labor and delivery, including inadequate planning for anesthesia complications that occur in the context of complex disability (Long-Bellil et al. 2017).

- **Knowledge of obstetrical practitioners.** Analysis of a 2015 national survey of women with physical disability found that 53% felt that their disability was an important factor in identifying an appropriate obstetrical care practitioner. However, 40% of respondents believed that their provider lacked understanding of how disability may affect pregnancy (Mitra, Akobirshoev, et al. 2017). In-depth interviews with obstetrical practitioners confirmed their lack of preparedness for providing maternity care to women with disability, as well as their unwillingness to provide this care, inaccessible practices, time and cost limitations, and insufficient disability-specific data to guide clinical practice (Mitra, Smith, et al. 2017).

- **Cancer care.** Surveillance, Epidemiology, and End Results (SEER) data suggest that women with disability under age 65 diagnosed with breast cancer between 1988 and 1999 were 20% less likely than non-disabled women to undergo breast-conserving surgery; of women who received such surgery, women with disability were 17% less likely to undergo radiotherapy. Furthermore, women with disability experience lower breast cancer survival rates (McCarthy et al. 2006). Similarly, analysis of SEER data identified that people under age 65 with disability diagnosed with non-small cell lung cancer between 1988 and 1999 were less likely than non-disabled persons to undergo surgical treatment (69% versus 82%) and experienced higher cancer-specific mortality (Iezzoni et al. 2008).

- **Physical accessibility of care.** Women with mobility disability who developed breast cancer described inaccessible exam tables, weight scales, and diagnostic equipment including mammography machines impeding their care (Iezzoni, Kilbridge, and Park 2010). Women with mobility disability receiving prenatal care also report inaccessible tables and weight scales (Iezzoni et al. 2015). A 2013 survey of 256 subspecialty practices found that 22% of practices could not accommodate a person with a mobility disability, 18% could not transfer a patient from a wheelchair to an exam table, and only 9% reported use of a height-adjustable exam table or lift device for transferring. The subspecialty with highest rate of inaccessibility was gynecology (Lagu et al. 2013).

- **Rural locations.** Analysis of 2002–2008 data from the Medical Expenditure Panel Survey found that women with mobility disability living in rural areas were less likely than non-disabled women to receive timely breast and cervical cancer screening (Horner-Johnson, Dobbertin, and Iezzoni 2015). Focus groups with people with disability living in rural areas identified barriers including old, inaccessible medical facilities, inadequate transportation, the need to travel to distant urban medical centers for specialty care, and difficulty identifying primary care physicians because of inadequate medical insurance (Iezzoni, Killeen, and O'Day 2006).
- **Satisfaction with care.** Focus groups with people with and without disability suggest that people with disability are less satisfied with care. Factors contributing to decreased satisfaction include poor coordination among providers, short visit times with providers, difficulties with insurance, physical access barriers (i.e., with facilities and transportation), and feeling invisible or being viewed as incompetent by providers (de Vries McClintock et al. 2016).

In 2013, the Agency for Healthcare Research and Quality released the National Healthcare Quality Report, which assessed progress toward quality improvement benchmarks for populations experiencing disparities. The report included people with basic activity limitations (defined as "problems with mobility, self-care, domestic life, or activities that depend on sensory functioning") and complex activity limitations (defined as "limitations experienced in work or in community, social, and civic life"). While 62% of quality benchmarks met improvement goals for people without activity limitations, only 36% improved for people with basic activity limitations and 21% improved for people with complex activity limitations (Agency for Healthcare Research and Quality 2014). Indeed, quality benchmark improvement rates were much lower for persons with disability than for racial/ethnic minorities: 42% of quality benchmarks met improvement goals for American Indians/Alaska Natives, as did 57% for Black, 57% for Asian, and 58% for Hispanic people (Agency for Healthcare Research and Quality 2014).

In our 2019–2020 nationwide survey of practicing physicians, we found that 68% of participants report that people with disability are treated unfairly in the USA healthcare system. Participants also believe that quality of care for people with disability is worse than care for non-disabled patients. However, this perception varies by disability type: 77% of physicians report that people with serious mental illness get worse quality of care, while 69% perceive substandard care for people with intellectual disability, 55% for people with hearing limitations, 56% for people with mobility limitations, and 54% for people with vision limitations.

II Contributors to Disparities and Implications for Solutions

Multiple factors contribute to the healthcare disparities experienced by people with disability, including social determinants of health, coexisting health problems and risk factors, and numerous considerations related to healthcare professionals, healthcare delivery systems, delivery settings, and financial access to care (e.g., health insurance coverage). Our survey of physicians nationwide found they commonly pointed to very practical constraints in caring for people with disability, including: lack of time during appointments (78%); lack of reimbursement for the additional time it takes to care for patients with disability (64%); lack of formal education/training about disability (77%); lack of funds to purchase special equipment (80%); and lack of physical space in their practices to accommodate patients with disability (69%).

These findings suggest that improving care for people with disability will require not only improving the training of physicians about caring for this population but also increasing resources – time, reimbursement, equipment, and physical space. Given concerns about high costs of care in the USA in general and intractable political debates about health insurance coverage, solving some of the barriers to equitable care for people with disability will be challenging. Below we consider selected factors contributing to healthcare disparities for people with disability and what needs to be done.

At the outset, however, it is important to note that individual people with disability – just like non-disabled people – have their own preferences for care based on their cultures and values and what is happening in their lives. Given their personal health situations, individuals need to balance competing risks. For example, although the USPSTF highly recommends screening colonoscopy, the physical demands of the preparatory bowel regimen may be too arduous or pose health risks for people with severe physical disability, especially for those without family histories of colon cancer. Therefore, we focus here on systems issues that affect populations.

III Social Determinants of Health

Disadvantages in social determinants of health contribute to healthcare disparities for people with disability, as noted above. To address these disparities, Healthy People 2020 recommended improving education, employment, social participation, access to assistive technologies and other supports, as well as using universal design concepts to remove accessibility barriers, promoting the inclusion of people with disability in public health data collection, health promotion activities, and expanding disability training opportunities for health professionals (Centers for Disease Control and Prevention 2016). It is beyond our scope here to detail the many steps necessary to make these massive social changes, other than to note that – under principles of universal design – changes that assist people with disability will likely benefit many non-disabled people as well.

IV Lack of Scientific Evidence to Inform Healthcare Decisions

Persons with disability are generally excluded from randomized, controlled clinical trials, which are the "gold standard" for assessing treatment options. Insufficient clinical evidence therefore often exists to guide clinical decision-making in diagnosing and treating persons with disability. People with disability may be excluded from clinical research for both explicit reasons (e.g., poorly justified exclusion criteria) and implicit reasons (e.g., inaccessible study documents, interventions, or other research measures) (Rios et al. 2016). A review examining rates of poorly justified exclusion criteria in 283 randomized control trials published in high-impact journals between 1994 and 2006 found that 81% excluded participants because of medical comorbidities, 11% of trials excluded participants because of physical disability or functional status, and 8% excluded participants because of intellectual disability (Van Spall et al. 2007). Another study that examined exclusion criteria across 300 randomized control trials published between 2007 and 2011 in high-impact medical journals found that over 90% of studies were designed to automatically exclude people with intellectual disability, even though at least 70% could have included people with intellectual disability with simple accommodations or other procedural modifications without compromising the research design (Feldman et al. 2014).

V Inaccessible Healthcare Delivery Settings and Medical Diagnostic Equipment

Despite the ADA, many aspects of healthcare delivery settings – and medical diagnostic equipment (which is technically not covered by ADA mandates) – remain physically inaccessible. Many disability disparities in health care are impossible to remedy without physical accessibility (e.g., height-adjustable tables, wheelchair-accessible scales, mammography machines). The national survey of actively practicing physicians found that only 18% "always" or "usually" used a wheelchair accessible weight scale to weight patients with mobility disability; 16% of physicians used the weight previously recorded in a patient's medical record, while 33% asked patients how much they weigh. Only 36% of physicians reported "always" or "usually" using automatic height adjustable exam tables when examining patients with mobility disability who could not independently transfer, and 39% of physicians "always" or "usually" required assistance from the person(s) accompanying the patient (under the ADA, requiring patients to provide their own assistants is illegal).

In January 2017, the Architectural and Transportation Barriers Compliance Board and U.S. Food and Drug Administration published Standards for Accessible Medical Equipment in the Federal Register (relating to exam tables and chairs, weight scales, gurneys or stretchers, diagnostic imaging equipment, and mammography machines). Despite this, accessible medical diagnostic equipment is not widely available across health settings. Indeed, the national physician survey found that 68% of physicians believe their practice is at risk of an ADA lawsuit because of problems providing reasonable accommodations for patients with disability. Implementation of these medical diagnostic equipment accessibility standards was stalled by failure to enact federal rules regarding these standards. In December 2017, the USA Department of Justice formally withdrew rulemaking that would have provided explicit regulations concerning the availability of accessible medical diagnostic equipment in healthcare settings (Iezzoni and Pendo 2018). Without such rulemaking, it is unclear how widely healthcare providers will voluntarily acquire equipment that meets federal accessibility standards.

Reasons for lack of widespread accessibility across health settings include concerns that accessible equipment and architectural changes may cost too much, especially for private practice settings or practices located in older buildings that may be difficult to modify. Contrary to assumptions about undue expense, most modifications cost less than expected, and clinical practices can view accessibility as an investment that allows them to serve the growing population of people with disability. However, if practices are truly inaccessible, the ADA requires them to make arrangements to provide equivalent services to the patient elsewhere (Agaronnik et al. 2019c).

The national survey highlights varying reasons for physical barriers in clinical settings. When asked about reasons for not being able to transfer patients with significant mobility disability onto exam tables or exam chairs, 57% described inadequate staffing as a reason; additional reasons included no height adjustable exam table or chair (66%), no lift device (86%), patient refusal to be transferred (83%), fear of injury to self or staff (73%), fear of injury to patient (91%), fear of legal liability or exposure (59%), and the additional time it takes (61%). In addition to these physical barriers, healthcare settings often fail to provide accommodations to ensure effective communication – as required under the ADA – for patients who are deaf or hard of hearing, are blind or have low vision, or have intellectual disability or other communication disorders. When asked about reasons for not providing communication accommodations, one study found that physicians worried about the additional time required to communicate with these patients, as well as discordance between patients' preferences for specific communication modalities and accommodations that physicians considered appropriate (Agaronnik et al. 2019b). Physicians clearly require more training about options for improving effective communication with patients – and their obligations to consider patients' preferences.

VI Physician Training

Medical education does not routinely address disability. Kirschner and Curry (2009) recommend six core competencies to incorporate into medical education to improve preparedness of physicians in caring for people with disability: (1) conceptual framework of disability in the context of human diversity, illness, and the lifespan; (2) skills for assessing level of disability, as well as functional consequences of illness and the implications of physical and social environments for treatment and management; (3) general principles for interacting with patients with disability (e.g., etiquette, language); (4) knowledge of interdisciplinary care teams as they relate to people with disability, as well as disability-specific resources available for patients; (5) legal requirements under the ADA, including key concepts of universal design as they relate to patient care; and (6) patient-centered care and understanding quality of life from patients' perspectives (Kirschner and Curry 2009).

Few practicing physicians in the USA identify as having a disability. Our national survey of practicing physicians indicates that only 4% report living with a significant limitation that requires accommodation(s) to perform their job as a physician. Despite abundant research suggesting that quality of care improves when core personal attributes, such as gender, race, or language, of physicians match those of their patient population, barriers remain for people with disability to pursue careers as healthcare providers. The requirement for medical students to meet technical standards is a substantial barrier for students with disability interested in medical training. Technical standards aim to ensure that the physician workforce can perform essential functions of the job, with or without reasonable accommodation, as required by Title I of the ADA.

However, technical standards for medical education are inconsistent across USA medical schools and other training settings. Furthermore, they do not necessarily reflect the essential functions that physicians are required to perform, depending on their chosen specialty. Unlike other professional careers such as law, where critical knowledge varies somewhat from state to state, core components of medical knowledge and the fundamentals of providing quality patient care are comparable across settings. Given the wide-ranging demands of different specialties, requiring all graduating students to meet uniform technical requirements no longer seems reasonable. To increase the representation of people with disability in medical schools, the profession should revisit technical standards requirements. It is important to recognize, however, that simply increasing the numbers of people with disability entering the medical profession will not eliminate disability disparities, given the systemic barriers to patients with disability in healthcare settings.

VII Lack of Knowledge about the ADA

In-depth interviews with practicing physicians reveal limited knowledge of legal obligations in providing equitable care for persons with disability in three key areas: deciding which accommodations practices should implement, refusing patients with disability, and holding patients accountable for cost of accommodations (Agaronnik et al. 2019c). The national survey found that only 14% of practicing physicians report knowing "a lot" about the ADA, 51% know "some," 28% know "a little," and 8% report knowing "nothing" about the ADA. When asked about who is responsible for determining reasonable accommodations for a patient being cared for in their practice, only 28% gave the correct response – that it is a collaborative effort between patients/family and physician(s) caring for the patient and/or practice staff/managers/administrators. When asked about who is responsible for paying for reasonable accommodations that patients with disability receive, 21% responded incorrectly that it is the responsibility of the patients/family or insurers/payers; these costs are the responsibility of the clinical practice. In-depth interviews with physicians found that some believe they can refuse a patient who they view as too challenging to accommodate. This is incorrect: under the ADA, physicians cannot refuse a patient based on disability, even if it may require extra time and resources to accommodate the patient (Agaronnik et al. 2019c).

VIII Erroneous Assumptions and Stigmatized Attitudes

Finally, erroneous assumptions about the lives and preferences of persons with disability and potentially stigmatized attitudes among providers about disability also contribute to healthcare disparities. For many healthcare providers, what has been called the "disability paradox" captures the rationale behind these erroneous assumptions "Why do many people with serious and persistent disabilities report that they experience a good or excellent quality of life when to most external observers these people seem to live an undesirable daily existence?" (Albrecht and Devlieger 1999). Findings from our national survey of actively practicing physicians found that 83% believe that people with significant disability experience worse quality of life compared to people without disability. These startling findings are in stark contrast with the way many people with disability perceive their quality of life.

These misconceptions about the quality of life of people with disability have influenced disability training approaches in medical education. One example is the use of disability simulation methods where students are required to mimic functional status limitations to understand the experience of living with various disabilities. These training techniques promote beliefs that living with disability causes distress, perpetuating fear and pity of this population. The absence of formal training on legal obligations under the ADA and requirements for providing equitable care compounds the lack of preparedness to care for a diverse disability population. Furthermore, medical education fails to include disability cultural competency training, including societal perceptions of disability and more specific considerations such as language usage (e.g., person-first versus identity-first language). Disability cultural competency may influence the quality of relationships between physicians and patients with disability (Agaronnik et al. 2019a).

Without adequate disability training in medical education, physicians may perpetuate implicit and explicit biases that affect quality of care. For example, the use of quantitative measures such as quality-adjusted life years (QALYs) to assess the costs/benefits of certain treatments and interventions is based on assumptions that quality of life decreases with increasing levels of disability. Clinical decisions about appropriate health interventions may therefore be biased by calculations falsely perceived as objective quantitative assessments. Times of crisis may amplify these biases, such as during the COVID-19 pandemic, when critical decisions potentially involved rationing of scarce medical resources and who should receive life-sustaining treatment.

IX Conclusion

The numbers of persons with disability will grow substantially in coming years, and virtually every health setting can expect to care for persons with disability – across the lifespan.

Nevertheless, three decades since the passage of the ADA, people with disability continue to face barriers for equitable care, including inaccessible health settings, inadequate physician training in caring for various aspects of disability, and misconceptions about quality of life that affect care quality. The ADA's legal mandates alone have not changed "the hearts" of healthcare providers in caring for patients with disability, given long-standing stigmatization of this population. Healthcare providers along the continuum of care must gain the knowledge and tools for accommodating patients with disability. Society should also facilitate access for people with disability to pursue health professions, including careers as physicians. Eliminating healthcare disparities for people with disability remains unfinished business.

References

PLEASE NOTE: Manuscripts reporting findings from our nationwide survey of physicians were being prepared as we wrote this chapter. Numerical results that appear in the published articles differ slightly from numbers reported here because of statistical weighting and final analytic decisions.

Agaronnik, Nicole, Eric G. Campbell, Julie Ressalam, and Lisa I Iezzoni. 2019a. "Exploring Issues Relating to Disability Cultural Competence among Practicing Physicians." *Disability and Health Journal* 12(3): 403–410.

———. 2019b. "Communicating with Patients with Disability: Perspectives of Practicing Physicians." *Journal of General Internal Medicine,* 34(7): 1139–1145.

———. 2019c. "Knowledge of Practicing Physicians about Their Legal Obligations When Caring for Patients with Disability." *Health Affairs (Project Hope)* 38(4): 545–553.

Agency for Healthcare Research and Quality. 2014. *2013 National Healthcare Quality Report.* Rockville, MD: US Department of Health and Human Services.

Albrecht, Gary L., and Patrick J Devlieger. 1999. "The Disability Paradox: High Quality of Life against All Odds." *Social Science & Medicine* 48(8): 977–988.

Andresen, Elena M., Jana J. Peterson-Besse, Gloria L. Krahn, Emily S. Walsh, Willi Horner-Johnson, and Lisa I. Iezzoni. 2013. "Pap, Mammography, and Clinical Breast Examination Screening among Women with Disabilities: A Systematic Review." *Women's Health Issues : Official Publication of the Jacobs Institute of Women's Health* 23(4): e205–e214.

Centers for Disease Control and Prevention. 1998. "Use of Cervical and Breast Cancer Screening among Women with and without Functional Limitations—United States, 1994-1995." *MMWR. Morbidity and Mortality Weekly Report* 47(40): 853–856.

———. 2016. *Healthy People 2020 — Disability and Health.* Department of Health and Human Services. https://www.healthypeople.gov/2020/topics-objectives/topic/disability-and-health.

———. 2018. "Disability and Health Data System." 2018. https://dhds.cdc.gov.

Dawes, Daniel E. 2020. *The Political Determinants of Health.* Baltimore, MD: Johns Hopkins University Press.

Feldman, M. A., J. Bosett, C. Collet, and P. Burnham-Riosa. 2014. "Where Are Persons with Intellectual Disabilities in Medical Research? A Survey of Published Clinical Trials." *Journal of Intellectual Disability Research : JIDR* 58(9): 800–809.

Gofine, Miriam, Thelma J. Mielenz, Sowmya Vasan, and Benjamin Lebwohl. 2018. "Use of Colorectal Cancer Screening among People with Mobility Disability." *Journal of Clinical Gastroenterology* 52(9): 789–795.

Horner-Johnson, Willi, Konrad Dobbertin, and Lisa I Iezzoni. 2015. "Disparities in Receipt of Breast and Cervical Cancer Screening for Rural Women Age 18 to 64 with Disabilities." *Women's Health Issues : Official Publication of the Jacobs Institute of Women's Health* 25(3): 246–253.

Iezzoni, Lisa I., Kerry Kilbridge, and Elyse R. Park. 2010. "Physical Access Barriers to Care for Diagnosis and Treatment of Breast Cancer among Women with Mobility Impairments." *Oncology Nursing Forum* 37(6): 711–717.

Iezzoni, Lisa I., Mary B Killeen, and Bonnie L. O'Day. 2006. "Rural Residents with Disabilities Confront Substantial Barriers to Obtaining Primary Care." *Health Services Research* 41(4 Pt 1): 1258–1275.

Iezzoni, Lisa I., Stephen G. Kurtz, and Sowmya R. Rao. 2015. "Trends in Mammography over Time for Women with and without Chronic Disability." *Journal of Women's Health (2002)* 24(7): 593–601.

———. 2016a. "Trends in Pap Testing over Time for Women with and without Chronic Disability." *American Journal of Preventive Medicine* 50(2): 210–219.

———. 2016b. "Trends in Colorectal Cancer Screening over Time for Persons with and without Chronic Disability." *Disability and Health Journal* 9(3): 498–509.

Iezzoni, Lisa I., Long H. Ngo, Donglin Li, Richard G. Roetzheim, Reed E. Drews, and Ellen P. McCarthy. 2008. "Treatment Disparities for Disabled Medicare Beneficiaries with Stage I Non-Small Cell Lung Cancer." *Archives of Physical Medicine and Rehabilitation* 89(4): 595–601.

Iezzoni, Lisa I., and Elizabeth Pendo. 2018. "Accessibility of Medical Diagnostic Equipment - Implications for People with Disability." *The New England Journal of Medicine* 378(15): 1371–1373.

Iezzoni, Lisa I., Sowmya Rao, Nicole Agaronnik, and Areej El-Jawahri. 2020. "Cross-Sectional Analysis of the Associations between Four Common Cancers and Disability." *Journal of the National Comprehensive Cancer Network* 18(8): 1031–1044.

Iezzoni, Lisa I., Amy J. Wint, Suzanne C. Smeltzer, and Jeffrey L. Ecker. 2015. "Physical Accessibility of Routine Prenatal Care for Women with Mobility Disability." *Journal of Women's Health* 24(12): 1006–1012.

Kirschner, Kristi L., and Raymond H. Curry. 2009. "Educating Health Care Professionals to Care for Patients with Disabilities." *JAMA* 302(12): 1334–1335.

Lagu, Tara, Nicholas S. Hannon, Michael B. Rothberg, Annalee S. Wells, K. Laurie Green, McAllister O. Windom, and Katherine R. Dempsey, et al. 2013. "Access to Subspecialty Care for Patients with Mobility Impairment: A Survey." *Annals of Internal Medicine* 158(6): 441–446.

Long-Bellil, Linda, Monika Mitra, Lisa I. Iezzoni, Suzanne C. Smeltzer, and Lauren D. Smith. 2017. "Experiences and Unmet Needs of Women with Physical Disabilities for Pain Relief during Labor and Delivery." *Disability and Health Journal* 10(3): 440–444.

McCarthy, Ellen P., Long H. Ngo, Richard G. Roetzheim, Thomas N. Chirikos, Donglin Li, Reed E. Drews, and Lisa I. Iezzoni. 2006. "Disparities in Breast Cancer Treatment and Survival for Women with Disabilities." *Annals of Internal Medicine* 145(9): 637–645.

Mitra, Monika, Ilhom Akobirshoev, Nechama Sammet Moring, Linda Long-Bellil, Suzanne C. Smeltzer, Lauren D. Smith, and Lisa I. Iezzoni. 2017. "Access to and Satisfaction with Prenatal Care among Pregnant Women with Physical Disabilities: Findings from a National Survey." *Journal of Women's Health (2002)* 26(12): 1356–1363.

Mitra, Monika, Karen M. Clements, Jianying Zhang, Lisa I. Iezzoni, Suzanne C. Smeltzer, and Linda M. Long-Bellil. 2015. "Maternal Characteristics, Pregnancy Complications, and Adverse Birth Outcomes among Women with Disabilities." *Medical Care* 53(12): 1027–1032.

Mitra, Monika, Lauren D. Smith, Suzanne C. Smeltzer, Linda M. Long-Bellil, Nechama Sammet Moring, and Lisa I. Iezzoni. 2017. "Barriers to Providing Maternity Care to Women with Physical Disabilities: Perspectives from Health Care Practitioners." *Disability and Health Journal* 10(3): 445–450.

Office of the Surgeon General, US Department of Health and Human Services. 2005. "The Surgeon General's Call to Action to Improve the Health and Wellness of Persons with Disabilities." Washington, DC.

Okoro, CA, ND Hollis, AC Cyrus, and S. Griffin-Blake. 2018. "Prevalence of Disabilities and Health Care Access by Disability Status and Type among Adults—United States, 2016." *MMWR Morbidity and Mortality Weekly Report* 67(32): 882–887.

Rios, Dianne, Susan Magasi, Catherine Novak, and Mark Harniss. 2016. "Conducting Accessible Research: Including People with Disabilities in Public Health, Epidemiological, and Outcomes Studies." *American Journal of Public Health* 106(12): 2137–2144.

Spall, Harriette G. C. Van, Andrew Toren, Alex Kiss, and Robert A. Fowler. 2007. "Eligibility Criteria of Randomized Controlled Trials Published in High-Impact General Medical Journals: A Systematic Sampling Review." *JAMA* 297(11): 1233–1240.

Vries McClintock, Heather F. de, Frances K. Barg, Sam P. Katz, Margaret G. Stineman, Alice Krueger, Patrice M. Colletti, and Tom Boellstorff et al.. 2016. "Health Care Experiences and Perceptions among People with and without Disabilities." *Disability and Health Journal* 9(1): 74–82.

Further Reading

Centers for Disease Control and Prevention. *Healthy People 2020 — Disability and Health*. Department of Health and Human Services. https://www.healthypeople.gov/2020/topics-objectives/topic/disability-and-health

Iezzoni, Lisa I. 2016. Stigma and Persons with Disabilities. In *Stigma and Prejudice: Touchstones in Understanding Diversity in Healthcare*, edited by Ranna Parekh and Ed W. Childs, pp. 3–21. New York: Springer International.

Iezzoni, Lisa I., and Bonnie L. O'Day. 2006. *More Than Ramps. A Guide to Improving Health Care Quality and Access for People with Disabilities*. New York: Oxford University Press.

Olkin, Rhoda. 1999. *What Psychotherapists Should Know about Disability*. New York: Guilford Press.

34

THE GOALS OF BIOMEDICAL TECHNOLOGY

Joseph A. Stramondo

I Introduction

While it is not the only plausible goal, one important goal of biomedical technology[1] is to advance equality of opportunity for social, political, and economic participation. However, as always, the devil is in the details. In particular, there is the detail of exactly how biomedical technology ought to advance the goal of equality of opportunity. For example, Norman Daniels has famously argued that people have a right to access a basic minimum of health care, including biomedical technology, because it advances this goal of equality of opportunity by maintaining or restoring their normal biological function (Daniels 2002). However, disability bioethicists have largely challenged Daniels' view because of this very idea that maintaining or restoring normal function is necessary for achieving such equality of opportunity. They argue that this feature of his view puts Daniels' argument at odds with the most fundamental assumptions of the disability rights movement: that equality of opportunity for the social, political, and economic participation of those with abnormal function is not only achievable, but it should be achieved (Amundson 2005; Amundson and Tresky 2007).

In this chapter, I will argue that disability bioethics can consistently maintain that biomedical technology should aim to advance equality of opportunity as a crucial goal without committing to the view that normal function is necessary for such equality of opportunity. This is because equality of opportunity can be advanced by biomedical technology that maintains, restores, or replaces normal function.

I begin by describing Daniels' widely accepted view that normal function is necessary for achieving equality of opportunity. Then, I argue that holding this view is actually paradoxical because it produces inequality of opportunity for disabled people when put into practice in the clinic or when held as a basic assumption for developing social policy. Next, I argue that this paradox can be resolved if one adopts the notion that greater equality of opportunity could be advanced equally well by biomedical technology that maintains, restores, or replaces normal function. Finally, I conclude with a discussion of whether there exists a moral obligation to advance the goal of equality of opportunity by developing and offering biomedical technology that replaces normal function and what the limit of that obligation might be.

DOI: 10.4324/9781003289487-45

II Biomedical Technology, Normal Function, and the Goal of Equality of Opportunity

Daniels argues that the need of individuals for normal functioning ought to justify a more equitable balance of social resources devoted to treating and preventing disease. To make the argument that more equitable access to health care deserves our attention as a matter of justice and not just kindness or decency, Daniels defines health care as distinct from and of greater moral importance than many other social goods. In this light, any defensible theory of healthcare justice must accord a special status to health care, since equitable access to it, in particular, promotes greater justice.

To establish this distinction and justification, Daniels turns to the notion of normal functioning,

> Specifically, the central function of health care is to maintain normal functioning. Disease and disability, by impairing normal functioning, restrict the range of opportunities open to individuals. Health care thus makes a distinct but limited contribution to equality of opportunity.
>
> *(Daniels 2002, 6)*

Daniels goes on to more fully flesh out what he means by equality of opportunity, as it relates to normal functioning,

> Specifically, by keeping people close to normal functioning, health care preserves for people the ability to participate in the political, social, and economic life of their society. It sustains them as fully participating citizens—normal collaborators and competitors—in all spheres of social life.
>
> *(7)*

In this way, Daniels places "normal function" in a position of primary importance for justifying why society ought to provide fairer access to health care: it is a necessary condition for achieving equality of opportunity. Within this framework, for a biomedical technology to achieve the goal of advancing equality of opportunity, it must maintain or restore normal function.

III Paradoxes Produced by Only Maintaining or Restoring Normal Function

Paradoxically, only developing and offering biomedical technology that maintains or restores normal function in order to advance equality of opportunity actually diminishes equality of opportunity for disabled people in at least two ways. First, if biomedical technology that maintains or restores normal function is all that is offered to a patient in a clinical setting, this disadvantages them at the individual level by narrowing the range of their options for how they exist in the world. Namely, it limits them to modes of function that most closely mimic normalcy. This diminishes their opportunities insofar as modes of function that mimic normalcy may not optimize how well a given person is able to achieve their goals. Second, and at a more fundamental level, a commitment to the notion that normal function is a necessary condition for equal social/political/economic participation maintains the disadvantages experienced by disabled people broadly. It does so by denying the possibility of achieving greater equality of opportunity via structural change that accommodates alternative, abnormal modes of function.

Anita Silvers argues that assuming normal function is required for equality of opportunity can actually thwart this goal by narrowing the range of choices a patient can make about biomedical technology, possibly removing those choices that would in fact maximize their access to the opportunities they value most. Namely, she argues that each patient should be seen as an individual with her own conception of the good life – a conception that should determine which treatment plan will maximize her equality of opportunity, regardless of whether it moves her closer to normal function or not (Silvers 2002).

To explain why, we must further unpack Silvers' distinction between mode and level of function. She observes how "we should notice that at least two aspects of functioning, the mode and the level, affect whether the performance of a function is normal" (Silvers 1998, 101). She defines the mode of a function as "the way it is accomplished," such as mobilizing on legs and feet rather than wheelchair tires. The level of function, however, is conceptually distinct and can be conceived as the speed or efficiency with which one accomplishes the function (101). Since mode and level of function come apart in both theory and practice, Silvers argues that we ought not prioritize achieving normalcy in the mode of function if it means we make sacrifices in level of function.

She argues that assuming normal function is necessary for equality of opportunity is an example of unfairly privileging a certain, narrow mode of functioning, while closing off choices that may better fit a person's life goals. Closing off certain modes of function diminishes a person's equality of opportunity. Silvers asks "Whose preferences about modes and levels of functioning justly take precedence in determining what medical services a patient should get – the patient's? the medical professional's? the dominant social group's?" (Silvers 2002, 236) and answers that our society tolerates pluralism in conceptions of the good in other contexts and we ought to extend such pluralism to disability as well. This motivates her call to embrace a concept of "multifunctionalism" in which we support "the equivalent value of various modes of performing such important human functions as mobilizing, socializing, acquiring information, and communicating" (236). Theories that justify the provision of health care only in terms of maintaining or restoring normal function cannot account for this sort of multifunctionalism, nor can they allow for the fullest range of choices in treatment options.

Silvers divorces the notion of normal function from the concept of equal opportunity range to show that multiple modes of function are compatible with an equal opportunity range, especially if those alternative modes of function are well supported within a given social environment. She suggests that we can still pursue equality of opportunity as an overarching goal of medicine – and, consequently, of biomedical technology – while respecting the pluralism that exists within the sorts of choices people may make regarding their mode of functioning. That is, individual people need various modes of functioning that open up different opportunities that they value uniquely in distinct social contexts. Silvers ultimately argues that, for disabled people, a mode of functioning that is closer to so-called normal function is many times actually less conducive to achieving the greatest possible access to a full range of opportunity. This is why she endorses the notion of multifunctionalism.

Silvers considers how this plays out in rehabilitative medicine because she sees this area of medicine as the largest threat to multifunctionalism (2002, 243). She offers the example of a college student who is coerced into receiving psychological counseling as a tradeoff for the continuation of service in a rehab medicine setting because he prefers to mobilize with a wheelchair rather than go through the arduous process of learning to use a prosthetic leg. She points out that this student chose his mode of functioning based on his particular environment – a large and often wheelchair accessible campus – because he values "moving efficiently more than moving in a species-typical mode" (243). For this student and many other disabled people, normal function does not equate with equality of opportunity but would, in fact, pose a significant barrier to this goal. Expending the time and effort to become adept at using a prosthetic leg would surely detract from his studies and social life, while offering very little additional opportunity on a college campus that is easily accessible by wheelchair.

This is not the only example of how assuming that normal function is necessary for equality of opportunity can actually diminish equality of opportunity. Another instance is the long history of conflict between those who would advocate for educating deaf children via divergent methods and the consequently divergent technologies. On the one hand, we have the techniques of oralism. This involves teaching deaf children to communicate with a combination of lip reading and any residual hearing they have, assisted or amplified by technologies like ear trumpets, hearing aids, and cochlear implants. This approach is premised on the idea that normal function, communication via speech and hearing, is necessary for a person to access the full range of opportunity and is, thus, the obvious path to achieving the goal of equality of opportunity via biomedical technology. In contrast, there is the approach championed by many residential Deaf schools that use the technologies of signed languages to educate young people, at the same time enculturating them into the Deaf community.[2] Like the student who preferred using a wheelchair to traverse his university campus, Deaf activists and scholars have argued that the oralist approach, while closer to normal function than signed languages like American Sign Language (ASL), actually diminishes the opportunities that can be accessed by deaf children because they produce a delay in the acquisition of general language skills that foster a full range of cognitive development (Hall et al. 2017). Accordingly, this commitment to a more normal mode of communicative function narrows a person's range of opportunities in much the same way that forcing prosthetics on someone who prefers a wheelchair narrows their opportunity range.

Of course, the prioritization of oralism in deaf education is as much a structural policy problem as it is a disadvantage for individual deaf children, so much so that it has been characterized as a form of oppression (Lane 1992). This framing brings us to the second, more fundamental, way in which a commitment to normal function as a necessary condition for equal social/political/economic participation maintains the disadvantages experienced by disabled people: it denies the very possibility that greater equality of opportunity can and should be achieved via structural change that accommodates alternative, abnormal modes of function.

That is, this commitment to the notion that normal function must be protected by health care because it is necessary for equality of opportunity holds an implicit criticism of some of the most fundamental goals of the disability movement, which aim at promoting equality of opportunity for disabled people who function abnormally. Ron Amundson articulates this well, arguing that bioethics, as a whole, is biased against the goals of the disability movement due to its failure to grasp and adopt the social model of disability. This model is the theoretical foundation of the Americans with Disabilities Act, which affords the strongest civil rights protection available to disabled people in the United States (Amundson 2005, 102).[3] Indeed, Amundson is not arguing for the truth of the social model as the only viable way to conceptualize anything about the causation or meaning of disability; he remains open to the idea that the medical and social models of disability may "present a false dichotomy, each attending to only one aspect of disability" (102). However, he contends that the medical and social models of disability are not merely explanatory, but deeply ideological in that

> contrasting causal accounts of the same phenomenon (here disability) serve and harm the interests of different groups of people. A causal account that depicts a social phenomenon as natural and inevitable (or changeable only at great cost) works to the advantage of the people who benefit from the phenomenon, and to harm of the people who are hurt by the phenomenon. When the same phenomenon is depicted as artificial and changeable, the reformist interests of those harmed by the phenomenon are served.
>
> *(103)*

In this same essay, Amundson also turns to Daniels's theory of healthcare justice as an example of a prominent mainstream theory in bioethics that builds on the problematic assumption that the harms of disability are natural and inevitable. The theory thus unfairly advantages those who benefit from these harms.[4] "The very purpose of the ADA is to remove barriers to opportunity that disadvantage 'abnormals,'" writes Amundson.

> To assume that 'normal opportunity range' is available only to a narrow range of body types is to assume that the Social Model is false and the ADA fruitless…. The assumed 'naturalness' of the linkage of normality to opportunity harms the interests of disabled people, just as the linkage of race and sex to opportunity has been harmful to other disadvantaged groups…. The notion that opportunity is by definition out of the reach of disabled people is rightly rejected by them, just as the same claims were rejected by women or minorities.
>
> *(Amundson 2005, 107–108)*

By binding normal function to equality of opportunity, Amundson charges that Daniels is creating a theory that is unavoidably at odds with the bedrock theoretical assumptions of the disability movement. In effect, Amundson is arguing that Daniels's theory is inescapably anti-disability rights because it assumes that normal function is required for equality of opportunity, while the disability movement works to divorce these concepts from each other in both thought and action. At best, Daniels's theory of healthcare justice fails to justify the provision of certain sorts of health care that promote equality of opportunity without normalizing function; at worst, his assumption that normal function is required for equal opportunity is implicitly an anti-disability rights political position (Amundson and Tresky 2007).

To be sure, we should not take issue with the idea that equalizing access to biomedical technology is justified, at least in part, by concerns about equality of opportunity. The examples referred to above such as adaptive mobility equipment like wheelchairs or services like ASL interpretation are all sought by disabled people largely because of their potential to provide greater access to social participation. Equal social, political, and economic participation is perhaps the overarching goal of the disability rights movement's decades-long struggle, and access to various forms of health care, especially biomedical technology, has been an essential component of that goal.

However, even if we grant that Daniels is correct that the provision of health care generally, and biomedical technology specifically, is important for promoting equality of opportunity, we are still left with the problem of determining how they do this in a way that is distinct from other social goods that serve this function, like public education, for instance. So, if it is not that health care, and thus biomedical technology, promotes equality of opportunity by maintaining or restoring normal function, there must be another way to specify how it is they will achieve this goal as health care and biomedical technology.

The solution to this dilemma is to recognize that health care and biomedical technology do not only promote equality of opportunity by maintaining or restoring normal function. Often, they promote equality of opportunity by replacing normal function with other modes of function that also can grant access to the social milieu.

IV Replacing Normal Function with Biomedical Technology to Achieve Equality of Opportunity

The apparent paradox that the egalitarian aims of biomedical technology instead thwart the goal of increased equality of opportunity can be resolved if we abandon a commitment to only doing this by maintaining or restoring normal function. Both of the ways in which biomedical technology might diminish equality of opportunity addressed above can be headed off if that technology, at least sometimes, replaces normal function as well.

By embracing multifunctionalism, clinicians might help patients achieve normalcy in their level of function by replacing the normal mode of function. In the case of the disabled college student Silvers describes, this would be done by giving priority to wheelchair use over bipedal locomotion via prosthetics. Indeed, as technology advances, it is possible there may be a proliferation of new technologies that open up opportunities for disabled people by improving their level of function without normalizing their mode of function. We can imagine that, in the future, rather than spending days, weeks, and years in rehabilitative therapy to gain "independence" in completing activities of daily living, a disabled person may be able to rely on brain-computer interface–equipped robotic attendants to accomplish these tasks swiftly and easily. In this way, biomedical technology might clearly advance the goal of equality of opportunity by replacing normal function.

Likewise, this intervention of replacing normal function with biomedical technology also addresses the second, more fundamental problem of how a commitment to normal function being a necessary condition for equal social participation maintains the disadvantages of disability by denying the possibility of achieving this via structural change that accommodates alternative, abnormal modes of function. At the theoretical level, it is clear how replacing normal function helps reconcile the provision of biomedical technology with the goal of achieving equality of opportunity for disabled people. While it may not be the only defining property of the concept of disability, completing tasks via an alternative, abnormal function seems to be a hallmark of what it means to be disabled. Thus, by replacing a normal function with an abnormal mode, biomedical technology advances equality of opportunity for disabled people not by erasing disability but by working in tandem with efforts to reform the social structure so that it is inclusive of disability. Further, we can notice that, in practice, many biotechnologies that advance the goal of equality of opportunity by replacing normal function can only do so in the context of a restructured social environment. This is sometimes a literal restructuring, such as the ramps and elevators that create access to opportunity for wheelchair users (Wasserman et al. 2018). Other times, it may be that a Deaf person is provided an ASL interpreter for their college class, doctor's appointment, or court hearing. This way, they access greater equality of opportunity via both the technology of a signed language and an intervention in the social world that allows for that technology to be effective.

V The Obligation to Replace Normal Function and Its Limits

Given the arguments presented above, we might ask whether there is a moral obligation to advance equality of opportunity via biomedical technology that replaces normal function. A detailed, comprehensive defense of the idea that there is such an obligation is beyond the scope of this chapter. However, I would suggest that it is worth exploring whether theories of health-care justice like Daniels', which maintain that there is a moral obligation to provide health care generally, can account for this sort of technology.[5]

What I have argued for implies an obligation to offer patients biomedical technologies that promote their equality of opportunity via abnormal modes of function and to do so without bias. Yet, by itself, this will not suffice to ensure that biomedical technology does not reproduce the inequality of opportunity it is intended to remedy. For this, we need to make certain that there is not a bias against replacing normal function in the development of new biomedical technologies. In other words, so long as we eliminate bias in the clinic as already discussed, we might allow individual patients to determine if their access to opportunity is advanced via maintaining, restoring, or replacing normal function in particular cases, but the specific technology that replaces normal function needs to be one of the choices available for this intervention to have any kind of impact. After all, choices about what kind of biomedical technology is designed and produced in the first place are made prior to any individual patient's choice about whether to adopt these technologies. Thus, we must embrace a commitment to multifunctionalism at both the clinical and policy levels of decision making. In practice, this will mean involving disabled people in the design of new biomedical technologies from their earliest conceptualization, rather than merely conducting market research for how to sell a product that has already been developed (Goering et al. 2020). It will also mean that there needs to be much greater parity in the public funding of the development of biomedical technologies that will maintain, restore, or replace normal function.

If there is such a set of obligations to develop and offer biotechnologies that maintain, restore, and replace normal function, we are left with the final difficulty of establishing a non-arbitrary limit to this obligation. This is a particularly thorny difficulty since we are wanting to include replacing normal function in our range of obligations. The reason is that Daniels' entire purpose for prioritizing the maintenance or restoration of normal function in the first place was to establish a non-arbitrary limit on the obligations he was demanding of us. Since there are many kinds of resources that help a person achieve equality of opportunity, Daniels needed a way to distinguish health care from all other sorts of resources if his theory was to be putatively about the obligations of healthcare justice. For this, he turned to the concept of normal function. Building on the work of Christopher Boorse (1977), Daniels argues that there is a natural, objectively normal way the human body is supposed to function and any deviation from that normalcy that nature has established is a state of disease. Health care, of course, is the practice of maintaining that normal disease-free state or restoring it, when it has been lost. So, Daniels hopes to establish an objective limit to the obligations of healthcare justice by endorsing an objective standard of what counts as healthy: normal function. The hope is that we can pretty easily identify what folks are owed by healthcare justice by discounting any intervention that provides functionality that is less than normal because it does not meet our obligation and dismissing any intervention that provides functionality that is greater than normal because it exceeds our obligation (Daniels 1985).

Endorsing the replacement of normal function as an important component of the goal of promoting equality of opportunity via biomedical technology is a fundamental challenge to Daniels' limit to the obligations of healthcare justice. So far, we have no way to know when we are not obligated to replace normal function. Indeed, it may seem that I am arguing that society is obligated to write disabled people a blank check for any and all biomedical technology that could benefit them.

However, this need not be the case. I have argued that, in order to avoid thwarting the very goal it sets out to achieve, biomedical technology must sometimes replace, as well as maintain or restore normal function. However, it would probably be more accurate to say that, in order to avoid paradox, biomedical technology must sometimes replace the normal mode of function. We can still establish a limit for what justice requires via the level of function provided by a type of biomedical technology. Of course, I don't have the space to offer a full defense of the idea here, but I would at least initially propose that any obligation to offer or fund the development of biomedical technology that replaces normal function could be limited, not by the normalcy of the mode of function it provides, but by how well it promotes equality of opportunity. For instance, there isn't much of an obligation to offer or fund the development of a biomedical technology that replaces a normal mode of function for a person in a way that far exceeds the normal level of function, such that the user is at a great advantage over non-disabled peers engaged in social, political, and economic encounters. This would not promote equality of opportunity, but quite the opposite.

Notes

1 In this chapter, I will use the term "biomedical technology" in its very broadest sense, encompassing any human-made device or practice that, by necessity, brings its user into contact with the institutions of biomedicine. Some examples would be obvious and intuitive like pharmaceuticals or prosthetics. However, others, like American Sign Language (ASL) interpretation, would be less so. Nevertheless, because of the control the institutions of medicine hold over access to these technologies I believe they are rightly characterized as biomedical. While ASL is a natural language and not intrinsically a "biomedical" technology any more so than any other language, one typically needs validation from a physician to get ASL *interpreting services* in a college classroom, for instance, and so these services bring the user into contact with the institutions of medicine and, thus, count as a biomedical technology on my view.
2 For a detailed discussion contextualizing attitudes toward biomedical technology via these educational approaches, see Scully (2008).
3 For selections of this piece, see Chapter 13, this volume.
4 For a full exploration of how certain people benefit from disability oppression, see Russell (2019).
5 For a closely related discussion of a possible moral obligation to provide disabled people with assistive technology that advances their equality of opportunity, see Stramondo (2020).

References

Amundson, Ron. 2005. "Disability, Ideology, and Quality of Life: A Bias in Biomedical Ethics." In *Quality of Life and Human Difference: Genetic Testing, Health Care, and Disability*, edited by David Wasserman, Jerome Bickenbach, and Robert Wachbroit, pp. 101–124. New York: Cambridge University Press.

Amundson, Ron, and Shari Tresky. 2007. "On a Bioethical Challenge to Disability Rights." *Journal of Medicine and Philosophy*. 32: 541–561.

Boorse, Christopher. 1977. "Health as a Theoretical Concept." *Philosophy of Science*, 44: 542–573.

Daniels, Norman. 1985. *Just Health Care*. Cambridge: Cambridge University Press.

Daniels, Norman. 2002. "Justice, Health, and Health Care." In *Medicine and Social Justice: Essays on the Distribution of Health Care*, edited by Rosamond Rhodes, Margaret P. Battin, and Anita Silvers, pp. 6–23. New York: Oxford University Press.

Goering, Sara and Eran Klein. 2020. "Neurotechnologies and Justice by, with, and for Disabled People." In *The Oxford Handbook of Philosophy and Disability*, edited by David Wasserman and Adam Cureton, pp. 616–632. New York: Oxford University Press.

Hall, Wyatte C., Leonard L. Levin, and Melissa L. Anderson. 2017. "Language Deprivation Syndrome: A Possible Neurodevelopmental Disorder with Sociocultural Origins." *Social Psychiatry and Psychiatric Epidemiology*, 52: 761–776.

Lane, Harlan. 1992. *The Mask of Benevolence: Disabling the Deaf Community*. New York: Knopf.

Russell. Marta. 2019. *Capitalism and Disability: Selected Writings by Marta Russell*, edited by Keith Rosenthal. Chicago, IL: Haymarket Books.

Scully, Jackie Leach. 2008. "Moral Bodies: Epistemologies of Embodiment." In *Naturalized Bioethics: Toward Responsible Knowing and Practice*, edited by Hilde Lindemann, Marian Verkerk, and Margaret Urban Walker, pp. 23–41. Cambridge: Cambridge University Press.

Silvers, Anita. 1998. "A Fatal Attraction to Normalizing: Treating Disabilities as Deviations from 'Species Typical' Functioning." In *Enhancing Human Traits: Ethical and Social Implications*, edited by Erik Parens, pp. 95–123. Washington, DC: Georgetown University Press.

Silvers, Anita. 2002. "Bedside Justice and Disability: Personalizing Judgment, Preserving Impartiality." In *Medicine and Social Justice: Essays on the Distribution of Health Care*, edited by Rosamond Rhodes, Margaret P. Battin, and Anita Silvers, pp. 235–247. New York: Oxford University Press.

Stramondo, Joseph. 2020. "The Right to Assistive Technology." *Theoretical Medicine and Bioethics* 24: 471–487.

Wasserman, David and Stephen M. Campbell. 2018. "A More 'Inclusive' Approach to Enhancement and Disability." In *The Ethics of Ability and Enhancement*, edited by Jessica Flanigan and Terry Price, pp. 25–38. Jepson Studies in Leadership. New York: Palgrave Macmillan.

Further Reading

Buchanan, Allen, Dan Brock, Norman Daniels, and Daniel Wickler. 2000. *From Chance to Choice: Genetics and Justice*. Cambridge: Cambridge University Press.

Ladner, Richard E. 2010. "Accessible Technology and Models of Disability." In *Design and Use of Assistive Technology: Social, Technical, Ethical, and Economic Challenges*, edited by Meeko Mitsuko K. Oishi, Ian M. Mitchell, H.F. Machiel Van Der Loos, pp. 25–32. New York: Springer.

Silvers, Anita. 2010. "Better than New! Ethics for Assistive Technologists." In *Design and Use of Assistive Technology: Social, Technical, Ethical, and Economic Challenge*, edited by Meeko Mitsuko K. Oishi, Ian M. Mitchell, H.F. Machiel Van Der Loos, pp. 3–16. New York: Springer.

Sparrow, Robert. 2010. "Implants and Ethnocide: Learning from the Cochlear Implant Controversy," *Disability & Society*, 25(4): 455–466.

Stramondo, Joseph. 2019. "The Distinction between Curative and Assistive Technology." *Science and Engineering Ethics*, 25(4): 1125–1145.

35

"WHY INSIST ON JUSTICE, WHY NOT SETTLE FOR KINDNESS?" KINDNESS, JUSTICE, AND COGNITIVE DISABILITY

Eva Feder Kittay

1 Including People with Cognitive Disabilities in Theories of Justice

One of the primary principles in bioethics is the principle of justice. In Tom Beauchamp and James Childress's classic formulation, the principle of justice is a set of norms "for fairly distributing benefits, risks, and costs" (2019). Theories of justice incorporate and systematize such norms. These norms are, by and large, created for the ideal of autonomous, independent, able-bodied, and equal rational agents. As a professor of philosophy teaching the various theories of justice, I found myself asking where my daughter fits in. This is because my daughter, Sesha, has a rare genetic variant on the PURA gene that is responsible for a number of very impactful impairments. She is very significantly disabled both cognitively and physically. People like Sesha simply weren't reflected in most theories of justice, and some explicitly exclude her. This raised for me a disturbing question: is it possible that my own daughter, a human being such as myself, should not be entitled to the same protections, rights, and distributive shares that are divvied up in a just society? In the context of bioethics, does her disability justify treating her differently (and worse) than I am treated?

But how can this be? Having been born in the USA where citizenship is a birthright, she is a citizen of the USA, and like all citizens in democratic states, she is presumed to be subject to and the beneficiary of justice. When that expectation is not met, we consider that situation to be unjust. Minimally, for example, I could secure for her some rights of US citizenship, for example, a USA passport; and were anyone to injure or cause her harm, they would no less be subject to a civil suit or criminal prosecution than had they harmed my nondisabled son. The law of the land included her, albeit as a dependent on myself and my husband.

Philosophical theories of justice, however, do not assume one has a birthright to justice. Instead subjects of justice are required to meet certain criteria. The most essential is personhood. A person, following Locke, is thought to be one who has reason and reflection and is the same self throughout time and place. This conception of personhood is equally operative in the sphere of bioethics. The extent to which my daughter can reason or reflect is not easily resolved. She has little control of her bodily movements and is unable to speak. Without these abilities, we cannot indisputably determine whether or not she can reason and reflect. Still, it has never occurred to me or our family that she wasn't a person—we talk about her, treat her and expect

DOI: 10.4324/9781003289487-46

others to treat her like a person, we presume that justice is due her: for example, that medical experimentation on her to benefit others is abhorrent, and that she should have the same access to medical treatments as anyone else.

When a theory of justice fails to include her, it fails to explain why she as a matter of course has the status of a citizen, and why the laws of the land pertain to her no less than to the rest of us. It fails to deliver on a promise that a just society is based on principles that apply to all who constitute it. The fulfillment of that promise has been long delayed for many, notably women of all races and classes, enslaved people, and people who have been colonized. And it has been far too long delayed for people with disabilities, especially for those with cognitive disabilities.

Yet at a conference on the impact of disability on philosophical concepts, a fellow philosopher— I imagine one without an intimate knowledge of people with cognitive disabilities—asked me the following: "Why insist on arguing that your daughter is due justice? Why is kindness not enough?" The question was motivated by the difficulty of including people with significant cognitive disabilities into most theories of justice. So before attempting to reply, it will be helpful to quickly rehearse some dominant theories of justice to make clear why and how these theories appear to exclude people who have mental disabilities of various sorts, especially those whom we presume have impaired cognitive functioning.

II What Is a Theory of Justice?

Justice, as we noted, enters into bioethics as one of four principles that should guide the actions and policies of the field of bioethics. The other three principles, autonomy, nonmaleficence, and benevolence, are largely ones that govern personal interactions between patients and providers. The principle of justice is about bioethical decisions that look at broader ethical issues such as the distribution of a scarce medical resource or the propriety of decisions that may have a disparate impact on different populations. Justice, while it can be an interpersonal virtue, is primarily thought of in social, political, or policy terms.

Current political philosophy is dominated by two different conceptions of justice. One is based on a social contract tradition that harkens back to Thomas Hobbes and has had many famous iterations, notably in John Locke, Jean-Jacques Rousseau, Immanuel Kant, and, most recently, John Rawls. The underlying idea is that social arrangements depend on the agreement of the governed about how to be governed. The second is based on utilitarian principles that legal, moral, and social policies should be directed at producing the greatest good for the greatest number. It rests largely on the work of Jeremy Bentham and his followers (Mill and Bentham 1987).

2.1 Utilitarian Theories and the Cognitively Disabled

Utilitarians believe that one can roughly gauge what sorts of actions, laws, or policies are best by considering what garners the greatest benefit for the greatest number. Disabled people, especially the cognitively disabled, frequently do not fare well under the utilitarian calculus. Consider the dilemmas that face medical personnel when medical resources are scarce, as has happened during surges during the COVID-19 pandemic. Policy makers have had to put in place policies that determine who gets an ICU bed and ventilator if demand outpaces supplies. The resulting triage schemes are meant to obtain the greatest benefit for the greatest number. Unfortunately, many triage protocols assume that the quality of life with a disability is worse than one without the disability. With this supposition, the calculus would determine that, given a choice between two similarly situated people (e.g., same age and same chance of survival) and a lone available ventilator, the resource should go to the nondisabled person because their remaining years would be of a higher quality (more benefit) than that of the disabled person (Nunez, this volume). The widespread, but nonetheless questionable, assumption that one who lives with an intellectual disability has a very low quality of life makes such triage protocols dangerous for this population. The same logic extends to other scarce resources or services.

A reason that utilitarian theories of justice, especially in an ableist society, are bad news for all, but especially cognitively, disabled people is that the good is aggregated. That is, what drives policies has to do with the overall benefit of the group rather than what is good for each individual. When the well-being of the disabled person is considered to be lesser (and more costly to maintain) than that of the nondisabled person, their interests come to count for less in an overall calculation of benefits (Campbell and Stramondo, 2017).

2.2 The Social Contract Tradition

Social contract theory does not see the good in such aggregated terms. In the contract tradition, we are believed to enter into an agreement with others for our mutual benefit: either to stave off a war of all against all (the view of Thomas Hobbes) or, more benignly, to cooperate to achieve goods we could not were we to act singly (the view of Locke, Rousseau, and Kant). On this second view, the contract not only secures our mutual advantage but also gains us the respect of our fellow human beings and gives us dignity.

The contemporary Hobbesian David Gauthier has maintained that there would be no rational reason for a bargainer to enter into a contract with a disabled person since the power between the two parties already favors the able person (1986). Crudely speaking, there is nothing in it for the nondisabled person.

John Rawls's theory, arguably the most influential today, reimagines the social contract (derived from Locke, Rousseau, and Kant) by positing an "original position" in which representatives who know nothing about the actual capacities and economic status of those whom they represent—"behind a veil of ignorance"—must chose principles of justice for rational and reasonable citizens in a well-ordered state. Rawls claims that two principles will be chosen. The first principle gives to all the most extensive liberties compatible with others having the same liberties. The second principle has two parts: (2a) it provides fair (not abstract) equality of opportunity for all and (2b) any inequalities in distribution should be to the advantage of those who are least well-off.

Rawls speaks of cooperation over a lifetime for he recognizes that we all have periods in which we are unable to function in the required manner. But he also maintains that those with life-long significant disabilities or costly medical requirements are not part of this ideal theory because the costs of maintaining them outweigh the contributions they would make. Furthermore, the situation is still worse for those with serious cognitive disabilities because Rawls maintained that to be a participant in a society governed by these two principles one needed to have two moral powers, to be rational (have a sense of one's one good) and to be reasonable (to have a sense of justice)—and it is not clear that people with significant cognitive disabilities have the needed capacities. While many have tried to show how social contract could be extended to disabled people, even to the severely cognitively disabled, it is not clear that anyone has succeeded (Beaudry 2021; cf. Simplican 2015).

2.2 Capability Theory

It is worth mentioning one other current theory of justice that more directly confronts the difficulties of including people with disabilities. It derives from Aristotle insofar as it aims at the ideal of flourishing. While the Aristotelian view of flourishing is tied to his own conception of what a life of virtue consists in, the way in which Martha Nussbaum takes up this conception is based on what she purports to be a list of capabilities drawn from what is thought to be "a truly human life." It is the task of the state to fashion constitutional principles, laws, and policies such that people who live within that state can achieve the functions that reflect this list of capabilities. Thus, if a person who is blind is to be able to exercise the capabilities that relate to an education, the state is obliged to ensure that educational opportunities are fully available to blind students, including, for example, texts set in Braille. The theory is very hospitable to the differing requirements of disabled people, but it is less clear how well the list of capabilities apply to people with very significant cognitive disabilities.

III Why Insist on Justice?

This sprint through important contemporary competing theories of justice is meant to give the reader an idea of how difficult it is to shoehorn people with disabilities into theories that have excluded them. It is this difficulty that gave rise to the question that is the title of this chapter.

Can kindness replace the work justice does and can we therefore settle for kindness? Kindness is indeed a very important virtue, one undertheorized in philosophy and undersold in moral theory. Amid the cruelty and loneliness experienced in the wake of the COVID-19 pandemic, a mantra frequently heard is "be kind." Everyone, save perhaps a vicious criminal or mass murderer, ought to be treated with kindness. Kindness and mercy temper justice, and perhaps there are times when we cannot even have claimed to treat a person justly, if we have been unkind to them.

Without question, the world would be a better place if kindness was the rule and cruelty was banished. But can it replace justice? As much as it is a challenge to find a theory of justice that does include people such as my daughter, I want to insist that the effort to search for or construct such a theory is necessary.

A world where only kindness reigned, untethered to enforceable norms, would be a world of angels, not humans. To justify the view that kindness is not enough, we will need to be explicit about what we expect from justice that kindness fails to deliver, and whether there is anything about people like my daughter that dictates that they should be excluded from just treatment and just principles.

As a partial answer to the second question, we might look at the situation of people with physical or sensory disabilities. Today it would be retrograde—if not outrageous—to ask, "why kindness is not enough" if my daughter had a paraplegia, sensory disability, or most any disability except one involving cognition. With a social model of disability in hand, we regard the rights of disabled people as civil rights analogous to those of African Americans, Indigenous people, women, or those who are LGBTQ.

Legislation such as the Americans with Disabilities Act (Silvers and Francis 2000) is intended to "level the playing field" for disabled people so that they may be full participating members of the society in which they live. Although there are questions about the adequacy of contractual theories of justice for the inclusion of disabled people, the argument from the social model is that the accommodations demanded are not responses to "special needs" but are being mindful of how our environment has been shaped to exclude people with disabilities and to remedy this societal failure. With accommodations in place, a disabled person is no less capable of sharing the benefits and burdens of social cooperation than a nondisabled person.

But when it comes to cognitive disability there is no leveling of the playing field. So totally are the structures and institutions of society aimed at those who have intellectual capacities and functions within a certain range—one defined to exclude those with serious cognitive disabilities—that the possibility of engaging in social cooperation is foreclosed—yielding the conclusion that these individuals are not owed justice. Hence the question, "why insist on justice, why is kindness not enough?" is reserved for people like my daughter—not all disabled people.

Not enough for what? I want to consider two parameters: well-being and self-respect/dignity, especially with respect to matters of bioethics.

IV Can Kindness Be Adequate in Assuring Well-Being?

The difference between kindness and justice can be seen in the relation of each to enforceable norms, most importantly to just laws. It is hoped and expected that one is kind to those we hold close, or those who are entrusted to our care. It is hoped, but not expected, that kindness is extended to fellow citizens and other distant strangers. The opposing vice to the virtue of kindness is cruelty, and cruelty can be prescribed by law. But kindness cannot be legislated, nor can anyone be legally liable for failing to be kind. This is because kindness is voluntary, while justice is obligatory.

And with that difference comes another. An evaluation of an action as kind is a judgment about the character of the person who acts kindly; a just act is evaluated by what is rightfully due to the other. For a disabled person who cannot care for themself in the most basic ways, kindness is neither necessary nor sufficient in getting one's needs met. Justice may not be sufficient—for the person needs not only justice but it also needs to be handled in a caring manner—but justice is necessary. Thus, while we may desire kindness, we need justice. While the impact of kind actions is significant, the impact of just or unjust actions is fundamental.

For the most part, the bonds of special relations motivate kindness. In the case of justice, special bonds and special relations do not play a role. Therefore, if anyone is not thought to be due justice, that individual must depend on those whose special bonds motivate their kindness. But even if we have such special bonds, that is, even in special relations, we can neither be assured of kindness nor do we have a *right* to demand it. Rights issue from the individual to whom a right is due. But kindness has to be initiated by the one who is exercising the virtue. We have little say about whether or not one will treat us kindly, even in special relations such as spousal, parental, or sibling relations. And surely those bereft of such special bonds can have no expectation of kindness from strangers—it is not a norm, much less an enforceable one.

For instance, a kind doctor will try to put a patient at ease, make another feel confident that the physician has their best interests at heart. Yet there is no public obligation to be kind. Being kind is not in a patient's bill of rights, and no social or political institutions assure that we are treated with kindness. However much we, as individuals, wish to be treated with kindness, if we are operating in a society whose institutions fail to be just, welfare is jeopardized for all who are affected by these institutions.

An extreme but important example comes from Nazi Germany (see Wilson, this volume). The hospitals that housed and presumably cared for people with mental disabilities were recruited by the Nazis for the eugenically motivated murder of the residents. Doctors and nurses were told that it was kinder to these residents not to have to live the awful lives they were condemned to live by virtue of their disabilities. Although they clothed their crimes in protestations of kindness, they knew enough to lie to parents and concerned family members, who were told that their loved one died a natural death due to some ailment. With these lies, family members were rendered powerless. Some well-meaning doctors doubtless deluded themselves into thinking that euthanizing these individuals was being kind to them. It is surely the case that actual kindness was not the operative motive in the killings, but the ease with which kindness could cover such crimes must make us suspect when we are told to be satisfied with kindness, and not worry about justice when we are talking about such vulnerable people. The Nazi program, which was known as T4, is an extreme example of how the failure of the protection of just institutions and laws was devastating—and no feigned or deluded motivations of kindness could mitigate the horror.

Without the protection of justice, people have no defense against others' indifference to their needs, and their vulnerability makes them tempting targets to abusers. Therefore, not only ought we not settle for kindness, we must insist on justice for people with significant cognitive disabilities (or indeed anyone whose needs are many and defenses are few), if we have concern for their well-being.

The last point draws our attention to the different ways in which kindness and justice operate in the political life of a society. The impact of kindness is amorphous because the demands of kindness are not well-defined or determinative. Kindness is highly context dependent, depending not only on the intentions of the kind person but also on the recipient's response. What is kind to one person in one context may be indifferent or even cruel to that person in another context, or it may be indifferent or cruel to another person in that same context. It is, in a word, subjective.

Justice, in contrast, is virtually, by definition, something that can be adjudicated by an impartial observer—at least, that is so when justice is realized. Even if one feels like one has been treated unjustly, there are, at least in principle, objective criteria to determine if one has, or has not, been treated unjustly.

For this reason, claims that we are kind are epistemically opaque in ways that just treatment is not. When we are dealing with people who lack communicative means of signifying what serves them well, we need to know, at minimum, that they are treated justly. Objective criteria may be especially important for people with significant cognitive disabilities who have difficulty being able to speak for themselves, and to tell us when, or how, they have been harmed.

4.1 Justice and Kindness with Respect to Dignity and Self-Respect

The scope of those to whom we can extend kindness is very broad. One can be kind to any sentient being we feel like being kind to. Justice, in contrast, can be given only to those to whom justice is due. We can extend our kindness to those we consider to be morally inferior. We are just to those who have claims on us because they are our moral equals.

Perhaps because we can choose to be kind to those we consider morally inferior, kindness more easily than justice can be used for perverse reasons. I act kindly to my old doddering aunt only because I hope that she will make me the principal beneficiary of her will. I am treating her as a mere means to my own ends. (Were she to suspect my motives, she wouldn't take kindly to my professed kindness.) One may object that this is not really kindness at all. But what if the aunt basks in what she takes to be kindness, never detects the deception, and *feels* herself treated kindly? Isn't that kindness enough? I do not believe that the answer is obvious.

Another manipulative move is to extend kindness knowing that the recipient thus becomes obligated to me, putting her in my power. I again use the other as a mere means, and also, whether intentionally or not, can make her feel demeaned and inferior—something that is quite unkind. Yet while she may feel demeaned, it is easy enough to imagine that she never blames the person who has apparently extended kindness to her.

In contrast, if I act justly, even if I do so for ulterior motives—for example I may want others to think well of me—my actions remain just. And the other is still well-served by being treated justly. I may observe that the just actions are not motivated by a good character, but the just action is still justice that benefits me. Manipulation and domination are forestalled—as the only obligation incurred, to act justly—is just what we are supposed to do in any case.

These points are telling: I can use a person as a mere means only if I do not regard them as an end-in-themself, as a moral equal. In contrast, I can only act justly if I am regarding the other as one who has ends of their own. When we are treated justly, treated as if we are due justice, then, however the other views us, we are being treated as a moral equal.

I suspect that when someone asks me why I insist on justice for my daughter, why kindness is not enough, my interlocutor is not convinced that my daughter and people with her disabilities are our moral equals. Perhaps this is an ungenerous rendering of the motive behind the question. Perhaps instead, the question is posed because the theories of justice that predominate in philosophical discussions make the requirements for full moral status too stringent for people like Sesha to meet. These conceptions of justice, after all, are constructed to be important to those for whom their self-respect depends on their ability to act autonomously, who view their rational agency as important. We demand justice from others because we can make autonomous claims against other equal autonomous individuals.

While it is relatively easy to see that lives of significantly cognitively disabled people can go better or worse depending on how they are treated, what is less clear is if they can receive the benefit of respect, dignity, and being regarded as a moral equal that being treated with justice confers on one. How would a person who cannot make autonomous claims, and who may not exhibit capacities we take to be necessary for self-respect and dignity, even understand that they are being viewed as inferiors? If they don't understand concepts of self-respect or dignity, then isn't it the case that we will not be withholding from them the benefits of recognizing them as equals deserving of justice?

We contrasted justice and kindness earlier by noting that kindness unlike justice does not rest on enforceable norms. But let us say, for argument's sake, that kindness (or decency) could be attached to strong social norms—even if they are not legally enforceable— and that these will, by and large, protect and serve the needs of the vulnerable people under consideration. In that case, could we not settle for this enhanced kindness? What would such a person lack if they were treated by such kindness instead of having a claim to just treatment? It is the respect and dignity that comes with knowing that one is a moral equal.

Maintaining that treating people with severe cognitive disabilities as moral inferiors who have no interest in the goods of dignity and self-respect, as long as they are treated kindly, is founded on the supposition that we know what happens in the minds of significantly cognitively disabled people. It assumes they lack thoughts coherent enough to feel the sting of unequal consideration, the injustice of having their preferences ignored, the value of their lives diminished. In fact, we simply do not know if this is so. To make such presumptions is epistemically unsound and is an arrogance borne of ignorance. Having lived with my daughter for over 50 years, I still am surprised to find that she understands something I would have thought impossible for her to understand. While my knowledge is meager, I have seen the hurt on a disabled young child's face when her baby brother gets priority in entering a play tent; I have seen a cognitively disabled person lower her head when people talk over her; I know of a women with Alzheimer's who overhears a philosophy discussion that says that people like her are not persons, and who, hurt and perplexed, murmurs, "I am a person…I am a person." Those who are around people with very significant cognitive disabilities encounter moments of surprise when it is clear that the person they thought didn't understand what is happening does something that indicates they surely understood some or all of it. We don't know, we can't know for certain, if conceptions such as dignity and self-respect, ideas for which they may not have fully articulated conceptions nonetheless are deeply meaningful. But we can know that when we treat them as if these conceptions didn't matter, we are unlikely to ever to see evidence that they do.

There is one further point. In the philosophical literature, an offense against dignity and self-respect is mostly framed in individualist terms. Yet others who are not directly targeted can feel offended, even damaged, by the moral demotion of others. Social movements often stress not only the harm done to the direct victims but also the moral harm done to those who share their lives—their families, friends, neighbors and even their fellow citizens.

The movement for Black liberation has not infrequently pointed out that the moral damage done to Black people redounds to the rest of society. We can hear Martin Luther King's (1963) resonant voice ring out, "Injustice anywhere is a threat to justice everywhere." More directly, the white privilege of white parents whose Black adopted child is faced with systemic racism does not protect them from a grave moral injury. And it is not only an abstract moral injury. In a society where structural racism is hard to escape, the white parent, like the Black parent, faces special hardships in fulfilling their parental responsibility to keep their child safe and provide them with opportunities for growth and social acceptance.

A responsible loving parent, faced with nearly insuperable obstacles to fulfill their parental responsibilities, finds their own self-respect at risk. It is no less true when the moral denigration is that of a disabled child. If I were to settle for kindness, I fail to see how I could fulfill my maternal obligations. And as one of those obligations is to assure Sesha's position in a social world, part of my maternal responsibility is to make sure others also accept her as a moral equal. Even if my daughter might not feel the sting of being treated as a moral inferior, my self-respect is undermined if I fail to make others recognize her full moral worth. The harm to me is a harm to her—a harm to her is a harm to me. The moral denigration of any human being ripples out, from the closest ties, to the more distant ones, and to the nation whose laws fail to guarantee justice for that person. "No one is free until we are all free."

In addition, if those left unprotected by justice are to depend on kin when available or the inconstant and scarce kindness of strangers, unless the interests of the cognitively disabled are recognized as a matter of justice, kin and kind strangers will be ill-equipped to meet the needs of the people they are to care for—thereby undermining even the premise that a normatively strong conception of kindness can do its work for those whom justice excludes. Justice is a guarantee that kindness will have what kindness needs to meet its concerns.

V Kindness, Justice, and an Ethic of Care

I have spent a great deal of ink promoting and attempting to articulate an ethic of care. Those familiar with an ethic of care and my own work might find it odd that I should argue that we need not only kindness but justice for those who are excluded or marginal in theories of justice. Kindness seems so much more easily aligned with an ethics of care than does justice. It would not be unreasonable to assume that a proponent of an ethic of care would be well satisfied with the demand that people with significant cognitive disabilities be treated with kindness and so not require justice.

To many ears "acting kindly" and "acting caringly" are nearly synonymous, while care and justice are often conceptualized as opposing ethical stances. Kindness, while a motivator and a highly desirable accompanying attitude to caring labor, is not sufficient for care understood in an ethically robust manner; nor is it necessary—other attitudes may suffice for genuine caring. And fewer and fewer care theorists hew to a dichotomous view of justice and care. On some views (including my own earlier view), justice and care are mutually necessary. True caring needs to take place in a background of justice and just institutions; justice in its fullest understanding needs to take concerns of care seriously.

On my current view, theories of justice and ethical theories occupy different but related planes: justice is primarily a political notion involving institutional structures; care is an ethical one that pertains to the character, actions, and intentions of individuals. Political theories of justice can be seen as aligned with particular ethical theories. Rawlsian justice as fairness presumes a Kantian moral framework (2005). Aristotelian conceptions of justice suppose ethical behavior is to be judged according to a virtue ethics. And so on. A theory of justice that envisions a society guided by an ethics of care remains to be developed. Such a theory, in my view, is needed, if we are to consider our condition not only as moral actors but also as moral patients—not only as those who are supposed to act morally but also as those to whom morally required action is directed. Most of us will at some time be moral agents and at other times moral patients, and at times both simultaneously. A few will not be moral agents, but will be moral patients with the same interests to preserve as those who are both moral agents and moral patients. A theory of justice that takes care as a principal moral orientation will be a theory that embraces a fulsome view of the human condition, one that understands that human dependency and vulnerability are inevitable, and that realizes that cognitive impairment or loss is as significant a part of human life as is cognitive prowess. In such a theory, all the reasons for excluding those with cognitive disabilities will fall away, and kindness will find its proper place.

References

Beauchamp, Tom, and James Childress. 2019. *Principles of Biomedical Ethics*, 8th edition. Oxford: Oxford University Press.

Beaudry, Jonas-Sébastien. 2021. *The Disabled Contract: Severe Intellectual Disability at the Margins of Justice and Morality*. Cambridge: Cambridge University Press.

Bentham, Jeremy and John Stuart Mills. 1987. *Utilitarianism and Other Essays*. New York: Penguin.

Campbell, Steven, and Joseph Stramondo. 2017. "The Complicated Relationship of Disability and Well-being." *Kennedy Institute of Ethics Journal* 27(2): 151–184.

Gauthier, David. 1986. *Morals by Agreement*. Oxford: Oxford University Press.

King, Jr., Martin Luther. 1963. "Letter from Birmingham Jail." California State University, Chico. https://www.csuchico.edu/iege/_assets/documents/susi-letter-from-birmingham-jail.pdf

Nunez-Landry, Lydia 2022. "Chronic Illness, Wellbeing, and Social Values." In *Disability and Bioethics Reader*, edited by Joel Michael Reynolds and Christine Wieseler. New York: Routledge.

Rawls, John. 2005. *A Theory of Justice*. Cambridge, MA: Belknap Press of Harvard University Press.

Silvers, Anita and Leslie P. Francis. 2000. *Americans with Disabilities: Exploring Implications of the Law for Individuals and Institutions*. London: Routledge.

Simplican, Stacy Clifford. 2015. *The Capacity Contract: Intellectual Disability and the Question of Citizenship*. Minneapolis, MN: University of Minnesota Press.

Wilson, Robert. 2022. "Eugenics, Disability, and Bioethics." In *Disability and Bioethics Reader*, edited by Joel Michael Reynolds and Christine Wieseler. New York: Routledge.

Further Reading

Beaudry, Jonas-Sébastien. 2021. *The Disabled Contract: Severe Intellectual Disability at the Margins of Justice and Morality.* Cambridge Cambridge University Press.

Carlson, Licia. 2009. *The Faces of Intellectual Disability: Philosophical Reflections.* Bloomington, IN: Indiana University Press.

Nussbaum, Martha. 2006. *Frontiers of Justice: Disability, Nationality, Species Membership (The Tanner Lectures on Human Values).* Cambridge, MA: Belknap Press of Harvard University Press.

Simplican, Stacy Clifford. 2015. *The Capacity Contract: Intellectual Disability and the Question of Citizenship.* Minneapolis, MN: University of Minnesota Press.

36

SELECTIONS OF *BRILLIANT IMPERFECTION*

Eli Clare

This chapter is a selection of excerpts from Eli Clare. *Brilliant Imperfection: Grappling with Cure* (2017), pp. 5–135.

I Prayers, Crystals, Vitamins

Strangers offer me Christian prayers or crystals and vitamins, always with the same intent—to touch me, fix me, mend my cerebral palsy, if only I will comply. They cry over me, wrap their arms around my shoulders, kiss my cheek. After five decades of these kinds of interactions, I still don't know how to rebuff their pity, how to tell them the simple truth that I'm not broken. Even if there were a cure for brain cells that died at birth, I'd refuse. I have no idea who I'd be without my tremoring and tense muscles, slurring tongue. They assume me unnatural, want to make me normal, take for granted the need and desire for cure.

Strangers ask me, "What's your defect?" To them, my body-mind just doesn't work right, defect being a variation of broken, supposedly neutral. But think of the things called defective—the mp3 player that won't turn on, the car that never ran reliably. They end up in the bottom drawer, dumpster, scrapyard. Defects are disposable and abnormal, body-minds or objects to eradicate.

Strangers pat me on the head. They whisper platitudes in my ear, clichés about courage and inspiration. They enthuse about how remarkable I am. They declare me special. Not long ago, a white woman, wearing dream-catcher earrings and a fringed leather tunic with a medicine wheel painted on its back, grabbed me in a bear hug. She told me that I, like all people who tremor, was a natural shaman. Yes, a shaman! In that split second, racism and ableism tumbled into each other yet again, the entitlement that leads white people to co-opt Indigenous spiritualities tangling into the ableist stereotypes that bestow disabled people with spiritual qualities. She whispered in my ear that if I were trained, I could become a great healer, directing me never to forget my specialness. Oh, how special disabled people are: we have *special* education, *special* needs, *special* spiritual abilities. That word drips condescension. It's no better than being defective.

Strangers, neighbors, and bullies have long called me *retard*. It doesn't happen so often now. Still, there's a guy down the road who, when he's drunk, taunts me as I walk by with my dog. But when I was a child, retard was a daily occurrence. Once, on a camping trip with my family,

I joined a whole crowd of kids playing tag in and around the picnic shelter. A slow, clumsy nine-year-old, I quickly became "it." I chased and chased but caught no one. The game turned. Kids came close, ducked away, yelling *retard*. Frustrated, I yelled back for a while. *Retard* became monkey. My playmates circled me. Their words became a torrent. "You're a monkey. Monkey. Monkey." I gulped. I choked. I sobbed. Frustration, shame, humiliation swallowed me. My body-mind crumpled. It lasted two minutes or two hours—I don't know. When my father appeared, the circle scattered. Even as the word *monkey* connected me to the nonhuman natural world, I became supremely unnatural.

All these kids, adults, strangers join a legacy of naming disabled people not quite human. They approach me with prayers and vitamins, taunts and endless questions, convinced that I'm broken, special, an inspiration, a tragedy in need of cure, disposable—the momentum of centuries behind them. They have left me with sorrow, shame, and self-loathing.

[...]

II The Restoration of Health

As an ideology seeped into every corner of white Western thought and culture, cure rides on the back of normal and natural. Insidious and pervasive, it impacts most of us. In response, we need neither a whole hearted acceptance nor an outright rejection of cure, but rather a broad based grappling.

The American Heritage Dictionary defines cure as the "restoration of health." Those three words seem simple enough, but actually *health* is a mire. Today inside white Western medicine, health ranges from individual and communal body-mind comfort to profound social control. Between these two poles, a multitude of practices exist. Health promotes both the well-being sustained by good food and the products sold by the multimillion-dollar diet industry. Health endorses both effective pain management for folks who live with chronic pain and the policed refusal to prescribe narcotic-based pain relief to people perceived as drug seeking. Health both saves lives and aggressively markets synthetic growth hormones to children whose only body-mind "problem" is being short.

Amidst these contradictions, I could try to determine who's healthy and who's not, acting as if there might be a single objective standard. I could struggle to clarify the relationship between health and disability. I could work, as many activists and healers do, to redefine health, moving toward theories and practices that contribute to the well-being of entire communities. But in using *The American Heritage Dictionary* definition as a springboard, I actually want to move away from this mire altogether and follow the word *restoration*.

★★★

To restore a house that's falling down or a tallgrass prairie ecosystem that's been devastated is to return it to an earlier, and often better, condition. In this return, we try to undo the damage, wishing the damage had never happened. Talk to anyone who does restoration work—carpenters who rebuild 150-year-old neglected houses or conservation biologists who turn agribusiness cornfields back to tallgrass prairie—and they'll say it's a complex undertaking. A fluid, responsive process, restoration requires digging into the past, stretching toward the future, working hard in the present. And the end results rarely, if ever, match the original state.[1]

Restoring a tallgrass prairie means rebuilding a dynamic system that has been destroyed by the near extinction of bison, the presence of cattle, and generations of agribusiness farming and fire suppression. The goal isn't to re-create a static landscape somehow frozen in time, but rather to foster dynamic interdependencies, ranging from clods of dirt to towering thunderheads, tiny microbes to herds of bison. This work builds on knowledge about and experience with an eight-thousand-year-old ecosystem, of which only remnants remain—isolated pockets of leadplants, milkweed, burr oaks, and switchgrass growing in cemeteries and on remote bluffs, somehow miraculously surviving. The intention is to mirror this historical ecosystem as closely as possible, even though some element is bound to be missing or different, the return close but not complete.

The process of restoration is simpler with a static object—an antique chair or old house. Still, if the carpenters aren't using axe-hewn timbers of assorted and quirky sizes, mixing the plaster with horse hair, building at least a few walls with chicken wire, using newspaper, rags, or nothing at all for insulation, then the return will be incomplete. It will be possibly sturdier and definitely more energy efficient, but different from the original house.

I circle back to the ideology of cure. Framing it as a kind of restoration reveals the most obvious and essential tenets. First, cure requires damage, locating the harm entirely within individual human body-minds, operating as if each person were their own ecosystem. Second, it grounds itself in an original state of being, relying on a belief that what existed before is superior to what exists currently. And finally, it seeks to return what is damaged to that former state of being.

But for some of us, even if we accept disability as damage to individual body-minds, these tenets quickly become tangled, because an original nondisabled state of being doesn't exist. How would I, or the medical industrial complex, go about restoring my body-mind? The vision of me without tremoring hands and slurred speech, with more balance and coordination, doesn't originate from my visceral history. Rather it arises from an imagination of what I should be like, from some definition of *normal* and *natural*.

[...]

III Defect

Defectiveness justifies cure and makes it essential. Across the centuries, how many communities have been declared inherently defective by white people, rich people, nondisabled people, men backed by medical, scientific, academic, and state authority? I ask this question rather than answer it, because any list I create will be incomplete. White women suffragists fighting for the right to vote were declared defective as a way of undercutting their demands. Black people kidnapped from Africa and enslaved in the Americas were declared defective as a way to justify and strengthen the institution of slavery. Immigrants at Ellis Island were declared defective and refused entry to the USA. Lesbians and gay men were declared defective and given hormones and shock treatments to cure their sexual desires. And today, police shoot homeless people, juries and judges sentence intellectually disabled Black men to death row, schools track Indigenous, Black, Latinx, poor, and disabled children into special ed—all of them deemed defective in one way or another. The list of peoples keeps growing, the damage deepening.[2]

Defectiveness wields incredible power because ableism builds and maintains the notion that defective body-minds are undesirable, worth less, disposable, or in need of cure. In a world without ableism, *defective*, meaning the "imperfection of a bodily system," would probably not even exist. But if it did, it would only be a neutral descriptor. However, in today's world where ableism fundamentally shapes white Western cultural definitions of normal and abnormal, worthy and unworthy, whole and broken body-minds, any person or community named defective can be targeted without question or hesitation for eradication, imprisonment, institutionalization. The ableist invention of defectiveness functions as an indisputable justification not only for cure but also for many systems of oppression.

★★★

Defective arcs repeatedly through history. Let me trace a single trajectory, starting in 1851, though I could begin nearly anywhere. Dr. Samuel Cartwright wrote in the *New Orleans Medical and Surgical Journal*: "It is this *defective* hematosis, or atmospherization of the blood, conjoined with a deficiency of cerebral matter in the cranium… which has rendered the people of Africa unable to take care of themselves" (emphasis added).[3] Using scientific language, Cartwright defended and justified slavery, casting Black people as inferior and racist stereotypes as medical truth. Defectiveness and deficiency lay at the center of his argument.

In the same article, he coined several "diseases of the mind": drapetomania, which he claimed led enslaved African Americans to run away, and dysaesthesia aethiopica, which led them to be lazy. These diagnoses not only turned Black resistance into illness but also allowed Cartwright to frame white power and control as cure: "The complaint [of *dysaesthesia aethiopica*] is easily curable… The best means… is, first, to have the patient well washed with warm water and soap; then, to anoint it all over in oil, and to slap the oil in with a broad leather strap."[4] Cartwright's sleight of hand is brutal. Enslaved Black people become patients and "it." The violence they endured becomes cure. The disabling nature of slavery is hidden away. Cartwright revealed in no uncertain terms the social control embedded in the declaration of defectiveness.

His words travel from 1851 to 1968, landing with white psychiatrists Walter Bromberg and Frank Simon, who pontificated: "The stress of asserting civil rights in the USA these past ten years and the corresponding nationalistic fervor of Afro-American nations… has stimulated specific reactive psychoses in American Negroes."[5] Cartwright's claims transform and yet stay the same, the 1851 "defective hematosis" twisting into the 1968 "specific reactive psychoses." Bromberg and Simon continued, "The particular symptomology we have observed, for which the term 'protest psychosis' is suggested, is influenced by… the Civil Rights Movement… and is colored by a denial of Caucasian values… This protest psychosis among prisoners is virtually a repudiation of 'white civilization.'"[6] In coining this new diagnosis "protest psychosis," cousin to schizophrenia, and declaring it widespread among Blacks who defied white supremacy, they, like Cartwright, framed resistance as pathology. They used defectiveness yet again to justify violence—this time the locking up of Black men in prisons and psychiatric facilities and drugging them with antipsychotic meds.

Bromberg and Simon's words travel from 1968 to 2014, landing in the grand jury testimony of the white police officer, Darren Wilson, who shot and killed the young, Black, unarmed Michael Brown in Ferguson, Missouri. In his testimony, Wilson recounts the altercation that happened moments before the shooting: "When I grabbed him, … I felt like a five-year-old holding onto Hulk Hogan [a six-foot-seven, three hundred-pound professional wrestler] … That's how big he felt and how small I felt."[7] There's no reflection of an adult man and a teenager of almost equal size—both of them six foot four, Brown weighing more and Wilson, the adult, armed and wielding the power of the state. Instead, Wilson creates a picture of a monstrously overpowering Black man. He continues, claiming at one point that the eighteen-year-old "had the most intense aggressive face. The only way I can describe it, it looks like a *demon*" (emphasis added). Wilson remembers that once he started shooting, Brown was "still coming at me, he hadn't slowed down…It looked like he was almost bulking up to run through the shots." Brown becomes in Wilson's story a monster, an embodiment of evil, a super human impervious to bullets.[8]

Unlike Cartwright, Bromberg, and Simon, Wilson doesn't characterize all Black people, wield diagnosis, or directly call Brown defective. Yet in painting him as an overpowering superhuman demon, Wilson calls on centuries of racism, his testimony joining with *drapetomania*, *dysaesthesia aethiopica*, and *protest psychosis*.

Cartwright and the rest use the ableist invention of defectiveness in order to explain and justify the practices of enslavement, imprisonment, institutionalization, and state violence. In essence, they fortify white supremacy by leveraging ableism.

★★★

Entire body-minds, communities, cultures are squeezed into defective. And then that single blunt concept turns, becoming defect. Bullies hurl it as an insult. Strangers ask it out of curiosity. Doctors note it in medical files. Judges and juries hear it in testimony. Scientists study it as truth. Politicians write it into policy. *Defect* and *defective* explode with hate, power, and control.

IV At the Center of Cure Lies Eradication

I play out an imaginary future in my head: disability has been cured. The medical-industrial complex has worked toward this moment for many decades. The visceral experiences named by thousands of diagnostic labels will soon cease to exist both in individual body-minds and collectively in the world. I think about myself and all the disabled people around me—acquaintances, friends, coworkers, neighbors, family members, lovers, activists, cultural workers. I think about what we offer the world—comedy, poetry, performance art, passionate activism, sexy films, important thinking, good conversation, fun. I think about who we are and the ways in which our particular body-minds have shaped us. Who would we be without disability?

Disability activist Harriet McBryde Johnson writes, "Are [disabled people] 'worse off'? I don't think so. Not in any meaningful sense. There are too many variables. For those of us with congenital conditions, disability shapes all we are. Those disabled later in life adapt. We take constraints that no one would choose and build rich and satisfying lives within them. We enjoy pleasures other people enjoy, and pleasures peculiarly our own. We have something the world needs."[9] In my imaginary future, we, or future generations like us, wouldn't exist. I feel neither triumph nor progress but loss.

★★★

At the center of cure lies eradication and the many kinds of violence that accompany it. On the surface, this claim appears hyperbolic. Many lives, including my own, depend on or have been made possible by cure and its technologies. As it supports and extends life, the restoration of health seems to be the opposite of eradication. But cure arrives in many different guises, connected to elimination and erasure in a variety of configurations.

In one permutation, the same medical-industrial complex that saved my mother and me would, if it could, eliminate cerebral palsy from both my individual body-mind and the world at large. In this guise, a multitude of visceral differences would cease to exist. They include both life threatening conditions (aids, malaria, smallpox, and many kinds of cancer, to name a few) and conditions deemed defects but that aren't necessarily lethal (autism, cerebral palsy, hearing voices, and the lasting impacts of spinal cord injuries, for example). The list of body-mind differences, illnesses, and so-called defects that the medical-industrial complex wants to eradicate goes on and on. This kind of elimination benefits some of us in significant ways—saving our lives or increasing our comfort. At the same time, it also commits damage, routinely turning body-minds into medical objects and creating lies about *normal* and *natural*.

In a second permutation, the medical-industrial complex focuses not on specific diseases and disorders but rather on the people who have these conditions. This kind of eradication is often intent on changing the future by manipulating the present. I think about disability-selective abortion. In today's world, the ideology of cure doesn't suggest that we round up everyone who has Down syndrome and eliminate them. Instead, genetic testing and counseling are paired with abortion, setting the scene for eradicating the future possibility of people with Down.

Every day doctors pressure pregnant people to undergo genetic testing, and counselors release the results and guide the course of the conversations that follow. As a result, prospective parents in the USA decide to abort about two-thirds of fetuses predicted to have Down.[10] This termination of pregnancy for the specific reason of not wanting a disabled child clearly manipulates the present. Eradication happens in this moment, but it also extends into a future that is no more than nine months away. In that future, one less person with Down syndrome exists. The choice of each individual parent stacks up until thousands of fetuses predicted to have Down are aborted every year. I'm less interested in the rightness or wrongness of these choices by themselves than in the distinct pattern they create when placed side by side, exposing the systemic desire to erase a whole group of people. This future-focused eradication is easy to shrug past, because many of us have been seduced into believing the need to eliminate disability and "defectiveness" is intuitively obvious.

In a third permutation, the resolve to eradicate particular body-mind conditions stops for nothing, including the possibility of death in the present. I think about the separation of conjoined twins. These surgeries are intensely risky and not always necessary for survival and well-being. Often the high-tech, hours-long medical procedures become media spectacles, with cameras following the families and filming the operations. In an ABC News story from 2015 about the separation of the infants Connor and Carter Mirabal, a nurse says, "Now they are truly boys, individuals," suggesting that a non-conjoined body-mind is a requirement for individuality, possibly even for personhood. Moments later one of the surgeons echoes her sentiments: "It felt good to see them in separate rooms. They seem like individuals now."[11] This emphasis on individuality underlines their belief in the superiority of one kind of body-mind over another. We never learn how Connor and Carter were actually doing before. Was this surgery essential for their survival? Or was it an exercise in eliminating what is deemed abnormal and defective, reshaping it to be normal?

In some separation surgeries, doctors intentionally sacrifice one of the twins in order to save the other, most often when neither will survive if they remain conjoined. This exact situation landed in the court system in the United Kingdom in 2000. Doctors at St. Mary's hospital in Manchester, England, wanted to pursue the separation of Gracie and Rosie Attard, a surgery that they knew would lead to Rosie's death. Their parents, Michaelangelo and Rina Attard, refused to give consent. The surgeons sued the Attards and won. In the legal decision, the judges' logic is revealing. One declared, "The operation would give [Rosie], even in death, bodily integrity as a human being."[12] Without apology, he justified the eradication of this disabled girl through an argument about personhood. In his logic, literal elimination of life becomes a cure.

★★★

In all three configurations, elimination of some kind—of a disease, of future existence, of present-day embodiments, of life itself—is essential to the work of cure. Sometimes these eradications result in benefit, but they can also cause individual death and the diminishment of whole groups of people. The violence that shadows these erasures could be framed as a mere side effect, or the unavoidable cost, of saving lives and normalizing body-minds.

But let me suggest a different framing: that this violence is something more inherent—a consequence, an impact, even an intent. I don't mean that each individual instance of cure is violent. Remember, the restoration of health arrives in many slippery guises. Rather I mean that as a widespread ideology centered on eradication, cure always operates in relationship to violence.

[...]

V Body-Mind Yearning

The desire for cure, for the restoration of health, is connected to loss and yearning. What we remember about our body-minds in the past seduces us. We wish. We mourn. We make deals. We desire to return to the days before immobilizing exhaustion or impending death, to the nights thirty years ago when we spun across the dance floor. We dream about the body-minds we once had before depression descended; before we gained twenty, fifty, a hundred pounds; before our hair turned gray. We ache for the evenings curled up in bed with a book before the ability to read vanished in an instant as a bomb or landmine exploded. We long for the time before pain and multigenerational trauma grabbed our body-minds.

We reach toward the past and dream about the future, feeling grief, envy, shame. We compare our body-minds to friends and lovers, models in *Glamour* and *Men's Health*. Photoshopped versions of humans hold sway. We find ourselves lacking. The gym, diet plan, miracle cure grip us. *Normal* and *natural* won't leave us alone. We remain tethered to our body-minds of the past, wanting to transport them into the future, imagining in essence a kind of time travel.

Even without a nondisabled past tugging at me, I too find myself yearning. Occasionally I wish I could step into the powerful grace of a gymnast or rock climber, but that wish is distant, fading away almost as soon as I recognize it. Sometimes in the face of a task I can't do, frustration overwhelms me, and I long for steady, nimble hands. But in those moments, I've learned to turn away from yearning and simply ask for help. At the same time, the longing I feel most persistently centers on body-mind change. As my wrists, elbows, and shoulders have grown stiff and sore, I've had to stop kayaking. It's a small loss in the scheme of things, but I do miss gliding on the rippling surface of a lake, the rhythm of my paddle dipping in and out of the water.

Cure is such a compelling response to body-mind loss precisely because it promises us our imagined time travel. But this promise can also devalue our present-day selves. It can lead us to dismiss the lessons we've learned, knowledge gained, scars acquired. It can bind us to the past and glorify the future. It can fuel hope grounded in nothing but the shadows of *natural* and *normal*. And when this time travel doesn't work or simply isn't possible, we need a thousand ways to process the grief prompted by body-mind loss.

★★★

Certainly our losses are real, but so is our adaptability. People living with body-mind conditions that grow more significant over time talk about drawing lines in the sand beyond which life would be intolerable. But as their body-minds change, they find their lines also shift. Reflecting on having multiple sclerosis, essayist Nancy Mairs writes:

> Everybody, well or ill, disabled or not, imagines a boundary of suffering beyond which, she or he is certain, life will no longer be worth living. I know that I do. I also know that my line, far from being scored in stone, has inched across the sands of my life: at various

times, I could not possibly do without long walks on the beach...; use a cane, a brace, a wheelchair; stop teaching; give up driving; let someone else put on and take off my underwear. One at a time... I have taken each of these (highly figurative) steps...I go on being, now more than ever, the woman I once thought I could never bear to be.[13]

What begins as loss or pure suffering frequently becomes ordinary and familiar over time. This transformation is another response to body mind loss.

[...]

VI Choosing Disability

Collectively in the white Western world, we go to such lengths to un choose disability. We wear seat belts. We don't dive into shallow water. We vaccinate against polio and measles. Certainly these actions are about avoiding death, but our avoidance quickly mashes into the un choosing of disability. Consider for instance public service announcements and advertisements that warn against unsafe and drunk driving. Many of them use disability as the cautionary tale, showing photos of tragic-appearing teenage boys in wheelchairs. One ad from Utah's "Zero Fatalities" campaign in 2009 reads: "Nothing kills more Utah teens than auto crashes. Not fazed? Okay, how does the thought of spending the rest of your life in a wheelchair grab you? Look, every year far too many Utah teens go from cool to crippled in a blink of an eye."[14] Disability and death are paired together, the first considered a more powerful argument against unsafe driving than the second.

We un-choose disability in hundreds of ways. We condone genetic testing for pregnant people and rarely question the ethics of disability selective abortion. Some pro-choice activists justify late-term abortions with talk about fetal abnormalities—or in plainer language, disability. We accept as a matter of course that sperm banks screen out donors with a whole host of body-mind conditions considered undesirable, including deafness, alcoholism, cystic fibrosis, depression, and schizophrenia. We walk to end breast cancer, run to end diabetes, bike to end multiple sclerosis, dump ice water on our heads to end also. We want to control how, when, and if disability and death appear in our lives.

★★★

In 1963 my mother was twenty-six, a newly married working-class student struggling through graduate school. Every day she answered to professors who believed women belonged not in the classroom but at home, tending children. In the spring of that year, she discovered she was pregnant with me. That pregnancy was unplanned. It completely changed the course of her life.

I grew up knowing she desperately didn't want a disabled child. She made that clear in a thousand ways. She was an intensely unhappy mother. Maybe she didn't want *any* children. Yet her grief, guilt, bitterness about my cerebral palsy was so distinct, so personal; at ages ten, eleven, twelve, I believed she didn't want *me*. I may have been right. But for sure, if she could have un-chosen disability, she would have.

★★★

There are also moments when disability is actively chosen. Prospective foster or adoptive parents fill out agency paperwork requesting a dis abled child—or more likely in the language of those bureaucracies, a "special needs" child. Pregnant people decide to keep their fetuses predicted to have Down syndrome. Or they decide against genetic testing altogether, letting the crapshoot of disability run its course unimpeded. Deaf people using alternative insemination to become pregnant seek out deaf sperm donors, wanting to increase their likelihood of having deaf children. Transabled people, sometimes called disability wannabes or amputee wannabes, feel a need to be disabled.[15] Many have sought out surgeons, planned self-amputations, or staged disabling events, manifesting their desire in actual disability. Or, unable to acquire a disability, they use crutches, braces, wheelchairs anyway.

How the world treats people who, in some fashion, choose disability reveals so much. When transabled people come out, putting words to their desire, they most often encounter revulsion, anger, disbelief. The medical-industrial complex pathologizes them, labeling their so-called troubled body-minds with the recently invented Body Identity Integrity Disorder. People who choose to increase the likelihood of having a disabled or deaf child are deemed categorically selfish and immoral. They're accused of burdening their children and sometimes publicly shamed by the media. People who forego genetic testing, deciding not to intervene in the possibility of disability, are seen as vaguely foolish. People who choose against selective abortion after a positive test for a variety of genetic conditions are frequently perceived as downright irresponsible. And people who adopt or foster disabled children—the world treats them as martyrs engaged in charity work. The act of choosing disability in the white Western world is never neutral, simply one choice among many, but rather pathologized, shamed, or sensationalized. In contrast, un-choosing disability is celebrated and framed as a collective imperative.

★★★

Beyond this binary of choosing and un-choosing lives the many ways we claim disability and chronic illness. We make peace. We accept. We celebrate. We let go. We find pride. We live with ambiguity. We face mortality. We reject pity and overcoming. We build community and grow accustomed to isolation. We seek interdependence. We turn away from expectations of hyperproductivity. We insist on what we know about our own body-minds. We learn to balance loss and pride. We deal with frustration and pain. I'm loath to define claims. Sometimes it lives near an active choosing of disability; other times it shares much in common with un-choosing; often it is laced with contradiction.

I know hard-of-hearing people who have thrown their hearing aids away and stopped struggling to be part of the non-deaf world. People who might have been able to walk again after their disabling accidents and chose to become wheelchair users. People who much prefer hearing voices or experiencing emotional highs and lows than managing the impacts and side effects of psychotropic drugs. To many non-deaf people, nondisabled walkies, and people without psych disabilities or psych labels, these choices seem unimaginable. But from the inside, they make all the sense in the world. They pave the way for finding community and connection. They allow for greater and easier mobility. They allow us to be ourselves.

★★★

On my forty-fifth birthday a friend writes me, "I'm so glad you were born crippled." She's a queer, disabled, white, working-class activist. We've organized together, sat together, struggled together. The word *crippled* makes me smile. In disability communities, many of us call each other crip, practicing the art of refashioning and reclaiming language full of hurt, but typically we veer away from *crippled*; it's too much. My friend uses that riskier word with affection. It contains a whirl of pain and centuries of history. The word *born* settles into me as inconvertible truth: I was indeed born forty-five years ago crippled. But glad is her gift to me, both surprise and revelation. *Glad* leans against un-choosing, my mother's dismay, a whole world of devaluing and eradication. *Glad* is more than uppity pride and stubborn resistance. *Glad* is matter-of-fact, unmovable in its conviction that the world needs disabled body-minds. *Glad* is a powerful claiming.

VII Airports and Cornfields

i

It's late spring in the San Francisco airport. I walk down a long concourse toward the plane that will take me home. I've been in the Bay Area for a long weekend with three hundred LGBTQ disabled people—queer crips as many of us like to call ourselves. I walk slowly, unable to keep up with the frenzied pace. People stream around me. A white businessman with a rainbow flag sticker on his briefcase hurries past an African American woman and her grandson; a Latino man speaking quiet Spanish into his cell phone stands next to a white teen joking in twangy English with her friends; an Asian American woman pushes her cleaning cart by, stop ping to empty the trash can. I know something is missing, but I don't know what. I let my exhaustion and images from the weekend roll over me. Suddenly I realize everyone around me has two arms and two legs. They're walking rather than rolling; speaking with their lips, not their hands, speaking in even, smooth syllables, no stutters or slurs. They have no canes, no crutches, no braces, no ventilators, no face masks, no oxygen tanks, no service dogs. Their faces don't twitch nor their hands flop; they don't rock back and forth. They hold their backs straight, and their smiles aren't lopsided. They move as if their body-minds are separate and independent from the others around them. For a split second, they all look the same.

★★★

That fleeting experience of sameness reminds me of monocultures— ecosystems that have been stripped, through human intervention, of a multitude of interdependent beings and replaced by a single species. I think of a wheat field with its orderly rows of one variety of grass, a clearcut forest replanted with one variety of tree.

I know there were many kinds of humans in the San Francisco airport. I was surrounded by differences created through race, language, citizenship, age, class, gender, sexuality, geography, spirituality, nationality, body-mind shape and size, and disability and chronic illness I didn't perceive. Yet, even with my recognition of human diversity, that moment at the airport when everyone looked the same has stayed with me.

ii

It's early autumn, and I step into an agribusiness cornfield. Rows envelop me, the whole world a forest of corn beginning to turn brown. Leaves and husks rattle overhead. I walk along the furrows between rows, step onto the mounds upon which the stalks grow. A repetition of the same plant fills the space. Nothing chirps or rasps, squawks or buzzes; the cicadas and grasshoppers have gone dormant for the season. I see no traces of grouse, pheasant, fox. If it were a rainy day, I'd see brown water running down the slight slope I'm standing on, washing the dirt away before my very eyes.

★★★

In a monoculture, a world of damage lies beneath the obvious sameness. During that autumn walk, I couldn't smell the pesticide residue, but it hung in the air I was breathing. I couldn't see the petroleum-based fertilizers in the dirt, but they were present in large quantities. I didn't know how depleted the earth was, each corn stalk sucking nutrients from the soil and giving nothing in return. Nor did I notice the six or seven inches of topsoil that had already been washed away in the last hundred and fifty years. I had no visceral awareness of all the invasive pests—the true armyworm, the European corn borer, the corn rootworm, among many others—that breed and eat with abandon in monoculture cornfields, which in turn force agri-business farmers to spray endless rounds of pesticides.

Simply put, monocultures do an immense amount of damage. So much labor and violence goes into creating and maintaining them. Their existence requires hundreds of eradications and removals.[16]

★★★

The history of agribusiness corn, soybeans, wheat, and beef haunts me. I return to an old Black-and-white photo, scratched and faded at the edges, taken in the 1870s[17] It starkly portrays the violence on which these monocultures were created. At the center looms a mountain of bison skulls—thousands upon thousands heaped on top of each other, maybe as many as 180,000. No single skull is distinct; instead they blur together, becoming a geometric pattern of bone. Soon they will be ground into fertilizer. Amidst these bones are two men. Both wear dark suits, and each stands with a foot resting on a skull that's been pulled out of the jumble. One is posed at the base of the pile; the other, twenty-five feet above him on top of the mountain. They make me shiver. They are braggarts—maybe bison hunters or government officials or land speculators. Their body-mind language proclaims, "Look, look at what I own."

My heart breaks and breaks again. Starting in the early 1800s, white hunters killed these big shaggy creatures indiscriminately, thirty million in less than a century. They left the carcasses to rot, took only tongues and skins with them to sell. Later white homesteaders collected the bones for fertilizer. The USA government encouraged this slaughter as one strategy among many to conquer the Indigenous peoples of the Great Plains. The Lakota medicine man John (Fire) Lame Deer (Lakota) described the connection between his nation and bison: "The buffalo was part of us, his flesh and blood being absorbed by us until it became our flesh and blood. Our clothing, our tipis, everything we needed for life came from the buffalo's body. It was hard to say where the animal ended and the man began."[18] So when, in 1867, Colonel Richard Irving Dodge commanded, "Kill every buffalo you can! Every buffalo dead is an Indian gone," he was calling for genocide.[19] Native peoples were starved, brutalized, killed, driven onto reservations.

White colonial settlers claimed the land as their own, dividing it into neat rectangles, fencing it, and establishing herds of cattle. The near eradication of the prairies started here. The grazing and migration patterns of bison had been integral parts of these ecosystems, whereas cows destroyed the grasses, giving nothing back. And then white farmers literally tore up the prairie with their plows. They planted monocultures of wheat, corn, and soybean. One hundred seventy million acres of tallgrass prairie used to exist in North America; seven million are left now. Today when we eat corn or steak produced on agribusiness farms in the Great Plains, we are connected all the way back to that mountain of skulls. Monocultures start with violence, removal, and eradication.

iii

The shadows, legacies, and ongoing realities of environmental destruction and genocide, incarceration and involuntary sterilization rise up. They haunt me. The desire for eradication runs so deep. It is revealed in specific moments, places, and histories—in a fleeting experience of sameness at the San Francisco airport, in an agribusiness cornfield before it's mowed for the winter, in a hundred-and-forty-year-old photo of a mountain of bison skulls. But the desire for eradication is also a pattern reaching across time and space. The un-choosing of disability fits into this pattern, one force among many, threatening to create a human monoculture.

Notes

1 For an overview of ecosystem restoration, see Jordan, The Sunflower Forest.
2 For more on the impact of ableism on white women's suffrage, slavery, immigration, and LGB identities, see, for example, Nielsen, A Disability History of the United States; Boster, African American Slavery and Disability; Barclay, "Mothering the 'Useless'"; Baynton, "Disability and the 192 Notes to Pages 24–46 Justification of Inequality in American History"; Kafer, "Compulsory Bodies."
3 Cartwright, "Report on the Diseases and Physical Peculiarities of the Negro Race," 693.
4 Cartwright, "Report on the Diseases and Physical Peculiarities of the Negro Race," 712.
5 Bromberg and Simon, "The 'Protest' Psychosis," 155.
6 Bromberg and Simon, "The 'Protest' Psychosis," 155.
7 All the quotes by Darren Wilson come from "State of Missouri v. Darren Wilson," Grand Jury Volume V, September 16, 2014, 212–28, https:// www.washingtonpost.com/apps/g/page/national/read-darren-wilsons -full-grand-jury-testimony/1472/.
8 My analysis of Darren Wilson's testimony springboards from Bouie, "Michael Brown Wasn't a Superhuman Demon."
9 Johnson, Too Late to Die Young, 207–8.
10 Natoli et al., "Prenatal Diagnosis of Down Syndrome." For more about the history of abortion and disability, see Reagan, Dangerous Pregnancies.
11 "Conjoined Twins Separated in Florida," video, abc News, May 12, 2015, http://abcnews.go.com/Health/conjoined-twins-separated-florida /story?id=30981266
12 Quoted in Dreger, One of Us, 103.
13 Mairs, Waist-High in the World, 121–22.
14 "Drive Stupid and Score Some Kickin' New Wheels," Don't Drive Stupid advertisement, accessed April 4, 2016, http://2.bp.blogspot.com /_iw4mpIACIU4/S3axcnaXowI/AAAAAAAACk/BjJhuz4DSmA/s1600 -h/dontdrivestupid-001.jpg
15 For more about transabled people, see Stevens, "Interrogating Transability"; Whole (dir. Melody Gilbert, 2003).
16 For more about environmental injustice and long-term processes that harm both the human and the nonhuman world, see, for example, Nixon, Slow Violence and the Environmentalism of the Poor.
17 "Bison Skull Pile," photograph, circa 1870 (Burton Historical Collection, Detroit Public Library), Wikimedia Commons, accessed April 4, 2016, https://commons.wikimedia.org/wiki/File:Bison_skull_pile-restored.jpg
18 Erdoes and Lame Deer, Lame Deer, Seeker of Visions, 269.
19 Smits, "The Frontier Army and the Destruction of the Buffalo," 328. For more about bison and Native peoples, see Jawort, "Genocide by Other Means."

References

Barclay, Jennifer. 2014. "Mothering the 'Useless': Black Motherhood, Disability, and Slavery." *Women, Gender, and Families of Color* 2, no. 2: 115–40.

Baynton, Douglas C. 2001. "Disability and the Justification of Inequality in American History." In *The New Disability History: American Perspectives*, edited by Paul K. Longmore and Lauri Umansky, pp. 33–57. New York: New York University Press.

Boster, Dea. 2013. *African American Slavery and Disability: Bodies, Property, and Power in the Antebellum South, 1800–1860.* New York: Routledge.

Bouie, Jamelle. 2014. "Michael Brown Wasn't a Superhuman Demon." *Slate*, November 24. http://www.slate.com/articles/news_and_politics/politics/2014/11/darren_wilson_s_racial_portrayal_of_michael _brown_as_a_superhuman_demon.html.

Bromberg, Walter, and Frank Simon. 1968. "The 'Protest' Psychosis: A Special Type of Reactive Psychosis." *Archives of General Psychiatry* 19, no. 2: 155–60.

Cartwright, Samuel A. 1851. "Report on the Diseases and Physical Peculiarities of the Negro Race." New Orleans Medical and Surgical Journal: 691–715.

Dreger, Alice Domurat. 2004. *One of Us: Conjoined Twins and the Future of Normal*. Cambridge, MA: Harvard University Press.

Erdoes, Richard, and John (Fire) Lame Deer. 1994. *Lame Deer, Seeker of Visions*. New York: Simon and Schuster.

Gilbert, Melody, dir. Whole. dvd. Saint Paul: Frozen Feet Films, 2003. Gonnerman, Jennifer. "The School of Shock." Mother Jones 32, no. 5 (2007): 36–90.

Jawort, Adrian. 2011. "Genocide by Other Means: U.S. Army Slaughtered Buffalo in Plains Indian Wars." Indian Country Today, May 9. http://indiancountrytodaymedianetwork.com/2011/05/09/genocide-other-means-us-army-slaughtered-buffalo-plains-indian-wars-30798.

Johnson, Harriet McBryde 2005. *Too Late to Die Young: Nearly True Tales from a Life*. New York: Picador.

Jordan, William R. 2003. *The Sunflower Forest: Ecological Restoration and the New Communion with Nature*. Berkeley: University of California Press.

Kafer, Alison. 2003. "Compulsory Bodies: Reflections on Heterosexuality and AbleBodiedness." *Journal of Women's History* 15, no. 3: 77–89.

Mairs, Nancy. 1997. *Waist-High in the World: A Life among the Nondisabled*. Boston: Beacon.

Natoli, Jaime L., Deborah L. Ackerman, Suzanne McDermott, and Janice G. Edwards. 2012. "Prenatal Diagnosis of Down Syndrome: A Systematic Review of Termination Rates (1995–2011)." *Prenatal Diagnosis* 32, no. 2: 142–53.

Nielsen, Kim E. 2012. *A Disability History of the United States*. Boston: Beacon.

Nixon, Rob. 2013. *Slow Violence and the Environmentalism of the Poor*. Cambridge, MA: Harvard University Press.

Smits, David D. 1994. "The Frontier Army and the Destruction of the Buffalo: 1865–1883." *Western Historical Quarterly* 25, no. 3: 312–38.

Stevens, Bethany. 2011. "Interrogating Transability: A Catalyst to View Disability as Body Art." *Disability Studies Quarterly 31,* no. 4. http://dsq-sds.org/article/view/1705/1755.

INDEX

Note: **Bold** page numbers refer to tables; *italic* page numbers refer to figures and page numbers followed by "n" denote endnotes.

For Product Safety Concerns and Information please contact our EU
representative GPSR@taylorandfrancis.com
Taylor & Francis Verlag GmbH, Kaufingerstraße 24, 80331 München, Germany

www.ingramcontent.com/pod-product-compliance
Lightning Source LLC
Chambersburg PA
CBHW081039220326
41598CB00038B/6927